Systems Behaviour

Second Edition

Edited by **John Beishon and Geoff Peters**
at the Open University

Published for
The Open University Press
by Harper & Row, Publishers London New York Hagerstown San Francisco

Acknowledgements

The editors would like to thank Professor Peter Checkland of Lancaster University, John Watt of Aston University and Ross Janes of City University for valuable suggestions for articles for this reader, and Mrs. Judy Baily and Mr. Caryl Hunter-Brown of the Open University for collating and collecting the ancillary material. They would also like to thank Michael Forster and Catherine Cahn at Harper & Row Ltd. for their patience and efficiency.

R. J. B
G. P

First published 1972
2nd edition 1976

Published by Harper & Row Ltd
28 Tavistock Street, London WC2E 7PN

Standard Book Number 06–318056–1 (cloth)
Standard Book Number 06–318057–X (paper)

Printed in Great Britain by
Robert MacLehose and Company Limited
Printers to the University of Glasgow

Preface

This book is a collection of papers and articles on systems. It arose out of the development of a course at the Open University on Systems Behaviour. In common with most Open University courses, the Systems Behaviour course adopts a multi-disciplinary and inter-disciplinary approach and it covers a range of topics not normally dealt with in a conventional university course. One of the problems in developing new courses is that suitable reference and back-up material is difficult to find and is usually spread across a wide variety of sources. This is particularly true of new areas such as Systems where there are few, if any, "standard" texts and most of the books available cover specific subject or technique areas. The Systems Behaviour course does not, for example, have any text books set as "required reading", for the reason that no book dealing with more than a few parts of the course can be found. So one of our main reasons for producing a book of readings is to gather together material on Systems which is scattered far and wide in the literature, often in journal articles or reports which are difficult to obtain.

Collections of readings suffer from a number of disadvantages: they are uneven in style and level, gaps in arguments and in the development of ideas occur, the coverage of a field is sometimes very patchy, and it is always possible to criticize the selection of material on one ground or another. Editors make valiant attempts to overcome these problems by providing linking passages and by careful editing where possible, but the results are not always satisfactory. However, one of the great advantages of readings from our point of view is that students can consult original sources and can compare and evaluate the different ideas and concepts presented. This is an essential part of our systems courses.

It is worth pointing out, however, that the final choice of articles is not always what one would like. Material which would logically appear at a certain point is sometimes left out because it is readily available elsewhere, or because it would make the book unduly long, or more simply because permission to reprint it is not forthcoming. We have selected articles which seem to us to be of general interest as well as of direct relevance to the Systems Behaviour course, and we have inserted recommendations for other reading at appropriate places in the linking texts between sections.

A final point – the articles we have selected supply essential background reading for students taking the Systems Behaviour course, hence the title of this Reader. The articles do not therefore deal exclusively with systems behaviour but cover a wider field of systems topics.

The Open University R. J. B.
G. P.

The Open University Systems Group

Professor John Beishon
Lyn Jones
John Martin
William Mayon-White
Dick Morris
John Naughton
Geoff Peters
Roger Spear
Alan Thomas
Dennis Bolton (visiting lecturer)
Ruth Carter (.. ..)

Research and Support Staff

George Barclay
Stephen Brown
Bob Drosdowski
Rita Daggett
Rob Forrester-Paton
Malcolm Henderson
Martin Lockett
Sue Parrott
Christopher Pym
Lindsay Read
Viv West
Bob Williams

The Open University Systems Courses

The Systems Group have produced four systems courses, each of half credit rating which entails 16 weeks' study totalling about 160-200 hours total study time. The courses have radio and television programmes as integral parts of the teaching material and two have home kits for experiments and practical activities which can be done in the students' homes. All students on systems courses attend a one week summer school.

Systems Behaviour T241

The idea of this course is to introduce students from many different backgrounds, scientists, technologists, social scientists, to the study and understanding of systems and systems behaviour. The course is at second level which approximates to a second year undergraduate course at a British University. The course contains eight modules of about two weeks work each covering a different case study in a correspondence text together with radio and TV programmes and computer exercises and home experiments where appropriate. Each case study is linked to the teaching of a specific technique or method in the systems area. The eight studies are:

Deep sea container berth

Air traffic control

Telephone system

Work group systems

Local government

The human respiratory system

Sheep farming-grass ecosystem

Economic case studies

The techniques developed include: data collection and statistical analysis, simulation and modelling both in digital and analogue forms and systems dynamics.

Systems Management T242

This is also a second level course and it concentrates on the organization as a system. It examines how such systems come to behave as they do and it draws on a number of case studies to illustrate its points. The course looks at the conventional management divisions of operations, man-management, and socio-economic aspects from a systems viewpoint studying the goal setting and goal conflicts, information and decision making, and the overall behaviour of these. It then integrates these views into a systems view of organizations and their behaviours. The course has 11 radio and TV programmes which develop case studies on such topics as British Rail's operations, kidney machines, blood analysis, coal winning, and common ownership firms.

Systems Modelling T341

This is a third level course which as the title implies, deals with the modelling of systems. The aim is to break down the mystique that surrounds the use of quantitative models and computer methods in the making of major decisions. The course concentrates on the problems of formulating models of complex systems and of interpreting the results they give. It draws on a range of case studies to illustrate the use of linear models and simulation modelling. Two units deal with modelling community decisions and large scale economic modelling (including world modelling).

Systems Performance: Human Factors and Systems Failures TD342

This is also a third level course and it deals comprehensively with analyses of the performance of systems concentrating particularly on the human element in systems and the problems of failures. It develops techniques for the analysis of failures such as failure modes and effects analysis and it critically deals with a wide range of systems failures. These cover catastrophic failures such as the Aberfan tip disaster, the Summerland fire and the Hixon level crossing train smash; and organizational studies in the field of science policy, mental health provisions and a major case study of the Bay Area Rapid Transit system in California. There are 16 TV and radio programmes which deal with these and other case studies including earthquakes and air crashes in depth. The course also studies concepts of reliability and safety in relation to the design of safe and effective systems. A unique home kit is included with the course which illustrates the range of problems which can arise in systems and shows how many of them can be avoided.

Contents

Introduction
Systems and Systems Behaviour

The concepts of systems have slowly emerged in the present century to assume a central importance in the thinking and approach of many scientists and technologists. The ideas are now having an increasing impact on the approach of social and political scientists and economists. In the slow development of knowledge about the physical and biological world the most significant advances have generally come from painstaking, detailed study and analysis of ever smaller areas of knowledge and application. The natural philosophy of earlier scientists gradually developed into the major natural sciences, physics, chemistry, and biology; these in turn have been increasingly subdivided into subjects such as nuclear physics, polymer chemistry, molecular biology, and so on. By the middle of the 20th century this subdivision process had chopped reality and the phenomena of the real world into hundreds of individual subjects and areas of professional expertise. Knowledge was compartmentalized and departmentalized into small and seemingly more manageable "chunks".

The success of analytical techniques, and in particular structural analysis, tended to concentrate attention on static and structural properties, and some writers have seen this as a reflection of the desire of men to live in well-ordered, static social and political environments. Victorian man might fit into this picture but not the more modern variety. At least two groups found this approach limiting and unhelpful: the mechanical and later the electrical engineer who wanted to build and run mechanisms and devices which *worked*, and the psychologist or ethologist who was interested less in the anatomy of the human or animal system than in the *behaviour* of the complete organism. The full history of the emergence and effects of the systems-thinking "revolution" has yet to be written although some of the ground has been covered ably by one of the pioneers in the field, L. von Bertalanffy.*

The impetus towards systems thinking and the systems approach has come partly from a recognition (some would say a belated recognition) of the complexity of behaviour which arises in both natural and man-made systems, and partly from the need to gain control over the more threatening outcomes of this behaviour. So one powerful stimulus to adopt a system approach arises from our attempts to predict and control the *behaviour* of systems instead of passively suffering from, or just reacting to, the often mysterious changes which occur in the surrounding physical, biological, social, economic, and political climate. Attempts to exercise control over systems, whether human or economic, have not been notably successful and, for example, there are real fears being expressed that control is no longer even possible over some of the ecosystems we have constructed or interfered with.

System problems often appear intractable since we know so little about systems, about systems analysis, about control over system behaviour, or about system design. Even the first step of recognizing the system we are trying to deal with, of drawing its boundaries, can present formidable difficulties. Most scientific disciplines have a large body of facts and knowledge to draw on which have been accumulated over years of testing and experimenting. Theories can be tested against this body of knowledge and gaps identified. But with systems behaviour we have surprisingly little in the way of studies or documented experience. The structure of a system is often known but its behaviour is another matter. It is unfortunate that we often have to wait for a system to fail before its possible behaviours are revealed and there are many vested interests engaged in seeing that full records of what happened are not readily available! There are other difficulties, of course: system malfunctioning often occurs suddenly and catastrophically and people are not going to start making detailed observations when all is falling about their ears. This is one reason why black box flight recorders are used on aircraft, so that a disaster can be analysed in detail after the event. The cost, a relatively small one, is that thousands of perfectly normal operations are recorded.

Systems thinking and the systems approach is now a growth area. Systems ideas appear in different guises in cybernetics, systems engineering, general system theory, operations research, systems analysis, computer systems,

* See Article 2.

and many other fields. New jargon abounds: open and closed systems, purposive systems, adaptive systems; the ideas are new and exciting and innovators rush to apply them to different problem areas. There must be many a bewildered manager, administrator, or scientist gazing at the brand-new, shiny, systems approach and wondering quite what hit him. Yet it is becoming clear that a systems approach is essential if we are to have any hope of coping with the complexity of modern life with its multi-national organizations, space modules, international monetary policies, high-speed transport systems, and so on.

This book of readings contains 19 different articles or extracts, all concerned with systems. They have been chosen to illustrate something of the systems approach itself and the way it can be applied to different subjects or activities. The articles are grouped into a number of sections to give some organization to the book but it is a feature of system design that it is always difficult to bring the components of a system into a relationship which is useful and yet can serve a number of different purposes. In a sense this book is a multi-purpose system. People will read separate articles for one purpose, a combination of articles for another, and some will want to read progressively through the book. An order of presentation which might suit a social science student might be inappropriate for a biology student, so our order is only one arrangement designed to serve one, fairly general, purpose. One of the messages of the systems approach is that very simple systems can nevertheless show remarkable complexity of behaviours. Our 19 articles could be arranged in 19! different ways, which is about $1{\cdot}2 \times 10^{17}$ ways. Enough for a different order for every person on Earth with a good few left over.

Ideally perhaps the book should be produced in loose-leaf form so that each reader could assemble, and re-assemble, the articles into any desired order, and also reject or add material to suit individual need. We have grouped the articles as follows:

Section I
Introductory material describing the system approach and defining systems concepts.

Section II
Two articles dealing with man-machine systems.

Section III
Five articles concerned with the impact and development of systems approaches to social systems and organizations.

Section IV
Three articles on biological and ecosystems.

Section V
One article dealing with implementation, one with social work and two on systems applications in geography.

We have provided linking passages where appropriate and references to other articles which could usefully be read at certain points or which complement the existing material. Since the articles cover such a wide range of scientific and organizational matters we have provided a glossary to assist people who are not familiar with all the different areas.

It is worth emphasizing two further points. This collection of articles reveals the way different people coming from different backgrounds and different base disciplines have interpreted and attempted to apply systems ideas. Not surprisingly there are often differences in the interpretation of the concepts or contradictory applications of the ideas. Sometimes these are unimportant and arise from different usage of terms. A rigorous set of definitions such as that proposed by Ackoff, (Article 5) is valuable in helping to clear up these problems. On other occasions however, the differences are more fundamental and reveal misunderstandings of the nature of basic ideas or deficiencies in the basic concepts themselves which limit their general application. The application of the ideas of open and closed systems in the organizational field is an example of an area where many kinds of confusion can arise. Reference to Bertalanffy (Article 2) reveals that the distinction is a fundamental one based on thermodynamical considerations. An organization could *never* have existed as a closed system, they have always been open systems. But the *idea* of closed and open systems encouraged sociologists to *think* of organizations, and to treat them as open systems which altered the way we analyzed organizational behaviour. One valuable function of this book is to help students make contrasts among different usages and to try to distill the common or core concept. The presentation of original material also enables students to realize the extent to which common systems ideas crop up in different guises and under different names in the various discipline areas covered.

A second point which arises from the first one, is that the language used in some fields is very difficult to understand by people from different backgrounds. The language of the sociologists, for example, as can be seen in the article by Buckley (Article 9) is particularly specialized and we have found that it presents serious difficulties to people with a technological or scientific background. Nevertheless it is important that students are encouraged to grapple with these different presentations because they will, hopefully, want to extend their reading on systems topics beyond the range of this book and they will then have to consult articles and books from the full range of disciplines as systems ideas spread and are incorporated in different theoretical and practical approaches.

One final point: this book has been produced in connection with an Open University course on systems but it does not in any sense represent either the views adopted in that course or by the systems group at the university. The book is an attempt to bring together the minimal library that a student should have available in studying courses in systems which cover a wide range of applications in the hard and soft science areas. So, just as any good library should, the book reflects a broad range of approaches and attitudes to systems and readers must be prepared to use their own critical judgements in assessing the material.

Section I
Introduction

The first five articles deal with introductory material about systems and systems concepts. Kast and Rosenzweig give a broad general introduction to a number of basic systems concepts, and then show how the systems approach can be applied to organizations. This article is followed by a more detailed, in depth review of General Systems Theory by one of the pioneers in this field, L. von Bertalanffy. His article gives some of the historical background to the emergence of systems ideas since the 1930s, and he also deals with some of the controversies and objections which have since arisen. This is an article written, as it were, from the insider's position and consequently will be difficult reading for anyone unfamiliar with the systems field. It is probably best to read it through, bypassing any difficult parts, and to come back to it again after reading other articles.

The next two articles by Peter Checkland and Gwilym Jenkins present the methodologies developed at Lancaster University for dealing with "soft" and "hard" systems respectively. These are particularly valuable contributions because they set out procedures and guidelines for tackling a very wide range of systems problems.

The article by Ackoff needs a more detailed introduction because although it is basically clearly written and Ackoff has attempted to produce rigorous definitions of systems concepts, we have found that some of the material can give rise to misunderstandings and confusion. It is worth spending time over this article because it raises several fundamental issues which need to be understood if systems ideas are to be applied successfully.

In any new and growing subject area problems of terminology and definition often predominate in the early stages of development and much time may be taken up in what, with hindsight, may turn out to be sterile or trivial debates on the meaning or status of concepts and terms. The establishment of an agreed vocabulary is, however, an essential step if effective communication is to be possible. Systems as a subject has been particularly prone to controversy and confusion about the meaning of its words. The emergence of systems thinking in various forms at different centres and arising from different base disciplines has been one of the main causes for the proliferation of terms and this has hindered the setting up of an agreed vocabulary. People starting from engineering, sociology or geography backgrounds are bound to draw on different terminology. The particular phenomenon or process being discussed may turn out to be a common, systems-based one but this may be obscured by these differences. Other problems arise when the terms from one discipline are uncritically transferred to another and applied inappropriately.

One of the most difficult problems for the newcomer to systems ideas is in understanding what systems thinking and the systems approach are actually all about. Ackoff does not discuss these ideas explicitly although he does by implication when he defines the various concepts involved. It is worth first making a distinction between *systematic* and *systemic*. The former is used to describe the situation when we have a method or follow a plan or an explicit and rational procedure. It is a useful aid to achieving results efficiently in a wide range of activities. *Systemic* however, is rather different. It is a term which has been coined to refer to the activity of using systems ideas especially in tackling problems. It can be defined strictly as "pertaining to system" but it is generally used to refer to the process or activity of treating things as systems or from a systems viewpoint. This, loosely speaking, means looking at situations, topics, problems, etc. as a complex of interacting parts which can be divided into specific systems, and within these, sub-systems, and if necessary into sub-sub-systems, and so on. Identification of these various systems is followed by an examination of the relationships among them and the flows of influences, materials and energy and the routes these take among and within the systems involved. There is much more to it than this of course as will become clear as readers work through articles in this book, but this is the basic starting point involved in systems thinking. It is important to emphasize that this process of identifying systems is not entirely an objective process since the purposes and interests of the person doing the identification will enter into the choice and identification of the systems concerned.

The idea of a (or the) systems approach is a development

of the use of systems thinking where a series of steps or guide lines are followed which assist with this conceptualization of things as systems and in identifying their properties and behaviours. At present, there is not a single, all-embracing systems approach which is applicable to all classes and types of systems thinking. Research workers have developed individual versions for specific purposes and then extended their range where possible. Although there are several forms of systems approach they have much in common as can be seen by comparing the two contained in the articles by Checkland and Jenkins (Articles 3 and 4). In their simplest form, system approaches consist of lists of questions which the would-be systems thinker can follow and apply to the system concerned. In their more sophisticated forms, the approaches embody more structuring and include iterative routes allowing more penetration and extension of the thinking. They may also include systems design or re-design sections.

System

After the idea of systems thinking and the more formalized system approach concept, the question of what a system itself is arises. In spite of the widespread use of the term there is as yet no generally agreed definition. An extensive analysis of the word has been carried out by Nehemiah Jordan in his article *Some thinking about "System"* in his book *Themes in Speculative Psychology* (Tavistock, 1970) which revealed a very confused state of affairs. A major cause of difficulty is due to a confusion between a system as it exists "objectively" out-there in the real world, for example, the solar system, and the *idea* of the system in a person's mind, for example, an 17th century astronomer's concept of the solar system. The point is perhaps clearer if we compare the 17th century astronomer's view with that of a present day astronomer. The solar system as we now conceive it has not changed essentially over the last few hundred years but the two astronomers had different systems in mind. Clearly the two ideas had much in common but they were not the same. One of the more dangerous consequences of the scientific approach is its tendency to encourage people to think that their perceptions of reality *are* real and that there is indeed an objective reality "out-there". We are not denying objectivity or reality but drawing attention to the *subjective aspect* of our perceptions of reality. A large number of objects and processes in the real world are sufficiently commonly perceived or experienced that there is no disagreement about their objective existence. But there are many other things in which subjective aspects are more important and which give rise to confusion. Systems concepts are particularly liable to these confusions because they tend to have strong objective and subjective parts. But the subjective is often unrecognized.

Most writers who have attempted to define system have ignored this subjective element or at least, not explicitly recognized it. A widely quoted definition of system, for example, states:

> a set of objects together with relationships between the objects and between their attributes connected or related to each other and to their environment in such a manner as to form an entirety or whole.

Ackoff defines system more simply as "a set of interrelated elements" and later on recognizes the objective and subjective aspect involved in this identification of the set. We have used a definition developed in our courses which has four parts:

A system is an assembly of parts where

1. the parts or components are connected together in an organized way
2. the parts or components are affected by being in the system and are changed by leaving it
3. the assembly does something
4. the assembly has been identified by a person as being of special interest.

This definition has advantages and disadvantages. It does exclude some things, by emphasizing the behavioural aspect, which definitions such as Ackoff's include; for example Ackoff includes a table as a static system. It does distinguish between an *assembly* and a *system* and it includes the subjective aspect *explicitly* which we regard as particularly important. It is not, however, *rigorous* in the way that Ackoff's definition is, but we have found that it is helpful in drawing attention to the important features of systems thinking for newcomers to the subject. A fuller discussion of this definition can be found in *Systems*, Unit 1 The Man-made World T100, (Open University Press).

Turning now to the definitions put forward by Ackoff, the first set (numbered 1 to 11) deal with the basic concepts. Although a case can be made for Ackoff's more rigorous version of "system" we find it rather contrary to the aim of *using* systems thinking to include static systems such as "table" as mentioned above, especially when a "compass" is also included in this category. As we have suggested above, an all-inclusive definition can be unhelpful when it distracts people from appreciating the particularly valuable emphasis that systems thinking places on the dynamic properties of systems. The section on *System Changes* introduces some useful definitions but many will find them difficult to appreciate on a first reading because Ackoff uses the words *reaction*, *response* and *act* in specialized ways which conflict with their normal, everyday usage. We prefer to distinguish between environmental events or changes which are *necessary and sufficient* for a system change to occur, and those which merely contribute to a system change, that is, are *necessary but not sufficient*. We would also favour a broader interpretation of "behaviour" (definition 15). In the strict sense a system's behaviour must consist of events which cause effects in other systems or we would not be able to detect the behaviour at all. We would prefer to include a *time dimension* in behaviour because there is a subjective aspect in apprehending "behaviour" just as there is in defining "system". Changes over very long time periods, such as millenia, may not be perceived or be worth regarding as behaviour in some contexts; at another extreme, the behaviour of a TV set "picture" will be different depending on the time scale used since what is "seen" by a camera is different from what is seen by a human observer.

Ackoff's behavioural classifications of *state-maintaining* and *goal-seeking* are useful but we find that the distinction between *purposive* and *purposeful* causes difficulty and is unhelpful in some respects. *Purposive* in Ackoff's terms is a special category of multi-goal seeking but in our view the concept of goals having a common property (Ackoff's distinction for purposive systems) is again a subjective one, or at least one with a strong subjective element, and the identification of a set or sub-set of goals as having a common property is not a feature of the system but of the external human observer. This is clear when we consider Ackoff's example of winning a game. The concept of a set of goals as a "winning" set is often purely arbitrary with no rational basis. Ackoff's use of the terms "goals" and "objectives" is also rather specialized and differs from the way these words are used in other articles and in the literature generally. We have been unable to find any general agreement on the way these words should or could be used and the distinction seems to us to be an unrealistic one anyway. If a distinction on the basis of the time scale concerned is needed we find it easier to speak of short- or long-term goals (objectives) and to specify the time scale. If the distinction is between a larger or longer term goal (objective) and subsidiary, intermediate or intervening goals which must be achieved if the longer term goal is to be reached eventually, the terms sub-goals or intermediate goals can be used. The distinction between attainable goals and unattainable ones (*ideals* in Ackoff's terms) seems to us to beg many philosophical questions but it would be inappropriate to discuss these here.

Finally we would like to draw attention to Ackoff's important distinction between *organism* and *organization*; confusion between these is frequently found in management and organizational literature which in our view can and does lead to many dangerous conclusions and recommendations for action. The understanding of the differences between the two concepts is vital and Ackoff makes a valuable contribution in the way he makes the distinction so clearly.

Associated reading

There are many general introductions to system concepts and most books with the term "systems" in the title include an opening section dealing with broadly similar ground.

A good straightforward coverage is given by C. W. Churchman in *The Systems Approach* (Dell Publishing Co.), 1968. Another useful general introduction is *System Engineering*, by Jenkins and Youle; *Systems Engineering: a unifying approach in industry and society* (C. A. Watts, 1971), which interprets systems engineering in the broader terms associated with the Lancaster University's System Engineering Department's approach. General concepts are dealt with in several books on cybernetics, e.g. Ross Ashby, *Introduction to Cybernetics* (Chapman and Hall, 1956); Stafford Beer, *Decision and Control* (Wiley, 1966); Van Court Hare Jnr in *Systems Analysis: a diagnostic approach* (Harcourt Brace Jovanovich, 1967) gives a more detailed introduction to the whole field of systems definition, analysis, and treatment, and we have included a chapter from his book later in these readings.

A full account of General Systems Theory is given in a book of that title by von Bertalanffy, *General system theory essay on its foundation and development* (Braziller, 1969). Other articles in this area can be found in General Systems Year Books published by the Society for General Systems Research edited by L. von Bertalanffy and A. Rapoport, 1956–. Some relevant extracts from several of the above books together with other more specialized material can be found in *System Thinking*, edited by F. E. Emery (Penguin Books), Ltd. 1969. A further collection of basic articles can be found in G. J. Klir *Trends in General*

System Theory (Wiley, New York, 1972). Ackoff's ideas are developed further in R. Ackoff and F. E. Emery *On Purposeful Systems* (Aldine Publishing Co. Chicago, 1972).

Several books by Geoffrey Vickers explore general applications of general system theory. See G. Vickers *Values systems and social process* (Basic Books, New York, 1968) and *Freedom in a rocking boat: changing values in an unstable society* (Penguin Press, London, 1970).

An introduction to systems ideas and their application to management can be found in *Management Systems: Conceptual considerations* by P. P. Schoderbek, A. G. Kefalas, and C. G. Schoderbek (Business Publications Inc. Dallas, Texas, 1975). Three Open University units cover introductory systems material: *Systems* (Unit 1, The Man-made World T100); *Human Activity Systems* (Unit 3, Systems Management T242); and *Systems and Failures* (Unit 1, Systems Performance: Human Factors and systems failures TD342); all published by the Open University Press.

1 The modern view: a systems approach

by F. E. Kast and J. E. Rosenzweig

Organization theory and management practice have undergone substantial changes in recent years. Traditional theory has been modified and enriched by informational inputs from the management sciences and the behavioral sciences. These research and conceptual endeavors have, at times, led to divergent findings. During the past decade, however, an approach has emerged which can serve as a basis of convergence – the systems approach, which facilitates unification in many fields of knowledge. It has been used in the physical, biological, and social sciences as a broad frame of reference. It can also be used as a framework for the integration of a modern organization theory. General systems theory, its pervasiveness, its relationship to organization theory, and its potential usefulness are discussed in this chapter via the following topics:

General Systems Theory
Pervasiveness of Systems Theory
Systems Approach and Organization Theory
Organization: An Open System in Its Environment
Organization: A Structured Sociotechnical System
Other Properties of Organizational Systems
Managerial Systems
Systems Concepts for Organization and Management

General Systems Theory

Over the past two decades the development of general systems theory has provided a basis for the integration of scientific knowledge across a broad spectrum.[1] *A system is an organized or complex whole: an assemblage or combination of things or parts forming a complex or unitary whole.* The term *system* covers a broad spectrum of our physical, biological, and social world. In the universe there are galaxial systems, geophysical systems, and molecular systems. In biology we speak of the organism as a system of mutually dependent parts, each of which includes many subsystems. The human body is a complex organism including, among others, a skeletal system, a circulatory system, and a nervous system. We come into daily contact with such phenomena as transportation systems, communication systems, and economic systems.

In considering the various types of systems in our universe, Kenneth Boulding provides a useful classification of systems which sets forth a hierarchy of levels as follows:

> 1. The first level is that of static structure. It might be called the level of *frameworks*; for example, the anatomy of the universe.
> 2. The next level is that of the simple dynamic system with predetermined, necessary motions. This might be called the level of *clockworks.*
> 3. The control mechanism or cybernetic system, which might be nicknamed the level of the *thermostat*. The system is self-regulating in maintaining equilibrium.
> 4. The fourth level is that of the "open system", or self-maintaining structure. This is the level at which life begins to differentiate from not-life: it might be called the level of the *cell.*
> 5. The next level might be called the genetic–societal level; it is typified by the *plant*, and it dominates the empirical world of the botanist.
> 6. The *animal* system level is characterized by increased mobility, teleological behavior, and self-awareness.
> 7. The next level is the *human* level, that is, of the individual human being considered as a system with self-awareness and the ability to utilize language and symbolism.
> 8. The *social system* or systems of human organization constitute the next level, with the consideration of the content and meaning of messages, the nature and dimensions of value systems, the transcription of images into historical record, the subtle symbolizations of art, music, and poetry, and the complex gamut of human emotion.
> 9. *Transcendental systems* complete the classification of levels.
>
> These are the ultimates and absolutes and the inescapable unknowables, and they also exhibit systematic structure and relationship.[2]

The first three levels in this hierarchy can be classified as physical or mechanical systems and provide the basis of knowledge in the physical sciences such as physics and astronomy. The fourth, fifth, and sixth levels are concerned with biological systems and are the interest of biologists, botanists, and zoologists. The last three levels are involved with human and social systems and are the concern of the social sciences as well as the arts, humanities, and religion.

General systems theory provides a basis for understanding and integrating knowledge from a wide variety of highly specialized fields. In the past, traditional knowledge has been along well-defined subject-matter lines. Bertalanffy * suggests that the various fields of modern science have had a continual evolution toward a parallelism of ideas. This parallelism provides an opportunity to formulate and develop principles which hold for systems in general. "In modern science, dynamic interaction is the basic problem in all fields, and its general principles will have to be formulated in General System Theory."[3] General systems theory provides the broad macro view from which we may look at all types of systems. "So has arisen systems theory – the attempt to develop scientific principles to aid us in our struggles with dynamic systems with highly interacting parts."[4]

Bertalanffy made another major contribution in setting forth a distinction between closed systems and open systems. Physical and mechanical systems can be considered as closed in relationship to their environment. Thus, the first three levels in Boulding's hierarchy are closed systems. On the other hand, biological and social systems are not closed but are in constant interaction with their environment. This view of biological and social phenomena as open systems has profound importance for the social sciences and organization theory. Traditional theory assumed the organization to be a closed system, whereas the modern approach considers it an open system in interaction with its environment. While the development of general systems theory has provided an overall conceptual view for dealing with all types of phenomena – physical, biological, and social – there have been many additional threads in intellectual development which have contributed to the development of the systems approach.

Pervasiveness of Systems Theory

In complex societies with rapid expansion of knowledge, the various scientific fields become highly differentiated and specialized. In many scientific fields, the concentration over the past several decades has been on analytical, fact-finding, and experimental approaches in highly specific areas. This has been useful in helping to develop knowledge and to understand the details of specific but limited subjects. At some stage, however, there should be a period of synthesis, reconciliation, and integration, so that the analytical and fact-finding elements are unified into broader, multi-dimensional theories.[5] There is evidence that every field of human knowledge passes alternately through phases of analysis and fact finding to periods of synthesis and integration. Recently systems theory has provided this framework in many fields – physical, biological, and social. Even more important, it provides a basis for communication between scientists in the various disciplinary areas – a problem of immense importance with the high degree of specialization of knowledge.[6]

The application of systems thinking has been of particular relevance to the social sciences. There is a close relationship between general systems theory and the development of functionalism in the social sciences, as described by Martindale:

> The essential unity of the social sciences has been revealed in the flooding of the functionalistic point of view across the boundaries of the special disciplines. This point of view has had both theoretical and methodological dimensions. Theoretically, it consists in the analysis of social and cultural life from the standpoint of the primacy of wholes or systems. Epistemologically, it involves analysis of social events by methods thought peculiarly adapted to the integration of social events into systems.
>
> The functionalistic point of view has been manifest in all of the social sciences from psychology through sociology, political science, economics, and anthropology to geography, jurisprudence, and linguistics.[7]

Although there are several connotations of the word "functionalism", its basic emphasis is upon systems of relationships and the integration of parts and subsystems into a functional whole.[8] Functionalism attempts to look at social systems in terms of structures, processes, and functions and attempts to understand the relationship between these components. It emphasizes that each element of a culture or social institution has a function in the broader system.

Functionalism, under the influence of the earlier works of A. R. Radcliffe-Brown and Bronislaw Malinowski, has become the framework for modern anthropology.[9] They pioneered the view that social customs, patterns of behavior, and institutions do not exist independently but must be considered in relationship to the total culture. All aspects of social life form a related whole, and society can best be

* [An article by Bertalanffy setting out the major ideas of General System Theory follows at p. 30. *Ed.*]

understood as an interconnected system. Thus each social action, such as a marriage ceremony or the punishment of a crime, has a function in the culture as a whole and contributes to the maintenance of the social structure.

In sociology, Talcott Parsons led in the adoption of functionalism and the general systems viewpoint.[10] Although Parsons acknowledges his debt to Pareto for the concept of systems in scientific theory, it is Parsons himself who has fully utilized the open-systems approach for the study of social structures.[11] He not only developed a broad social system framework but also related his ideas to the organization. Many of his concepts relating to the structure and processes of social systems will be used later in this book.

In the field of psychology, the systems approach has achieved prominence. The various types of behaviorism in psychological theory have given way to the holism of gestalt psychology and field theory. The very word *gestalt* is German for configuration or pattern.[12] "The Gestaltists early adopted the concept of system, which is more than the sum of its components, and which determines the activity of these components."[13] Kurt Lewin was among the first to apply to tenets of gestalt psychology to the field of individual personality. He found that purely psychological explanations of personality were inadequate and that sociocultural forces had to be taken into account. He viewed personality as a dynamic system, influenced by the individual's environment. Harry Stack Sullivan, in his *Interpersonal Theory of Psychiatry*, went even further in relating personality to the sociocultural system. He viewed the foundation of personality as an extension and elaboration of social relationships. A further extension of psychology to give greater consideration to broader interpersonal and social systems is seen in the rapidly expanding field of social psychology.[14]

Modern economics has increasingly used the systems approach. Equilibrium concepts are fundamental in economic thought, and the very basis of this type of analysis is consideration of subsystems of a total system. Economics is moving away from static equilibrium models appropriate to closed systems toward dynamic equilibrium considerations appropriate to open systems. Leontief and his followers in industrial input–output analysis utilize the systems approach. "Considered from the point of view of the input-output scheme any national economy can be described as a system of mutually interrelated industries or – if one prefers a more abstract term – interdependent economic activities. The interrelation actually consists in the more or less steady streams of goods and services which directly or indirectly link all the sectors of the economy to each other."[15]

The very foundation of the discipline of cybernetics is based upon a systems approach.[16] It is primarily concerned with communication and information flow in complex systems. Although cybernetics has been applied primarily to mechanistic engineering problems, its model of feedback, control, and regulation has a great deal of applicability for biological and social systems as well.

Another similar point of view permeating many of the social and physical sciences is the concept of holism which is closely related to functionalism and the systems approach. Holism is the view that all systems – physical, biological, and social – are composed of interrelated subsystems. The whole is not just the sum of the parts, but the system itself can be explained only as a totality. Holism is the opposite of elementarism, which views the total as the sum of its individual parts. The holistic view is basic to the systems approach. In traditional organization theory, as well as in many of the sciences, the subsystems have been studied separately, with the view to later putting the parts together into the whole. The systems approach emphasizes that this is not possible and that the starting point has to be with the total system.

The foregoing discussion has attempted to show how the systems approach and associate views have become the operating framework for many physical and social sciences. We agree with Chin, who says:

> Psychologists, sociologists, anthropologists, economists, and political scientists have been "discovering" and using the system model. In so doing, they find intimations of an exhilarating "unity" of science, because the system models used by biological and physical scientists seem to be exactly similar. Thus, the system model is regarded by some system theorists as universally applicable to physical and social events, and to human relationships in small and large units.[17]

It is important for the student of organization and management to recognize that the developing body of knowledge and applications of the systems approach to complex organizations is but a part of the broad trend in many of the physical and social sciences and that this field is part of a pervasive stream of thought. Furthermore, understanding that organization theory can be put in the context of general systems theory allows for a growing community of interest and understanding with widely diverse disciplines. We will now look more closely at the direct relationship between the systems approach and organization theory.

Systems Approach and Organization Theory

Traditional organization theory used a highly structured, closed-system approach. Modern theory has moved toward the open-system approach. "The distinctive qualities of modern organization theory are its conceptual-analytical base, its reliance on empirical research data, and, above all, its synthesizing, integrating nature. These qualities are framed in a philosophy which accepts the premise that the only meaningful way to study organization is as a system."[18]

Chester Barnard was one of the first management writers to utilize the systems approach.[19] Herbert Simon and his associates viewed the organization as a complex system of decision-making processes. Simon has ranged widely in seeking new disciplinary knowledge to integrate into his organization theories. However, the one broad consistency in both his research and writings has been the utilization of the systems approach. "The term 'systems' is being used more and more to refer to methods of scientific analysis that are particularly adapted to the unraveling of complexity."[20] He not only emphasizes this approach for the behavioral view of organizations but also stresses its importance in management science.

The systems approach has been advocated by a number of other writers in management science. Churchman and his associates were among the earliest to emphasize this view. "The comprehensiveness of O. R.'s aim is an example of a 'systems' approach, since 'system' implies an interconnected complex of functionally related components. Thus a business organization is a social or man-machine system."[21] Although the systems approach has been adopted and utilized in management science, the models typically used are closed in the sense that they consider only certain variables and exclude from consideration those not subject to quantification.

The sociologist George Homans uses systems concepts as a basis for his empirical research on social groups. He developed a model of social systems which can serve as an appropriate basis for small groups and also for larger organizations.[22] In his view, an organization is comprised of an external environmental system and an internal system of relationships which are mutually interdependent. There are three elements in a social system. *Activities* are the tasks which people perform. *Interactions* occur between people in the performance of these tasks, and *sentiments* develop between people. These elements are mutually interdependent.

Philip Selznick utilizes structural functional analysis and the systems approach in his studies of organizations. The institutional leader is concerned with the adaptation of the organization to its external systems. The organization is a dynamic system, constantly changing and adapting to internal and external pressures, and is in a continual process of evolution. The organization is a formal system influenced by the internal social structure and subject to the pressure of an institutional environment. "Cooperative systems are constituted of individuals interacting as wholes in relation to a formal system of coordination. The concrete structure is therefore a resultant of the reciprocal influences of the formal and informal aspects of organization. Furthermore, this structure is itself a totality, an adaptive 'organism' reacting to influences upon it from an external environment."[23] Selznick used this systems frame of reference for empirical research on governmental agencies and other complex organizations.

The systems approach has not only been used by many students of organization theory in the United States but has also provided the model in other countries. The group of social scientists associated with Tavistock Institute of Human Relations in London is one of the strongest proponents of the open-systems approach. As a result of a number of research studies in the mining, textile, and manufacturing industries in England and other countries, this group developed the concept of the sociotechnical system.[24] They also stressed that the organization is an open system in interaction with its environment.

The systems approach has also been adopted by social psychologists as a basis for studying organizations. Using open-systems theory as a general conceptual scheme, Katz and Kahn present a comprehensive theory of organization.[14] They suggest that the psychological approach has generally ignored or has not dealt effectively with the facts of structure and social organization, and they use systems concepts to develop an integrated model.

There are numerous examples of the utilization of the systems approach at operational levels. For example, the trend toward automation involves implementation of these ideas. Automation suggests a self-contained system with inputs, outputs, and a mechanism of control. Automated production systems for processing of materials are becoming increasingly important in many industries. Another phase which has been automated is information flow. With the introduction of large-scale, electronic data processing equipment, information processing systems have been developed for many applications. Physical distribution systems have received increasing attention. The concepts of logistics, or material management, have been used to emphasize the flow of materials through distribution channels.

The systems approach has been utilized as a basis of organization for many of our advanced defense and space programs. Program management is geared to changing managerial requirements in research, development, procurement, and utilization. With the new, complex programs such as ballistic missiles and advanced space programs it became impossible to think of individual segments or parts of the program as separate entities, and it was necessary to move to a broader systems approach.[25] In many other types of governmental projects, which require the integration of many agencies and activities – transportation problems, pollution control, and urban renewal, for example – the systems approach is being used.

These examples of the increasing trend in adapting the systems approach to modern organization theory and management practice are by no means exhaustive; they merely illustrate current developments. However, they are sufficient to indicate that increasing attention is being given to the study of organizations as complex systems. This modern view treats the organization as a system of mutually dependent parts and variables, which is part of the whole system of society. Modern organization theory and general systems theory are closely related. Many systems concepts taken from the investigation of other types of physical, biological, and social systems are meaningful to the study of organizations.

Organization: An Open System in Its Environment

Systems can be considered in two ways: (1) closed or (2) open and in interaction with their environments. This distinction is important in organization theory. Closed-system thinking stems primarily from the physical sciences and is applicable to mechanistic systems. Many of the earlier concepts in the social sciences and in organization theory were closed-system views because they considered the system under study as self-contained. Traditional management theories were primarily closed-system views concentrating only upon the internal operation of the organization and adopting highly rationalistic approaches taken from physical science models. The organization was considered as sufficiently independent so that its problems could be analyzed in terms of internal structure, tasks, and formal relationships – without reference to the external environment.

A characteristic of all closed systems is that they have an inherent tendency to move toward a static equilibrium and entropy. Entropy is a term which originated in thermodynamics and is applicable to all physical systems. It is the tendency for any closed system to move toward a chaotic or random state in which there is no further potential for energy transformation or work. "The disorder, disorganization, lack of patterning, or randomness of organization of a system is known as its *entropy*."[26] A closed system tends to increase in entropy over time, to move toward greater disorder and randomness.

Biological and social systems do not fall within this classification. The open-system view recognizes that the biological or social system is in a dynamic relationship with its environment and receives various inputs, transforms these inputs in some way, and exports outputs. The receipt of inputs in the form of material, energy, and information allows the open system to offset the process of entropy. These systems are open not only in relation to their environment but also in relation to themselves, or "internally" in that interactions between components affect the system as a whole. The open system adapts to its environment by changing the structure and processes of its internal components.[27]

The organization can be considered in terms of a general open-system model, as in Figure 1.1. The open system is in

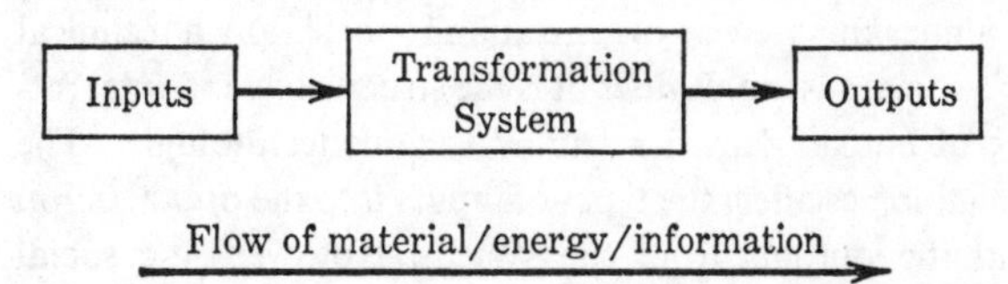

Figure 1.1 General model of organization as an open system.

continual interaction with its environment and achieves a "steady state" or dynamic equilibrium while still retaining the capacity for work or energy transformation. The survival of the system, in effect, would not be possible without continuous inflow, transformation, and outflow. In the biological or social system this is a continuous recycling process. The system must receive sufficient input of resources to maintain its operations and also to export the transformed resources to the environment in sufficient quantity to continue the cycle.

For example, the business organization receives inputs from the society in the form of people, materials, money, and information; it transforms these into outputs of products, services, and rewards to the organizational members sufficiently large to maintain their participation. For the business enterprise, money and the market provide a mechanism of the recycling of resources between the firm and its environment. The same kind of analysis can be made for all types of social organizations. Open-system views provide the basis for the development of a more comprehensive organization theory.

Organization: A Structured Sociotechnical System

In addition to being considered as an open system in interaction with its environment, the organization can also be viewed as a structured sociotechnical system. This view of the organization is set forth by Trist and his associates at the Tavistock Institute. Technology is based upon the tasks to be performed and includes the equipment, tools, facilities, and operating techniques. The social subsystem is the relationship between the participants in the organization. The technological and social subsystems are in interaction with each other and are interdependent. "Trist's concept of the socio-technical system arose from the consideration that any production system requires both a technological organization – equipment and process layout – and a work organization – relating those who carry out the necessary tasks to each other. Technological demands limit the kind of work organization possible, but a work organization has social and psychological properties of its own that are independent of technology."[24]

Under this view an organization is not simply a technical or a social system. Rather, it is the structuring and integrating of human activities around various technologies. The technologies affect the types of inputs into the organization and the outputs from the system. However, the social system determines the effectiveness and efficiency of the utilization of the technology.

Technical subsystems are determined by the task requirements of the organization and vary widely. The technical subsystem for the manufacturing of automobiles differs significantly from that in an oil refinery or in an electronics or aerospace company. Similarly, the task requirements and technology in a hospital are substantially different from those in a university. The technological subsystem is shaped by the specialization of knowledge and skills required, the types of machinery and equipment involved, and the layout of facilities.

Technology frequently prescribes the type of human inputs required. For example, an aerospace company requires the employment of many scientists, engineers, and other highly trained people. Technology also is a prime factor in determining the structure and relationships between jobs.

In addition to the technical subsystem, every organization has within its boundaries a psycho-social subsystem, which consists of the interactions, expectations and aspirations, sentiments, and values of the participants. However, it must be emphasized that these two subsystems, the technical and the social, cannot be looked at separately but must be considered in the context of the total organization. Any change in the technical subsystem will have repercussions on the social subsystem and conversely.

The organization *structure* can be considered as a third subsystem intermeshed between the technical and the social subsystems. The task requirements and technology have a fundamental influence upon the structure. Structure is concerned with the ways in which the tasks of the organization are divided into operating units and with the coordination of the units. In the formal sense, structure is set forth by the organization chart, by positions and job descriptions, and by rules and procedures. It also concerns the pattern of authority, communications, and work flow. In a sense, the organization structure provides for formalization of relationships between the technical and the psycho-social subsystems. However, it should be emphasized that this linkage is by no means complete and that many interactions and relationships occur between the technical and the psycho-social subsystems which bypass the formal structure.

One way of visualizing the organization as a structured sociotechnical system is shown in Figure 1.2. The goals and

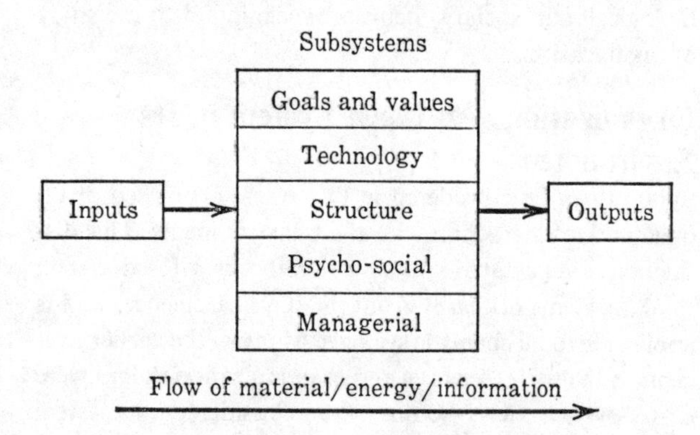

Figure 1.2 Organization as a sociotechnical system.

values, as well as the technical, structural, psycho-social, and managerial subsystems are shown as integral parts of the overall organization. This figure is an aid to understanding the evolution in organization theory. Traditional management theory emphasized the structural and managerial subsystems and was concerned with developing principles. The human relationists and behavioral scientists emphasized the psycho-social subsystem and focused their attention on motivation, group dynamics, and other related factors. The management science school emphasized the economic–technical subsystem and techniques for quantifying decision-making and control processes. Thus each approach to organization and management has emphasized particular primary subsystems, with little recognition of the importance of the others. The modern approach views the organization as a structured, sociotechnical system and considers *each* of the primary subsystems *and* their interactions.

Other Properties of Organizational Systems

The systems approach emphasizes that an organization has a number of interacting subsystems and can only be considered in a holistic or synergistic framework. A number of other characteristics of organizational systems can be identified.

Contrived systems

Social organizations are not natural like mechanical or biological systems; they are contrived. They have structure, but the structure of events rather than of physical components, and it cannot be separated from the processes of the system. The fact that social organizations are contrived by human beings suggests that they can be established for an infinite variety of objectives and do not follow the same life-cycle pattern of birth, maturity, and death as biological systems. Katz and Kahn say:

> Social structures are essentially contrived systems. They are made of men and are imperfect systems. They can come apart at the seams overnight, but they can also outlast by centuries the biological organisms which originally created them. The cement which holds them together is essentially psychological rather than biological. Social systems are anchored in the attitudes, perceptions, beliefs, motivations, habits, and expectations of human beings.[14]

Recognizing that the social organization is a contrived system cautions us against making an exact analogy between it and physical or biological systems.

Boundaries

The view of the organization as an open sociotechnical system suggests that there are boundaries which separate it from the environment. The concept of boundaries helps us understand the distinction between open and closed systems. The closed system has rigid, impenetrable boundaries, whereas the open system has permeable boundaries between itself and a broader supersystem. "Boundaries are the demarcation lines or regions for the definition of appropriate system activity, for admission of members into the system, and for other imports into the system. The boundary constitutes a barrier for many types of interaction between people on the inside and people on the outside, but it includes some facilitating device for the particular types of transactions necessary for organizational functioning."

The boundaries set the "domain" of the organization's activities. In a physical, mechanical, or biological system the boundaries can be identified. In a social organization, the boundaries are not easily definable and are determined primarily by the functions and activities of the organization. It is characterized by rather vaguely formed, highly permeable boundaries. Frequently, in the study of social organizations, where to draw the boundaries is a matter of convenience and strategy. Thus, in the study of a small work group, we may artificially establish the boundary to include only the activities of the immediate group and may consider interactions with other groups as outside these boundaries. Or, we might set our boundaries to include an entire department, division, company, industry, or total economic system. The boundaries of a social organization are often quite flexible and changeable over time, depending upon its activities and functions.

One of the key functions within any organization is that of boundary regulation between systems. A primary role of management is serving as a linking pin or boundary agent between the various subsystems to ensure integration and cooperation.[28] Furthermore, an important managerial function is that of serving as boundary agent between the organization and environmental systems.

The concept of *interface* is useful in understanding boundary relationships. An interface may be defined as the area of contact between one system and another. Thus, the business organization has many interfaces with other systems: suppliers of materials, the local community, prospective employees, unions, customers, and state, local, and federal governmental agencies. There are many transactional processes across systems boundaries at the interface involving the transfer of energy, materials, people, money, and information.

Hierarchy of systems

In general, all systems – physical, biological, and social – can be considered in a hierarchical sense. A system is composed of subsystems of a lower order and is also part of a supersystem. Thus, there is a hierarchy of the components in the system. Large organizations are almost universally hierarchical in structure. People are organized into groups; groups are organized into departments; departments are organized into divisions; divisions are organized into companies; and companies are part of an industry and economy. Many general systems writers have concluded that this hierarchical relationship is paramount in all types of systems. "Hierarchical subdivision is not a characteristic that is peculiar to human organizations. It is common to virtually all complex systems of which we have knowledge. [There are] strong reasons for believing that almost any system of sufficient complexity would have to have the rooms-within-rooms structure that we observe in actual human organizations. The reasons for hierarchy go far beyond the need for unity of command or other considerations relating to authority."[29]

The hierarchical structure is not only related to levels but is based upon the need for more inclusive clustering or combination of subsystems into a broader system, in order to coordinate activities and processes. In complex organizations there is a hierarchy of processes as well as structure.

Negative entropy

Closed physical systems are subject to the force of entropy which increases until eventually the entire system stops. The tendency toward maximum entropy is a movement to disorder, complete lack of resource transformation, and death. In a closed system, the change in entropy must always be positive. However, in the open biological or social system, entropy can be arrested and may even be transformed to negative entropy – a process of more complete organization and ability to transform resources. This is possible because in open systems and resources (material, energy, and information) utilized to arrest the entropy process are imported from the external environment. "Living systems, maintaining themselves in a steady state, can avoid the increase of entropy, and may even develop towards states of increased order and organization."[30] Obviously, for the biological system, this process of negative entropy is never perfect. The organism lives and grows for a period of time but is subject to deterioration and death. The contrived, or social, organization which can continue to import new human components and other resources in order to continue its functioning may be capable of indefinitely offsetting the entropy process. However, the only way in which the organization can offset entropy is by continually importing material, energy, and information in one form or another, transforming them, and redistributing resources to the environment.

The steady state or dynamic equilibrium

The concept of *steady state* is closely related to that of negative entropy. A closed system must eventually attain an equilibrium state with maximum entropy – death or disorganization. An open system, however, may attain a state where the system remains in dynamic equilibrium through the continuous inflow of material, energy, and information. This is called a steady state. This relationship between negative entropy and the steady state for living organisms and social systems is suggested by Emery and Trist.

> In contradistinction to physical objects, any living entity survives by importing into itself certain types of material from its environment, transforming these in accordance with its own system characteristics, and exporting other types back into the environment. By this process the organism obtains the additional energy that renders it "negentropic"; it becomes capable of attaining stability in a time-independent steady state – a necessary condition of adaptability to environmental variance.[31]

The steady state for the open system, as contrasted to the closed system subject to entropy, occurs while the system can still maintain its functions and perform effectively. Under this concept, an organization is able to adapt to changes in its environment and to maintain a continual steady state. An analogy can be seen in a biological system. The human body is able to maintain a steady state of body temperature in spite of wide variations in the environmental temperature. Obviously, there are limits to the degree to which the biological organism or the social organization can maintain a steady state in response to environmental changes. Massive environmental changes may be so great that it is impossible for the system to adapt. The organism dies, or the social organization is disbanded.

The steady state has an additional meaning; within the organizational system, the various subsystems have achieved a balance of relationships and forces which allows the total system to perform effectively. In biological organisms, the term *homeostasis* is applied to the organism's steady state. For social organizations, it is not an absolute steady state but rather a dynamic or moving equilibrium, one of continual adjustment to environmental and internal forces. The social organization will attempt to accumulate a certain "slack" of resources which helps it to maintain its equilibrium and to mitigate some of the possible variations in the inflow and environmental requirements.

Feedback mechanisms

The concept of feedback is important in understanding how a system maintains a dynamic equilibrium. Through the process of feedback, the system continually receives information from its environment which helps it adjust.

The concept of feedback has been used in looking at a number of biological phenomena. The maintenance of homeostasis, or the balance in a living organism, depends on a continual feedback of information to that organism from its environment. For example, the cooling of the blood from a drop in external temperatures stimulates certain centers in the brain which activate heat-producing mechanisms of the body, and the organism's temperature is monitored back to the center so that temperature is maintained at a steady level. Man uses principles of feedback in many of his physical activities. For example, in riding a bicycle, he receives feedback in regard to direction and balance which cause him to take corrective actions. Feedback can be both positive and negative, although for our purposes the most important consideration is that of negative feedback. Negative feedback is informational input which indicates that the system is deviating from a prescribed course and should readjust to a new steady state. Feedback is of vital importance in the complex organization which must continually receive informational inputs from its environment. Management is involved in interpreting and correcting for this information feedback. This is a vital part of the organizational control function.[32]

Adaptive and maintenance mechanisms

Systems must have two mechanisms which are often in conflict. First, in order to maintain an equilibrium, they must have maintenance mechanisms which ensure that the various subsystems are in balance and that the total system is in accord with its environment. The forces for maintenance are conservative, and attempt to prevent the system from changing so rapidly that the various subsystems and total system become out of balance. Second, adaptive mechanisms are necessary in order to provide a *dynamic* equilibrium, one which is changing over time. Therefore, the system must have adaptive mechanisms which allow it to respond to changing internal and external requirements.

Some forces within the social organization are geared to the maintenance of the system, and other forces and subsystems are geared to adaptation. These counteracting forces will often create tensions, stresses, and conflicts which are natural and should not be considered as totally dysfunctional.[33] Katz and Kahn describe the importance of maintenance and adaptive mechanisms for social organizations.

> If the system is to survive, *maintenance substructures* must be elaborated to hold the walls of the social maze in place. Even these would not suffice to insure organizational survival, however. The organization exists in a changing and demanding environment, and it must adapt constantly to the changing environmental demands. *Adaptive structures* develop in organizations to generate appropriate responses to external conditions.[14]

Growth through internal elaboration

In the closed system subject to the laws of physics, the system moves toward entropy and disorganization. In contrast, open systems appear to have the opposite tendency and move in the direction of greater differentiation and a higher level of organization. Bertalanffy points to the continual elaborations of biological organisms: "In organic development and evolution, a transition toward states of higher order and differentiation seems to occur. The tendency toward increasing complication has been indicated as a primary characteristic of the living, as opposed to inanimate, nature."[34]

This same process appears to hold true for most social systems. There is a tendency for them to elaborate their activities and to reach higher levels of differentiation and organization. An examination of certain attributes of complex organizations may help explain this tendency. Complex social organizations are made up of many subsystems, some of which have excess capacity or resources which create a continual pressure toward growth. Furthermore, social organizations will often try to encompass within their boundaries additional activities in order to limit uncertainties and to ensure their survival. The business organization may use vertical integration in order to ensure a continual source of raw materials. The pattern of conglomerate diversification and mergers by many corporations in the United States is another indication of this process. In many cases, these mergers result from product innovation and technological breakthroughs which provide opportunities for the organization to extend its boundaries into new areas. Or it may be attributed to an imbalance of managerial and technical skills which are seeking outlets for their activities and creativity. An indication of this elaboration has been the expansion of many of our large corporations into international activities, significantly increasing the boundaries of their operations.

There is also a tendency for complex organizations to achieve greater differentiation and specialization among internal subsystems. The increased number of specialized departments and activities in complex business organizations is readily apparent. The great proliferation of departments, courses, and subject matter in universities is another example of differentiation and elaboration.

Equifinality of open systems

In physical systems there is a direct cause and effect relationship between the initial conditions and the final state. Biological and social systems operate differently. The concept of *equifinality* says that final results may be achieved with different initial conditions and in different ways. This view suggests that the social organization can accomplish its objectives with varying inputs and with varying internal activities. Thus, the social system is not restrained by the simple cause and effect relationship of closed systems.

The equifinality of social systems has major importance for the management of complex organizations. The closed-system cause and effect relationship adopted from the physical sciences would suggest that there is *one best way* to achieve a given objective. The concept of equifinality suggests that the manager can utilize a varying bundle of inputs into the organization, can transform these in a variety of ways, and can achieve satisfactory output. Extending this view further suggests that the management function is not necessarily one of seeking a rigid optimal solution but rather one of having available a variety of satisfactory solutions to his decision problems.

The foregoing are a few of the characteristics of open systems. To the student who is initially exposed to some of these concepts, they may seem complicated. Much of our educational experience emphasizes closed-system approaches – mathematics and the physical sciences, for example. The open-system view, with the properties set forth in the previous sections, is pertinent for organization theory.

Managerial Systems

Having looked at the characteristics of organizations as open sociotechnical systems, we now turn to a more detailed consideration of managerial systems. Parsons provides a useful framework. He suggests that there are three managerial levels in the hierarchical structure of complex organizations: the technical or production level, the organizational (managerial) level, and the institutional or community level.[35]

The *technical* system is involved with the actual task performance in the organization. In the business firm, the technical functions involve the actual production and distribution of the products or services – the task performance activities of the organization. The technical system is not just involved with physical work but includes many types of technical activities utilizing knowledge. For example, research and development, production control, market research, operations research, and most accounting functions are part of the technical system. The teacher performs a technical task in the school, the doctor a technical task in the hospital, and scientists a technical task in the research laboratory. In complex organizations, many of the technical tasks are performed by professionals and highly trained experts, as well as by skilled and unskilled employees.

The second level, the *organizational*, coordinates and integrates the task performance of the technical system. A primary function of management at this level is to integrate the input of material, energy, and information to the technical level.

The *institutional* level is involved in relating the activities of the organization to its environmental system. The organization must continually receive supporting inputs from the society in order to carry on its transformation activities.

The managerial system spans the entire organization by directing the technology, organizing people and other resources, and relating the organization to its environment. However, there are basic differences in the orientation of the managerial system at these different levels. The technical level is concerned primarily with economic–technical rationality and tries to create certainty by "closing the technical core" to many variables. Thompson says, "Under norms of rationality, organizations seek to seal off their core technologies from environmental influences. Since complete closure is impossible, they seek to buffer environmental influences by surrounding their technical cores with input and output components."[36] The closed-system view is applicable to the "technical core" of the organization.

By contrast, at the institutional level the organization faces the greatest degree of uncertainty in terms of inputs from its environment over which it has little or no control. Therefore, management at this level should have an open-system view and concentrate on adaptive and/or innovative strategies. The organizational manager operates between the technical core and the institutional level and serves to mediate and coordinate the two. This level transforms the uncertainty of the environment into the economic-technical rationality necessary for input into the technical core. Figure 1.3 shows the organization as a composite system using these concepts. In describing this figure, Petit says:

> The technical level has a boundary that does not seal it off entirely from the firm's environment but does have a

high degree of closure. The organizational level has less closure and consequently is more susceptible to the intrusion of external elements. The institutional level has a highly permeable boundary and therefore is strongly affected by uncontrollable and unpredictable elements in the environment. Inputs enter the firm, are transposed into outputs in the technical subsystem, and then disposed of in the environment.[37]

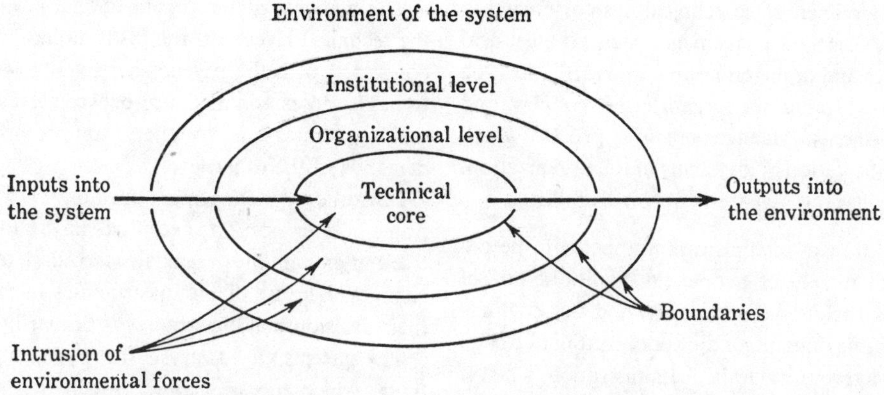

Environment of the system
Institutional level
Organizational level
Inputs into the system
Technical core
Outputs into the environment
Boundaries
Intrusion of environmental forces

Figure 1.3 The firm as a composite system (Thomas A. Petit, "A Behavioral Theory of Management", *Academy of Management Journal,* December, 1967, p. 346).

Thus the managerial system in the organization involves three levels – the technical core activities, intraorganizational interactions, and inter-institutional relationships. In many organizations these roles are separated theoretically. For example, in the university, the board of regents is thought of as fulfilling the institutional role, whereas the president, deans, and department heads are involved with organizational aspects. The professors, under this concept, perform the technical functions. In a hospital, the board of trustees performs the institutional role, the hospital administrator's staff is involved with organizational aspects, while the doctors, nurses, and other specialists perform the technical functions. Theoretically, in the business organization the board of directors relates the institution to its environment, upper and middle management deal with organizational aspects, and other employees perform the technical tasks. However, this distinction is not clear-cut in any of these organizations. For example, the president of a corporation usually has both institutional and organizational roles.

Parsons makes an interesting observation regarding these three levels. He stresses that they frequently are performed relatively independently and that therefore it is not realistic to talk of a "line authority" extending from the top or institutional level, through the organization level, and down to the technical level. "I have emphasized the relative independence of the three level-types of organization, an independence that constitutes my main objection to the continuous line-authority picture of formal organization. This relative independence means that there is, at each linkage point, a range of possible *different* types of articulation."[38] He suggests that there is a break in the line authority between these three levels.

This view represents a major departure from the traditional scalar chain of command. There is a break between the organizational level and the technical level, particularly in the case of professional personnel. The manager cannot, because of his limited knowledge in a specific area, exert absolute authority over professional personnel. He must rely upon them to provide specialized technical expertise which he cannot develop himself. The manager can veto recommendations of the highly trained specialist; however, he cannot propose alternatives. Thus, it is necessary to develop means of articulation and adjustment at the boundaries between these various levels.

However, this does not mean that these different managerial levels can operate independently. Quite the contrary, they are interdependent. For example, the institutional level must perform effectively if the organization is to receive the necessary inputs for the technical level. Also, the technical level must produce outputs efficiently to ensure that the organization receives environmental support.

Role of manager

The view of the organization as a sociotechnical system suggests a substantially different role for management than in traditional theory. In the traditional theory, the emphasis was upon economic–technical rationality. This closed-system view was appropriate for the technical level but not for the organizational and institutional levels. The human relations revolution did bring into focus the psycho-social subsystem but neglected the technical, structural, and environmental aspects. The management science approach adopted a closed-system view, focusing on the techniques of managerial decision making.

The view of an open sociotechnical system creates a more difficult role for the management system. It must deal with uncertainties and ambiguities and, above all, must be concerned with adapting the organization to new and changing requirements. Management is a process which spans and links the various subsystems of the organization. Thompson describes the management role as follows:

> The basic function of administration appears to be co-alignment, not merely of people (in coalition) but of institutionalized action – of technology and task environment into a viable domain, and of organizational design and structure appropriate to it. Administration, when it works well, keeps the organization at the nexus of the several necessary streams of action. Paradoxically, the administrative process must reduce uncertainty but at the same time search for flexibility.[36]

It is useful to consider the different types of managers necessary at the three levels in the managerial system. Petit suggests a basis for this classification in terms of the task performed, point of view, technique employed, time horizon, and decision-making strategy (see Figure 1.4).

The task at the technical level is primarily economic–technical and is concerned with efficiency of output with a given technology. Managers tend to adopt a closed-system view and utilize the methods of scientific management and management science. They have a task orientation with a short time perspective and utilize computational decision-making strategies.

The manager at the organizational level must integrate the technical level with the institutional level. He is more concerned with the psycho-social system of the organization. He serves as a mediator between levels. The decision-making strategy is to effect compromise between the demands of the other levels.

Institutional management must have a broad conceptual frame of reference. Top executives have a long-run perspective and prepare the organization to adapt to environmental changes. Because of the uncertainties in the environment, their decision-making strategy is primarily judgmental.

The systems view suggests that management faces situations which are dynamic, inherently uncertain, and frequently ambiguous. Management is not in full control of all the factors of production as suggested by traditional theory. It is strongly restrained by many environmental and internal forces. The technical system, the psycho-social system, and the environmental system all constrain the management system. Sayles outlines the role of management under the systems approach:

Figure 1.4 The managerial system: technical, organizational, and institutional levels

TYPE OF MANAGER	TASK	VIEWPOINT	TECHNIQUE	TIME HORIZON	DECISION-MAKING STRATEGY
Technical	Technical rationality	Engineering	Scientific management, operations research	Short run	Computational
Organizational	Coordination	Political	Mediation	Short run and long run	Compromise
Institutional	Deal with uncertainty, relate organization to environment	Conceptual and philosophical	Opportunistic surveillance, negotiate with environment	Long run	Judgmental

SOURCE: Adapted from Thomas A. Petit, "A Behavioral Theory of Management", *Academy of Management Journal*, December, 1967, p. 349.

A systems concept emphasizes that managerial assignments do not have these neat, clearly defined boundaries; rather, the modern manager is placed in a network of mutually dependent relationships. . . . The one enduring objective is the effort to build and maintain a predictable, reciprocating system of relationships, the behavioral patterns of which stay within reasonable physical limits. But this is seeking a moving equilibrium, since the parameters of the system (the division of labor and the controls) are evolving and changing. Thus the manager endeavors to introduce regularity in a world that will never allow him to achieve the ideal . . . Only managers who can deal with uncertainty, with ambiguity, and with battles that are never won but only fought well can hope to succeed.[39]

One of the most pervasive functions of management at all levels is decision making. Figure 1.4 indicates that the decision-making strategy differs for various managerial levels. At the technical level, closed-system approaches are appropriate. However, at the organizational and institutional levels, open-system decision-making approaches are necessary.

Summary

The systems approach provides an integrative framework for modern organization theory and management practice. General systems theory includes concepts for integrating knowledge in the physical, biological, and social sciences.

There is a close relationship between general systems theory and the development of functionalism in the social sciences. Functionalism emphasizes integration of parts and subsystems into a functional whole. It has been used as a primary frame of reference in anthropology and sociology. In the field of psychology, the systems concept has achieved prominence. Modern economics also uses this approach, particularly in dynamic equilibrium analysis and input-output studies. The discipline of cybernetics is founded on the systems approach, focusing on communication and information flow in complex systems. Although it has been applied primarily to mechanistic systems, its model of feedback, control, and regulation has applicability for social systems as well.

The systems approach is directly related to organization theory. Traditional theory used closed-system thinking. Modern theory has moved toward considering the organization as an open system interacting with its environment. In contrast to the closed or mechanical system, the open system is not subject to the process of entropy – it can maintain a dynamic equilibrium by importing material, energy, and information from its environment.

The organization can be viewed as a structured sociotechnical system with five primary components – goals and values, and technical, structural, psycho-social, and managerial subsystems.

There are several key characteristics of organizational systems. They are not natural, like physical or biological systems, but are *contrived.* There are *boundaries* which separate the organization from its environment. In general, a system is composed of subsystems of a lower order and is also part of a supersystem; there is a *hierarchy* of systems. In open biological or social systems, entropy can be arrested and may even be transformed to *negative entropy* – a process of more complete organization. The concept of *steady state* is closely related to that of negative entropy. The organization is able to adapt to changes in its environment and to maintain a continual dynamic equilibrium.

The concept of *feedback* is important in understanding how a system maintains a steady state. Through the process of feedback, it continually receives information from its environment which helps it to adjust. A system must have both *adaptive and maintenance mechanisms.* The forces for maintenance are conservative and attempt to prevent the system from changing so rapidly that the various subsystems become out of balance. In contrast, adaptive mechanisms are necessary in order to provide for change. Open systems display *growth through internal elaboration.* They tend to move in the direction of greater differentiation and to a higher level of organization. Finally, open systems have the characteristic of *equifinality* – objectives may be achieved with varying inputs and in different ways.

There are three levels in the managerial system of complex organizations; technical, organizational, and institutional. The *technical level* is involved with actual task performance. The *organizational level* integrates the technical and institutional levels. The *institutional level* relates the activities of the organization to its environment. The view of the organization as a sociotechnical system creates a different role for the manager. He must integrate and balance the various subsystems and their activities in the environmental setting.

References

1 The name "general systems theory" and many of the basic concepts were set forth by the biologist Ludwig von Bertalanffy. For a general discussion of his views, see "The Theory of Open Systems in Physics and Biology", *Science*, Jan. 13, 1950, pp. 23–9; and

"General System Theory: A New Approach to Unity of Science", *Human Biology*, December 1951, pp. 302–61.

2 Boulding, Kenneth E. (1956) "General Systems Theory: The Skeleton of Science", *Management Science*, April, pp. 197–208.

3 von Bertalanffy, Ludwig (1952) *Problems of Life*, John Wiley & Sons, Inc., New York, p. 201. On page 176 he stresses this view: "If we survey the various fields of modern science, we notice a dramatic and amazing evolution. Similar conceptions and principles have arisen in quite different realms, although this parallelism of ideas is the result of independent developments, and the workers in the individual fields are hardly aware of the common trend. Thus, the principles of wholeness, of organization, and of the dynamic conception of reality become apparent in all fields of science."

4 Ashby, W. Ross (1964) in Mesarovic, Mihajlo D. (ed.), *Views on General Systems Theory*, John Wiley & Sons, Inc., New York, p. 166.

5 Eddington suggests that attention has been focused more and more on overall systems as frames of reference for analytical work in various areas. This synthesizing process "marked a reaction from the view that everything to which science need pay attention is discovered by microscopic dissection of objects. It provided an alternative stand-point in which the centre of interest is shifted from the entities reached by the customary analysis (atoms, electric potentials, etc.) to qualities possessed by the system as a whole, which cannot be split up and located – a little here and a little bit there." Eddington, Sir Arthur (1958) *The Nature of the Physical World*, The University of Michigan Press, Ann Arbor, Mich., pp. 103–4.

6 For an overview of the integrative possibilities of the systems approach, see Walter Buckley (ed.) (1968) *Modern Systems Research for the Behavioral Scientist*, Aldine Publishing Company, Chicago.

7 Martindale, Don (1965) *Functionalism in the Social Sciences*, Monograph 5, American Academy of Political and Social Science, February, pp. viii–ix.

8 Robert K. Merton discusses various connotations of the word *function* in *Social Theory and Social Structure*, rev. ed., The Free Press of Glencoe, New York, 1957, pp. 20–25.

9 Radcliffe-Brown, A. R. (1952) *Structure and Function in Primitive Society*, Cohen & West, London, and Bronislaw Malinowski (1960) *A Scientific Theory of Culture*, Oxford University Press, New York.

10 Talcott Parsons uses the systems approach in much of his writings. His *The Social System*, The Free Press of Glencoe, New York, 1951, presents a comprehensive treatise on his views.

11 For a view of Pareto's works, see Lawrence J. Henderson (1935) *Pareto's General Sociology*, Harvard University Press, Cambridge, Mass.

12 "A *gestalt* is an organized entity or whole in which the parts, though distinguishable, are interdependent; they have certain characteristics produced by their inclusion in the whole, and the whole has some characteristics belonging to none of the parts. The gestalt thus constitutes a 'unit segregated from its surroundings', behaving according to certain laws of energy distribution. It is found throughout human behaviour as well as in physiological and physical events and is thus a fundamental aspect of scientific data." Gould, Julius and Kolb, William L. (eds.) (1964) *A Dictionary of the Social Sciences*, The Free Press of Glencoe, New York, p. 287.

13 Whitaker, Ian (1965) "The Nature and Value of Functionalism in Sociology", in *Functionalism in the Social Sciences*, Monograph 5, American Academy of Political and Social Science, February, pp. 137–8.

14 Katz, Daniel and Kahn, Robert L. (1966) *The Social Psychology of Organizations*, John Wiley & Sons, Inc. New York, is an example of the movement of social psychology into broader systems of analysis.

15 Leontief, Wassily *et al.* (1953) *Studies in the Structure of the American Economy*, Oxford University Press, New York, p. 8.

16 Wiener, Norbert (1948) *Cybernetics*, John Wiley & Sons, Inc., New York.

17 Chin, Robert (1961) "The Utility of System Models and Developmental Models for Practitioners", in Warren G. Bennis, Kenneth D. Benne, and Robert Chin (eds.), *The Planning of Change*, Holt, Rinehart & Winston, Inc., New York, p. 202.

18 Scott, William G. (1967) *Organization Theory*, Richard D. Irwin, Inc., Homewood, Ill., pp. 122–3.

19 Barnard, Chester I. (1938) *The Functions of the Executive*, Harvard University Press, Cambridge, Mass.

20 Simon, Herbert A. (1964) "Approaching the Theory of Management", in Harold Koontz (ed.), *Toward a Unified Theory of Management*, McGraw-Hill Book Company, New York, pp. 82–3.

21 Churchman, C. West, Ackoff, Russell L. and Arnoff, E.

Leonard (1957) *Introduction to Operations Research*, John Wiley & Sons, Inc., New York, p. 7.

22 Homans, George C. (1950) *The Human Group*, Harcourt, Brace & World, Inc., New York.

23 Selznick, Philip (1948) "Foundations of the Theory of Organization", *American Sociological Review*, February, pp. 23–35.

24 Emery, F. E. and Trist, E. L. (1960) "Socio-technical Systems", in C. West Churchman and Michel Cerhulst (eds.), *Management Sciences: Models and Techniques*, Pergamon Press, New York, vol. 2, pp. 83–97; and Rice, A. K. (1963) *The Enterprise and Its Environment*, Tavistock Publications, London.

25 For a discussion of the evolution of this approach in military and space programs, see Kast, Fremont E. and Rosenzweig, James E. (1965) "Organization and Management of Space Programs", in Ordway, Frederick I. III (ed.), *Advances in Space Science and Technology*, Academic Press, Inc., New York, vol. 7, pp. 273–364.

26 Miller, James G. (1965) "Living Systems: Basic Concepts", *Behavioral Science*, July, p. 195.

27 Buckley, Walter, "Society as a Complex Adaptive System", in Buckley, *op. cit.*, pp. 490–1 (see Article 9).

28 This point is made by Rensis Likert in *New Patterns of Management*, McGraw-Hill Book Company, New York, 1961. In his interaction-influence system, he recommends the overlapping-group form of organization in which a "linking-pin function" is performed to integrate activities of the various subsystems in the organization.

29 Simon, Herbert A. (1960) *The New Science of Management Decision*, Harper & Row, Publishers, Incorporated, New York, pp. 40–2.

30 von Bertalanffy, Ludwig (1956) "General System Theory", in *General Systems*, Yearbook of the Society for the Advancement of General Systems Theory, vol. I, p. 4.

31 Emery, F. E. and Trist, E. L. (1965) "The Causal Texture of Organizational Environments", *Human Relations*, February, p. 2.1.

32 The concept of negative feedback is fundamental to the discipline of cybernetics. Mechanisms of feedback are basic to the purposeful behavior in man-made machines, as well as in living organisms and social systems. In discussing the cybernetic hypothesis, Wisdom says, "We may describe machines that embody negative feed-back mechanisms as proceeding by 'trial and error', or as 'error-compensating', or – best – as 'self-correcting'; and we may define 'a simple negative feed-back mechanism' as a mechanism by which part of the input-energy of a machine is utilized at intervals to impose a check on the output-energy. The basic hypothesis of cybernetics is that the chief mechanism of the central nervous system is one of negative feed-back." Wisdom, J. O. (1956) "The Hypothesis of Cybernetics", *General Systems*, Yearbook of the Society for the Advancement of General Systems Theory, vol. I, p. 112.

33 Robert Chin says, "The presence of tensions, stresses or strains, and conflict within the system often are reacted to by people in the system as if they were shameful and must be done away with. Tension reduction, relief of stress and strain, and conflict resolution become the working goals of practitioners but sometimes at the price of overlooking the possibility of increasing tensions and conflict in order to facilitate creativity, innovation, and social change." Chin, *op. cit.*, p. 204.

34 von Bertalanffy, Ludwig, (1950) "The Theory of Open Systems in Physics and Biology", *Science*, Jan. 13, p. 26.

35 Parsons, Talcott (1960) *Structure and Process in Modern Societies*, The Free Press of Glencoe, New York, pp. 60–96. Parsons calls the middle level in the managerial hierarchy "the managerial level". Actually management is involved in each of the three levels, and for our purpose it is more appropriate to call this "the organizational level".

36 Thompson, James D. (1967) *Organizations in Action*, McGraw-Hill Book Company, New York, p. 24.

37 Petit, Thomas A. (1967) "A Behavioral Theory of Management", *Academy of Management Journal*, December, p. 346.

38 Parsons, *Structure and Process in Modern Societies*, p. 95.

39 Sayles, Leonard (1964) *Managerial Behavior*, McGraw-Hill Book Company, New York, pp. 258–9.

2 General system theory – a critical review

by Ludwig von Bertalanffy

Since creative thought is the most important thing that makes people different from monkeys, it should be treated as a commodity more precious than gold and preserved with great care. – A. D. Hall, A Methodology for Systems Engineering.

It is more than 15 years since the writer has first presented, to a larger public, the proposal of a General System Theory.* Since then, this conception has been widely discussed and was applied in numerous fields of science. When an early reviewer found himself "hushed into awed silence" by the idea of a General System Theory, now in spite of obvious limitations, different approaches and legitimate criticism, few would deny the legitimacy and fertility of the interdisciplinary systems approach.

Even more: The systems concept has not remained in the theoretical sphere, but became central in certain fields of applied science. When first proposed, it appeared to be a particularly abstract and daring, theoretical idea. Nowadays "systems engineering" "research", "analysis" and similar titles have become job denominations. Major industrial enterprises and government agencies have departments, committees or at least specialists to the purpose; and many universities offer curricula and courses for training.

Thus the present writer was vindicated when he was among the first to predict that the concept of "system" is to become a fulcrum in modern scientific thought. In the words of a practitioner of the science [R. L. Ackoff]:

> In the last two decades we have witnessed the emergence of the "system" as a key concept in scientific research. Systems, of course, have been studied for centuries, but something new has been added. . . . The tendency to study systems as an entity rather than as a conglomeration of parts is consistent with the tendency in contemporary science no longer to isolate phenomena in narrowly confined contexts, but rather to open interactions for examination and to examine larger and larger slices of nature. Under the banner of *systems research* (and its many synonyms) we have also witnessed a convergence of many more specialized contemporary scientific developments. . . . These research pursuits and many others are being interwoven into a cooperative research effort involving an ever-widening spectrum of scientific and engineering disciplines. We are participating in what is probably the most comprehensive effort to attain a synthesis of scientific knowledge yet made.

This, however, does not preclude but rather implies that obstacles and difficulties are by no means overcome as is only to be expected in a major scientific reorientation. A reassessment of General Systems Theory, its foundations, achievements, criticisms and prospects therefore appears in place. The present study aims at this purpose.

According to the Preface to the VIth volume of *General Systems* by Meyer, the greatest number of enquiries made asks for "new statements describing the method and significance of the idea". Another central theme is "the organismic viewpoint". As one of the original proponents of the *Society for General Systems Research* and founders of the organismic viewpoint in biology, the author feels obliged to answer this challenge as well as readily admitted limitations of his knowledge and techniques permit.

1 The rise of interdisciplinary theories

The motives leading to the postulate of a general theory of systems can be summarized under a few headings.

1. Up to recent times the field of science as a nomothetic endeavor, i.e., trying to establish an explanatory and predictive system of laws, was practically identical with theoretical physics. Few attempts at a system of laws in nonphysical fields gained general recognition; the biologist would first think of genetics. However, in recent times the biological, behavioral and social sciences have come into their own, and so the problem became urgent whether an expansion of conceptual schemes is possible to deal with fields and problems where application of physics is not sufficient or feasible.

2. In the biological, behavioral and sociological fields, there exist predominant problems which were neglected in classical science or rather which did not enter into its considerations. If we look at a living organism, we observe an amazing order, organization, maintenance in continuous change, regulation and apparent teleology. Similarly, in

* This article first appeared in 1962.

human behavior goal-seeking and purposiveness cannot be overlooked, even if we accept a strictly behavioristic standpoint. However, concepts like organization, directiveness, teleology, etc., just do not appear in the classic system of science. As a matter of fact, in the so-called mechanistic world view based upon classical physics, they were considered as illusory or metaphysical. This means, to the biologist for example, that just the specific problems of living nature appeared to lie beyond the legitimate field of science.

3. This in turn was closely connected with the structure of classical science. The latter was essentially concerned with two-variable problems, linear causal trains, one cause and one effect, or with few variables at the most. The classical example is mechanics. It gives perfect solutions for the attraction between two celestial bodies, a sun and a planet, and hence permits one to exactly predict future constellations and even the existence of still undetected planets. However, already the three-body problem of mechanics is unsolvable in principle and can only be approached by approximations. A similar situation exists in the more modern field of atomic physics. Here also two-body problems such as that of one proton and electron are solvable, but trouble arises with the many-body problem. One-way causality, the relation between "cause" and "effect" or of a pair or few variables cover a wide field. Nevertheless, many problems particularly in biology and the behavioral and social sciences, essentially are multivariable problems for which new conceptual tools are needed. Warren Weaver, cofounder of information theory, had expressed this in an often-quoted statement. Classical science, he stated, was concerned either with linear causal trains, that is, two-variable problems; or else with unorganized complexity. The latter can be handled with statistical methods and ultimately stems from the second principle of thermodynamics. However, in modern physics and biology, problems of organized complexity, that is, interaction of a large but not infinite number of variables, are popping up everywhere and demand new conceptual tools.

4. What has been said are not metaphysical or philosophic contentions. We are not erecting a barrier between inorganic and living nature which obviously would be inappropriate in view of intermediates such as viruses, nucleoproteins and self-duplicating units in general which in some way bridge the gap. Nor do we protest that biology is in principle "irreducible to physics" which also would be out of place in view of the tremendous advances of physical and chemical explanation of life processes. Similarly, no barrier between biology and the behavioral and social sciences is intended. This, however, does not obviate the fact that in the fields mentioned we do not have appropriate conceptual tools serving for explanation and prediction as we have in physics and its various fields of application.

5. It therefore appears that an expansion of science is required to deal with those aspects which are left out in physics and happen to concern just the specific characteristics of biological, behavioral, and social phenomena. This amounts to new conceptual models to be introduced. Every science is a model in the broad sense of the word, that is a conceptual structure intended to reflect certain aspects of reality. One such model is the system of physics – and it is an incredibly successful one. However, physics is but *one* model dealing with certain aspects of reality. It needs not to have monopoly, nor is it *the* reality as mechanistic methodology and metaphysics presupposed. It apparently does not cover all aspects and represents, as many specific problems in biology and behavioral science show, a limited aspect. Perhaps it is possible to introduce other models dealing with aspects outside of physics.

These considerations are of a rather abstract nature. So perhaps some personal interest may be introduced by telling how the present author was led into this sort of problem.

When, some 40 years ago, I started my life as a scientist, biology was involved in the mechanism–vitalism controversy. The mechanistic procedure essentially was to resolve the living organism into parts and partial processes: the organism was an aggregate of cells, the cell one of colloids and organic molecules, behavior a sum of unconditional and conditioned reflexes, and so forth. The problems of organization of these parts in the service of maintenance of the organism, of regulation after disturbances and the like were either by-passed or, according to the theory known as vitalism, explainable only by the action of soul-like factors, like hobgoblins as it were, hovering in the cell or the organism – which obviously was nothing less than a declaration of bankruptcy of science. In this situation, I was led to advocate the so-called organismic viewpoint. In one brief sentence, it means that organisms are organized things and, as biologists, we have to find out about it. I tried to implement this organismic program in various studies on metabolism, growth, and biophysics of the organism. One way in this respect was the so-called theory of open systems and steady states which essentially is an expansion of conventional physical chemistry, kinetics and thermodynamics. It appeared, however, that I could not stop on the way once taken and so I was led to a still further generalization which I called "General System Theory". The idea goes back for some considerable time – I presented it first in 1937 in Charles Morris' philosophy seminar at the University of Chicago. However, at this time theory was in bad reputation in biology, and I was afraid of what Gauss, the mathematician, called the "clamor of the Boeotians". So I left my drafts in the drawer, and it was only after the war that my first publications in this respect appeared.

Then, however, something interesting and surprising happened. It turned out that a change in intellectual climate had taken place, making model building and abstract generalizations fashionable. Even more: quite a number of scientists had followed similar lines of thought. So General Systems Theory, after all, was not isolated or a personal idiosyncrasy as I have believed, but rather was one within a group of parallel developments.

Naturally, the maxims enumerated above can be formulated in different ways and using somewhat different terms. In principle, however, they express the viewpoint of the more advanced thinkers of our time and the common ground of system theorists. The reader may, for example, compare the presentation given by Rapoport and Horvath which is an excellent and independent statement and therefore shows even better the general agreement.

There is quite a number of novel developments intended to meet the goals indicated above. We may enumerate them in a brief survey:

(1) Cybernetics, based upon the principle of feedback or circular causal trains providing mechanisms for goal-seeking and self-controlling behavior.

(2) Information theory, introducing the concept of information as a quantity measurable by an expression isomorphic to negative entropy in physics, and developing the principles of its transmission.

(3) Game theory, analyzing in a novel mathematical framework, rational competition between two or more antagonists for maximum gain and minimum loss.

(4) Decision theory, similarly analyzing rational choices, within human organizations, based upon examination of a given situation and its possible outcomes.

(5) Topology or relational mathematics, including non-metrical fields such as network and graph theory.

(6) Factor analysis, i.e., isolation by way of mathematical analysis, of factors in multivariable phenomena in psychology and other fields.

(7) General system theory in the narrower sense (G.S.T.), trying to derive from a general definition of "system" as complex of interacting components, concepts characteristic of organized wholes such as interaction, sum, mechanization, centralization, competition, finality, etc., and to apply them to concrete phenomena.

While systems theory in the broad sense has the character of a basic science, it has its correlate in applied science, sometimes subsumed under the general name of Systems Science. This development is closely connected with modern automation. Broadly speaking, the following fields can be distinguished:

Systems Engineering, i.e., scientific planning, design, evaluation, and construction of man–machine systems;

Operations research, i.e., scientific control of existing systems of men, machines, materials, money, etc.

Human Engineering, i.e., scientific adaptation of systems and especially machines in order to obtain maximum efficiency with minimum cost in money and other expenses.

A very simple example for the necessity of study of "man–machine systems" is air travel. Anybody crossing continents by jet with incredible speed and having to spend endless hours waiting, queuing, being herded in airports can easily realize that the physical techniques in air travel are at their best, while "organizational" techniques still are on a most primitive level.

Although there is considerable overlapping, different conceptual tools are predominant in the individual fields. In systems engineering, cybernetics and information theory, also general system theory are used. Operations research uses tools such as linear programming and game theory. Human engineering, concerned with the abilities, physiological limitations and variabilities of human beings, includes biomechanics, engineering psychology, human factors, etc., among its tools.

The present survey is not concerned with applied systems science; the reader is referred to Hall's book as an excellent textbook of systems engineering. However it is well to keep in mind that the systems approach as a novel concept in science has a close parallel in technology. The systems viewpoint in recent science stands in a similar relation to the so-called "mechanistic" viewpoint, as stands systems engineering to physical technology.

All these theories have certain features in common. *Firstly*, they agree in the emphasis that something should be done about the problems characteristic of the behavioral and biological sciences, but not dealt with in conventional physical theory. *Secondly*, these theories introduce concepts and models novel in comparison to physics: for example, a generalized system concept, the concept of information compared to energy in physics. *Thirdly*, these theories are particularly concerned with multivariable problems, as mentioned before. *Fourthly*, these models are interdisciplinary and transcend the conventional fields of science. If, for example, you scan the *Yearbooks* of the *Society for General Systems Research*, you notice the breadth of application: Considerations similar or even identical in structure are applied to phenomena of different kinds and levels, from networks of chemical reactions in a cell to populations of animals, from electrical engineering to the social sciences. Similarly, the basic concepts of cybernetics stem from certain special fields in modern technology. However, starting with the simplest case of a thermostat which by way of feedback maintains a certain temperature and advancing to servomechanisms and automation in modern technology, it turns out that similar schemes are applicable to many biological phenomena of regulation or behavior. Even more, in many instances there is a formal correspondence or isomorphism of general principles or even of special laws. Similar mathematical formulations may apply to quite different phenomena. This entails that general theories of systems, among other things, are labor-saving devices: A set of principles may be transferred from one field to another, without the need to duplicate the effort as has often happened in science of the past. *Fifthly*, and perhaps most important: Concepts like wholeness, organization, teleology and directiveness appeared in mechanistic science to be unscientific or metaphysical. Today they are taken seriously and as amenable to scientific analysis. We have conceptual and in some cases even material models which can represent those basic characteristics of life and behavioral phenomena.

An important consideration is that the various approaches enumerated are not, and should not be considered to be monopolistic. One of the important aspects of the modern changes in scientific thought is that there is no unique and all-embracing "world system". All scientific constructs are models representing certain aspects or perspectives of reality. This even applies to theoretical physics: far from being a metaphysical presentation of ultimate reality (as the materialism of the past proclaimed and modern positivism still implies) it is but one of these models and, as recent developments show, neither exhaustive nor unique. The various "systems theories" also are models that mirror different aspects. They are not mutually exclusive and often combined in application. For example, certain phenomena may be amenable to scientific exploration by way of cybernetics, others by way of general system theory; or even in the same phenomenon, certain aspects may be describable in the one or the other way. Cybernetics combine the information and feedback models, models of the nervous system net and information theory, etc. This, of course, does not preclude but rather implies the hope for further synthesis in which the various approaches of the present toward a theory of "wholeness" and "organization" may be integrated and unified. Actually, such further syntheses, e.g., between irreversible thermodynamics and information theory, are slowly developing.

The differences of these theories are in the particular model conceptions and mathematical methods applied. We therefore come to the question in what ways the program of systems research can be implemented.

2 Methods of general systems research

Ashby has admirably outlined two possible ways or general methods in systems study:

> Two main lines are readily distinguished. One, already well developed in the hands of von Bertalanffy and his co-workers, takes the world as we find it, examines the various systems that occur in it – zoological, physiological, and so on – and then draws up statements about the regularities that have been observed to hold. This method is essentially empirical. The second method is to start at the other end. Instead of studying first one system, then a second, then a third, and so on, it goes to the other extreme, considers the set of all conceivable systems and then reduces the set to a more reasonable size. This is the method I have recently followed.

It will easily be seen that all systems studies follow one or the other of these methods or a combination of both. Each of these approaches has its advantages as well as shortcomings.

1. The first method is empirico-intuitive; it has the advantage that it remains rather close to reality and can easily be illustrated and even verified by examples taken from the individual fields of science. On the other hand, the approach lacks mathematical elegance and deductive strength and, to the mathematically minded, will appear naïve and unsystematic.

Nevertheless, the merits of this empirico-intuitive procedure should not be minimized.

The present writer has stated a number of "system principles", partly in the context of biological theory and without explicit reference to G.S.T., partly in what emphatically was entitled an "Outline" of this theory. This was meant in the literal sense: It was intended to call attention to the desirability of such field, and the presentation was in the way of a sketch or blueprint, illustrating the approach by simple examples.

However, it turned out that this intuitive survey appears to be remarkably complete. The main principles offered such as wholeness, sum, centralization, differentiation, leading part, closed and open system, finality, equifinality, growth in time, relative growth, competition, have been used in manifold ways (e.g., general definition of system; types of growth; systems engineering; social work). Excepting minor variations in terminology intended for clarification or due to the subject matter, no principles of similar significance were added – even though this would be highly desirable. It is perhaps even more significant that this also applies to considerations which do not refer to the present writer's work and hence cannot be said to be unduly influenced by it. Perusal of studies such as those by Beer and Kremyanskiy on principles, Bradley and Calvin on the network of chemical reactions, Haire on growth or organizations, etc., will easily show that they are also using the "Bertalanffy principles".

2. The way of deductive systems theory was followed by Ashby. A more informal presentation which summarizes Ashby's reasoning lends itself particularly well to analysis.

Ashby asks about the "fundamental concept of machine" and answers the question by stating "that its internal state, and the state of its surroundings, defines uniquely the next state it will go to". If the variables are continuous, this definition corresponds to the description of a dynamic system by a set of ordinary differential equations with time as the independent variable. However, such representation by differential equations is too restricted for a theory to include biological systems and calculating machines where discontinuities are ubiquitous. Therefore the modern definition is the "machine with input": It is defined by a set S of internal states, a set I of input and a mapping f of the product set $I \times S$ into S. "Organization", then, is defined by specifying the machine's states S and its conditions I. If S is a product set $S = \pi_i T_i$, with i as the parts and T is specified by the mapping f. A "self-organizing" system, according to Ashby, can have two meanings, namely: (1) The system starts with its parts separate, and these parts then change toward forming connections (example: cells of the embryo, first having little or no effect on one another, join by formation of dendrites and synapses to form the highly interdependent nervous system). This first meaning is "changing from unorganized to organized". (2) The second meaning is "changing from a bad organization to a good one" (examples: a child whose brain organization makes it fire-seeking at first, while a new brain organization makes him fire-avoiding; an automatic pilot and plane coupled first by deleterious positive feedback and then improved). "There the organization is bad. The system would be 'self-organizing' if a change were automatically made" (changing positive into negative feedback). But "*no machine can be self-organizing in this sense*" (author's emphasis). For adaptation (e.g., of the homeostat or in a self-programming computer) means that we start with a set S of states, and that f changes into g, so that organization is a variable, e.g., a function of time $\alpha(t)$ which has first the value f and later the value g. However, this change "cannot be ascribed to any cause in the set S; *so it must come from some outside agent, acting on the system* S *as input*" (our emphasis). In other terms, to be "self-organizing" the machine S must be coupled to another machine.

This concise statement permits observation of the limitations of this approach. We completely agree that description by differential equations is not only a clumsy but, in principle, inadequate way to deal with many problems of organization. The author was well aware of this emphasizing that a system of simultaneous differential equations is by no means the most general formulation and is chosen only for illustrative purposes.

However, in overcoming this limitation, Ashby introduced another one. His "modern definition" of system as a "machine with input" as reproduced above, supplants the general system model by another rather special one: the cybernetic model, i.e., a system open to information but closed with respect to entropy transfer. This becomes apparent when the definition is applied to "self-organizing systems". Characteristically, the most important kind of these has no place in Ashby's model, namely, systems organizing themselves by way of progressive differentiation, evolving from states of lower to states of higher complexity. This is, of course, the most obvious form of "self-organization", apparent in ontogenesis, probable in phylogenesis, and certainly also valid in many social organizations. We have here not a question of "good" (i.e., useful, adaptive) or "bad" organization which, as Ashby correctly emphasizes, is relative on circumstances; increase in differentiation and complexity – whether useful or not – is a criterion that is objective and at least on principle amenable to measurement (e.g., in terms of decreasing entropy, of information). Ashby's contention that "no machine can be self-organizing", more explicitly, that the "change cannot be ascribed to any cause in the set S" but "must come from some outside agent, an input" amounts to exclusion of self-differentiating systems. The reason that such systems are not permitted as "Ashby machines" is patent. Self-differentiating systems that evolve toward higher complexity (decreasing entropy) are, for thermodynamic reasons, possible only as open systems, i.e., systems importing matter containing free energy to an amount over-compensating the increase in entropy due to irreversible processes within the system ("import of negative entropy"). However, we cannot say that "this change comes from some outside agent, an input"; the differentiation within a developing embryo and organism is due to its internal laws of organization, and the input (e.g., oxygen supply which may vary quantitatively, or nutrition which can vary qualitatively within a broad spectrum) makes it only possible energetically.

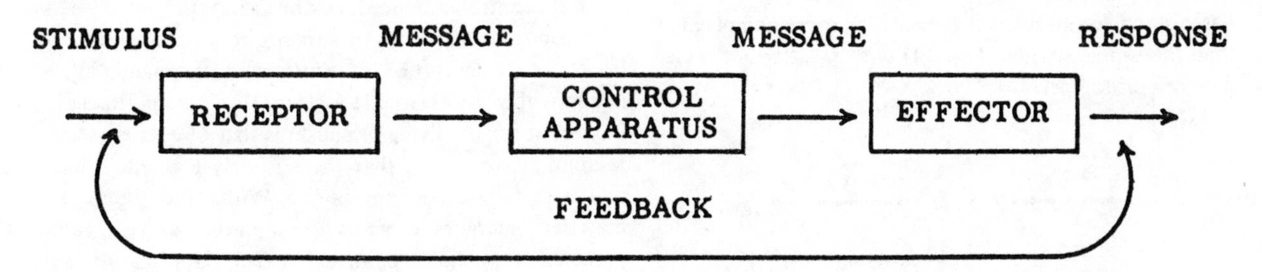

Figure 2.1 Simple feedback model.

The above is further illustrated by additional examples given by Ashby. Suppose a digital computer is carrying through multiplications at random; then the machine will "evolve" toward showing even numbers (because products even × even as well as even × odd give numbers even), and eventually only zeros will be "surviving". In still another version Ashby quotes Shannon's Tenth Theorem, stating that if a correction channel has capacity H, equivocation of the amount H can be removed, but no more. Both examples illustrate the working of closed systems: The "evolution" of the computer is one toward disappearance of differentiation and establishment of maximum homogeneity (analog to the Second Principle in closed systems); Shannon's Theorem similarly concerns closed systems where no negative entropy is fed in. Compared to the information content (organization) of a living system, the imported matter (nutrition, etc.) carries not information but "noise". Nevertheless, its negative entropy is used to maintain or even to increase the information content of the system. This is a state of affairs apparently not provided for in Shannon's Tenth Theorem, and understandably so as he is not treating information transfer in open systems with transformation of matter.

In both respects, the living organism (and other behavioral and social systems) is not an Ashby machine because it evolves toward increasing differentiation and inhomogeneity, and can correct "noise" to a higher degree than an inanimate communication channel. Both, however, are consequences of the organism's character as an open system.

Incidentally, it is for similar reasons that we cannot replace the concept of "system" by the generalized "machine" concept of Ashby. Even though the latter is more liberal compared to the classic one (machines defined as systems with fixed arrangement of parts and processes), the objections against a "machine theory" of life remain valid.

These remarks are not intended as adverse criticism of Ashby's or the deductive approach in general; they only emphasize that there is no royal road to General Systems Theory. As every other scientific field, it will have to develop by an interplay of empirical, intuitive and deductive procedures. If the intuitive approach leaves much to be desired in logical rigor and completeness, the deductive approach faces the difficulty of whether the fundamental terms are correctly chosen. This is not a particular fault of the theory or of the workers concerned but a rather common phenomenon in the history of science; one may, for example, remember the long debate as to what magnitude – force or energy – is to be considered as constant in physical transformations until the issue was decided in favor of $mv^2/2$.

In the present writer's mind, G.S.T. was conceived as a working hypothesis; being a practicing scientist, he sees the main function of theoretical models in the explanation, prediction, and control of hitherto unexplored phenomena. Others may, with equal right, emphasize the importance of axiomatic approach and quote to this effect examples like the theory of probability, non-Euclidean geometries, more recently information and game theory, which were first developed as deductive mathematical fields, and later applied in physics or other sciences. There should be no quarrel about this point. The danger, in both approaches, is to consider too early the theoretical model as being closed and definitive – a danger particularly important in a field like general systems which is still groping to find its correct foundations.

3 Homeostasis and open systems

Among the models mentioned, cybernetics in its application as homeostasis, and G.S.T. in its application to open systems lend themselves most readily for interpretation of many empirical phenomena. The relation of both theories is not always well understood, and hence a brief discussion is in place.

The simplest feedback scheme can be represented as follows (Fig. 2.1). Modern servomechanisms and automation, as well as many phenomena in the organism, are based upon feedback arrangements far more complicated than the simple scheme (Fig. 2.1) but the latter is the elementary prototype.

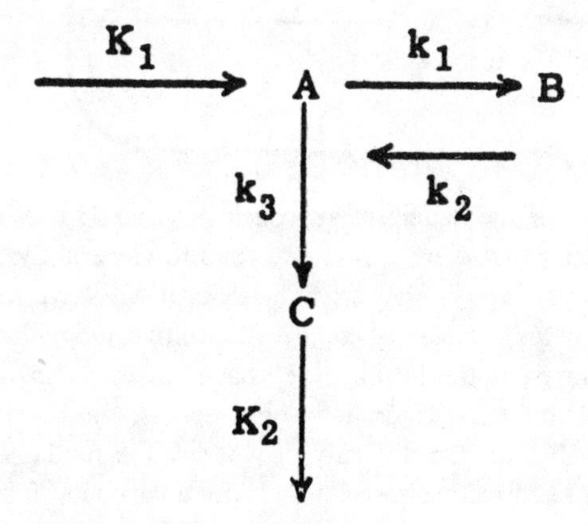

Figure 2.2 Model of a simple open system. The component A is introduced into the system and transformed, in a reversible reaction, into B; it is catabolized in an irreversible reaction, into C which eventually is excreted. K_1, K_2 are constants of import and export, respectively; k_1, k_2, k_3 are reaction constants. The model approximately corresponds, for example, to protein turnover in an animal organism, A representing amino acids, B proteins, and C products of excretion.

In application to the living organism, the feedback scheme is represented by the concept of homeostasis.

Homeostasis, according to Cannon, is the ensemble of organic regulations which act to maintain the steady states of the organism and are effectuated by regulating mechanisms in such a way that they do not occur necessarily in the same, and often in opposite, direction to what a corresponding external change would cause according to physical laws. The simplest example is homeothermy. According to Van't Hoff's rule in physical chemistry, a decrease in temperature leads to slowing down of the rate of chemical reactions, as it does in ordinary physico-chemical systems and also in poikilothermic animals. In warm-blooded animals, however, it leads to the opposite effect, namely, to an increase of metabolic rate, with the result that the temperature of the body is maintained constant at approximately 37°C. This is effectuated by a feedback mechanism. Cooling stimulates thermogenic centers in the brain thalamus which "turn on" heat-producing mechanisms in the body. A similar feedback pattern is found in a great variety of physiological regulations. Regulation of posture and the control of actions in animals and man toward a goal are similarly controled by feedback mechanisms.

In contradistinction to cybernetics concerned with feedback arrangements, G.S.T. is interested in dynamic interaction within multivariable systems. The case particularly important for the living organism is that of open systems. It amounts to saying that there is a system into which matter is introduced from outside. Within the system, the material undergoes reactions which partly may yield components of a higher complexity. This is what we call anabolism. On the other hand, the material is catabolized and the end products of catabolism eventually leave the system. A simple model of an open system is indicated in Figure 2.1.

A few main characteristics of open as compared to closed systems are in the fact that, appropriate system conditions presupposed, an open system will attain a steady state in which its composition remains constant, but in contrast to conventional equilibria, this constancy is maintained in a continuous exchange and flow of component material. The steady state of open systems is characterized by the principle of equifinality; that is, in contrast to equilibrium states in closed systems which are determined by initial conditions, the open system may attain a time-independent state independent of initial conditions and determined only by the system parameters. Furthermore, open systems show thermodynamic characteristics which are apparently paradoxical and contradictory to the second principle. According to the latter, the general course of physical events (in closed systems) is toward increasing entropy, leveling down of differences and states of maximum disorder. In open systems, however, with transfer of matter import of "negative entropy" is possible. Hence, such systems can maintain themselves at a high level, and even evolve toward an increase of order and complexity – as is indeed one of the most important characteristics of life processes.

The open-system model also has a wide application. According to its character, it is particularly applicable to phenomena showing non-structural, dynamic interaction of processes, such as those of metabolism, growth, metabolic aspects of excitation, etc.

Speaking more generally, living systems can be defined as hierarchically organized open systems, maintaining themselves, or developing toward a steady state. Disease is the life process regulating toward normalcy after disturbance, owing to the equifinality of biological systems and with the assistance of the physician. In this way, the *vis medicatrix naturae* of old is divested of its metaphysical paraphernalia; it is not a vitalistic agent but an expression of the dynamics of living systems, maintaining and reestablishing, so far as possible, the steady state.

In this way, the theory of open systems accounts for basic characteristics of the living organism which have baffled physicists, biologists, and philosophers, and appeared to be violations of the laws of physics, explainable only by vitalistic factors beyond the competence of science and scientific explanation.

Thus "feedback" and "open systems" are two models for biological and possibly behavioral phenomena. It should be made clear that the term "homeostasis" can be used in two ways. It is either taken in the original sense as proferred by Cannon and illustrated by examples like maintenance of body temperature and other physiological variables by feedback mechanisms – or else the term is often used as a synonym for organic regulation and adaptation in general. This is a question of semantics. However, it is a wise rule in the natural sciences to use terms in the sense originally attached to them by their authors. So I propose to use the word homeostasis in its narrower but well-defined sense, and this has important consequences, as it reveals certain limitations which are often forgotten.

As was already emphasized, regulations of the homeostasis or feedback type are abundant in the mature higher organism. However, it is clear from the scheme (Fig. 2.1) or any other flow diagram that feedback represents a machine-like arrangement, that is, an order of processes based upon fixed arrangements and representing linear, though circular, causal trains. The primary phenomena of organic regulation, e.g. the regulations in early embryonic development, in regeneration, etc., appear to be of a different nature. It seems that the primary regulations in the organism result from dynamic interaction within a unitary open system that reestablishes its steady state. Superimposed upon this by way of progressive mechanization are secondary regulatory mechanisms governed by fixed structures especially of the feedback type.

Although the homeostasis model transcends older mechanistic models by acknowledging directiveness in self-regulating circular processes, it still adheres to the machine theory of the organism. This also applies to a second aspect. An essential component of the mechanistic view is a utilitarian conception which is deeply connected with the economic outlook of the 19th and early 20th centuries. This is well known, for example, in the history of Darwinism: Struggle for existence and survival of the fittest are a biological version of the economic model of free competition. This utilitarian or economic viewpoint also prevails in the concept of homeostasis: The organism is essentially envisaged as an aggregate mechanism for maintenance of minimum costs. However, there seem to be plenty of non-utilitarian structures and functions in the living world.

The concept of homeostasis also retains a third aspect of the mechanistic view. The organism is essentially considered to be a reactive system. Outside stimuli are answered by proper responses in such a way as to maintain the system. The feedback model (Fig. 2.1) is essentially the classical stimulus-response scheme, only the feedback loop being added. However, an overwhelming amount of facts shows that primary organic behavior as, for example, the first movements of the fetus, are not reflex responses to external stimuli, but rather spontaneous mass activities of the whole embryo or larger areas. Reflex reactions answering external stimuli and following a structured path appear to be superimposed upon primitive automatisms, ontogenetically and phylogenetically, as secondary regulatory mechanisms. These considerations win particular importance in the theory of behavior as we shall see later on.

In this sense, it appears that in development and evolution dynamic interaction (open system) precedes mechanization (structured arrangements particularly of a feedback nature). In a similar way, G.S.T. can logically be considered the more general theory; it includes systems with feedback constraints as a special case, but this assertion would not be true vice versa. It need not be emphasized that this statement is a program for future systematization and integration of G.S.T. rather than a theory presently achieved.

4 Criticism of general system theory

A discussion of G.S.T. must take account of the objections raised, both to clarify misunderstanding and to utilize criticism for improvement.

A "devastating" criticism of "General Behavior Systems Theory" by Buck would hardly deserve discussion were it not for the fact that it appeared in the widely read *Minnesota Studies in the Philosophy of Science*, a leading publication of modern positivism. In passing, it should be noted that the lack of interest in, or even hostility of logical positivists against, G.S.T. is a rather remarkable phenomenon. One would expect that a group whose program is "Unified Science" should be concerned with a novel approach to this problem, however immature it may still be. The opposite is the case; no contribution or even pertinent criticism came forward from these quarters. The reason is not difficult to see. Abandoning the debatable but challenging position of Logical Positivism and replacing it by a rather tame "Empirical Realism", modern positivists have come back to what is generally agreed among modern scientists, avoiding commitments which trespass and would imply an adventure of thought. It needs to be said that modern positivism has been a singularly sterile movement. It is paradoxical that the declared "philosophers of science" have neither contributed any empirical research nor new idea to modern science – while professional or half-time philosophers who were justly censored for their "mysticism", "metaphysics", or "vitalism", indubitably did. Eddington and Jeans in physics, Driesch in biology, Spengler in history are but a few examples.

Buck's critique is not directed against the present author but against J. G. Miller and his Chicago group. Its essence is in the "So what?" argument: Supposing we find an analogy or formal identity in two "systems", it means nothing. Compare, for example, a chessboard and a mixed dinner party; a general statement expressing the alternation of black and white squares on the one hand, and of men and women on the other can be made. "If one is tempted to say 'All right, so they're structurally analogous, so what?' my answer is 'So, nothing.' " In the same vein, Buck pokes fun at some of Miller's more hazardous comparisons, such as the behavior of slime molds and Londoners during the *blitz*. He asks, "What are we to conclude from all this? That Londoners are a form of slime mold? That myxamoebae are a sort of city dweller?" Or "if no conclusion, why bother with the analogy at all?"

As proof of the emptiness of analogies Buck offers the example of a scientist, A, who finds a formula for the rate of formation of frost in a refrigerator; of another, B, formulating the rate of carbon deposit in an automobile motor; and a "general systems theorist," C, who notices that both formulas are the same. The similarity of mathematical expressions and models is, according to Buck, "sheer coincidence" – it does not prove that a refrigerator is an automobile or vice versa, but only that both are "systems" of some sort. This, however, is a meaningless statement; for

> One is unable to think of anything, or of any combination of things, which could not be regarded as a system. And, of course, a concept that applies to everything is logically empty.

Regardless of the question whether Miller's is a particularly felicitous presentation, Buck has simply missed the issue of a general theory of systems. Its aim is not more or less hazy analogies; it is to establish principles applicable to entities not covered in conventional science. Buck's criticism is, in principle, the same as if one would criticize Newton's law because it draws a loose "analogy" between apples, planets, ebb and tide and many other entities; or if one would declare the theory of probability meaningless because it is concerned with the "analogy" of games of dice, mortality statistics, molecules in a gas, the distribution of hereditary characteristics, and a host of other phenomena.

The basic role of "analogy" – or rather of isomorphisms and models in science – has been lucidly discussed by Ashby. Hence a few remarks in answer to Buck will suffice.

The "So what?" question mistakes a method which is fundamental in science although – like every method – it can be misused. Even Buck's first example is not a meaningless pseudoproblem; in the "analogy" of chessboard and dinner party topology may find a common structural principle that is well worth stating. Generally speaking, the use of "analogy" (isomorphism, logical homology) – or, what amounts to nearly the same, the use of conceptual and material models – is not a half-poetical play but a potent tool in science. Where would physics be without the analogy or model of "wave", applicable to such dissimilar phenomena as water waves, sound waves, light and electromagnetic waves, "waves" (in a rather Pickwickian sense) in atomic physics? "Analogies" may pose fundamental problems, as for example, the analogy (logically not dissimilar from that of chessboard and dinner party) of Newton's and Coulomb's law which raises the question (one of the most basic for "Unified Science") of a general field theory unifying mechanics and electrodynamics. It is commonplace in cybernetics that systems which are different materially; e.g., a mechanical and an electrical system, may be formally identical; far from considering this as a meaningless So what? the researcher has to work out the common structure (flow diagram), and this may be of incomparable value for practical technology.

A similar lack of understanding is manifest in the criticism of the system concept. By the same token ("One is unable to think of anything" which would not show the properties in question) mechanics would have to be refused as "logically empty" because every material body shows mass, acceleration, energy, etc. In the following paragraphs of his paper, Buck has some glimpse of this truism, but he soon comes back to ridiculing Miller's use of "analogies".

Although Buck justly criticizes certain unfortunate formulations, his misunderstanding of the basic problems involved makes one wonder how his essay found its way into a treatise on "Philosophy of Science".

At an incomparably higher level stands the criticism by the Soviet authors, Lektorsky and Sadovsky. The writers give a sympathetic and fair presentation of Bertalanffy's G.S.T. sketching diligently its gradual evolution from "organismic biology" and the theory of open systems. In view of the above criticism by Buck, the following quotation is of interest:

> Bertalanffy emphasizes the idea that a general system theory is not an investigation of hazy and superficial analogies. . . . Analogies as such have little value, since differences can always be found among phenomena as well as similarities. Bertalanffy declares that the kind of isomorphism with which general system theory is concerned is a consequence of the fact that in some respects corresponding abstractions and conceptual models can be applied to different phenomena.

"We can only welcome [the] goal [of G.S.T.]," write Lektorsky and Sadovsky, "i.e., the attempt to give a general definition of the concept of 'organized system', to classify logically various types of systems and to work out mathematical models for describing them . . . Bertalanffy's theory of organization and of organized complexes is a special scientific discipline. At the same time it certainly fulfills a definite methodological function" (i.e., avoiding duplication of effort in various disciplines by a single formal apparatus). "Its mathematical apparatus can be utilized for analyzing a comparatively large class of systemed objects of interest to biologists, chemists, biochemists, biophysicists, psychologists and others."

The criticism of the Russian authors is directed against imperfections of G.S.T. which, unfortunately, cannot be denied: "Bertalanffy's definition is rather a description [not pretending to precision] of the class of events which we may call systems than a strictly logical definition." "The description contains no trace of logical elegance." "Elementary methods of analysis and synthesis are insufficient for the analysis of systems." Fairly enough the authors concede that "The flaws we have noted speak only for the fact that general system theory, like any scientific theory, should develop further and in the process of development should strive for more adequate reflection of the objects of investigation."

The "main flaws of the theory", according to Lektorsky and Sadovsky, are in the lack of "methodology" (i.e., presumably of rules to establish and to apply system principles) and in considering G.S.T. "a philosophy of modern science". With respect to the first item, the present study is devoted to just this problem. The second point is a misunderstanding. G.S.T. in its present form is one – and still very imperfect – model among others. Were it completely developed, it would indeed incorporate the "organismic" world view of our time, with its emphasis on problems of wholeness, organization, directiveness, etc., in a similar way as when previous philosophies have presented a mathematical world view (philosophies *more geometrico*), a physicalistic one (the mechanistic philosophy based upon classical physics), etc., corresponding to scientific development. Even then, this "organismic" picture would not claim to be a "Nothing-but" philosophy: It would remain conscious that it only presents certain aspects of reality (richer and more comprehensive than previous ones, as corresponds to the advance of science), but never exhaustive, exclusive or final.

According to the authors, Marxist–Leninist philosophy "formulates a series of most important methodological principles of analysis of complex systems"; Soviet scientists "attempt to give a general definition of the notion of systems and to obtain a classification". Difficulties in international communication make it unfortunately impossible to the present writer to evaluate these claims.

Another criticism backed by the same *weltanschauung* is that of Kamaryt. The main arguments are:

1. Underestimation of the structural and morphologic aspects of organization of the theory of open systems (and implicitly in G.S.T.). The theory of open system does not "solve" the problem of life, its origin and evolution which is successfully attacked in modern biochemistry, submicroscopic morphology, physiological genetics, etc. The reply to this is that the functional and processual aspect has been emphasized in the theory, particularly in contradistinction to structural, homeostatic mechanisms. But neither the importance of the latter is denied, nor of course the specificity of the material basis of life. "Morphology and physiology are different and complementary ways of studying the same integrated object." If one wishes, this may be called a "dialectic unity of structure and function".

2. Neglect of "qualitative specifity" of biological open system and of the specific "chemodynamics" of the first.

The reply is: Thermodynamic considerations (of machines, chemical reactions, organisms, etc.) permit balance statements regarding the system as a whole, without entering into, or even knowing partial reactions, components, organization, etc., in detail. Hence part of the "theory of open systems" is concerned with such over-all balances of the system as a whole. If, however, the theory is applied to individual processes such as formation of proteins, behavior of tracers in the organism, ionic steady states, etc., the "specifity" of the respective components enters as a matter of course.

5 Advances of general system theory

The decisive question is that of the explanatory and predictive value of the "new theories" attacking the host of problems around wholeness, teleology, etc. Of course, the change in intellectual climate which allows us to see new problems which were overlooked previously, or to see problems in a new light, is in a way more important than any single and special application. The "Copernican Revolution" was more than the possibility somewhat better to calculate the movement of the planets; general relativity more than an explanation of a very small number of recalcitrant phenomena in physics; Darwinism more than a hypothetical answer to zoological problems; it was the changes in the general frame of reference that mattered. Nevertheless, the justification of such change ultimately is in specific achievements which would not have been obtained without the new theory.

There is no question that new horizons have been opened up but the relations to empirical facts often remain tenuous. Thus, information theory has been hailed as a "major breakthrough" but outside the original technological field contributions have remained relatively scarce. In psychology, they are so far limited to rather trivial applications such as role learning, etc. When, in biology, DNA is spoken of as "coded information" and of "breaking the code" when the structure of nucleic acids is elucidated, this is more a *façon de parler* than added insight into the control of protein synthesis. "Information theory, although useful for computer design and network analysis, has so far not found a significant place in biology." Game theory, too, is a novel mathematical development which was considered to be comparable in scope to Newtonian mechanics and the introduction of calculus; again, "the applications are meager and faltering" (the reader is urgently referred to Rapoport's discussions on information and game theory which admirably analyze the problems here mentioned). The same is seen in decision theory from which considerable gain in applied systems science was expected; but as regards the much-advertised military and business games, "there has been no controlled evaluation of their performance in training, personnel selection, and demonstration".

A danger in recent developments should not remain unmentioned. Science of the past (and partly still the present) was dominated by one-sided empiricism. Only collection of data and experiments were considered as being "scientific" in biology (and psychology); "theory" was equated with "speculation" or "philosophy", forgetting that a mere accumulation of data, although steadily piling up, does not make a "science". Lack of recognition and support for development of the necessary theoretical framework and unfavorable influence on experimental research itself (which largely became an at-random, hit-or-miss endeavor) was the consequence. This has, in certain fields, changed to the contrary in recent years. Enthusiasm for the new mathematical and logical tools available has led to feverish "model building" as a purpose in itself and often without regard to empirical fact. However, conceptual experimentation at random has no greater chances of success than at-random experimentation with biological, psychological, or clinical material. In the words of Ackoff, there is the fundamental misconception in game (and other) theory to mistake for a "problem" what actually is only a mathematical "exercise". One would do well to remember the old Kantian maxim that experience without theory is blind, but theory without experience a mere intellectual play.

The case is somewhat different with cybernetics. The model here applied is not new; although the enormous development in the field dates from the introduction of the name, Cybernetics, application of the feedback principle to physiological processes goes back to R. Wagner's work nearly 40 years ago. The feedback and homeostasis model has since been applied to innumerable biological phenomena and – somewhat less persuasively – in psychology and the social sciences. The reason for the latter fact is, in Rapoport's words that

> usually, there is a well-marked negative correlation between the scope and the soundness of the writings. . . . The sound work is confined either to engineering or to rather trivial applications; ambitious formulations remain vague.

This, of course, is an ever-present danger in all approaches to general systems theory: doubtless, there is a new compass of thought but it is difficult to steer between the scylla of the trivial and the charybdis of mistaking neologisms for explanation.

The following survey is limited to "classical" general system theory – "classical" not in the sense that it claims any priority or excellence, but that the models used remain in the framework of "classical" mathematics in contradistinction to the "new" mathematics in game, network, information theory, etc. This does not imply that the theory is merely application of conventional mathematics. On the contrary, the system concept poses problems which are partly far from being answered. In the past, system problems have led to important mathematical developments such as Volterra's theory of integro-differential equations, of systems with "memory" whose behavior depends not only on actual conditions but also on previous history. Presently important problems are waiting for further developments, e.g., a general theory of non-linear differential equations, of steady states and rhythmic phenomena, a generalized principle of least action, the thermodynamic definition of steady states, etc.

It is, of course, irrelevant whether or not research was explicitly labeled as "general system theory". No complete or exhaustive review is intended. The aim of this unpretentious survey will be fulfilled if it can serve as a sort of guide to research done in the field, and to areas that are promising for future work.

Open systems

The theory of open systems is an important generalization of physical theory, kinetics and thermodynamics. It has led to new principles and insight, such as the principle of equifinality, the generalization of the second thermodynamic principle, the possible increase of order in open systems, the occurrence of periodic phenomena of overshoot and false start, etc. The possibility of measuring organization in terms of entropy ("chain entropy" of high molecular compounds showing a certain order of component molecules) deserves further attention.

The extensive work done cannot be reviewed here. . . . It should be briefly mentioned, however, that apart from theoretical developments, the field has two major applications, i.e., in industrial chemistry and in biophysics.

The applications of "open systems" in biochemistry, biophysics, physiology, etc., are too numerous to permit more than brief mentioning in the present study. The impact of the theory follows from the fact that the living organism, the cell as well as other biological entities essentially are steady states (or evolving toward such states). This implies the fundamental nature of the theory in the biological realm, and a basic reorientation in many of its specialties. Among others, the theory was developed and applied in such fields as e.g., the network of reactions in photosynthesis, calculation of turnover rates in isotope experiments, energy requirements for the maintenance of body proteins, transport processes, maintenance of ion concentrations in the blood, radiation biology, excitation and propagation of nerve impulses, and others. The organism is in a steady state not only with respect to its chemical components, but also to its cells; hence the numerous modern investigations on cell turnover and renewal have also to be included here. Beside the work already cited, results and impending problems in biophysics and related fields may be found in Netter (1959).

There are certainly relations between irreversible thermodynamics of open systems, cybernetics, and information theory, but they are still unexplored. First approaches to these problems are those by Foster, Rapoport and Trucco, and by Tribus. Another interesting approach to metabolizing systems was made by Rosen (1960) who instead of conventional reaction equations, applied "relational theory" using mapping by way of block diagrams.

Beyond the individual organism, systems principles are also used in population dynamics and ecologic theory. Dynamic ecology, i.e., the succession and climax of plant populations, is a much-cultivated field which, however, shows a tendency to slide into verbalism and terminological debate. The systems approach seems to offer a new viewpoint. Whittacker has described the sequence of plant communities toward a climax formation in terms of open systems and equifinality. According to this author, the fact that similar climax formations may develop from different initial vegetations is a striking example of equifinality, and one where the degree of independence of starting conditions and of the course development has taken appears even greater than in the individual organism. A quantitative analysis on the basis of open systems in terms of production of biomass, with climax as steady state attained, was given by Patten.

The open-system concept has also found application in the earth sciences, geomorphology and meteorology, drawing a detailed comparison of modern meteorological concepts and Bertalanffy's organismic concept in biology. It may be remembered that already Prigogine in his classic mentioned meteorology as one possible field of application of open systems.

Growth-in-time

The simplest forms of growth which, for this reason, are particularly apt to show the isomorphism of law in different fields, are the exponential and logistic. Examples are, among many others, the increase of knowledge of number of animal species, publications on drosophila, of manufacturing companies. Boulding and Keiter have emphasized a general theory of growth.

The theory of animal growth after Bertalanffy (and others) – which, in virtue of using overall physiological parameters ("anabolism", "catabolism") may be subsumed under the heading of G.S.T. as well as under that of biophysics – has been surveyed in its various applications.

Relative growth

A principle which is also of great simplicity and generality concerns the relative growth of components within a system. The simple relationship of allometric increase applies to many growth phenomena in biology (morphology, biochemistry, physiology, evolution).

A similar relationship obtains in social phenomena. Social differentiation and division of labor in primitive societies as well as the process of urbanization (i.e., growth of cities in comparison to rural population) follow the allometric equation. Application of the latter offers a quantitative measure of social organization and development, apt to replace the usual, intuitive judgments. The same principle apparently applies to the growth of staff compared to total number of employees in manufacturing companies.

Competition and related phenomena

The work in population dynamics by Volterra, Lotka, Gauss and others belongs to the classics of G.S.T., having first shown that it is possible to develop conceptual models for phenomena such as the "struggle for existence" that can be submitted to empirical test. Population dynamics and related population genetics have since become important fields in biological research.

It is important to note that investigation of this kind belongs not only to basic but also to applied biology. This is true of fishery biology where theoretical models are used to establish optimum conditions for the exploitation of the sea (survey of the more important models: Watt). The most elaborate dynamic model is by Beverton and Holt developed for fish populations exploited in commercial fishery but certainly of wider application. This model takes into account recruitment (i.e., entering of individuals into the population), growth (assumed to follow the growth equations after Bertalanffy), capture (by exploitation), and natural mortality. The practical value of this model is illustrated by the fact that it has been adopted for routine purposes by the Food and Agriculture Organization of the United Nations, the British Ministry of Agriculture and Fisheries and other official agencies.

Richardson's studies on armaments races, notwithstanding their shortcomings, dramatically show the possible impact of the systems concept upon the most vital concerns of our time. If rational and scientific considerations matter at all, this is one way to refute such catch words as *Si vis pacem para bellum.*

The expressions used in population dynamics and the biological "struggle for existence", in econometrics, in the study of armament races (and others) all belong to the same family of equations. A systematic comparison and study of these parallelisms would be highly interesting and rewarding. One may, for example, suspect that the laws governing business cycles and those of population fluctuations according to Volterra stem from similar conditions of competition and interaction in the system.

In a non-mathematical way, Boulding has discussed what he calls the "Iron Laws" of social organizations: the Malthusian law, the law of optimum size of organizations, existence of cycles, the law of oligopoly, etc.

Systems engineering

The theoretical interest of systems engineering and operations research is in the fact that entities whose components are most heterogeneous – men, machines, buildings, monetary and other values, inflow of raw material, outflow of products and many other items – can successfully be submitted to systems analysis.

As already mentioned, systems engineering employs the methodology of cybernetics, information theory, network analysis, flow and block diagrams, etc. Considerations of G.S.T. also enter. The first approaches are concerned with structured, machine-like aspects (yes-or-no decisions in the case of information theory); one would suspect that G.S.T. aspects will win increased importance with dynamic aspects, flexible organizations, etc.

Personality theory

Although there is an enormous amount of theorizing on neural and psychological function in the cybernetic line based upon the brain–computer comparison, few attempts have been made to apply G.S.T. in the narrower sense to the theory of human behavior. For the present purposes, the latter may be nearly equated with personality theory.

We have to realize at the start that personality theory is at present a battlefield of contrasting and controversial theories. Hall and Lindzey have justly stated: "All theories of behavior are pretty poor theories and all of them leave much to be desired in the way of scientific proof" – this being said in a textbook of nearly 600 pages on "Theories of Personality".

We can therefore not well expect that G.S.T. can present solutions where personality theorists from Freud and Jung to a host of modern writers were unable to do so. The theory will have shown its value if it opens new perspectives and viewpoints capable of experimental and practical application. This appears to be the case. There is quite a group of psychologists who are committed to an organismic theory of personality, Goldstein and Maslow being well-known representatives. Biological considerations may therefore be expected to advance the matter.

There is, of course, the fundamental question whether, first, G.S.T. is not essentially a physicalistic simile, inapplicable to psychic phenomena; and secondly whether such model has explanatory value when the pertinent variables cannot be defined quantitatively as is in general the case with psychological phenomena.

1. The answer to the first question appears to be that the systems concept is abstract and general enough to permit application to entities of whatever denomination. The notions of "equilibrium", "homeostasis", "feedback", "stress", etc., are no less of technologic or physiological origin but more or less successfully applied to psychological phenomena. System theorists agree that the concept of "system" is not limited to material entities but can be applied to any "whole" consisting of interacting "components". . . . Systems engineering is an example where components are partly not physical and metric.

2. If quantitation is impossible, and even if the components of a system are ill-defined, it can at least be expected that certain principles will qualitatively apply to the whole *qua* system. At least "explanation on principle" (see below) may be possible.

Bearing in mind these limitations, one concept which may prove to be of a key nature is the organismic notion of the organism as a spontaneously active system. In the present author's words:

> Even under constant external conditions and in the absence of external stimuli the organism is not a passive but a basically active system. This applies in particular to the function of the nervous system and to behavior. It appears that internal activity rather than reaction to stimuli is fundamental. This can be shown with respect both to evolution in lower animals and to development, for example, in the first movements of embryos and fetuses.

This agrees with what von Holst has called the "new conception" of the nervous system, based upon the fact that primitive locomotor activities are caused by central automatisms that do not need external stimuli. Therefore, such movements persist, for example, even after the connection of motoric to sensory nerves had been severed. Hence the reflex in the classic sense is not the basic unit of behavior but rather a regulatory mechanism superimposed upon primitive, automatic activities. A similar concept is basic in the theory on instinct. According to Lorenz, innate releasing mechanisms (I.R.M.) play a dominant role, which sometimes go off without an external stimulus (in-vacuo or running idle reactions): A bird which has no material to build a nest may perform the movements of nest building in the air. These considerations are in the framework of what Hebb called the "conceptual C.N.S. of 1930–1950". The more recent insight into activating systems of the brain emphasizes differently, and with a wealth of experimental evidence, the same basic concept of the autonomous activity of the C.N.S.

The significance of these concepts becomes apparent when we consider that they are in fundamental contrast to the conventional stimulus-response scheme which assumes that the organism is an essentially reactive system answering, like an automaton, to external stimuli. The dominance of the S–R scheme in contemporary psychology needs no emphasis, and is obviously connected with the *zeitgeist* of a highly mechanized society. This principle is basic in psychological theories which in all other respects are opposite, for example, in behavioristic psychology as well as in psychoanalysis. According to Freud it is the supreme tendency of the organism to get rid of tensions and drives and come to rest in a state of equilibrium governed by the "principle of stability" which Freud borrowed from the German philosopher, Fechner. Neurotic and psychotic behavior, then, is a more or less effective or abortive defense mechanism tending to restore some sort of equilibrium (according to D. Rappaport's analysis of the structure of psychoanalytic theory: "economic" and "adaptive points of view").

Charlotte Buhler, the well-known child psychologist, has aptly epitomized the theoretical situation:

> In the fundamental psychoanalytic model, there is only one basic tendency, that is toward *need gratification* or *tension reduction*. . . . Present-day biologic theories emphasize the "spontaneity" of the organism's activity which is due to its built-in energy. The organism's autonomous functioning, its "drive to perform certain

movements" is emphasized by Bertalanffy. . . . These concepts represent *a complete revision of the original homeostasis principle* which emphasized exclusively the tendency toward equilibrium. It is the original homeostasis principle with which psychoanalysis identified its theory of discharge of tensions as the only primary tendency. (Emphasis partly ours.)

In brief, we may define our viewpoint as "Beyond the Homeostasis Principle":

(1) The S–R scheme misses the realms of play, exploratory activities, creativity, self-realization, etc.;
(2) The economic scheme misses just specific, human achievements – the most of what loosely is termed "human culture";
(3) The equilibrium principle misses the fact that psychological and behavioral activities are more than relaxation of tensions; far from establishing an optimal state, the latter may entail psychosis-like disturbances, as e.g., in sensory-deprivation experiments.

It appears that the S–R and psychoanalytic model is a highly unrealistic picture of human nature and, in its consequences, a rather dangerous one. Just what we consider to be specific human achievements can hardly be brought under the utilitarian, homeostasis, and stimulus–response scheme. One may call mountain climbing, composing of sonatas or lyrical poems "psychological homeostasis" – as has been done – but at the risk that this physiologically well-defined concept loses all meaning. Furthermore, if the principle of homeostatic maintenance is taken as a golden rule of behavior, the so-called well-adjusted individual will be the ultimate goal, that is a well-oiled robot maintaining itself in optimal biological, psychological and social homeostasis. This is a *Brave New World* – not, for some at least, the ideal state of humanity. Furthermore, that precarious mental equilibrium must not be disturbed. Hence in what somewhat ironically is called progressive education, the anxiety not to overload the child, not to impose constraints and to minimize all directing influences – with the result of a previously unheard-of crop of illiterates and juvenile delinquents.

In contrast to conventional theory, it can safely be maintained that not only stresses and tensions but equally complete release from stimuli and the consequent mental void may be neurosogenic or even psychosogenic. Experimentally this is verified by the experiments with sensory deprivation when subjects, insulated from all incoming stimuli, after a few hours develop a so-called model psychosis with hallucinations, unbearable anxiety, etc. Clinically it amounts to the same when insulation leads to prisoners' psychosis and to exacerbation of mental disease by isolation of patients in the ward. In contrast, maximal stress need not necessarily produce mental disturbance. If conventional theory were correct, Europe during and after the war, with extreme physiological as well as psychological stresses, should have been a gigantic lunatic asylum. As a matter of fact, there was statistically no increase either in neurotic or psychotic disturbances, apart from easily explained acute disturbances such as combat neurosis.

We so arrive at the conception that a great deal of biological and human behavior is beyond the principles of utility, homeostasis and stimulus-response, and that it is just this which is characteristic of human and cultural activities. Such new look opens new perspectives not only in theory, but in practical implications with respect to mental hygiene, education, and society in general.

What has been said can also be couched in philosophical terms. If existentialists speak of the emptiness and meaninglessness of life, if they see in it a source not only of anxiety but of actual mental illness, it is essentially the same viewpoint: that behavior is not merely a matter of satisfaction of biological drives and of maintenance in psychological and social equilibrium but that something more is involved. If life becomes unbearably empty in an industrialized society, what can a person do but develop a neurosis? The principle which may loosely be called spontaneous activity of the psychophysical organism, is a more realistic formulation of what the existentialists want to say in their often obscure language. And if personality theorists like Maslow or Gardner Murphy speak of self-realization as human goal, it is again a somewhat pompous expression of the same.

Theoretical history

We eventually come to those highest and ill-defined entities that are called human cultures and civilizations. It is the field often called "philosophy of history". We may perhaps better speak of "theoretical history", admittedly in its very first beginnings. This name expresses the goal to form a connecting link between "science" and the "humanities"; more in particular, between the "social sciences" and "history".

It is understood, of course, that the techniques in sociology and history are entirely different (polls, statistical analysis against archival studies, internal evidence of historic relics, etc.). However, the object of study is essentially the same. Sociology is essentially concerned with a temporal cross-section as human societies *are*; history with the "longitudinal" study how societies *become* and develop. The object and techniques of study certainly justify practical differentiation; it is less clear, however, that they justify fundamentally different philosophies.

The last statement already implies the question of constructs in history, as they were presented, in grand form, from Vico to Hegel, Marx, Spengler and Toynbee. Professional historians regard them at best as poetry, at worst as fantasies pressing, with paranoic obsession, the facts of history into a theoretical bed of Procrustes. It seems history can learn from the system theorists, not ultimate solutions but a sounder methodological outlook. Problems hitherto considered to be philosophical or metaphysical can well be defined in their scientific meaning, with some interesting outlook at recent developments (e.g., game theory) thrown into the bargain.

Empirical criticism is outside the scope of the present study. For example, Geyl and many others have analyzed obvious misrepresentations of historical events in Toynbee's work, and even the non-specialist reader can easily draw a list of fallacies especially in the later, Holy-Ghost inspired volumes of Toynbee's *magnum opus*. The problem, however, is larger than errors in fact or interpretation or even the question of the merits of Marx's, Spengler's or Toynbee's theories; it is whether, in principle, models and laws are admissible in history.

A widely held contention says that they are not. This is the concept of "nomothetic" method in science and "idiographic" method in history. While science to a greater or less extent can establish "laws" for natural events, history, concerned with human events of enormous complexity in causes and outcome and possibly determined by free decisions of individuals can only describe, more or less satisfactorily, what has happened in the past.

Here the methodologist has his first comment. In the attitude just outlined, academic history condemns constructs of history as "intuitive", "contrary to fact", "arbitrary", etc. And, no doubt, the criticism is pungent enough vis-à-vis Spengler or Toynbee. It is, however, somewhat less convincing if we look at the work of conventional historiography. For example, the Dutch historian, Peter Geyl, who made a strong argument against Toynbee from such methodological considerations, also wrote a brilliant book about Napoleon, amounting to the result that there are a dozen or so different interpretations – we may safely say, *models* – of Napoleon's character and career within academic history, all based upon "fact" (the Napoleonic period happens to be one of the best documented) and all flatly contradicting each other. Roughly speaking, they range from Napoleon as the brutal tyrant and egotistic enemy of human freedom to Napoleon the wise planner of a unified Europe; and if one is a Napoleonic student (as the present writer happens to be in a small way), one can easily produce some original documents refuting misconceptions occurring even in generally accepted, standard histories. You cannot have it both ways. If even a figure like Napoleon, not very remote in time and with the best of historical documentation, can be interpreted contrarily, you cannot well blame the "philosophers of history" for their intuitive procedure, subjective bias, etc., when they deal with the enormous phenomenon of universal history. What you have in both cases is a conceptual model which always will represent certain aspects only, and for this reason will be one-sided or even lopsided. Hence the construction of conceptual models in history is not only permissible but, as a matter of fact, is at the basis of any historical interpretation as distinguished from mere enumeration of data, i.e., chronicle or annals.

If this is granted, the antithesis between idiographic and nomothetic procedure reduces to what psychologists are wont to call the "molecular" and "molar" approach. One can analyze events within a complex whole – individual chemical reactions in an organism, perceptions in the psyche, for example; or one can look for over-all laws covering the whole such as growth and development in the first or personality in the second instance. In terms of history, this means detailed study of individuals, treaties, works of art, singular causes and effects, etc., or else over-all phenomena with the hope of detecting grand laws. There are, of course, all transitions between the first and second considerations; the extremes may be illustrated by Carlyle and his hero worship at one pole and Tolstoy (a far greater "theoretical historian" than commonly admitted) at the other.

The question of a "theoretical history" therefore is essentially that of "molar" models in the field; and this is what the great constructs of history amount to when divested of their philosophical embroidery.

The evaluation of such models must follow the general rules for verification or falsification. First, there is the consideration of empirical bases. In this particular instance it amounts to the question whether or not a limited number of civilizations – some twenty at the best – provide a sufficient and representative sample to establish justified generalizations. This question and that of the value of proposed models will be answered by the general criterion; whether or not the model has explanatory and predictive value, i.e., throws new light upon known facts and correctly foretells facts of the past or future not previously known.

Although elementary, these considerations nevertheless are apt to remove much misunderstanding and philosophical fog which has clouded the issue.

1. As had been emphasized, the evaluation of models should be simply pragmatic in terms of their explanatory and predictive merits (or lack thereof); *a priori* considerations as to their desirability or moral consequences do not enter.

Here we encounter a somewhat unique situation. There is little objection against so-called "synchronic" laws, i.e., supposed regularities governing societies at a certain point in time; as a matter of fact, beside empirical study this is the aim of sociology. Also certain "diachronic" laws, i.e., regularities of development in time, are undisputed such as, e.g., Grimm's law stating rules for the changes of consonants in the evolution of Indo-Germanic languages. It is commonplace that there is a sort of "life-cycle" – stages of primitivity, maturity, baroque dissolution of form and eventual decay for which no particular external causes can be indicated – in individual fields of culture, such as Greek sculpture, Renaissance painting or German music. Indeed, this even has its counterpart in certain phenomena of biological evolution showing, as in ammonites or dinosaurs, a first explosive phase of formation of new types followed by a phase of speciation and eventually of decadence.

Violent criticism comes in when this model is applied to civilization as a whole. It is a legitimate question – Why often rather unrealistic models in the social sciences remain matters of academic discussion, while models of history encounter passionate resistance? Granting all factual criticism raised against Spengler or Toynbee, it seems rather obvious that emotional factors are involved. The highway of science is strewn with corpses of deceased theories which just decay or are preserved as mummies in the museum of history or science. In contrast, historical constructs and especially theories of historical cycles appear to touch a raw nerve, and so opposition is much more than usual criticism of a scientific theory.

2. This emotional involvement is connected with the question of "Historical Inevitability" and a supposed degradation of human "freedom". Before turning to it, discussion of mathematical and non-mathematical models is in place.

Advantages and shortcomings of mathematical models in the social sciences are well known. Every mathematical model is an oversimplification, and it remains questionable whether it strips actual events to the bones or cuts away vital parts of their anatomy. On the other hand, so far as it goes, it permits necessary deduction with often unexpected results which would not be obtained by ordinary "common sense".

In particular, Rashevsky has shown in several studies how mathematical models of historical processes can be constructed.

On the other hand, the value of purely qualitative models should not be underestimated. For example, the concept of "ecologic equilibrium" was developed long before Volterra and others introduced mathematical models; the theory of selection belongs to the stock-in-trade of biology, but the mathematical theory of the "struggle for existence" is comparatively recent, and far from being verified under wildlife conditions.

In complex phenomena, "explanation on principle" by qualitative models is preferable to no explanation at all. This is by no means limited to the social sciences and history; it applies alike to fields like meteorology or evolution.

3. "Historical inevitability" – subject of a well-known study by Sir Isaiah Berlin – dreaded as a consequence of "theoretical history", supposedly contradicting our direct experience of having free choices and eliminating all moral judgment and values – is a phantasmagoria based upon a world view which does not exist any more. As, in fact, Berlin emphasizes, it is founded upon the concept of the Laplacean spirit who is able completely to predict the future from the past by means of deterministic laws. This has no resemblance with the modern concept of "laws of nature". All "laws of nature" have a statistical character. They do not predict an inexorably determined future but probabilities which, depending on the nature of events and on the laws available, may approach certainty or else remain far below it. It is nonsensical to ask or fear more "inevitability" in historical theory than is found in sciences with relatively high sophistication like meteorology or economics.

Paradoxically, while the cause of free will rests with the testimony of intuition or rather immediate experience and can never be proved objectively ("Was it Napoleon's free will that led him to the Russian Campaign?"), determinism (in the statistical sense) can be proved, at least in small-scale models. Certainly business depends on personal "initiative", the individual "decision" and "responsibility" of the entrepreneur; the manager's choice whether or not to expand business by employing new appointees, is "free' in precisely the sense as Napoleon's choice whether or not to accept battle at Austerlitz. However, when the growth curve of industrial companies is analyzed, it is found that "arbitrary" deviations are followed by speedy return to the normal curve, as if invisible forces were active. Haire states that "the return to the pattern predicted by earlier growth suggests the operation of *inexorable forces* operating on the social organism" (our emphasis).

It is characteristic that one of Berlin's points is "the fallacy of historical determinism (appearing) from its utter inconsistency with the common sense and everyday life of looking at human affairs". This characteristic argument is of the same nature as the advice not to adopt the Copernican system because everybody can see that the sun moves from morning to evening.

4. Recent developments in mathematics even allow to submit "free will" – apparently the philosophical problem most recalcitrant against scientific analysis – to mathematical examination.

In the light of modern systems theory, the alternative between molar and molecular, nomothetic and idiographic approach can be given a precise meaning. For mass behavior, system laws would apply which, if they can be mathematized, would take the form of differential equations of the sort of those used by Richardson mentioned above. Free choice of the individual would be described by formulations of the nature of game and decision theory.

Axiomatically, game and decision theory are concerned with "rational" choice. This means a choice which "maximizes the individual's utility or satisfaction", that "the individual is free to choose among several possible courses of action and decides among them at the basis of their consequences", that he "selects, being informed of all conceivable consequences of his actions, what stands highest on his list", he "prefers more of a commodity to less, other things being equal", etc. Instead of economical gain, any higher value may be inserted without changing the mathematical formalism.

The above definition of "rational choice" includes everything that can be meant by "free will". If we do not wish to equate "free will" with complete arbitrariness, lack of any value judgment and therefore completely inconsequential actions (like the philosopher's favorite example: It is my free will whether or not to wiggle my left little finger) it is a fair definition of those actions with which the moralist, priest, or historian is concerned: free decision between alternatives based upon insight into the situation and its consequences and guided by values.

The difficulty to apply the theory even to simple, actual situations is of course enormous; so is the difficulty in establishing over-all laws. However, without explicit formulation, both approaches can be evaluated in principle – leading to an unexpected paradox.

The "principle of rationality" fits – not the majority of human actions but rather the "unreasoning" behavior of animals. Animals and organisms in general do function in a "ratio-morphic" way, maximizing such values as maintenance, satisfaction, survival, etc.; they select, in general, what is biologically good for them, and prefer more of a commodity (e.g., food) to less.

Human behavior, on the other hand, falls far short of the principle of rationality. It is not even necessary to quote Freud to show how small is the compass of rational behavior in man. Women in a supermarket, in general, do not maximize utility but are susceptible to the tricks of the advertiser and packer; they do not make a rational choice surveying all possibilities and consequences; and do not even prefer more of the commodity packed in an inconspicuous way to less when packed in a big red box with attractive design. In our society, it is the job of an influential specialty – advertisers, motivation researchers, etc. – to *make* choices irrational which essentially is done by coupling biological factors – conditioned reflex, unconscious drives – with symbolic values.

And there is no refuge by saying that this irrationality of human behavior concerns only trivial actions of daily life; the same principle applies to "historical" decisions. That wise old mind, Oxenstierna, Sweden's Chancellor during the Thirty Years' War, has perfectly expressed this by saying: *Nescis, mi fili, quantilla ratione mundus regatur* – you don't know, my dear boy, with what little reason the world is governed. Reading newspapers or listening to the radio readily shows that this applies perhaps even more to the 20th than the 17th century.

Methodologically, this leads to a remarkable conclusion. If one of the two models is to be applied, and if the "actuality principle" basic in historical fields like geology and evolution is adopted (i.e., the hypothesis that no other principles of explanation should be used than can be observed as operative in the present) – then it is the statistical or mass model which is backed by empirical evidence. The business of the motivation and opinion researcher, statistical psychologist, etc., is based upon the premise that statistical laws obtain in human behavior; and that, for this reason a small but well-chosen sample allows for extrapolation to the total population under consideration. The generally good working of a Gallup poll and prediction verifies the premise – with some incidental failure like the well-known example of the Truman election thrown in, as is to be expected with statistical predictions. The opposite contention – that history is governed by "free will" in the philosophical sense (i.e., rational decision for the better, the higher moral value or even enlightened self-interest) is hardly supported by fact. That here and there the statistical law is broken by "rugged individualists" is in its character. Nor does the role played in history by "great men" contradict the systems concept in history; they can be conceived as acting like "leading parts", "triggers" or "catalyzers" in the historical process – a phenomenon well accounted for in the general theory of systems.

5. A further question is the "organismic analogy" unanimously condemned by historians. They combat untiringly the "metaphysical", "poetical", "mythical" and thoroughly unscientific nature of Spengler's assertion that civilizations are a sort of "organisms", being born, developing according to their internal laws and eventually dying. Toynbee takes great pains to emphasize that he did not fall into Spengler's trap – even though it is somewhat difficult to see that his civilizations, connected by the biological relations of "affiliation" and "apparentation", even (according to the latest version of his system) with a rather strict time span of development, are not conceived organismically.

Nobody should know better than the biologist that civilizations are no "organism". It is trivial to the extreme that a biological organism, a material entity and unity in space and time, is something different from a social group consisting of distinct individuals, and even more from a civilization consisting of generations of human beings, of material products, institutions, ideas, values, and what not. It implies a serious underestimate of Vico's, Spengler's (or any normal individual's) intelligence to suppose that they did not realize the obvious.

Nevertheless, it is interesting to note that, in contrast to the historians' scruples, sociologists do not abhor the "organismic analogy" but rather take it for granted. For example, in the words of Rappoport and Horvath:

> There is some sense in considering a real organization as an organism, that is, there is reason to believe that this comparison need not be a sterile metaphorical analogy, such as was common in scholastic speculation about the body politic. Quasibiological functions are demonstrable in organizations. They maintain themselves; they sometimes reproduce or metastasize; they respond to stresses; they age, and they die. Organizations have discernible anatomies and those at least which transform material inputs (like industries) have physiologies.

Or Sir Geoffrey Vickers:

> Institutions grow, repair themselves, reproduce themselves, decay, dissolve. In their external relations they show many characteristics of organic life. Some think that in their internal relations also human institutions are destined to become increasingly organic, that human cooperation will approach ever more closely to the integration of cells in a body. I find this prospect unconvincing (and) unpleasant. [N.B. so does the present author.]

And Haire:

> The biological model for social organizations – and here, particularly for industrial organizations – means taking as a model the living organism and the processes and principles that regulate its growth and development. It means looking for lawful processes in organizational growth.

The fact that simple growth laws apply to social entities such as manufacturing companies, to urbanization, division of labor, etc., proves that in these respects the "organismic analogy" is correct. In spite of the historians' protests, the application of theoretical models, in particular, the model of dynamic, open and adaptive systems to the historical process certainly makes sense. This does not imply "biologism", i.e., reduction of social to biological concepts, but indicates system principles applying in both fields.

6. Taking all objections for granted – poor method, errors in fact, the enormous complexity of the historical process – we have nevertheless reluctantly to admit that the cyclic models of history pass the most important test of scientific theory. The predictions made by Spengler in the *Decline of the West*, by Toynbee when forecasting a time of trouble and contending states, by Ortega y Gasset in the *Uprise of the Masses* – we may as well add *Brave New World* and *1984* – have been verified to a disquieting extent and considerably better than many respectable models of the social scientists.

Does this imply "historic inevitability" and inexorable dissolution? Again, the simple answer was missed by moralizing and philosophizing historians. By extrapolation from the life cycles of previous civilizations nobody could have predicted the Industrial Revolution, the Population Explosion, the development of atomic energy, the emergence of underdeveloped nations, and the expansion of Western civilization over the whole globe. Does this refute the alleged model and "law" of history? No – it only says that this model – as every one in science – mirrors only certain aspects or facets of reality. Every model becomes dangerous only when it commits the "Nothing-but" fallacy which mars not only theoretical history, but the models of the mechanistic world picture, of psychoanalysis and many others as well.

We have hoped to show in this survey that General System Theory has contributed toward the expansion of scientific theory; has led to new insights and principles; and has opened up new problems that are "researchable", i.e., are amenable to further study, experimental or mathematical. The limitations of the theory and its applications in their present status are obvious; but the principles appear to be essentially sound as shown by their application in different fields.

References

Ackoff, R. L. (1959) "Games, decisions, and organizations". *General Systems* **IV**, 145–50.

—— (1960) "Systems, organizations, and interdisciplinary research", *General Systems* **V**, 1–8.

Arrow, K. J. (1956) "Mathematical models in the social sciences", *General Systems* **I**, 29–47.

Ashby, W. R. (1958a) "General systems theory as a new discipline", *General Systems* **III**, 1–6.

—— (1958b) *An Introduction to Cybernetics*, 3rd impr., Wiley, New York.

—— (1962) "Principles of the self-organizing system", in H. von Foerster, G. W. Zopf, Jr. (eds.), *Principles of Self-organization*, Pergamon Press, New York, pp. 255–78.

Attneave, F. (1959) *Application of Information Theory to Psychology*, Holt, New York.

Beer, St. (1960) "Below the twilight arch. A mythology of systems", *General Systems* **V**, 9–20.

Bell, E. (1962) "Oogenesis", C. P. Raven (review), *Science* **135**, 1056.

von Bertalanffy, L. (1947) "Vom Sinn und der Einheit der Wissenschaften", *Der Student*, Wien, 2, No. 7/8.

—— (1949) "Zu einer allgemeinen systemlehre", *Biologia Generalis* **19**, 114–29.

—— (1950) "An outline of general system theory", *Brit. J. Philos. Sci.* **1**, 134–65.

—— (1953) *Biophysik des fliessgleichgewichts*, (Transl. by W. Westphal), Vieweg, Braunschweig.

—— (1956) "General system theory", *General Systems* **I**, 1–10.

—— (1956) "A biologist looks at human nature", *Scientific Monthly* **82**, 33–41.

—— (1952) *Problems of Life. An evaluation of Modern Biological and Scientific Thought*. Torchbook edition, Harper, New York, 1960a.

—— (1960b) "Principles and theory of growth", in W. W. Nowinski (ed.), *Fundamental Aspects of Normal and Malignant Growth*, Elsevier, Amsterdam, pp. 137–259.

—— (1933) *Modern Theories of Development. An Introduction to Theoretical Biology*. Torchbook edition, Harper, New York, 1962.

von Bertalanffy, L., Hempel, C. G., Bass, R. E., and Jonas, H. (1951) "General system theory: A new approach to unity of science", *Human Biol.* **23**, 302–61.

Beverton, R. J. H. and Holt, S. J. (1957) "On the dynamics of exploited fish populations", *Fishery Investigation*, Ser. II, vol. XIX, H.M.S.O. London.

Boulding, K. E. (1953) *The Organizational Revolution*, Harper, New York.

—— (1956) "Toward a general theory of growth", *General Systems* **I**, 66–75.

Bradley, D. F. and Calvin, M. (1956) "Behavior: Imbalance in a network of chemical transformation", *General Systems* **I**, 56–65.

Bray, J. R. (1958) "Notes toward an ecology theory". *Ecology* **9**, 770–6

Bray, H. G. and White, K. (1957) *Kinetics and Thermodynamics in Biochemistry*, Academic Press, New York.

Buck, R. C. (1956) "On the logic of general behavior systems theory", in H. Feigel and M. Scriven (eds.), *Minnesota Studies in the Philosophy of Science*, vol. I, Univer. of Minnesota Press, Minneapolis, pp. 223–38.

Buhler, Ch. (1959) "Theoretical observations about life's basic tendencies", *Amer. J. Psychother.* **13**, 501–81.

Chorley, R. J. "Geomorphology and general systems theory", in press.

Dost, R. H. (1953) *Der Blutspiegel. Kinetik der Konzentrationsabläufe in der Körperflussigkeit*. Thieme, Leipzig.

Egler, F. E. (1953) "Bertalanffian organismicism", *Ecology* **34**, 443–6.

Feigl, H. (1956) "Some major issues and developments in the philosophy of science of logical empiricism", in H. Feigl and M. Scriven (eds.), *Minnesota Studies in the Philosophy of Science*, vol. I, Univer. of Minnesota Press, Minneapolis, pp. 3–37.

Foster, C., Rappoport, A., and Trucco, E. (1957) "Some unsolved problems in the theory of non-isolated systems", *General Systems* **II**, 9–29.

Gessner, F. (1952) "Wieviel Tiere bevölkern die Erde?", *Orion*, 33–5.

Geyl, P. (1957) *Napoleon for and Against*, Cape, London.

—— (1958) *Debates with Historians*, Meridian Books, New York.

Haire, M. (1959) "Biological models and empirical histories of the growth of organizations", in M. Haire (ed.),

Modern Organization Theory, Wiley, New York, pp. 272–306.

Hall, A. D. (1962) *A Methodology for Systems Engineering*, Nostrand, Princeton.

Hall, A. D. and Fagen, R. E. (1956) "Definition of system", *General Systems* **I,** 18–28.

Hall, C. S. and Lindzey, G. (1957) *Theories of Personality*, Wiley, New York.

Hayek, F. A. (1955) "Degrees of explanation", *Brit. J. Philos. Sci.* **6,** 209–25.

Hearn, G. (1958) *Theory Building in Social Work*, Univer. of Toronto Press, Toronto.

Hersh, A. H. (1942) "Drosophila and the course of research", *Ohio J. of Science* **42,** 198–200.

Holt, S. J. "The application of comparative population studies to fisheries biology – an exploration", in E. D. Le Cren and M. W. Holdgate (eds.), *The Exploitation of Natural Animal Populations*, Blackwell, Oxford.

Kamaryt, J. (1961) "Die Bedeutung der Theorie des offenen Systems in der gegenwaertigen Biologie", *Deutsche Z. fuer Philosophie* **9,** 2040–59.

Keiter, F. (1951–2) "Wachstum und Reifen im Jugendalter", *Koelner Z. fuer Soziologie* **4,** 165–74.

Kment, H. (1959) "The problem of biological regulation and its evolution in medical view", *General Systems* **IV,** 75–82.

Kremyanskiy, V. I. (1960) "Certain peculiarities of organisms as a 'system' from the point of view of physics, cybernetics, and biology", *General Systems* **V,** 221–30.

Lektorsky, V. A. and Sadovsky, V. N. (1960) "On principles of system research (related to L. Bertalanffy's general system theory)", *General Systems* **V,** 171–9.

McClelland, Ch. A. (1958) "Systems and history in international relations. Some perspectives for empirical research and theory", *General Systems* **III,** 221–47.

Meyer, R. L. (1961) Preface, *General Systems* **VI, III–IV.**

Miller, J. G. *et al.* (1953) Symposium. "Profits and problems of homeostatic models in the behavioral sciences", *Chicago Behavioral Sciences Publications No. 1.*

Naroll, R. S. and von Bertalanffy, L. (1956) "The principle of allometry in biology and the social sciences", *General Systems* **I,** 76–89.

Netter, H. (1959) *Theoretische Biochemie*, Springer, Berlin.

Oppenheimer, R. (1956) "Analogy in science", *Amer. Psychol.* **11,** 127–35.

Patten, B. C. (1959) "An introduction to the cybernetics of the ecosystem: The trophic-dynamic aspect", *Ecology* **40,** 221–31.

Prigogine, I. (1947) *Etude thermodynamique des phénomènes irréversibles*, Dunod, Paris.

Rapoport, A. (1956) "The promise and pitfalls of information theory", *Behav. Sci.* **1,** 303–15.

—— (1957) "Lewis F. Richardson's mathematical theory of war", *General Systems* **II,** 55–91.

—— (1959) "Critiques of game theory", *Behav. Sci.* **4,** 49–66.

—— (1960) *Fights, Games, and Debates*, Univer. of Mich. Press, Ann Arbor.

Rapoport, A. and Horvath, W. J. (1959) "Thoughts on organization theory and a review of two conferences", *General Systems* **IV,** 87–93.

Rappaport, D. (1960) "The structure of psycho-analytic theory", *Psychol. Issues* **2,** Monogr. 6, pp. 39–64.

Rashevsky, N. (1952) "The effect of environmental factors on the rates of cultural development", *Bull. Math. Biophysics* **14,** 193–201.

Rosen, R. (1960) "A relational theory of biological systems", *General Systems* **V,** 29–55.

Schulz, G. V. (1951) "Energetische und statistische Voraussetzungen fuer die Synthese der Makromolekuele im Organismus", *Z. Elektrochem. u. angew. phys. Chemie* **55,** 569–74.

Thompson, J. W. (1961) "The organismic conception in meteorology", *General Systems* **VI,** 45–9.

Toynbee, A. J. (1961) *A Study of History, Vol. XII. Reconsiderations*, Oxford Univer. Press, New York.

Tribus, M. (1961) "Information theory of the basis for thermostatics and thermodynamics", *General Systems* **VI,** 127–38.

Vickers, G. (1957) "Control, stability and choice", *General Systems* **II,** 1–8.

Watt, K. E. F. (1958) "The choice and solution of mathematical models for predicting and maximizing the yield of a fishery", *General Systems* **III,** 101–21.

Weaver, W. (1948) "Science and complexity", *American Scientist* **36,** 536–644.

Whittaker, R. H. (1953) "A consideration of climax theory: The climax as a population and pattern", *Ecol. Monographs* **23,** 41–78.

Weiss, P. (1962) "Experience and experiment in biology", *Science* **136,** 468–71.

Wiener, N. (1948) *Cybernetics*, Wiley, New York.

Zacharias, J. R. (1957) "Structure of physical science", *Science* **125,** 427–8.

Reprinted from von Bertalanffy, L. (1962) "General System Theory – A Critical Review", in *General Systems* **VII,** 1–20.

3 Towards a systems-based methodology for real-world problem solving

*by P. B. Checkland**

Introduction

The notions of "systems analysis" of complex problem situations and "a systems approach" to problem solving have become so modish that few are brave enough publicly to reject them and to argue for a reductionist rather than a holistic approach. One reason for this must be the sheer *generality* of claims made for using "a systems approach". The complex problems of organizations private and public, and the yet more difficult problems of society as a whole, are so obviously multi-faceted and contain so many connections that it is "obvious" that we must somehow embrace "the whole problem" in seeking to solve it, lest improvements in one area produce effects elsewhere which are inimical to the whole.

But constant lip service to "a systems approach", and exhortations to use it, do not hide the relative lack of determined persistent efforts both to define what a systems approach consists of and to use it in tackling real-world problems. It is not helpful to make claims for an approach unless it is possible to detail the concept sufficiently for interested persons to grasp it and use it in tackling problems of their own. A recent international symposium on "A Systems Approach to Management", for example, generated much useful discussion,[1] but absolutely absent from it was any account of how a systems approach to management might actually be made manifest in the management of an organization, let alone the "management" of society.

If "a systems approach" is to become more than an easily accepted but somewhat irritating concept, there is a need for expressions of it which eliminate any difference between what it is and how we may use it; the need is for accounts of systems-based *methodologies* which describe a systems approach as a way of analysing and hence trying to solve real-world problems.

In a previous paper[2] it has been suggested that the challenge facing those seeking to develop a systems approach is to develop methodologies appropriate across the spectrum from "the relatively 'hard' systems involving industrial plants characterized by easy-to-define objectives, clearly defined decision-taking procedures and quantitative measures of performance . . . (to) . . . 'soft' systems in which objectives are hard to define, decision-taking is uncertain, measures of performance are at best qualitative and human behaviour is irrational".

This paper is an attempt to begin to meet that challenge. It presents a general methodology which uses systems ideas to find a structure in apparently unstructured "soft" problems, and hence leads to action to eliminate, alleviate or solve the problem, or provides an orderly way of tackling "hard" problems. The methodology was evolved in the course of an action research programme of client-sponsored systems studies during the period 1969–1971, and refined in further projects in 1972 and 1973. It was derived from experience in real-world problem situations and "tested" in further problems. Tested ideas are rare, and although it is impossible in logic to establish accurately the worth or otherwise of a methodology such as this (a point discussed below), the users of it have found its guidelines helpful.

The methodology is expounded here by means of some account of projects which led to its formulation and reformulation. The aim is to hit a level between the merely anecdotal: "This happened . . .", and the dubiously academic: "Questionnaires sent to 100 firms showed that . . .".

Section 1 argues the need for ways of using systems ideas which are at once precise yet vague – precise in the sense that the ideas can actually be used to initiate and guide action, vague in the sense that the methodology must not be seen as, must not become, a technique. This section reviews the development of the systems movement and the strands of thinking within it which have led to previous expositions of systemic methodologies. Section 2 reviews the organizational context in which the present action research programme was carried out, and explains

* Peter Checkland is Professor and Head of the Department of Systems Engineering at the University of Lancaster, England.

its methods; Section 3 describes some experiences which contributed directly to development of the methodology; Section 4 expounds the methodology itself; Section 5 describes further experience gained in attempting to use it subsequent to its initial formulation; and Section 6 discusses some of the problems and issues raised by the idea of a useable methodology for tackling real-world problems.

Busy, practical men with no time for nonsense might turn straight to Section 4 where they might think they have found a recipe. But that would be a pity, since the main aim of the paper is both to offer a methodology which is of practical use in real-world problems, and at the same time gain the reader's acceptance that the methodology is not a how-to-do-it handbook and cannot be applied in any automatic or mechanistic way.

1 The context of the work

The systems movement

The work described in this paper has been carried out within "the systems movement"; that is to say the author's conscious sense of location has been that his work has been a small part of the effort by workers in many different disciplines who are loosely united by the concept "system".

The systems *movement* (and any other word would imply a structure or unity of outlook which does not exist) is concerned ultimately with the concept of systemic wholeness, and the working out of the implications of the concept in any (or every) one of the arbitrary divisions of human knowledge which we presently know as "disciplines". There are systems thinking scientists, technologists, economists, managers, management scientists, psychologists, sociologists, anthropologists, geographers, political scientists, historians, philosophers, artists . . . and many more; and there is also a small group of people working in the area whose interest is in systems concepts as such.

The systems movement represents an attempt to be holistic, and to find out the consequences of being holistic, in any area of endeavour. The only unifying element is the notion "system", and this is the movement's paradigm, in Kuhn's sense,[3] that a paradigm in science is an achievement or set of achievements which a scientific community "acknowledges as supplying the foundation for its further practice", achievements which "attract an enduring group of adherents away from competing modes of scientific activity" and are "sufficiently open-ended to leave all sorts of problems for the redefined group of practitioners to resolve".

The systems movement believes that the concept "system" can provide a source of explanations of many different kinds of observed phenomena – which are beyond the reach of reductionist science. And there is no doubt that achievements so far are sufficiently open-ended to leave many problems to be resolved!

Although there is a long philosophical tradition of systems thinking, explicit statement of the systems stance as a conscious paradigm would seem to stem from Bertalanffy's work in the 1930's on biological organisms as open systems which creatively maintain a degree of organization by continuous exchange of energy, materials and information with their environment. Bertalanffy[4] went on to point out the relevance of his treatment to systems of many different kinds, and it is his exposition which has led to the systems movement having a conscious identity and at least a partial language, as well as organizations and publications.

Within this young tradition, the present work has been carried out. It is concerned with attempts to use the systems paradigm to analyse, elucidate and help to solve real-world problems, and is part of a wider attempt to understand and develop systems concepts by using them in this way.[5] The work is thus concerned with problems of "management", in the broad sense that much human activity involves planning, doing and monitoring, and that some aspects of what is seen as "a problem" are likely to be a mismatch between intention or expectation, and outcome. Thus while the work has nothing to contribute towards understanding of a human activity of the kind: falling in love, it is relevant to any human activity which is *purposeful.* (It need hardly be stated, one hopes, that it is not restricted to "management" interpreted in a class sense.)

Methodologies for using systems ideas

Within the systems movement defined above two main strands of thinking may be identified.

The first is that which derives from the work of Bertalanffy, and is formally identified in the phrase General System(s) Theory (G.S.T.).

The second is that which grew up around the development and subsequent widespread use of computers, and with the application of computers to problems in technology. Here the concern was both the engineering of large complex systems containing many different components brought together to achieve some explicit objective, and also the organization of all the different activities which must be successfully integrated if a technological project

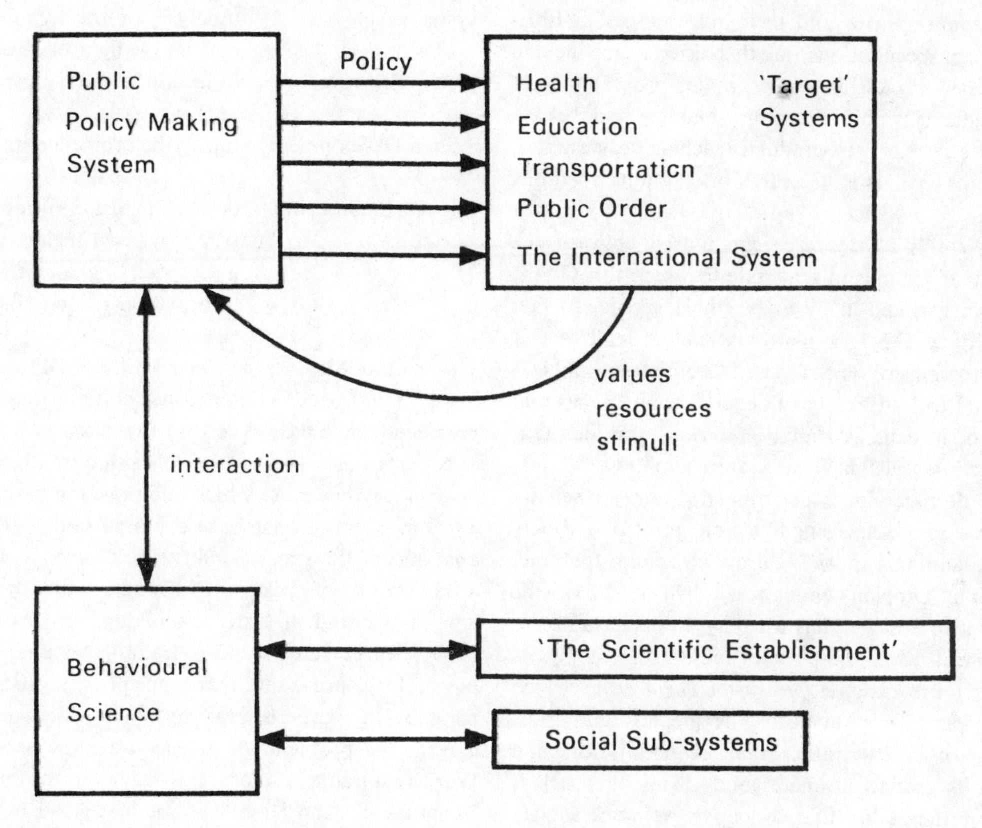

Figure 3.1. A picture version of Dror's GST model of policy making and behavioural science.

is to be efficiently accomplished; this we may identify as the Systems Analysis approach.

G.S.T. has not been much concerned with the development of tools for use in practical problems. Its concern has been mainly with obtaining an understanding of the fundamental nature of systems, broadly defined, and with tracing isomorphisms in systems whose content is at first sight very different. As Boulding describes it[6]

> General Systems Theory is a name which has come to describe a level of theoretical model-building which lies somewhere between the highly generalized constructions of pure mathematics and the specific theories of the specialized disciplines.

When the concepts of G.S.T. are used to provide a methodological approach to a practical problem there is obviously an immense difficulty in building a bridge between the theory and the problem; usually the bridge collapses into the gap between the two. We may briefly illustrate the difficulties by considering an interesting paper by Dror[7] ambitiously concerned with a General Systems approach to the problem of improving public policy-making by making use of the behavioural sciences. In some happy phrases which can make us all blush, Dror describes the present situation as a "mix-up between reliable factual knowledge, axiomatic assumptions, provisional theories, conceptual taxonomies, doubtful hypotheses and various types of value judgements . . .". He then goes on, "using a very simple version of systems theory", to suggest the model I have converted in outline into Figure 3.1. Something called the "Public Policy Making System" produces policies on a number of "Target Systems" such as Education, Health, Transportation, etc. and is itself affected by feedback from those areas in the form of values, resources and stimuli. The Policy Making System has interactions with the discipline "Behavioural Science", which itself also interacts with "The Scientific Establishment" and with many "social sub-systems". Consideration of this as a totality then leads to the beguiling, defensible, but very dangerous conclusion that "what is required is not some incremental change here or there, but fargoing redesign in the two relevant systems and their interchanges". "Illustrative redesign specifications" are then listed which include creation of new professional roles, major programmes of structural change within the policy-making and behavioural-science systems, new graduate teaching programmes, and significant changes in "research orientations, methods and subjects"!

Clearly there is a very large gap indeed between the present state of affairs and that state to which this G.S.T. analysis points, a gap likely to be unbridgeable. In fact the generality of G.S.T. is almost bound to lead to this kind of Utopianism; G.S.T. used methodologically is almost bound to lead to what Popper,[8] in his magisterial demolition of it, calls *Utopian engineering* – the idea that "rational action must have a certain aim", that "only when (the) ultimate aim is determined . . . only when we are in possession of something like a blueprint of a society at which we aim . . . can we . . . draw up a plan for practical action". Utopian engineering, which seeks the establishment of an ideal, has led only, Popper claims, to "the use of violence in the place of reason". He contrasts it with the more rational *piecemeal engineering* which searches for and fights against "the greatest and most urgent evils of society rather than searching for, and fighting for its greatest ultimate good". This approach is surely nearer the reality that, wherever we want society to go, we have no choice but to start from its present state, since we have no detailed knowledge for the large scale Utopian social experiments with which G.S.T. is likely to seduce us.

The second strand of systems thinking, the Systems Analysis strand, is more concerned with methodology, as it is bound to be given its base in engineering, the art of making things happen. Many methodologies for creating complex systems have been proposed, and some of them have been put to use during the last twenty years.

In the early 1960s A. D. Hall[9] described a methodology for translating scientific discoveries into realized applications which meet human needs, his account deriving from Bell Telephone Laboratories' experience. He envisages, and illustrates, the following sequence:

Problem Definition (essentially definition of a *need*)
↓
Choice of Objectives (a definition of physical needs and of the value system within which they must be met)
↓
Systems Synthesis (creation of possible alternative systems)
↓
Systems Analysis (analysis of the hypothetical systems in the light of objectives)
↓
System Selection (selection of the most promising alternative)
↓
System Development (up to the prototype stage)
↓
Current Engineering (system realization beyond prototype stage and including monitoring, modifying and feeding back information to design).

This methodology is conceived with hardware engineering in mind, and similar considerations have been expounded from a base in control engineering.[10]

An obvious extension of this kind of thinking is to systems beyond those which comprise the hardware itself, and this extension has been explored both in the military area and in the process industries. Optner,[11] as editor of a collection of papers concerned with the "systems analysis" strand of systems thinking, records that in the 1950s "Professionals with training in the social and physical sciences and management specialists with experience in weapons evaluation and procurement were agreed: the problems of defence after the Second World War required a more far-reaching problem-solving technology". And Hitch[12] in the first paper in this collection describes the essential elements as: an objective, alternative systems by which it may be accomplished, costs and resources of each alternative, a model (linking system, objectives, environment and resources), and a criterion for choosing the optimal alternative. These elements constantly recur as the components of the methodology developed within the RAND Corporation during the 1950s.

In the introduction to a book on the application of systems analysis to the planning of defence policy, edited by Quade and Boucher,[13] the RAND approach is summarized as follows:

> One strives to look at the entire problem, as a whole, in context, and to compare alternative choices in the light of their possible outcomes. Three sorts of enquiry are required, any of which can modify the others as the work proceeds. There is a need, first of all, for a systematic investigation of the decision-makers objectives and of the relevant criteria for deciding among the alternatives that promise to achieve these objectives. Next, the alternatives need to be identified, examined for feasibility, and then compared in terms of their

effectiveness and cost, taking time and risk into account. Finally an attempt must be made to design better alternatives and select other goals if those previously examined are found wanting.

Again there is emphasis on generating alternatives and comparing them using criteria relevant to the system objective or to the need the system is intended to fill. Obviously the lacuna here is the absence of guidance on how to generate alternatives. Even if alternatives are obvious, logically there is a blatant possibility that some unthought-of alternative would have given a better solution than any of those considered, and no amount of "brain-storming" or "lateral thinking" during an actual study can remove this defect.

One methodology within the "systems analysis" tradition has been described which avoids explicit steps of the kind: "generate alternative systems"; it is that described by Jenkins in the first issue of this Journal.[14]*

This methodology I interpret as follows. It starts from an organizational definition of system as a complex grouping of human beings and machines for which there is an overall objective. The procedure is then to select that system the engineering of which will solve the problem whose existence initiates the activity. The system is placed in a systems hierarchy, objectives and criteria of performmance are defined, and the chosen system is designed (via model building, simulation and optimization), implemented and reappraised in operation. Here the "generate alternatives" problem has to be solved by the initial examination of the problem and the choice of the system to be engineered. Logically, of course, what subsequently comes to be recognized as a wrong choice may be made; but the whole activity is conceived as a team activity aimed at helping a manager, so it is at least implicit that he will himself profoundly influence the choice of system to be engineered to solve his problem, even if that problem-solving task is itself conceived as a professional team activity.

As the methodology described below evolved during the action research programme, it diverged from those described above, and for interesting reasons which are discussed later. But the starting point was the methodology described by Jenkins, and the explicit aim was to explore the implications of trying to apply it to ill-structured, fuzzy, "soft" problem situations. What has finally emerged can be viewed as a general systems methodology which is perhaps nearer to the G.S.T. tradition than the Systems Analysis tradition, but which under special circumstances (those of a structured problem) can sharpen up into a version which would be at home in the latter tradition.

* Reprinted as Article 4 in this book.

Before describing the work which led to formulation of the methodology it is useful briefly to examine some of the requirements of an ordered procedure for tackling ill-structured, or unstructured problems.

The requirements of a methodology for tackling unstructured problems

The common characteristic of the systems analysis methodologies described above is that they make use of the kind of thinking which is natural to design engineers, whose role is to provide an efficient means of meeting a defined need. The design engineer's situation is one in which *what* is required has been defined, and he must examine *how* it can be provided. His skill and flair are directed to providing ingenious possible answers to the question: How can it be done? The best design engineer is the man who generates the most efficient and ingenious alternatives, the man, for example, who first designs a bridge with arch and keystone where previous beam-based structures have required closely-spaced support pillars.

Much skill is called for in engineering design, and much resolution in converting design into realized artifact, but the relevant point here is that the design engineer's problem is essentially a *structured* one. Methodologies stemming from engineering thinking inevitably postulate a certain kind of structure in the situation in which the methodology will be used: either a need can be defined, or a system's objectives can be stated unequivocally, thus enabling the system to be engineered to achieve them. The need- and objective-defining are taken as given, in the sense that need-defining and objective-defining systems are not part of the problem situation.

Basically these methodologies derive from the model that sees human behaviour as goal-seeking. But this model, useful though it is in many situations, is inappropriate in many real-world problems which are seen as "problems" precisely because there is no agreement on needs, objectives, measures of performance, etc; and the lack of agreement may not be due simply to lack of understanding or lack of information – it may be fundamental, and for two different reasons. The lack may be due to incompatible ways of viewing the problem, incompatible weltanschauungen, or it may be that any goal-seeking model itself imposes false structure on the problem situation by seeing it as a matter of *ends* and *means* rather than ongoing relationships through time. Vickers[15] has argued most cogently against the dominance of "objectives" over "relationships" in our conscious thinking, and it is in the context of what he terms shared *appreciative systems* "carved out by our interests, structured by our expectations and evaluated by our standards of judgement" that any methods of using systems ideas to tackle real-world problems must be developed.

Vickers' insight is that our human experience develops within us "readinesses to notice particular aspects of our situation, to discriminate them in particular ways and to measure them against particular standards of comparison, which have been built up in similar ways". These *readinesses* are organized more or less coherently into "an appreciative system". New experience is conditioned by these readinesses and itself modifies them; such circular relations constitute the common facts of social life, but we fail to see this clearly because of our concentration on linear causal chains and on the notion of goal seeking.

I take this concept of an appreciative system to be the most useful description of the context of "problems" in the real-world, and one we must seek to use in spite of the greater simplicities of the goal-seeking model.

If we are to make progress, then, we must avoid both content-free methodologies derived in a flabby way from General Systems Theory, and over-precise goal-oriented formulations stemming from Systems Analysis. We need simultaneously to be precise enough to provide guidelines which can actually be used, and vague enough to remain problem-oriented, avoiding distorting the problem into a particular structure just because we would know how to tackle it if it came to us in that form.

Finally before describing the research and its outcome, it is necessary to make some general remarks about the need for methodology, since the idea that we need to develop systems-based methodology relevant to tackling real-world problems is not an everyday concept, and may seem a strange or unnecessary one. Scientists and engineers do not show much concern about methodology, why should we agonize about the importance of it in systems work?

A methodology I take to be an explicit, ordered, non-random way of carrying out an activity. As such it is independent of the *content* of the activity and can be considered separately from content, although in practice this is not easy to achieve: most people, asked to describe how they tackled a problem will quickly slip into describing the content of the problem itself. The reason is that methodology, once adopted, tends to become invisible; it becomes, in both a logical and a behavioural sense, simply the way the activity is carried out, and is taken for granted. Most practising scientists and engineers show little interest in methodology because the pattern of their activity, both in its logic and its displayed behaviour has long been established. The four-year degree course in Chemistry I took at Oxford contained nothing at all about the methodology of science. Although I now find this astonishing I did not notice it at the time. The course was concerned with the content of chemistry; it was implicitly assumed that the research activity of the fourth year would be carried out in the same way as researches already described in the literature, with practical guidance gained by working alongside more senior researchers. The methodology of carrying out chemical research had become invisible.

In using a systems approach it is likely that methodology will eventually cease to be a major object of study, and the attention of most practitioners will be concentrated wholly on problem content. At present, while hopefully bringing about improvements in the problem situations themselves, it is more important to use groups of related studies to establish methodology itself.

2 The research method

Action research in systems at Lancaster

The research on which this paper is based was carried out as part of an on-going action research programme in the Department of Systems Engineering at Lancaster University, a programme which may well be unique in form and content. The programme is associated with Masters, Doctoral and Post-Doctoral work and involves carrying out mainly client-sponsored systems studies in real-world problem situations. In starting it in 1966, G. M. Jenkins' uniquely bold concept was that a fruitful way to explore systems concepts would be to tackle actual problem situations with a project team consisting of a faculty member, a full-time mature postgraduate student (for whom the project would be an exceptionally rich learning experience) and a member of the client organization which poses the problem.

More than a hundred such projects have now been carried out.

Systems ideas are the basis of project thinking, and the projects themselves, as well as seeking practical improvements in the problem situation, provide a growing body of experience which is a source both of insight into systems ideas and methodologies for using systems ideas. A methodology derived tentatively from a number of projects can hopefully be further tested and refined in subsequent projects: that is the pattern followed in this work.

Obviously there are problems which stem from the "action" nature of the research, and this point requires some discussion.*

The concept of action research arises first in the behavioural sciences. Forster[16] identifies Kurt Lewin as the father of the approach in the 1940's and the prime ideational source as Lewin's Gestaltist view of "the limitations of studying complex real social events in a laboratory, the artificiality of splitting out single behavioural elements from an integrated system". Forster arrives at a definition by augmenting that of Rapaport[17]. Forster's version is as follows:

> A type of applied social research differing from other varieties in the immediacy of the researcher's involvement in the action process and the intention of the parties, although with different roles, to be involved in a change process of the system itself. It aims to contribute both to the practical concerns of people in an immediate problematic situation and to the goals of social science by joint collaboration within a mutually acceptable ethical framework.

This definition fits the Lancaster systems projects in that they are involved in "a change process in the system itself" as a means to both practical action and an experience relevant to the research aim of developing systems concepts. The problem with action research arises from the fact that it cannot be wholly directed. The essential nature of action research is revealed by asking the question: "What would action research in physics be like?" The impossibility of that concept shows that action research is a possibility only in a situation in which the elements are social interactions. (You cannot do action research on magnetism because there is no way in which you can intervene and affect or change the physical laws). The point is that, being concerned with intervention in purposeful systems, the action researcher, unlike the researcher in experimental science or technology can express his research aims as hopes but cannot with certainty design them into his "experiments". He is prepared to react to whatever happens in the research situation; he has to follow wherever the situation leads him or stop the research. In the present work an attempt was made to tackle ill-defined problems by using systems concepts, to explore the difficulties encountered and to propose and test ways in which the systems concepts were used. The problems, though very diverse in content were treated as a single group and lessons from the ones first tackled modified the approach used in later ones. In this situation the use of action research projects as the vehicle for the research is made legitimate because, and only because, the research aim is the very general one of developing a methodology for tackling "soft" ill-structured problems; hence *any* problems which meet that criterion are suitable for the purpose whatever other characteristics they may have.

* I am grateful to my colleague Dr. T. R. Barnett for helping sharpen my ideas on action research.

3 Some systems projects

Three project experiences

Some aspects of three systems projects are now summarized. They were all characterized by having initial definitions of "the problem" which were vague and unstructured. The client organizations had no doubt that serious problems existed, but could not or (sensibly) would not define them in precise terms. Since a problem defined is a problem solved, in the sense that any precise formulation of a problem implies its solution, these projects were suitable ones for examining means of tackling "soft" problems.

The first project was in a textile firm having 1,000 employees. It had ceased to be profitable the year before the study and had newly recruited some senior managers with large-firm experience. One of them, the Marketing Director, initiated the systems project, probably, we felt, as a means of bringing pressure bear on the Production side, whose planning and scheduling performance he regarded as inadequate. It was at once apparent that this was correct, but equally apparent was the fact that perfecting the production planning system would have a negligible effect on the Company's main aim, which was now nothing less than survival. There were at least a dozen significant candidates for the role of "the problem" including the inadequacy of financial information, the absence of any Company planning, poor production-marketing collaboration, inadequate control of nine regional stores, no quality control system, inadequate information flows to senior managers, and low staff morale. In the end progress was made by *choosing to view the Company in a particular way and working out the consequences of that view*. The Company was taken to be a system whose survival was threatened unless it could quickly – say within a year – improve its most basic short-term operation, namely winning customer orders and meeting them efficiently. The Company was therefore taken to be a system which generates and accepts customer orders and meets them sufficiently quickly and efficiently to ensure an increasing flow of them. The sequence of *decisions which recur* in such a system was isolated and a conceptual model was constructed by postulating a system component to take each decision. The sequence itself then led to a system structure, as has been described previously (Checkland and Griffin).[18] The concept led to the creation of two units for order processing, one customer-oriented, the other production-oriented, and to the engineering of the necessary information flows. Four subsequent projects then sought to bring the reality gradually nearer to the new concept.

The second project was in the engineering industry, in a large firm concerned with the design and construction of a small number of very large and very complex objects. A perspicacious director, concerned with the problems of integrating highly specialist expertise into the enterprize as as whole, set up the study with a remit which was deliberately, and usefully, vague: its theme was to be "information flow for decision making", and there was a specific suggestion from some of the specialist middle managers that the *interface* between Design and Production Planning should be examined as a particular problem area. Major elements in the overall problem situation were: an incredibly complex functional organization within which decision-taking was diffuse and which contained a number of project "task forces" which had lived on beyond the end of their intended life; a "project" organization which cut across the functional organization but which was purely a reporting, not a managing function; a senior management which (perhaps because of these characteristics) continually became involved in low-level day-to-day engineering problems; and a middle management remarkably dedicated to the success of the total enterprize, showing much more commitment to it, in fact, than is usually found in the process industries. Because of the middle management's dedication, unworkable procedures did actually work. Here again the thinking hinged on deciding the basic nature of the organization. It was decided to view it as a system to carry out a gigantic task – that of converting physical and abstract resources into the final complex object within a certain time, at a certain cost and while meeting various technical and safety constraints.

Within this frame of reference a conceptual model of the necessary system to carry out such a task was compared with the existing functional organization and project administration. General conclusions were drawn and these were then illustrated by examining a specific example. Specific illustration was centred not on the flow across the Design/Production Planning interface (which would have automatically assumed this to be a required part of the *structure* of the system) but on a comparison of how one specific part of the complex object *would have* been provided within the conceptualized system and *was actually* provided within the existing procedures. The illustration was apt in that it showed in microcosm the operation of the system as a whole.

The third project was in the publishing/printing industry, being specifically concerned with consumer and technical magazine publishing and printing. The work was sponsored by a small central group who were engaged in building a corporate planning system in a divisionalized Company; its audience was a committee chaired by the Managing Director and having as members the executive heads of the different divisions. The project problem was exemplified rather than posed. It was exemplified by the stated difficulty of answering questions of the kind "What is the appropriate mix of production facilities within a printing division over the next five years, given the rapidly changing market, the spread of printing contracts between our own and external printers, and the general overall need to improve Company profitability"? (The division structure was a typical "business area" organization, with each division encouraged to be autonomous. The management consultants who had suggested it said in their report that the divisional chief executives "can be held truly accountable for their own profit contribution or cost. However, they cannot be entirely free to take action . . . Before setting out on a major course of action . . . directors must seek approval and coordination from higher management". They did not, alas, say how that "coordination" should be provided. Asking: How can it be provided? is one way of stating the problem).

It did not seem useful to identify "the system" as "the corporate planning system"; instead, analysis of existing arrangements gradually moved over into consideration of the basic nature of the business and conceptualization of a system to carry it out. Here the basic business can be viewed as a system concerned with the provision of consecutive "images" to an audience which must be assessed continuously. The work of the system is carried out by two different groups whose technical interests are different (publishers and printers) and who constitute two very different social systems. In the end the conceptualization led to definition of a range of possible changes to the existing structure, the range covering different "amounts of change", and an assessment of the difficulty of implementing them.

An emerging methodology

What generalizations do these experiences illustrate?

All these projects included, as every systems study must, an *analysis* phase, a finding out of what exists at present in the problem situation. In methodologies appropriate to "hard" systems this analysis is "in systems terms", and the art of the analysis is to view the problem situation, without distorting it, as a hierarchy of systems, in order that "the system" (i.e. "the system to seek to engineer") can be selected. There is an implication here that the systems hierarchy will already exist in at least rudimentary form. This will certainly be the case in most "fairly hard" systems studies. If the study is concerned with, say, the management information needed to run a chlorine plant, we may be sure that there does exist at least a rudimentary version of such an information system – the plant simply could not run without it – just as there are bound to be existing versions of plant subsystems concerned with other necessary activities: production, costing, maintenance, industrial relations, etc.

But in softer systems it is much less obvious that the situation can be analyzed as an existing hierarchy of (albeit imperfect) systems. To pursue such an analysis may well distort the picture of the problem situation and may lead to a confused compromise between analysis and design in which the picture built up is neither an account of what exists nor a proposal concerning what should exist. This is difficult to avoid if existing arrangements fall short of being purpose-built systems – as most "existing arrangements" do.

Similarly any concentration on definition of objectives implies that an agreed definition of objectives for the systems in the hierarchy will not be too difficult to obtain, and will be useful once obtained; this is then the lead-in to systems design: the design is such that these objectives may be achieved. But in soft systems such definitions of objectives may be impossible to obtain and if obtained may not be useful. It may be that obtainable definitions are "public" ones which conceal the "real" objectives – perhaps even from the people who profess them; or it may be that objectives of a soft system are genuinely inexpressible; or it may be, remembering Vickers' distinction between goal-seeking and relationship-maintaining, that it is simply inappropriate to try to define objectives, objectives being finite goals different in kind from the deeper underlying ongoing purposes of a soft system which, though it may be other things as well, will be a social system. And social systems are not usefully regarded as goal-seeking.

In soft systems like those of the three studies under discussion it was found most useful to make the analysis stage a building up of *the richest possible picture* of the situation being studied without pressing the analysis in systems terms. Such an analysis (guidelines for which are discussed below) enables selection to be made of a *viewpoint* from which to study further the problem situation, and enables some relevant systems (relevant, that is, to improving the problem situation) to be named tentatively.

In each of the three studies described above an early feeling of helplessness led to asking, not: What system needs to be engineered? but: *What is the fundamental nature of (notional) systems which from the analysis phase seem relevant to the problem?* In the textile firm the sheer multiplicity of problems forced the emergence of the view that the Company as a whole had to be encompassed, and that it had to be viewed as an order-processing system which needed to improve within a year in order to survive. In the engineering firm the useful lack of a precise problem definition and the suggestion that a particular organizational interface be examined led to a realization that the present position of such interfaces is arbitrary, and irrelevant to the nature of the enterprize, which is to use one of many possible organizational structures to accomplish a huge task. In the publishing/printing firm the analysis suggested that the problems were concerned centrally with the notionally autonomous divisions, different in activity and attitudes, which were nevertheless closely entwined in the overall activity of the enterprise. A view of the overall activity which brought this out was required.

In each of these projects (and a number of others) analysis led to this kind of *root definition* of relevant systems. From such definitions *conceptual models* of notional systems whose fundamental nature is embodied in the root definition were derived (relatively rigorous ways of doing this have since been developed and are discussed below). These models, which were not normative, but rather provided a means of organizing the systems thinking, then gave something which could be set alongside the actual situation in a *formal comparison* which enabled *definition of a range of possible changes*.

In the projects under discussion it so happened that structural, rather than procedural changes were debated, and in two cases, the textile firm and the publishing/printing firm, new organizational units were agreed as both desirable and feasible. (The engineering firm, whose historical pride was in the technical excellence of its functional groups, was not prepared to contemplate the possible changes which the comparison stage defined.)

It follows from the above that any *design* and *implementation* is, in general terms, the design and implementation of an agreed change rather than of a system. Design of a system emerges as a special case, appropriate when the problem definition is sufficiently sharp to enable objectives and measures of performance to be defined and hence to allow the thinking to be in terms of model building, simulating performance and optimizing.

Out of the three projects illustrated here and six similar ones, there emerged during 1969, 1970 and 1971 the outline of a methodology whose crux was the root definition and conceptualization of relevant systems. Attempts to use it (and understand it!) in subsequent projects since then have enabled it to be expressed in clearer terms and have enabled some of its implications to be revealed. It departed from "systems analysis" methodologies (as defined previously) firstly in a more open-ended analysis phase not carried out in systems terms, secondly in introducing the root definition/conceptualization stage, and thirdly in being forced to abandon the idea of *designing a system* except in the special circumstances of either a well-defined problem with explicit objectives and measures of performance, or a "green field" situation in which conceptualization can be used as the first phase of the design activity, a phase concerned with defining "what" before design itself gets down to defining "how".

The next section expounds the methodology and further illustrates it. Meanwhile we may summarize it in outline as in Figure 3.2.

4 The methodology

Introduction

The methodology assumes that we are concerned to tackle problems which arise in human activity systems. In a previous paper[2] it was argued that the basic proposition of a systems approach is that it is "reasonable and useful" to view the universe as a complex of interacting systems.

We may identify *natural* systems, from atoms to galaxies; and one of them, the human being, can create *designed* systems, whether abstract (a philosophy, for example) or concrete (a machine) which can be used in *human activity systems*. Quoting[2]: "Human activity systems comprise the area in which objectives can be originated"; in other words, they contain purposeful elements, and it is this which justifies the development of problem-solving methodology: the knowledge that there exists a will to understand and improve the situations we find ourselves in. The methodology now described can be viewed as a means of using systems ideas to structure problem situations in order that that will can be exercised.

Analysis

It has already been argued that the analysis phase should not be in systems terms unless the problem situation is a relatively structured one. If it is unstructured then seeking out "systems" at the start almost always leads to the identification of organizational groupings – departments or functions – as systems, which may or may not be appropriate. An organizational grouping is a particular "how"; the analysis needs to be in terms of "what".

Guidelines for analysis found adequate in a number of studies, including those described above, are illustrated in Figure 3.3.

The problem situation will exist within a number of environments and/or wider systems with which it will interact. Most of the environments will be what Emery and Trist call "turbulent fields".[19] The problem situation will itself be turbulent, but it will contain elements of "structure" which are relatively static, and elements of "process" which are dynamic, the former existing as the framework within which the latter operate. In the publishing/printing project the environments included an owner-company of which the publishing business was one business area, a particular information/entertainment market, a changing technology and two different social systems. "Structure" was present in the form of an organization broken down into subdivisions defined according to product type and functional activity. The "process" was the set of activities concerned with publishing and printing as a business activity and the activity of staying in that business in the long term.

Structure may be examined in terms of physical layout (which often has an effect on the "process" and on the relationship between the two, the "climate"), hierarchy, reporting structure and the pattern of communications both formal and informal. The *process* may, in any organizational context, be analyzed in terms of the basic activities: planning to do something, doing it, monitoring both

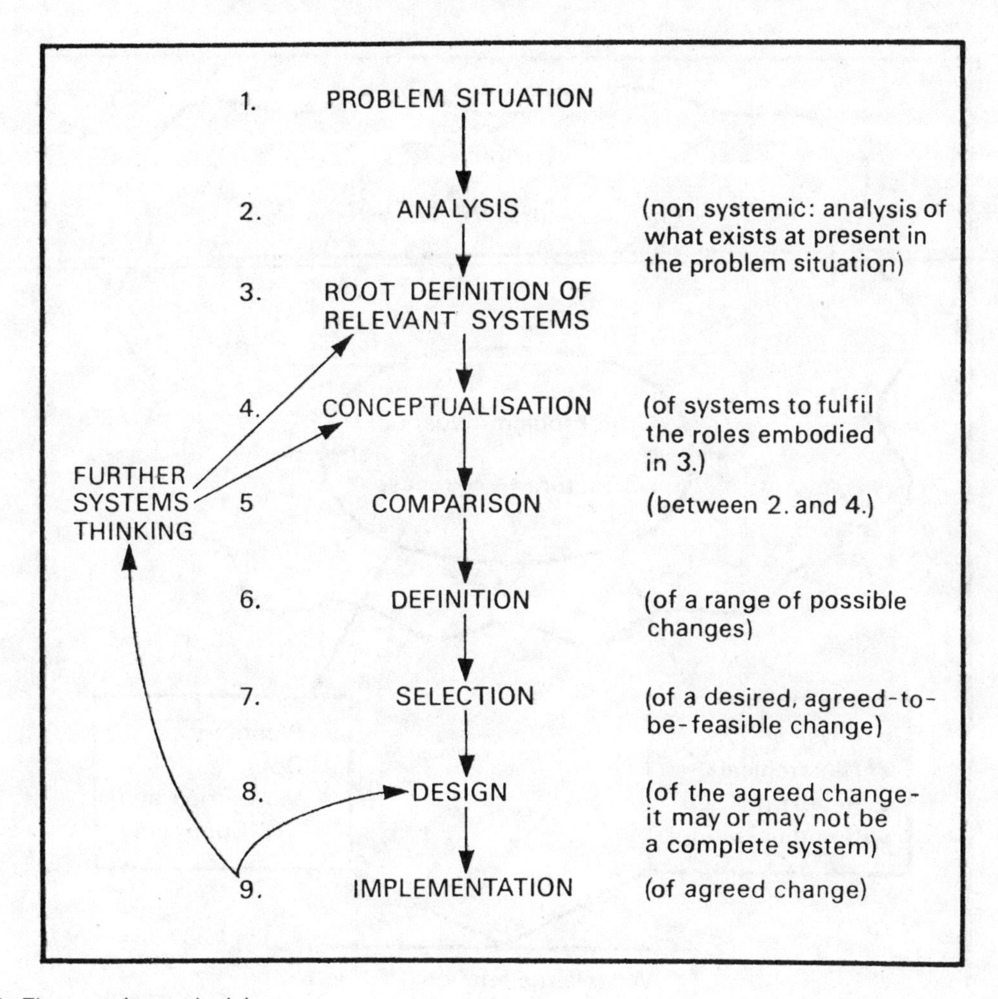

Figure 3.2. The emerging methodology.

how well it is done and its external effects, and taking action to correct deviations.

The relationship between structure and process, the "climate", is a core characteristic of any problem situation, and especially so if the problem is in an organization. In the engineering industry project a technology-based functional structure matched ill with the process operations of a task-system, but managed to survive (with ever-increasing complexity) because of the attitudes of those middle managers dedictated to the process. This was the core characteristic of this problem situation, the end point of the analysis.

It is difficult to know when to finish the analysis phase – even if only temporarily! But the analysis should not be regarded as complete until at least the following questions can be answered convincingly and in some depth.

Within and/or around the problem situation:

1. What *resources* are deployed in what *operational processes* under what *planning procedures* within what *structure*, in what *environments and wider systems*, *by whom?*

2. How is this resource deployment *monitored* and *controlled?*

Root definition of relevant systems

The analysis may be taken to be complete, at least on a first iteration, when it is possible to postulate a root definition of the basic nature of the system or systems thought to be relevant to the problem situation. This should be a condensed representation of the system(s) in

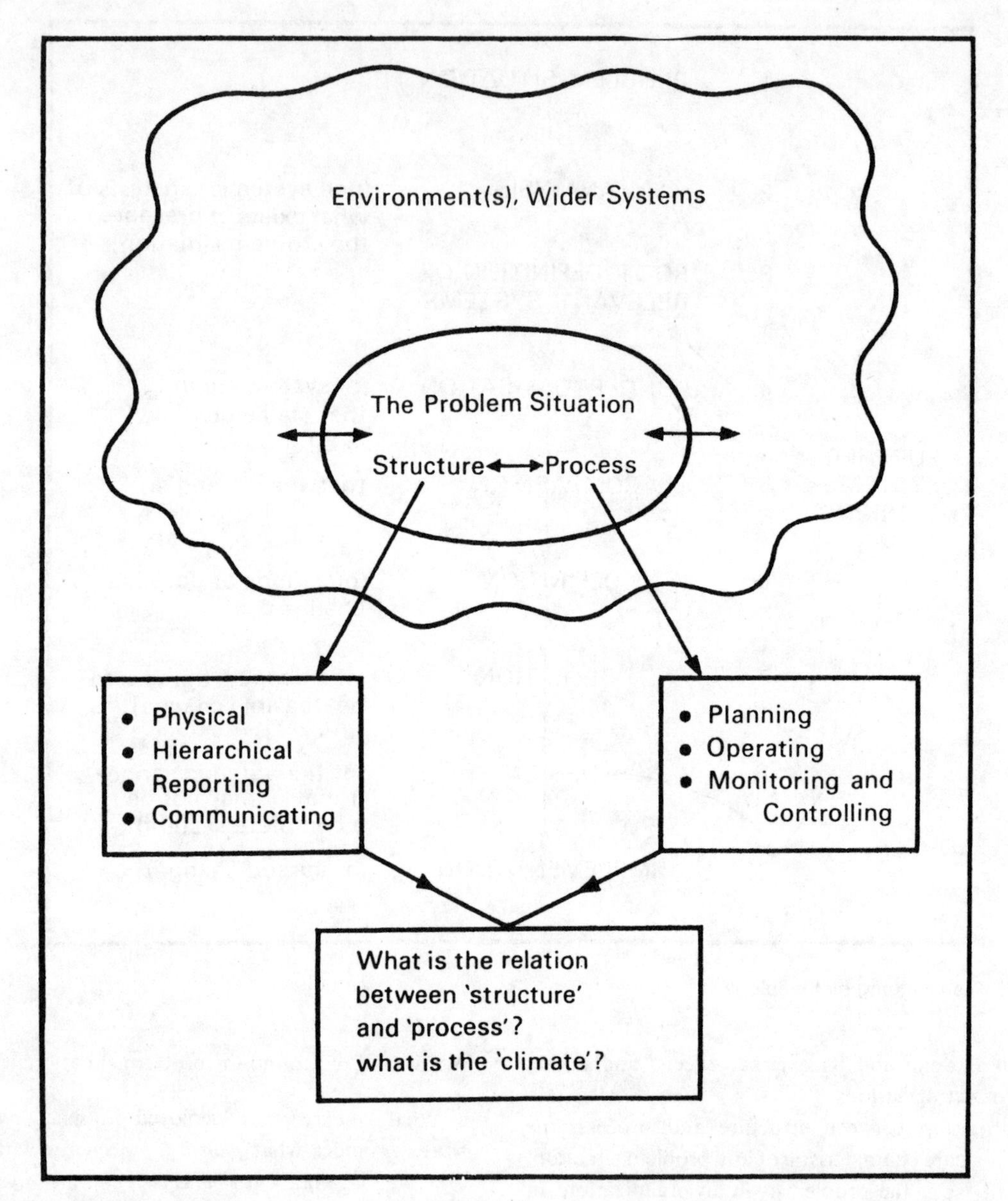

Figure 3.3. Guidelines for analysis.

its most fundamental form. Much will depend upon this definition and it needs to be pondered carefully; its aim is to capture insight.

In the case of the textile industry project the analysis revealed a firm whose crucial objective was survival. Hence the root definition of the relevant system was taken to be an order-generating and order-processing system, the aim being a quick improvement (within about a year) in the level of customer service.

In the publishing industry project the conclusion of the analysis was that the problems here were embedded in a rather more complicated system, one crucially concerned with assessing potential audiences, providing that audience with successive images, and ensuring that doing this as a business operation earned a surplus; Figure 3.4 summarizes the concept.

This root definition derived from the realization that the editorial activity and the printing activity were, in essence,

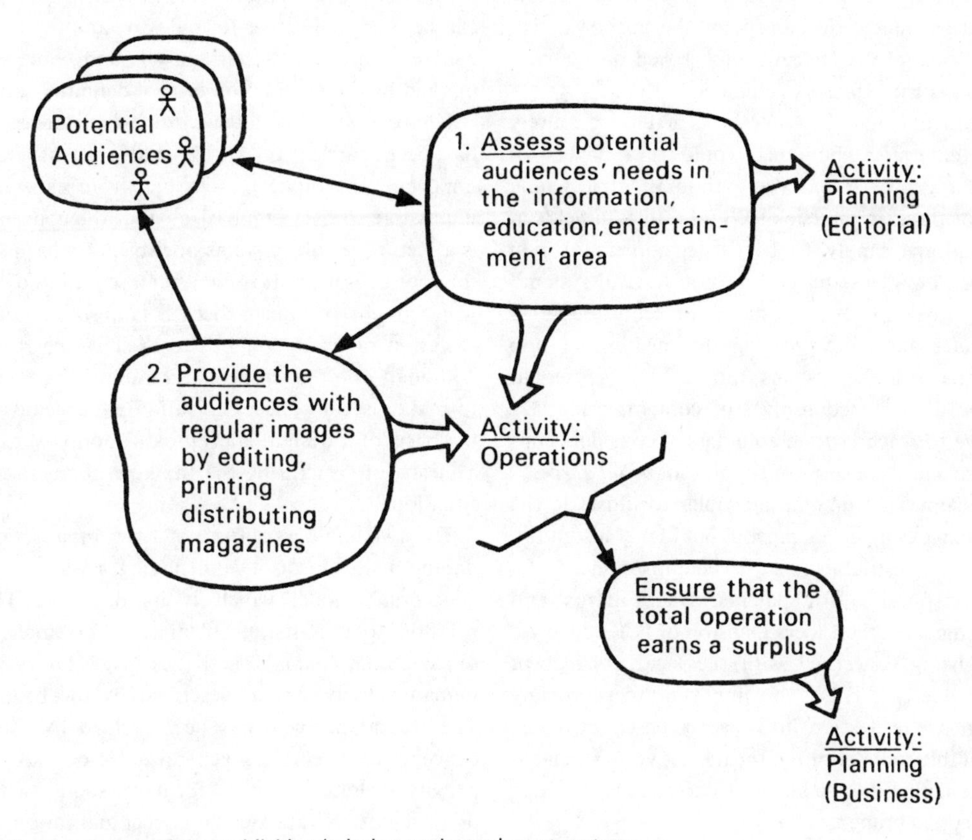

Figure 3.4. Root defintion of the publishing/printing project relevant system.

not the "marketing" and "production" activities of typical commercial enterprizes but were the activities, valued for their own sake, of two different social systems in which business considerations were of low importance. ("I was spending all my time reading 'Management Today' instead of doing the editing I love" wrote a disgruntled magazine editor recently, complaining about the increasingly professional approach of her company). In such a system purely business considerations tend to be neglected, need to be built into the system, but are inappropriate to the kind of people involved in editing and printing.

For those who concocted it the "assess potential audiences' needs" and "provide regular images" elements in this root definition embodied the fact that a magazine editor genuinely feels a very close (almost mystical) rapport with his audience, and has to assess sensitively its changing attitudes. It is a fact that magazines for pubescent girls, for example, need to be edited by young women who can identify with teenagers and dream their dreams, rather than by cynical 50-year-old males. Such editors are unlikely to treat magazine publishing primarily as a commercial enterprise: hence the form of the root definition.

There will of course never be a demonstrably "correct" root definition, only a range of possibilities, some glib and shallow, some full of insight. What is needed is a penetrating definition, derived from the richness of the analysis, which is revealing to those involved in the day-to-day workings of the systems concerned.

The nature of root definitions may be further illustrated by a few somewhat extreme examples.

We could formulate root definitions of a church around several different concepts: perhaps "a social welfare system", which downgrades the core of supernaturalism upon which most churches insist, or as "a ritual-organizing system", or as a "belief-maintenance system", or – building in a different weltanschauung – as a "system to provide support in the face of unanswerable questions". In an actual situation it might be profitable to explore the implications of several such root definitions, covering as these do both what a participant might want the system to be as well as what an outside observer might take it to be. A postgraduate student developed for the author a rich conceptual model of the Jewish church based on the root definition "a system to preserve ethnic identity".

A pop festival could be conceptualized as a purely commercial enterprise, but this would miss layers of meaning at a deeper level. Many thousands of young people do not travel to the Isle of Wight to live in fields in squalid conditions simply to be the customers of pop music sellers. A systems study of a "pop festival system" would have to begin with some such root definition as "a system to celebrate a life-style via the medium of pop music". Similarly if the systems approach were used to conceptualize an intended festival of contemporary jazz it would have to reach beyond commercial considerations to the politico-racial concerns of the current *avant garde*.

These are somewhat fanciful examples to illustrate the concept. A final example, less fanciful but highly insightful, comes from an industrial firm. A. B. Thompson, head of a department responsible for patents, trade marks and licence agreements in the Fibers Division of ICI, conceives his role as being concerned with the legal concept of intellectual property.[20] This is very important to a company operating on a world scale in a science-based business. Clearly a suitable root definition for the activity associated with this role is that it is "a system to *acquire, protect and exploit* intellectual property".

All of the above examples are necessarily general for illustrative purposes. In any particular systems project it is extremely important that the root definition should be specific to the particular situation of that project. To imagine that there can be once-and-for-all definitive root definitions applicable, say, to any distribution system, any marketing system, or any quality control system would be to make the same mistake as is made in those management science text-books which imagine that problems recur, and that they can therefore deal with "the inventory problem", "the investment problem", "the depot location problem", etc. chapter by chapter.

Conceptualization

Conceptualization is the process of building conceptual models of the system or systems which are relevant to the problem situation and for which we have root definitions. The purpose is to create, in the light of the problem situation but in a mood of detachment from it, something which can be compared, formally and specifically, with the picture built up in the analysis phase. It is the part of the methodology in which systems thinking can be used in a rather formal way, and there should be a conscious break between analysis and conceptualization, marked by the postulation of root definitions.

Before describing the way in which conceptualization has been carried out in 20-odd systems studies, it is important to emphasize – in fact from experience so far impossible to over-emphasize – that the conceptual model is a picture, as objective as possible, of what is implied by the root definition. It is *not* a representation of the "ideal" which might be approached; it is absolutely *not* a representation of what "ought" to exist in the real situation. Although in a "green field" situation it has been possible to make use of conceptualization as a means of defining the basis of a design, conceptualization is *not* design, it is a means of structuring the thinking about fuzzy problem situations.

The two main problems of conceptualization concern finding a way to do it, and finding a way to validate the conceptual model which is its outcome. The present solution to the first problem is to assemble in correct sequence the *minimum* activities which are *necessary* in a human activity system described by the root definition. The second problem has been tackled by using systems concepts to assemble a generalized model of any human activity system, called a "formal system", which can be used, if not to validate the conceptualization, at least to ensure that it is not basically deficient.

Conceptualization starts from the root definition and asks: What would the activities have to be in a human activity system which meets the requirements of this definition? The aim is to include every activity implied by the definition, but no extra ones, on the principle of Occam's Razor. When a tentative sequence has been assembled it is useful to annotate it with major inputs and outputs. When this has been done we may ask of each activity: What decisions would have to be taken in carrying out this activity? And then at the next level of consideration: What information would the decision-taker require? Once these steps have been completed it is useful to consider the structure implied by the model so far: How might the activities be grouped? An example of this kind of activity is given in[18]. (It is of course possible that answering these questions may lead to reformulation of the root definition).

It is important in building the concept not to slip into describing what happens to exist in the real-world problem situation. The art of conceptualization lies in cleaving to the implications of the root definition while at the same time being not unaware of the reality; the art lies in finally achieving a bridgeable gap between the conceptual model and the problem situation.

The process may be illustrated by reference again to the publishing/printing project. The root definition (Figure 3.4) sees a relevant system as one concerned to:

assess a potential audience
provide it with successive images
ensure that the total operation earns a surplus.

It was decided to examine its implications in a system which begins with the "assess" activity, and produces a single consumer magazine.

The first activity (see Figure 3.5) is that of assessing the audience response to a sequence of issues. This is the crucial editorial activity of forecasting a kind of material which will serve to establish that editor-reader rapport which is the essence of this business and a prime value of the social system which the editor inhabits.

Given that assessment, material for the issues must now be selected or commissioned, as must advertising matter. This must also fit in with the assessment of the audience, and must be in sufficient quantity to meet those financial requirements which will have to be fed in from outside, from the Business Planning sub-system which the root definition implies.

Assembly and appraisal of a particular issue follows, and when satisfactory to the editor becomes "copy" which has now to be transformed into a form (plate or photographic image) which can be used for image duplication.

There will here be interaction between what the editor desires and what the printer's technologists can provide. Reappraisal may be necessary.

Printing follows agreement on satisfactory copy and from here the activity becomes technical: printing, collating, binding. Finally the magazine is distributed to the potential audience.

This operational sequence takes place with the production of each issue and involves the two activities of editing and printing. Logically the next activity to consider is of a different kind: the monitoring of the success of the operational activities by the other sub-system, that concerned with the operations as a business activity. This we may add to the operational diagram by representing it as an operation concerned with the whole of it, as in Figure 3.6.

The "monitor" activity itself will consist of appraising resources, making a plan to use those resources, comparing performance with plan and replanning. This sequence can

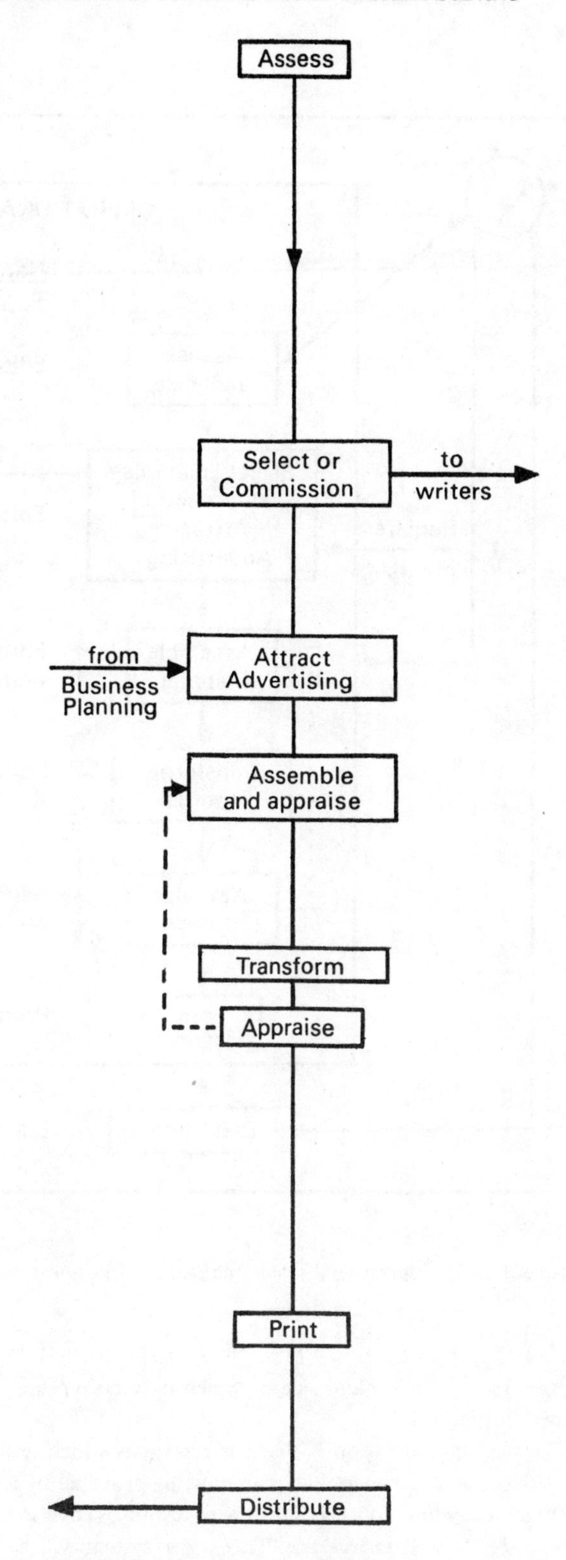

Figure 3.5. An operational system to create one magazine.

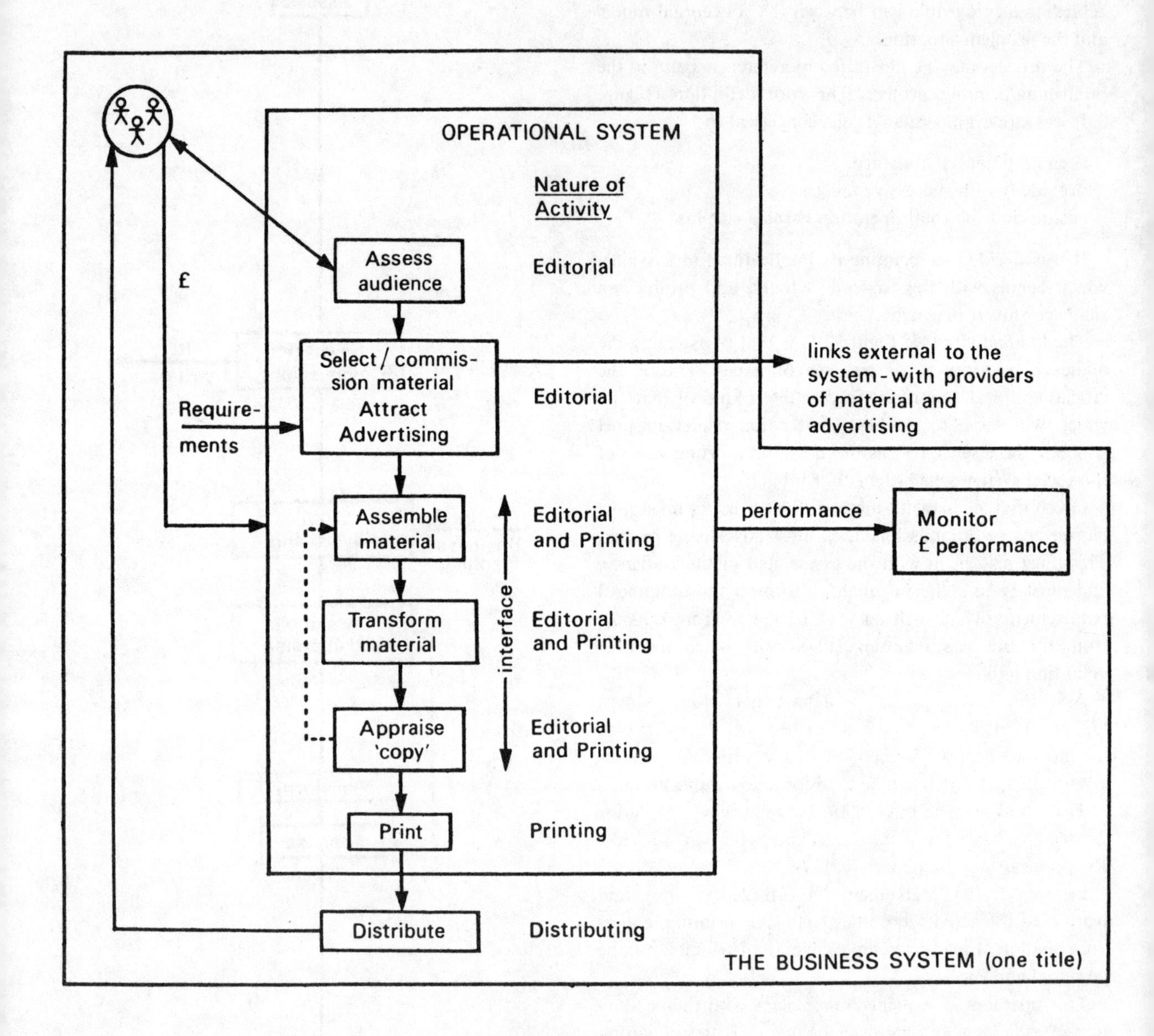

Figure 3.6. The operational system as part of a business system.

usefully replace the box "Monitor £ performance" in Figure 3.6, the right-hand side of which now becomes as shown in Figure 3.7.

The next consideration is that the resources which are the subject of this (single title) plan must be provided by a wider system whose concern is the use of resources covering other titles as well as this one. This wider system will be concerned with:

(a) providing resources for this title, or not providing them, depending upon its assessment of the total resources available,

(b) the relative merits (subjectively and/or quantitatively assessed) of different plans, and

(c) other more general considerations which in this business might be of the kind "In spite of its profitability we will not publish pornography".

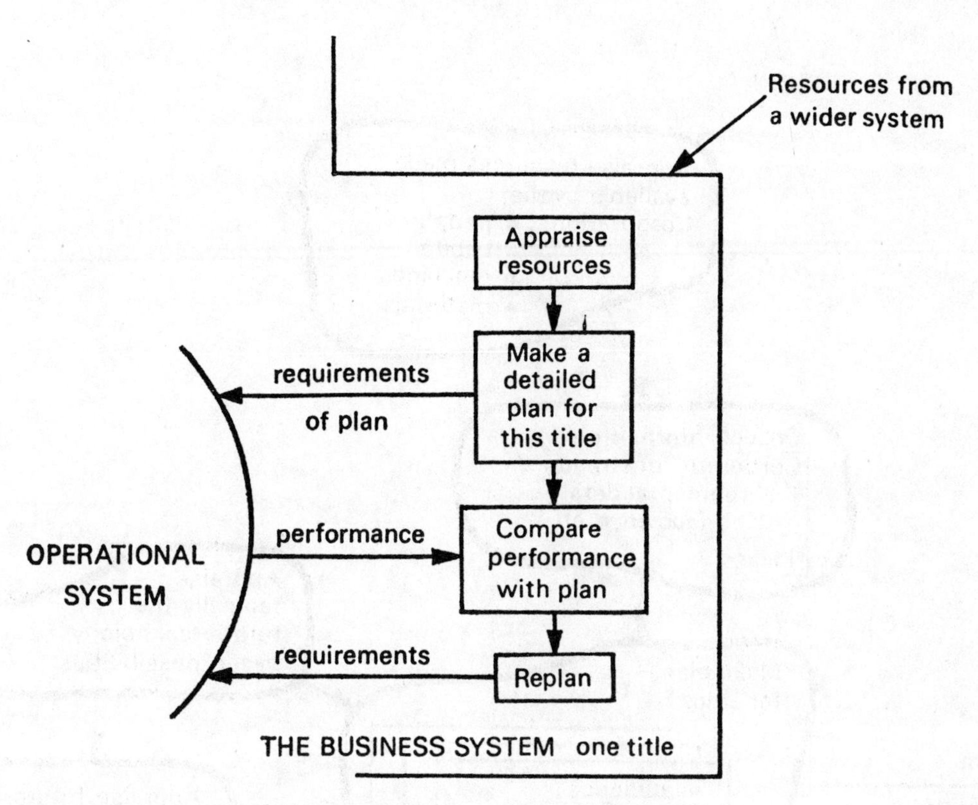

Figure 3.7. Expansion of the "Monitor" activity.

The conceptualization may now proceed in two directions.

At a more detailed level it is now possible to examine each activity in the diagrams above and answer the questions:

What information is needed to take this action?
What is its source?
Where is it received?
When is it sent? With what frequency? In what form?

The chart containing the answers to these questions and the diagrams above could then be used to draw up diagrams of

(a) operational and business system structures, and
(b) the information flows within those structures.

At a higher level, attention now turns to the wider system whose existence was a necessary postulate when considering the "monitor" activity of the business system for one title. This system will be concerned with appraising the *total* resources available, collecting information relevant to overall plans, making individual-title outline plans, appraising whether these require extra resources not available within the company or leave resources unused, allocating resources to individual operational systems, monitoring performance and controlling via modifications to plans. In addition, a planning system at this level must also be concerned with long-term survival, and must contain a sub-system concerned with long-term appraisal of changes both internal and external. Internal changes will include investment in other resources, or divestment, and external changes will include changes in the nature of the market for successive "images", in the entertainment/education business. All this may be represented as in Figure 3.8.

In the event, the crux of this analysis turned out to be the central box of Figure 3.8, in which a central body can take decisions concerning particular titles – especially how and where to print them – in the light of overall knowledge of the total use of resources. It was this aspect which was highlighted in the "Compare" phase.

The activity described above consists of the first stages of conceptualization of a "formal" system for publishing

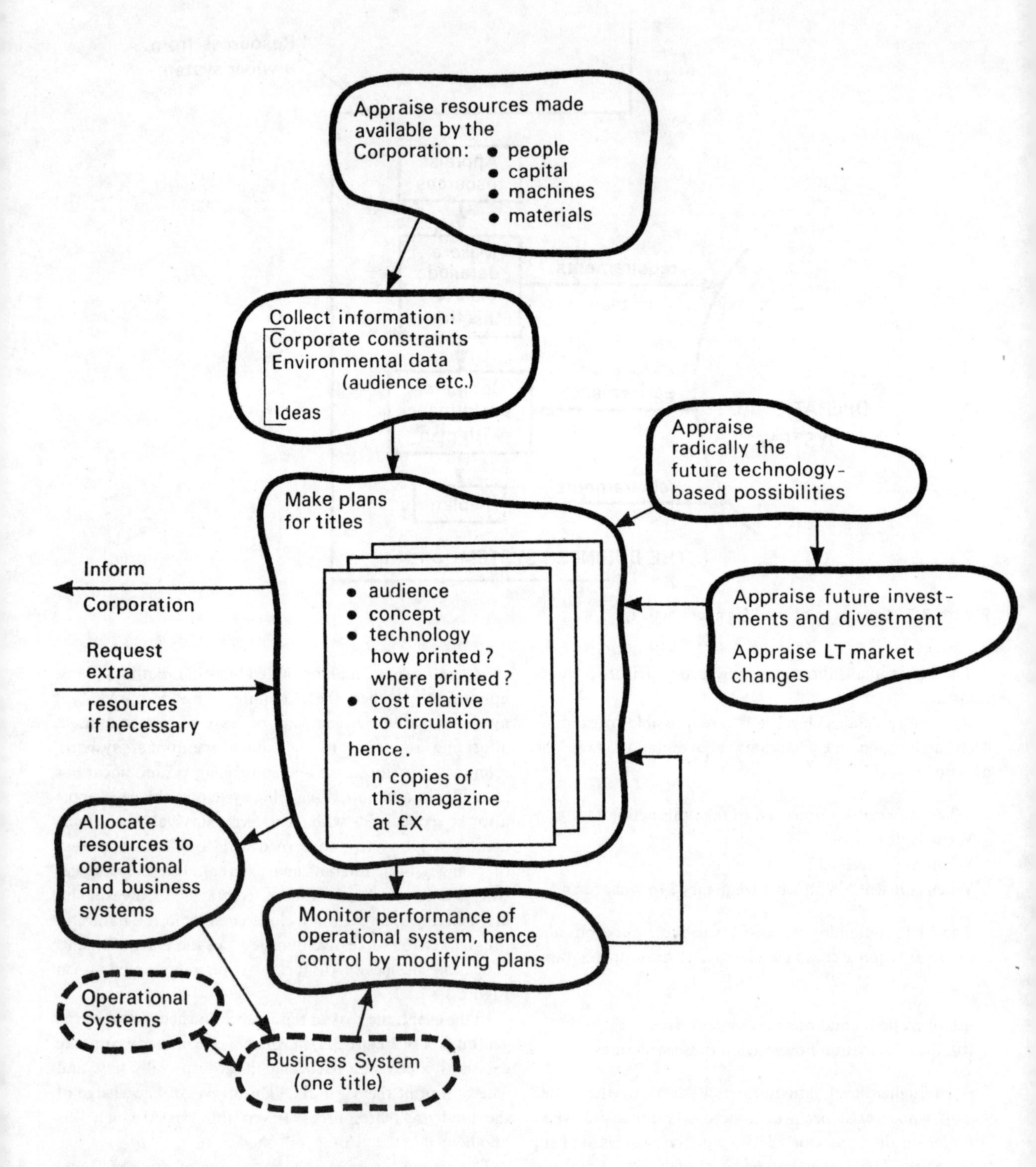

Figure 3.8. The overall business system.

magazines as a business activity. How can we have any certainty that it is "valid"? In absolute terms we cannot achieve that certainty, but it has been found useful to develop a simple general model of a purposeful human activity system and to use it as a means of ensuring that as the conceptualization proceeds the conceptual model does not lack some vital aspect or component. The emphasis has been on usability rather than sophistication, but obviously anyone may guide and validate a conceptualization by means of whatever systems concepts he finds helpful – I am thinking of material like Ackoff's compendium of systems concepts[21] or Beer's cybernetic model of an organization based on analogy with the central nervous system.[22]

In the present work it has been found useful to use a simple model which is a compilation of components which have to be present if an (open) human activity system is to be capable of purposeful activity, the latter defined to cover "maintaining relationships" as well as "achieving objectives". The model extends Jenkins' earlier "summary of properties of systems"[14] and follows Churchman's "Anatomy of System Teleology".[23]

The model, and two illustrations of it are as follows. Figure 3.9 is a picture-summary.

S is a "formal system" if, and only if:	Examples: Industrial Firm	Examples: University
(i) S has an objective, a mission, a definition of a final desirable state, or an ongoing purposes.	Exploitation of particular technological skills in a particular market? Social responsibilities? Survival?	Discovery, preservation and transfer of knowledge?
(ii) S has a measure of performance.	Financial performance? Profit? Employee relations? Public image?	Class hours? Public image? Publications? Ability to attract students and staff?
(iii) S has sub-components which are themselves systems (with objectives, measures of performance, sub-systems, etc.)	Divisions? Departments? Projects? (because Departments often are not related to mission) Functions? (e.g. Manpower Planning)	Academic departments? Functions? "Schools"? Individuals?
(iv) S has sub-components which interact, which show a degree of *connectivity* such that effects and actions can be transmitted through the system.	Divisions, Departments, Projects, Functions interacting by flows of money? materials? energy? information? decisions? instructions?	Academic departments, Functions, interacting by flows of money? information? opinions? advice?
(v) S exists in wider systems and/or environments with which it interacts (inputs and outputs). Boundaries are defined by the area within which the decision takers (vii) can cause action to be taken.	The economy? Government policy? The market (consumer preferences)?	U.G.C. policy? D.E.S. policy? Public opinion? Student opinion? Future employers?

	Examples: Industrial Firm	Examples: University
(vi) S has resources, both physical and, through the human components, abstract.	Money? Men? Materials? Machines?	Money? Men? Materials? Machines?
(vii) S contains a decision taker and a decision taking process. (Action is caused to be taken – which requires information flows via (iv)).	Various levels of management? Unions? Employees at all levels?	Council? Senate? H.O.D's? Staff? Students?
(viii) S has some guarantee of continuity, is not ephemeral, will recover stability after disturbance ('long-run stability').	Government subsidies? Society's basic need for the products?	Society's commitment to enlightenment? Immunity from political attack? Shared beliefs?

Use of the model consists of asking, of the conceptual model, questions based on it, for example: What are the definable sub-systems and are the influences on them of their environments taken into account in the activities of the system?

Comparison and definition of possible changes

Judgement must be exercised as to when to finish a conceptualization and move to an explicit comparison of it with what exists in the problem situation. In the publishing/printing project the nature and level of the problems was such that the conceptual model was not pressed as far as the detailed examination of information flows implied by the activities in the model. In the textile company on the other hand the conceptualization included a detailed specification of the required information flows including their source, content, recipient, frequency and form. Similarly the comparison may be in general terms: how does the mode of operation and structure of the conceptual model compare with what exists in the real world as it emerged from the analysis? Or it may be highly specific; in a later stage of work in the textile company, when attention turned to the distribution system, a conceptualization was used to define a written set of detailed specific questions which were then answered by reference to present arrangements.

At whatever level to detail it is carried out, the comparison is done in order to reveal possible changes which could "improve" the problem situation.

In the publishing/printing company the comparison revealed about half a dozen major areas in which it was useful to debate the differences between the model and the reality. For example major implications of the model were:

- that important decisions concerning environment forecasts, magazine specification, resource acquisition and allocation, control of the business in the light of developments affecting many titles, and the decision where to print were taken within one system which itself assembled information relevant to those decisions
- that the summation of individual plans would be related to total resources available within the group.
- that the printer's activity was explicitly linked to his role in the magazine publishing system, and was not an autonomous profit generator.
- that individual printing contracts would not be negotiated individually.
- that an 'R and D' activity is concerned with the future of the business not just of the technology.

Discussion of all these, and comparison with present arrangements led to insight into the reasons for present arrangements being what they were, and enabled formulation of seven possible significant changes which could

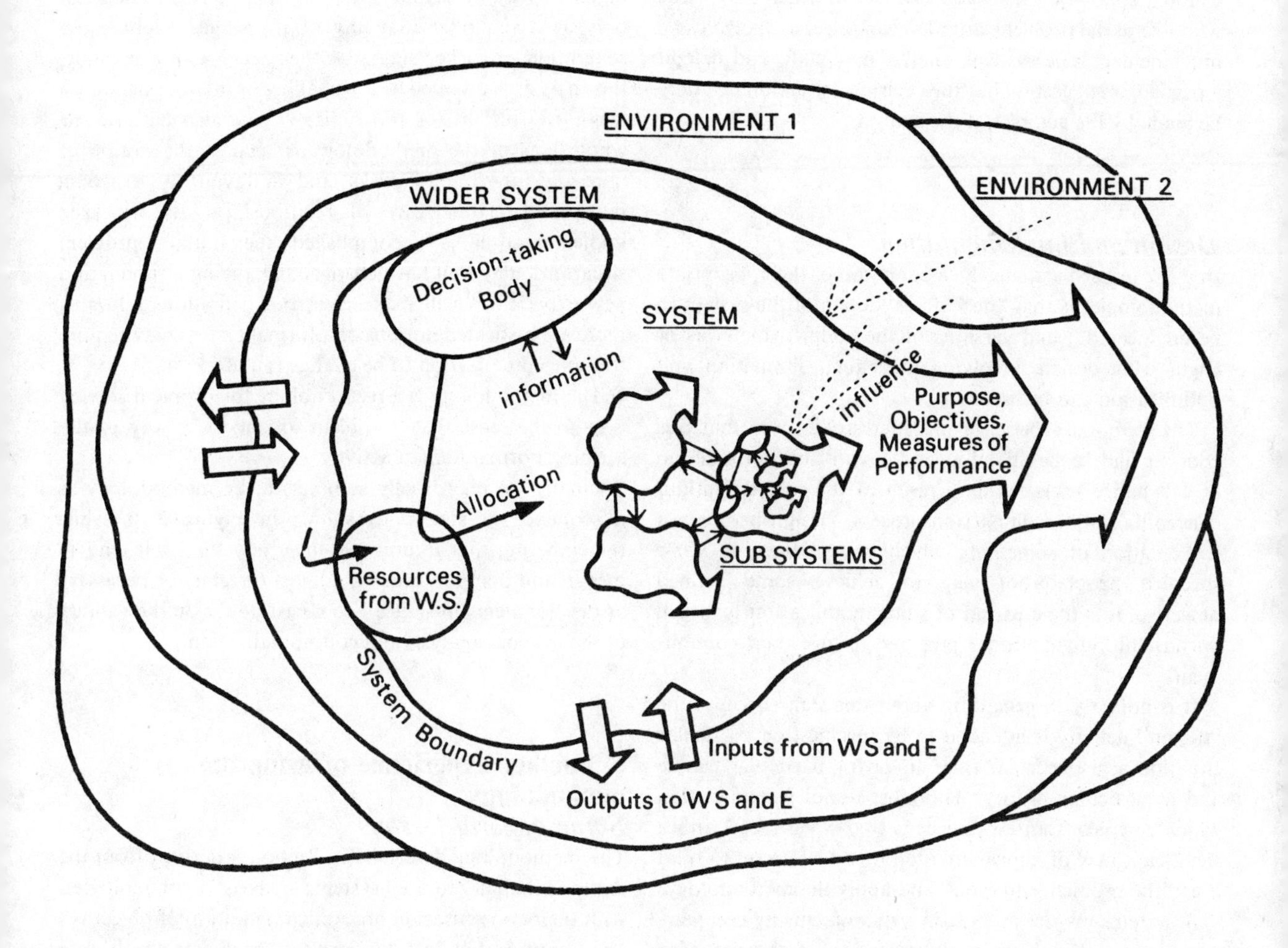

Figure 3.9. A simple model of a purposeful human activity system—a "formal system".

hopefully improve the problem, which, interestingly, had still never been stated in any precise way. "Changes" ranged from major structural changes in the organization and its management to the appointment of "title managers" at printers with the specific task of liaison with the publishing and editorial groups concerned with a particular magazine or group of magazines. Each potential change was assessed in terms of the *kind* of change involved (structural, procedural, attitudinal, or a change in the environment) and the subjectively assessed *amount* of change, and estimates were made of whether the changes would be acceptable to middle managers given the attitudes revealed during the interviews of the analysis phase.

In the event the managing director's committee accepted from the central planning group, with whom we were working, a proposal for a new organizational unit concerned with the "where to print?" decision.

Two points should be reiterated. Firstly, in the Comparison stage of the study the conceptual model is there to help tackle the prob lemsolving in a structured way, not to prescribe what "ought" to exist. Conceptual models tend to be ahistoric, unless the root definition can capture something of the previous history of the situation – which is difficult – and in any case such models have an antiseptic inhuman air. Implemented conceptual models would require inhuman beings to operate them. Secondly, the Comparison stage consists of a debate in which only those who live in the problem situation can select a change to be implemented. The systems analyst may state and defend a particular suggestion but the decision for action can only be made by the actors, not the analyst.

Design and implementation

In work in hard systems the assumption of the appropriate methodologies is that "design" is "design of the system to be engineered", and versions of the codified methods of engineering design, involving modelling, simulation and optimization can be used.

The author's experience in softer studies was that this concept had to be abandoned; "design" becomes design of a "change" selected as a result of the root definition/conceptualization/comparison process. Design here is not the creation of something which will perform in some specified people-proof way to achieve some defined objective, it is the creation of some modification to which purposeful individuals are prepared to give their commitment.

It is not easy to generalize about this stage because the "design" activity is intended to be specific to a particular situation at a particular time, involving particular people and a particular history. The most useful generalization which emerges from experience is that when faced with a significant task of implementation it can be useful to treat *it* as "the problem situation" and apply the methodology!

A systems project in 1972 in a manufacturing company was concerned with the implementation of the use of a rather sophisticated management tool (related to production planning and product assembly) which had been developed in previous studies. It was found helpful to view "implementation" as an activity system of limited life in which the decision-taking body would at first be those who developed the management tool and later would gradually become the managers who would make day-to-day use of it; as soon as these managers became the decision takers in a system which makes use of the tool the temporary "implementation" system would cease to exist and the developers of the tool could leave the stage. Progress was made by viewing implementation in this way and then using the conceptualization process as described above, in the light of existing management modes, to plan the introduction of the management tool.

Appraisal and reiteration

Including an Appraisal stage in the methodology is a dignified way of saying "Start again". It is included not only as a reminder that any of the stages might cause reiteration of earlier stages – either because, for example, the analysis is revealed as too shallow or the root definition lacking in insight (or too radical) – but also because the whole bias of the methodology is against the notion of once-and-for-all finite tasks and in favour of on-going purposeful maintenance of relationships. By the time implementation is accomplished the initial "problem situation" may well have changed. Changing attitudes and new experiences will make new root definitions relevant; more sophisticated notions of a formal system may require the conceptualization to be changed; and so on.

The methodology is a methodology for problem-solving only in the sense that "problem solving" is a way of describing normal human activity.

Finally we may briefly summarize the methodology as a sequence of stages, as shown in Figure 3.10, while remembering that in any real-life study they will tend to merge, and that iteration will almost certainly be necessary as developments in stages 3 to 6 cast doubt on the validity of the original analysis and conceptualization.

5 Further experience of using the methodology

Some general lessons

The methodology described in Section 4 derived from the author's inability to use "systems analysis" methodologies, with their concentration on explicit definition of objectives and "optimization" of "designs" on problems which were so woolly and unstructured that considering them in those terms rapidly caused the onset of a state described by Wittgenstein's phrase: "The methods pass the problems by". Attempts to remain at grips with the problems and yet draw out generalizations from a number of experiences led to the first formulation of the new methodology during nine systems studies carried out in 1969, 1970 and 1971. Increasingly precise formulation of it since then has enabled the author and colleagues to try to use it in more than twenty further studies. Some of the lessons learnt are summarized in this section.

The main lesson is that believing one has understood this way of tackling ill-structured problems is much easier than actually using it. There is a holding back involved in it, a refusal to entertain possible solutions until the analysis

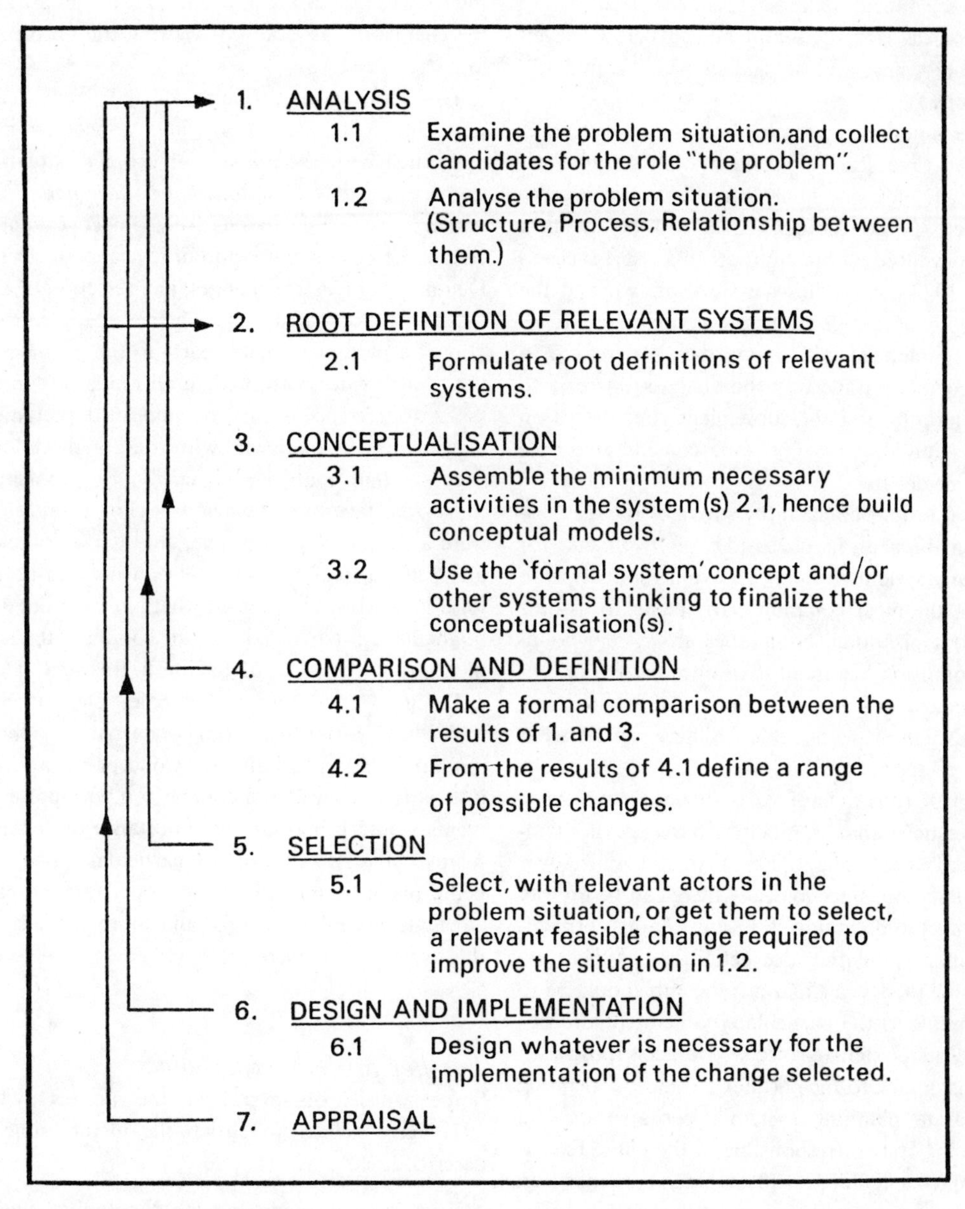

Figure 3.10. A summary of the methodology.

and the hard thinking which follows it have been completed, which accords ill with the most popular form of problem solving, namely to find some element in the problem situation which is striking, and to leap from that to possible solutions, subsequently, perhaps, taking into account a few of the interactions which obtrude. A management consultant visiting one of the companies in which we were working expressed amazement at the thoroughgoing nature of the analysis phase. "We try to start implementing something in the first week", he said, "in order to convince the client we are active". This outlook, commercially understandable in a consultant, unfortunately applies frequently to people trying to solve their own problems. Taking some kind of action is undoubtedly easier than thinking, and there is a common temperament, action rather than insight-oriented, which finds application of the methodology very difficult indeed. (It is not at all surprising that when Descartes' friends found him in bed at eleven o'clock in the morning, and he said he was "Thinking", they were furious.)

In fact there is nothing in the methodology which links its phases to a time scale: implementation might start within a week. If only one hour is available for conceptualization then the conceptual model will be the best which can be produced within an hour. In a short successful project in 1972 analysis took a week or two and the root definition, conceptualization and comparison stages were then all completed within one day; whereas in a major study currently underway the analysis phase is to occupy three months and the subsequent root definition of relevant systems may occupy some considerable time and involve wide debate.

Other general lessons concern the stages of the methodology itself. In the analysis phase it is often difficult for anyone with any systems training to avoid an analysis in systems terms, the most common error being to assume that internal organizational boundaries are system boundaries. This confusion has more than once led to conceptualization being a tidied-up description of what exists, rather than a rigorous-as-possible building up of what must follow from the root definition.

Formulation of root definitions is obviously not easy, but experience shows that it is better to accept one temporarily and return to refine it later, than to agonize over the initial formulation, since logically there can be no way of knowing that the definition is in any sense "correct". There are as many potential root definitions as there are ways of choosing to view a situation; the aim should be to choose one specific to the particular problem situation of the study. The best systems projects have been those with root definitions unique to the particular situation, defining not "a corporate planning system", perhaps, but "a strategic planning system (responsible to the Chief Executive) whose implemented plans will enable the organization to decide within two years whether or not it would be advantageous to move from marketing product W in market X, to marketing a service Y which involves the use of W, in market Z".

The greatest difficulty with conceptualization, apart from the intrinsic difficulties of deciding logically what activities in what sequence are necessary and sufficient to meet the implications of the root definition, lies in the temptation to regard it as a means of defining "the ideal". This is a temptation also for whoever it is who wants a solution to the problem, and it is worth special efforts to ensure that the Conceptualization and Compare stages are accepted by the client only as a way of viewing the problem situation in a structured way.

This difficulty is probably related to the "what/how" problem. Root definitions and conceptualizations are concerned with the "what" of problem situations; what elements do they contain, *what is their nature?* These are matters susceptible to relatively objective analysis. They enable the powerful weapon of logical thought to be used. Definition of feasible changes, and selection of a change to make in the actual situation, however, are concerned with "how" a particular client reacts to the problem situation; the "how" can accommodate all kinds of constraints, all kinds of deference to history, ignorance, prejudice or plain irrationality. Experience with the methodology so far suggests that many "problems" *are* problems because those who face them tenaciously cling to a "how" as if it were a "what". For example, consider a company which, given its particular business, has a need to produce technical innovations (that is a "what" in their situation). They might have a particular way of coping with that requirement which consists of having a Research Department, a Design Department and a Sales Department (that is their "how"). But that is only one particular "how" out of innumberable possibilities. Conceptualization of "an innovation system" and comparison with present arrangements would bring out the importance of "what" and the relative unimportance of any particular "how". The point being made is that this distinction is not an easy one to elucidate in real situations, and more work is needed on this aspect of the methodology.

Further project experiences

Some aspects of several of the projects which since 1971 have helped to enrich the methodology are now described.

(i) A project in an industrial company which manufactures synthetic fibres illustrated the value of the notion in the "formal system" model (see Section 4) that a system boundary is defined by the area within which a decision-taking body can exercise authority.

The problem situation was one in which the fibre manufacturer made a yarn which was sold to the textile trade for further processing; the processed yarn was bought by weavers or knitters and finally reached the consumer in a garment. Complaints at the garment stage or earlier obviously tended to be passed back down the processing chain and ultimately lodged with the fibre manufacturer.

Given the nature of this total yarn-to-garment system it is clear that it is logical for the management of the fibre manufacturer to be the decision-taker of a smaller system whose boundary coincides with the organizational boundary between himself and the yarn processor *only* if there exists a test which can be made on the original yarn which will indicate predictively that the yarn will be satisfactory at all stages through to the final garment. Such a test ensures that the yarn which passes it will make an acceptable garment as long as the yarn processing, knitting and garment making are done properly. Equally, if the yarn passes the test, faulty goods subsequently must be due to mistakes outside the fibre manufacturer's system. Unfortunately *no such test exists*; so that unless the market is extremely buoyant, there is an inherent weakness in the fibre manufacturer's position. The obvious solution, from a systems point of view, is for the fibre manufacturer to make himself the decision-taker of a system which includes the steps beyond yarn manufacture. Those who know this industry will be aware that this has been happening in recent years.

(ii) A rather unusual use of the methodology occurred in a project in which a remarkably open-minded professional consultant asked for a systems analysis of one of his completed assignments which both he and his client felt had had an unsatisfactory outcome.

Here the analysis was historical, involving interviews with both consultant and client, and the root definition and conceptualization stages consisted of an examination of what would have happened if consultant and client had consciously and mutually engineered a (temporary) system to tackle the client's problem. This was different from what had actually happened, and the differences suggested ways in which the consultant should change his mode of operation.

(iii) Most of the early projects were carried out in user-supported organizations (using Vickers' distinction between user-supported and public-supported,[15b] which is more insightful than the more usual "private sector" and "public sector"). Here there will be relatively easy-to-define aims for the organization and identifiable decision-taking bodies. Some of the more recent work has been in public-supported organizations where aims, decision-taking and control are more diffuse. Here it is more likely that the outcome of the systems work will be a structure for a debate.

Work in a hospital, for example, was concerned with an ill-defined problem connected with "the use of resources consequent upon medical decisions". Two groups are primarily concerned: doctors who would like to be able to take medical decisions as if the resources were infinite, and administrators who are primarily concerned with the limited nature of the resources and the fact that they are allocated to the hospital at certain time intervals.

Here the methodology was used to initiate and structure discussion between medical staff and administrators on the problems of resource allocation. Agreed experiments on resource-use planning came out of these discussions.

(iv) Beyond the diffuse problems of public-supported organizations are the yet more diffuse social problems which are not confined within organization boundaries. If you are a member of a government "think tank" you are likely to be given the problem of "the future of the National Health Service", or "juvenile delinquency", problems with societal rather than organizational boundaries. Can the methodology help?

Tentative findings are that it can be useful to tease out and formulate root definitions *as if* they could be made manifest in an organization. For example, if we were to examine the question of the future of the N.H.S. in the United Kingdom in these terms one useful strand in the examination might well be to recognize that what we have at present can be viewed as a National Ill-health Service, concerned with treating those who have become ill. A root definition of a *Health* Service might be more concerned with defining the possible state of health of a population, given the state of medical knowledge, measuring the present state of health, and formulating plans to bring the latter nearer to what is possible. The World Health Organization is actually doing some work on these lines.[24]

The approach might be a useful part of an appraisal of the N.H.S., but this is speculation: the project has not been done! One non-organization-based project which has been started is to consider the problem of the use of the Northern Uplands of England and Southern Scotland. This is an area which is of interest to many different groups for different reasons: farmers, local authorities, national Government, land owners, commercial interests, the military, amenity interests, etc. The problem is being examined by conceptualizing a *notional* "system to manage the uses of the Northern Uplands". Such a system will never be embodied in an organization but conceptualizing it does provide a way into the analysis which enables the different aspects of the problem to be marshalled in a hopefully neutral manner. The outcome is as yet uncertain, but this kind of approach will be vindicated if it achieves the modest aim of a survey which is "better than piecemeal".

6 Conclusion

The idea of a methodology for problem solving in the real world is a curious one. The nature and status of such a methodology is not obvious, as may be illustrated by considering what it is not.

It is not a technique. A technique is a procedure which if applied correctly in a suitable situation will with certainty produce a known result: there are *techniques* for making a cricket ball swing, for launching space rockets, for handling queueing problems mathematically, and for myriad other activities. The methdology does not have that kind of certainty.

It is not a science. A science is characterized in practice by paradigms which define meaningful experiments,[3] and by the repeatability of experimental results; and in logic by the progress which can be made by the refutation of hypotheses, as Popper establishes.[25] The methodology does not measure up to this prescription. Although it is based on the systems paradigm, we cannot obtain or expect repeatable results in purposeful systems, and the idea of making progress via refutation is equally inapplicable. Thus if a reader tells the author "I have used your methodology and it works", the author will have to reply "How do you know that better results might not have been obtained by an *ad hoc* approach?" If the assertion is: "The methodology does not work" the author may reply, ungraciously but with logic, "How to do you know the poor results were not due simply to your incompetence in using the methodology?"

If the methodology is not a technique and is not a science, what then is it? Some further negative statements are helpful:

It is not optimum-seeking; it seeks what are agreed by concerned people to be "improvements". The powerful but unreasonable hunger for "optimum" solutions in human activity systems will remain unsatisfied, happily, just because those systems are purposeful, that is to say, human.

It is not utopian; it does not *require* any definition of ultimate goal or objective, only an elucidation of ongoing purpose.

It is hence not a methodology of system design, only of conceptualization, and design of changes.

In summary: the methodology tries to face the fact that problem solving is dependent upon problem structuring. It provides through systems ideas a way of seeing diffuse, ill-structured problems in a patterned way, and tries to do so without distorting the problems in the way that application of techniques usually does. It seeks not to be reductionist but to provide a conceptual framework within which many different aspects of problem situations can be accommodated. It enables those who have a concern for a problem to make decisions – incremental or radical – aimed at solving, alleviating or eliminating it.

The methodology has been found to be helpful in a wide range of problems of many different types; but given the nature of it even that is an assertion which cannot be *proved*.

Acknowledgements

The author is extremely grateful to all of those people, colleagues, students and managers, with whom he has discussed the methodology as it has evolved.

Firstly I am grateful to my colleague Gwilym Jenkins who with boldness and drive conceived and established the action research programme which was the vehicle for this work.

I am also especially grateful to those who struggled with me on the systems studies which are the source of this paper: R. H. Anderton, T. R. Barnett, D. H. Brown, J. H. Collins, C. H. Pogson, L. Watson, I. Woodburn, D. G. W. Allen, D. J. Brown, R. Griffin, G. R. Hewitt, N. G. Jarman, R. W. Keen, G. L. Moss, W. P. Murray, D. C. Nevin, D. J. Scott, G. Severn, A. R. Thomas, D. I. Thomas, A. M. Waugh.

Finally I am grateful to Ross Barnett for his constructive criticisms of the first version of the paper, which enabled a number of improvements to be made.

References

1 Checkland, P. B., "A Systems Approach to Management: a discussion of the St. Gallen Symposium", *J. Sys. Eng.*, **3**, No. 1, 1972.

2 Checkland, P. B., "A Systems Map of the Universe", *J. Sys. Eng.*, **2**, No. 2, 1971.

3 Kuhn, T. S., "The Structure of Scientific Revolutions", University of Chicago Press, 1962,

4 Bertalanffy, L. von, "General System Theory", Braziller, 1968.

5 (a) Jenkins, G. M., "Systems and their Optimisation", Inaugural Lecture, University of Lancaster, 1967.

(b) Checkland, P. B., "Systems and Science, Industry and Innovation", Inaugural Lecture, University of Lancaster, 1969, and *J. Sys. Eng.*, **1**, No. 2, 1970.

6 Boulding, K. E., "General Systems Theory – the Skeleton of Science", *Management Science*, **2**, No. 3, 1956.

7 Dror, Y., "A General Systems Approach to Uses of Behavioural Sciences for better Policymaking", Rand Paper P. 4091.

8 Popper, K. R., "The Open Society and its Enemies", Routledge, 1945.

9 Hall, A. D., "A Methodology for Systems Engineering", Van Nostrand, 1962.

10 Chestnut, H., "Systems Engineering Methods", Wiley, 1967.

11 Optner, S. L. (Ed.), "Systems Analysis", Penguin Books, 1973.

12 Hitch, C., "An Appreciation of Systems Analysis", Rand Corporation, 1955.

13 Quade, E. S. and Boucher, W. I. (Eds.), "Systems Analysis and Policy Planning: Applications in Defence", Elsevier, 1968.

14 Jenkins, G. M., "The Systems Approach", *J. Sys. Eng.*, **1**, No. 1, 1969.

15 (a) Vickers, G., "Value Systems and Social Process", Tavistock, 1968.

(b) Vickers, G., "Freedom in a Rocking Boat", Allen Lane, the Penguin Press, 1970.

16 Forster, M., "An Introduction to the Theory and Practice of Action Research in Work Organisations", *Human Relations*, **25**, No. 6.

17 Rapaport, R. N., "Three Dilemmas in Action Research", *Human Relations*, **23**, No. 6.

18 Checkland, P. B. and Griffin, R., "Management Information Systems: A Systems View", *J. Sys. Eng.*, **1**, No. 2, 1970.

19 Emery, F. E. and Trist, E. L., "The Causal Texture of Organisational Environments", *Human Relations*, **18**, No. 1, 1965.

20 Thompson, A. B., Lecture at Lancaster University, 1971.

21 Ackoff, R. L., "Towards a System of System Concepts", *Management Science*, **17**, No. 11, 1971.

22 Beer, S., "Brain of the Firm", Allen Lane, The Penguin Press, 1972.

23 Churchman, C. W., "The Design of Inquiring Systems", Basic Books, 1971.

24 Litsios, S., "General Introduction to Organisation and Strategy of Health Services Research". Paper to a meeting of the Scientific Group on Research in Epidemiology and Communications Science; WHO, Geneva, October, 1970.

25 (a) Popper, K. R., "The Logic of Scientific Discovery", Hutchinson, 1959.

(b) Popper, K. R., "Conjectures and Refutations", Routledge, 1963.

Reprinted from Checkland, P. B. (1972) "Towards a system-based methodology for real-world problem solving" *Journal of Systems Engineering*, **3**, 2.

4 The systems approach

by Gwilym M. Jenkins

Summary

The objective of this paper is to discuss the philosophy underlying a systems approach to the solution of problems. The most important conclusion may be summarized as follows:

1. A *piecemeal approach* to problems within firms and in local and national government is *no longer good enough* if firms and nations are to compete, and indeed to collaborate, efficiently.
2. This is so because technology, firms, organizations and affairs in general are becoming *increasingly complex* and because policy decisions *increasingly require the expenditure of large sums of money* – so that the consequences of *bad decision making* are becoming *increasingly costly*.
3. A *systems approach* to problems demands that a *piecemeal approach* is replaced by an *overall approach*. Systems Engineering is the science of designing complex systems, by the efficient use of resources in the form of *Men*, *Money*, *Machines* and *Materials*, so that the individual sub-systems making up the overall system can be designed, fitted together, checked and operated so as to achieve the *overall objective* in the most efficient way.
4. The rapid development of Systems Engineering during the last few years has been stimulated not only by the increasing complexity of businesses but also by the increasing potential of large analogue, digital and hybrid computers which enable an overall *model* of the system to be *optimized*.
5. One of the greatest benefits of Systems Engineering is that it exerts a *unifying influence* on management by tying together the many specialist techniques needed to solve complex problems.
6. The systems engineer is seen as the *generalist* who always takes an overall view and who always takes particular care to ensure that the system objectives are correct, are communicated to all concerned, and are achieved with maximum efficiency.
7. One of the most important consequences of the systems approach is that it highlights the fact that *fundamental changes* are needed in the way that both individuals and organizations go about their work. In particular, it demands that problem solving needs to be carried out on a more *interdisciplinary* basis and that many firms and organizations need to be organized in a more *integrated* way than at present.
8. Thus, Systems Engineering is seen as a *key factor* in improving *management practice*, and hence, in making big improvements to the efficiency of firms and organizations.
9. Finally, the paper draws attention to the *urgent need* at this point in time to inject *systems thinking* at all levels into industry, commerce, and into local and national government.

The paper is in three parts. Part A is concerned with answering the questions "What is Systems Engineering?" and "What is a Systems Engineer?" Part B answers the question "How does a Systems Engineer go about solving any problem?" *

A. The nature and objectives of Systems Engineering

A.1. Systems and their properties

The expression "All Systems Go" is now an established part of popular jargon. It means that the overall system, consisting of millions of electronic components, making up a space rocket and hundreds of men, making up the management and technical teams, has been designed in such a way that each component and human being is ready to play its *designed role* efficiently in making the rocket achieve its predetermined objective. Such an impressive feat of engineering and project management calls for sophisticated systems engineering skills. Before discussing these skills in

* [The original article contained a further part discussing the question, "How can systems thinking help to improve efficiency within firms and other organizations?" *Ed.*]

greater detail, it is useful by way of introduction to say what is meant by a system.

The notion that it is useful to regard such diverse entities as a domestic water heater, an industrial plant, a company, a space rocket, a hospital, a port and the entire regional government set-up of a country as *systems*, must surely go down as a very important contribution to twentieth-century thought. That the word is not new is seen from its Greek origin "systema", which derives from "syn" meaning "together" and "histemi", which means "to set". A typical dictionary definition of a system would read "A *plan* or scheme according to which things are *connected* into a *whole*", as in a system of philosophy or in the solar system. Thus, the key words, *plan*, *connected* and *whole*, which will recur throughout this paper, are present even in popular definitions of the word system.

Although a system has a well-established popular meaning, from the point of view of systems engineering a more precise and extended definition is needed [1]. In the following discussion, six important properties of systems are listed and illustrated by considering a relatively simple system in the form of a chemical plant.

Systems as complex groupings of human beings and machines

A chemical plant usually consists of a very large number of different items of equipment, together with stocks of raw materials, intermediate products and finished products and also services in the form of water, steam and electricity. To operate the plant, a plant manager is required, and he will need to be assisted by several foremen and process workers, usually working three eight-hour shifts per day. For efficient running of the plant, these "line personnel" must be backed up by a host of technical and commercial "service personnel", for example maintenance engineers, research chemists accountants and salesmen. Thus, the first property of a system is that it is *a complex grouping of human beings and machines*.

Sub-systems and flow-block diagrams

An important characteristic of systems is that they may be broken down into *sub-systems*. The way in which this breaking down is done in any particular situation will depend on the nature of the system being studied and therefore on the extent to which detail is important. To decide on the amount of sub-system detail needed may require an analysis in depth of the system and its interactions.

A convenient and readily understood way of displaying how systems may be broken down into sub-systems is provided by a *flow-block diagram*, as indicated in Figure 4.1. This shows a simplified flow-block diagram of a chemical plant for making acrolein (C_3H_4O) by the catalytic oxidation of propylene (C_3H_6) * [2]. The diagram displays the individual sub-systems or process units, making up the whole plant, as *blocks* and the links or *flows* between them, as arrows. In the present example, the flows refer to materials and energy. More generally, the flows between individual sub-systems may refer to Money, Materials, Energy, Information or Decisions.

Flow-block diagrams provide an invaluable tool for helping to clarify one's thinking about a particular system. It is remarkable how much light can be shed on a complex problem by the mere act of constructing a flow-block diagram.

Sub-systems interact with each other

The overall efficiency of the acrolein plant depends on the correct functioning of all the sub-systems shown in Figure 4.1. This is because the plant units have *interacting tasks* to perform. In addition, some material is *recycled* from one part of the plant to an earlier stage and this in turn produces further interactions. An example of a simple interaction is provided by the fact that the more concentrated the acrolein leaving the absorption column, the less is the demand on the distillation column; thus the design of the absorption and distillation columns may be balanced against each other. Similarly, a small reactor will be cheaper but may give a lower conversion to acrolein and hence may demand more recycling of unconverted raw materials and also bigger absorption and distillation columns. At the same time the catalyst is expensive and has a limited life; it can be shown that this has an effect on the most economic size of the reactor tubes. Because of these interactions, and many more besides, it is impossible to arrive at the best design of the plant, or even at the best design for an individual piece of equipment, by considering each item separately. This brings us to the third general property of systems, namely that *the individual sub-systems interact with each other*. The performance of a given sub-

* In the process propylene and air are mixed with recycle gas and passed into a tubular reactor where they react over a catalyst to produce acrolein plus other gases. The outlet gases are cooled and then passed on to an absorption column where some gases are absorbed in water, the undissolved gases from the top of the column being recycled to the reactor. The liquid from the bottom stream of the absorber is passed into the distillation column where the acrolein is distilled and removed at the top of the column and the liquid (mostly water) from the bottom of the column is then returned to the absorption column for further use.

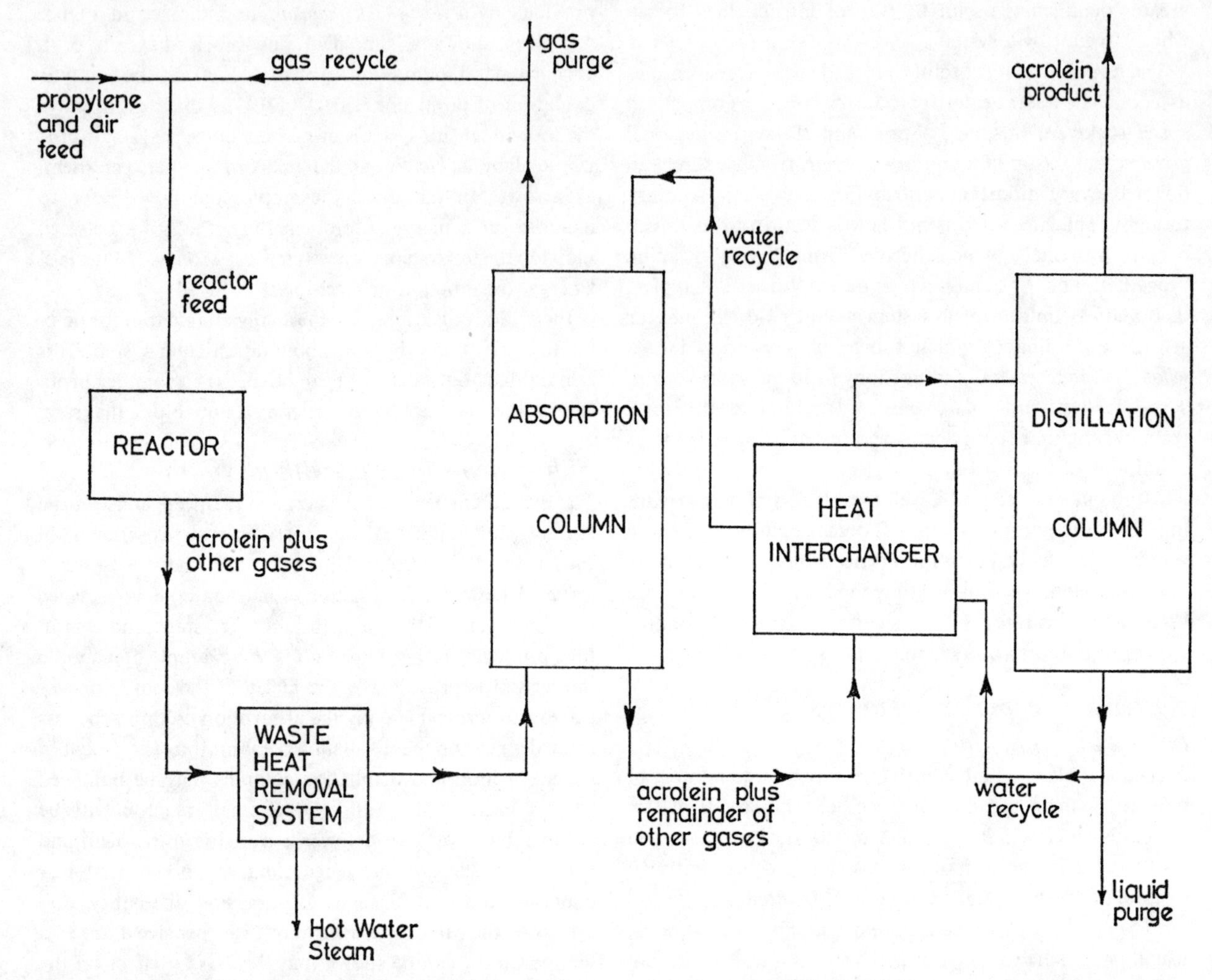

Figure 4.1 Simplified flow-block diagram of an acrolein plant as an example of a simple system with interacting sub-systems.

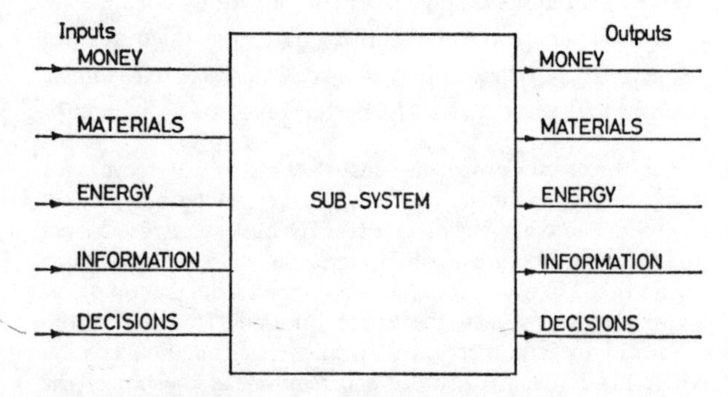

Figure 4.2 A sub-system as some process which transforms input flows of money, materials, energy, information and decisions into corresponding output flows.

system interacts with the performance of other sub-systems and hence it cannot be designed in isolation from these other sub-systems.

In general, a sub-system may be regarded as some process which *transforms* certain *input* flows of money, materials, energy, information or decisions into corresponding *outputs*, as shown in Figure 4.2.

It is a fact that the outputs from one sub-system provide the inputs for other sub-systems which is responsible for the interactions between the various sub-systems. Understanding the detailed nature of these interactions is one of the primary concerns of the Systems Engineer.

Systems form part of hierarchies of systems

The process of breaking down a system into sub-systems can be taken further as we look at the system in greater detail. Thus, at a later stage in the design of a chemical process, it will be necessary to specify individual plant units in much greater detail and to include their instrumentation and control equipment. The breaking down process could be continued even further until one was concerned with minute detail such as the frequency with which pumps should be serviced or even the ordering of the grease for the maintenance mechanic's grease gun!

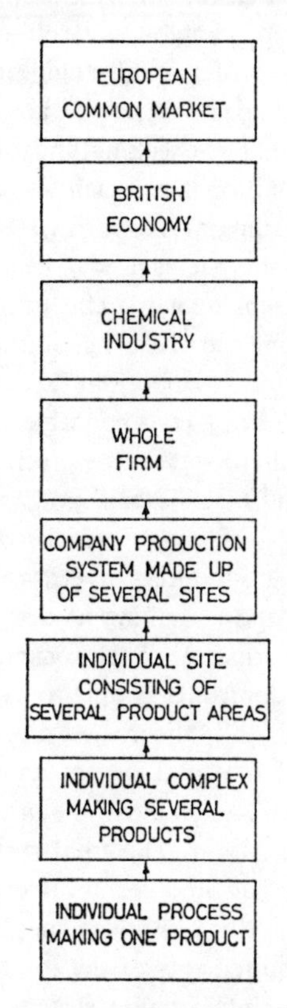

Figure 4.3 Hierarchy of systems associated with the wider system of which an individual plant forms part.

In turn, as indicated in Figure 4.3, the plant being studied may be part of a complex of plants making several products from common feedstocks and using common and interlocking utility supplies. Again, the complex may be one of several plants and complexes on a single large industrial site. In turn, the production facilities of the entire company may consist of several manufacturing sites *to which* raw materials have to be transported, *at which* production has to be planned on a company-wide basis and *from which* finished products have to be delivered to marketing areas. If the whole firm is to operate efficiently, it must *coordinate* and *integrate* the functions of planning and investment, research, design and development, production and equally importantly, selling and marketing. This can be done efficiently only if proper account is taken of other firms, especially those in the chemical industry. In turn the chemical industry, along with other industries, forms part of an even wider system, the British Economy of this country, possibly itself to form part of an even wider system, the European Common Market.

Thus, the fourth property of systems is that *they form part of a hierarchy of systems.* There will usually be strong interactions between the various systems which occur at the same level of a hierarchy and between systems at different levels in the hierarchy. However, the systems "at the top" are the most important because they exert considerable influence on the systems lower down in the hierarchy.

Systems have conflicting objectives

The fifth property of a *system is that it should have an objective.* Getting the objectives right is not an easy matter but it is the key to successful systems design. For example, in the case of designing a plant, is the objective "to minimize capital costs" or "to minimize operating costs" or "to achieve the highest possible safety standards" or "to maximize reliability" or "to maximize the ease of maintenance" or what? In fact, for a given system, it is not difficult to write down a long list of possible objectives. However, these objectives will usually tend to be in conflict with each other. For example, lower capital costs result in higher operating costs, higher safety and reliability standards increase both capital and operating costs but may increase throughput and profit, and so on.

In general, *all systems have conflicting objectives,* so that some form of compromise is essential. Therefore, a balance or *trade-off* must be sought between the conflicting objectives if the best overall result is to be obtained. Thus, in the present example, a compromise between the conflicting requirements of low capital costs, low operating costs, high safety standards, high reliability and ease of maintenance must be obtained by calculating their contribution to some overall objective, such as the financial rate of return on the plant over its expected future lifetime. Reaching the best compromise between conflicting objectives poses many important questions which usually require detailed investigation at the beginning of a systems study.

Systems must be designed to be able to achieve their objectives

The sixth, and most important, property of a system is that it must be *designed* in such a way that it is capable of achieving its overall objective. Achieving the overall objective may be a difficult and complicated question and may involve analysis, planning and designing over a long time period. This is what Systems Engineering is all about. Before going on to discuss its role in greater detail, a summary is given of the main properties of systems.

Summary of properties of systems

1. A system is a *complex grouping* of human beings and machines.
2. Systems may be broken down into *sub-systems*, the amount of sub-system detail depending on the problem being studied. *Flow-block diagrams* provide a readily understood way of describing these sub-systems.
3. The *outputs* from a given sub-system provide the *inputs* for other sub-systems. Thus the performance of a given sub-system *interacts* with the performance of other sub-systems and hence can not be studied in isolation.
4. The system being studied will usually form part of a *hierarchy* of such systems. The systems at the top are very important and exert considerable influence on the systems lower down.
5. To function at all, a system must have an *objective*, but this is influenced by the wider system of which it forms part. Usually, systems have multiple objectives which are in conflict with one another, so that an *overall objective* is required which effects a compromise between these conflicting objectives.
6. To function at maximum efficiency, a system must be *designed* in such a way that it is capable of achieving its overall objective in the best way possible.

A.2. The four M's and Systems Engineering

A system has been defined as any complex grouping of human beings and machines with a definite objective, such as an entire industrial plant, a whole firm or a rocket system, which is made up of a very large number of electronic components, each affecting the overall performance of the rocket. *Systems Engineering is the science of designing complex systems in their totality* to ensure that the component sub-systems making up the system are designed, fitted together, checked and operated in the most efficient way [1], [3]. It is not a new discipline, since its history is deeply rooted in good industrial design practice. However, it brings a new emphasis on *overall performance*, as opposed to the performance of individual parts of the system.

An important central feature of Systems Engineering is the building of *quantitative models* so that some overall measure of the performance of the system can be *optimized*. One of the advantages of the systems approach is that it is possible to look at entirely different problems coming from different areas of technology and business in a way that emphasizes their common features when regarded as systems.

The word "engineer" in Systems Engineering is used in the everyday sense of "designing, constructing and operating works of public utility", a definition which includes a much wider range of people than would be recognized as engineers by the professional engineering institutions! Hence, Systems Engineering is the activity of planning, designing, constructing, checking and operating complex systems. This definition has much in common with the original Greek meaning of the word system, namely "to set together". Many sub-systems need to be integrated together for the whole system to work effectively. Thus, systems engineering is the science of setting, or knitting together, or engineering, systems so that jointly they perform more efficiently in pursuit of a common objective.

It is now possible to state more precisely what is meant by the expression "All systems go", mentioned in the introduction. "All systems go" means that in the launch controller's view, each piece of electronic equipment is working properly and is making its designed contribution to the overall performance of the rocket. To reach a state where the launch controller is able to say with confidence that "All systems go", the rocket designers must have previously defined the total system and its objective. Its objective may be to launch a capsule into a predetermined path to reach the Moon. The total system consisting of rocket, launching pad and technical crew must then be designed to achieve this objective. The designers will need to analyse the requirements of the total system and then break them down into smaller systems capable of finer definition until every component and human being *has its planned place and role* in a suitably defined subsidiary system. Nothing less is acceptable from a systems engineering point of view.

The four M's

Perhaps a simpler way of saying all this is that Systems Engineering is concerned with the *optimal use of resources* of all kinds. The major resources are the four M's, namely Men, Money, Machines and Materials or, as once translated by an earthy Yorkshire executive, Blokes, Brass, Gadgets and Muck! It has been objected by some that Systems Engineering will result in the regimentation of people to such an extent that they may become technological robots. On the contrary – a systems approach to many problems in British industry during the last twenty-five years would have shown that too much emphasis was being placed on Money, Machines and Materials and not enough on Blokes. Greater emphasis on people would have achieved the dual result of making firms much better places in which to work and also of increasing efficiency, profits and benefits all round.

Reasons for the rapid development of Systems Engineering

Systems Engineering is becoming increasingly important because of the increasing *complexity* of business and of modern affairs in general. If all that were concerned was running a small family business, or the local parish council, there would be no particular need for Systems Engineering. However, many problems nowadays are far more complicated. Governments are forced to consider future national energy requirements and to balance up supply and demand of oil, coal, natural gas, hydro-electric power and atomic power. City administrations have to control vast financial resources in the areas of health, sanitation, education and transport. Large oil companies operate huge refineries where plant units are strung together in series and in parallel so that the performance of any one affects the efficiency of the whole. Many separate systems need to come together if a satisfactory solution is to be obtained for such complex problems. Thus, the science of knitting together separate systems is rapidly becoming an important branch of science – the science of Systems Engineering.

Alongside the increasing complexity of affairs, the ability to perform the sometimes very complicated calculations needed for systems engineering studies has been made possible by rapid developments in the development of large digital, analogue and hybrid computers. Such computers have now become indispensable weapons in the system engineer's armoury since they enable him to explore in considerable detail the economics of different ways of operating a system and to select those ways that are most efficient, resulting in optimization of the system.

Interdisciplinary approach

In addition to providing a method by which complex problems, activities and organizations can be analysed, Systems Engineering also provides a framework within which to tie together many separate and possibly divergent disciplines, which otherwise might fail to make an effective contribution to the overall optimization of the problem. Thus Systems Engineering is a team activity and brings together specialists with such diverse backgrounds as natural science, engineering, mathematics, statistics, economics, accountancy and behavioural science. By contrast, the Systems Engineer himself is a generalist, a man trained to think in terms of an overall approach to problem solving, of getting the objectives right and seeing that they are achieved efficiently. As such, it is essential that he is able to liaise and *communicate* effectively with the various specialists whose advice is essential and to stimulate their creativity within this interdisciplinary approach. In fact, the role of the Systems Engineer is very much like that of the general practitioner whose main concern is with the general health of his patient but who, from time to time, will call in specialists for guidance. The systems team will contain specialists and Systems Engineers but the main job of the Systems Engineer is to sort out what is happening, and why, and how it can be done better. Then, together with the specialists, he ensures that the agreed objectives are realized as efficiently as possible in minimum time and at minimum cost and that a good case is presented to the decision makers who will eventually have to sanction the implementation of the designed system.

Disasters that could have been avoided with Systems Engineering

That there is urgent need to apply a systems approach throughout industry is highlighted by the following examples of the consequences of a piecemeal approach which the writer has seen during the last few years:

1. A plant translated too quickly from laboratory stage to full-scale plant, without a proper systems study, failed to operate at all on the large scale and had to be re-designed at considerable expense.
2. A plant, which had been engineered excellently, was built but was written off immediately and did not manufacture a single ton of product. This was because the firm's assessment of the market had been at fault and was outstripped by the assessment of a rival company.
3. A large integrated plant complex lost a great deal of money during the first two years of its life because plant reliability and raw material availability had not been assessed properly.
4. A fibre manufacturer responded quickly to an increase in demand by installing additional spinning capacity without ensuring that its raw material supply was adequate and so lost money by tying up valuable capital resources.

Mistakes of this kind, which are so obvious in hindsight, are caused by a piecemeal approach to problems. It is such disasters that the disciplined approach of Systems Engineering is designed to prevent and can prevent.

Success stories that went with good Systems Engineering

By contrast, the following represent some successful applications of systems engineering which the author has seen at close quarters during the past few years:

1. A plant designed using a systems engineering approach was estimated to be at least 10% cheaper than a plant designed by conventional methods [2].
2. A systems study leading to the installation of an on-line computer on a paper making machine led to an increase in profitability of 9% and the cost of the computer was recovered within two years [4].
3. Short term production planning of an olefines complex resulted in savings of the order of £200,000 per annum [5, 6].
4. A systems study of a petrochemical plant resulted in improvements to the process and in savings of approximately £80,000 per annum at a total cost of £6,000, including the cost of systems effort [7].

Such examples could be multiplied several times over and testify to the efficiency of the extra discipline instilled by a systems approach. Detailed stages in the development of a systems engineering project will be described later. Figure 4.4 and the following discussion summarize the main sequence of events in a typical systems study.

Summary of stages in a systems approach to problems

1. *Systems analysis.* Systems Engineering starts with a common-sense analysis of what is going on, and why, and whether it might be done better. Then the system and its objectives have to be defined and data gathered about its likely performance.

2. *Systems design* (*or systems synthesis*). First, the future environment of the system has to be *forecast*. Then a quantitative *model* has to be built and used to *simulate* or explore a number of different ways of operating the system, finally choosing the system or systems which are in some sense "best", thus optimizing the system.

3. *Implementation.* The results of the system study must be presented and approval sought for their implementation. The optimized system will then have to be built, that is suitable hardware and/or software constructed. The project will require careful planning at this stage to ensure that the full benefits of the system approach are realized. After construction, the system will need to be checked for performance, reliability, etc.

4. *Operation.* A point will be reached when the system will need to be handed over to those who have to operate the system on a routine basis. This is where great care is needed to avoid misunderstanding and inefficiency and probably represents *the area which is least well done in any project*. Finally the effectiveness of the operational system will need to be assessed, and if unsatisfactory, the system "tuned", or reoptimized, to operate in an environment which may turn out to be different from that for which it was designed.

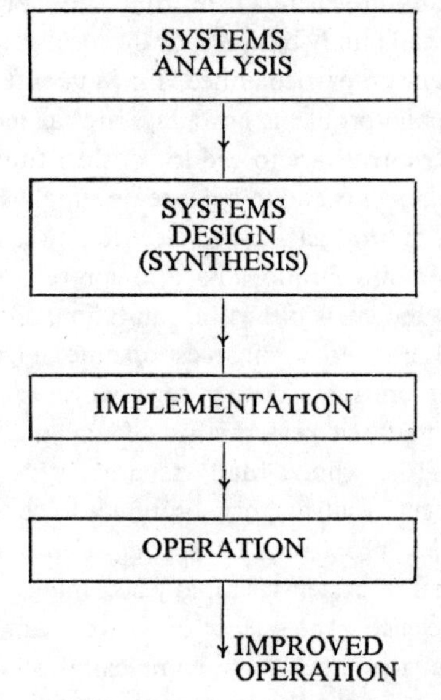

Figure 4.4 Broad stages in the development of a systems engineering project.

Benefits which follow from using a Systems Engineering approach

Several benefits follow from applying a systems approach. The following are the most important:

1. The majority of systems in industry and government have not been designed at all – like Topsy, they have just happened! New systems get added on to old systems and the result becomes a mess. An example of this is provided by the piecemeal development of an industrial site where plant after plant is added without a full examination of the consequences in terms of overall efficiency. Even if there were well defined objectives originally, these change gradually in time without any consequent change in the

system. Hence, the most important benefit of taking an overall systems approach is *that it affords an opportunity to stand back, take a good look at the system and to start formulating new objectives.*

2. Getting the objectives right is the most important part of any study. If the objectives are wrong, most of the subsequent effort is going to be wasted. In this context, it is easy to delude oneself into thinking that a well-defined mathematical objective is relevant. It may be that the real objective is more difficult to define and more vague. Nevertheless, it must be brought out into the open, no matter how subjective certain features of it turn out to be. Hence the second benefit of Systems Engineering is that by its disciplined method of attack on problems, it focuses attention on the important issues – *the correct objective will tend to be brought to the foreground as opposed to an irrelevant objective which might be kept in the background.* Once the objective has been formulated and agreed, it should be explained carefully to all those involved with the system, *down* to the men who will have to operate the system and *up* to the senior managers whose support is necessary both for implementation and efficient operation. Because of his overall approach, the systems engineer is the best person to define *what* should be optimized and hence is the best man to carry out the optimization.

3. There is still a great tendency in industry and in government to base decisions on *guesswork*. For example, the use of guesswork in the design of industrial plants has often led in the past to large safety factors being used in the design, resulting in excessive capital and operating costs. The third important benefit of Systems Engineering is that it *replaces guesswork by model building and optimization.* Careful consideration is then given to the consequences of alternative ways of designing the system. However, it should be emphasized that model building and optimization, if they are to be effective, must supplement and enhance intuition, judgement and inventiveness and not replace them!

4. The fourth important benefit is that by taking an overall view, *problems will be tackled in their correct order of importance.* Time will not be wasted in exploring avenues which, although interesting in themselves are not important to the realization of the overall objective. For example, there is little point in worrying about what is happening on the 17th plate of a distillation column in a chemical plant if the 17th plate has no effect on the overall economic performance of the plant. By contrast, the objectives of Systems Engineering are to pin-point those areas which are cost sensitive and to see that these are studied carefully.

5. Related to the fourth benefit is the fact that if simple devices and techniques result in big improvements, *these will be applied as soon as possible*, leaving the last 5% or 10%, say, of the improvements to the development of more sophisticated techniques later on, provided, of course, that the effort involved in getting this last 10% of improvement is worthwhile and could not more usefully be spent elsewhere. In other words, a good systems engineer should be applying Systems Engineering to find out where he should be doing Systems Engineering!

A.3. The origins and scope of Systems Engineering

It is claimed by Schlager in [8] that the term "Systems Engineering" was probably first used in the Bell Telephone Laboratories in the early 1940s. The widespread development of Systems Engineering since that time is demonstrated by references [9] to [29], which represent a few of the many books and articles written on the subject during the last 15 years. Systems Engineering was born and has been developed in two quite distinct areas – in military and space technology systems during and since World War II and also in industrial problems, particularly in the oil, chemical and power generation industries.

Systems Engineering in military and space applications has been dominated by the U.S.A. and has fitted in very well with the practice whereby government agencies specify the overall system requirements and then sub-contract major sub-systems to individual firms. By contrast, in the non-military area, British development, although as yet in its infancy, compares very favourably with that in other countries. This is especially true in relation to the application of systems thinking to the activities of a whole firm.

Engineering and Systems Engineering

Differences of opinion seem to exist as to the scope and nature of Systems Engineering. There are many engineers who would argue that good engineering and Systems Engineering are synonymous. For example, Affel in [22] writes "I still find it hard to convince myself that there is a difference between systems engineering and just good engineering". In a similar vein, J. M. English [29] says "Since engineers have always been concerned with developing systems, they always have been doing systems engineering or (synonymously) engineering design. . . . Design is the function which characterises the engineer." The author would agree that "Design" is the central activity in engineering (and unfortunately an activity which is sadly neglected in many university engineering departments) and, therefore, that good engineering and good systems engineering go hand-in-hand. However, in the author's opinion, such a narrow view of Systems Engineering seems to miss two very important points.

The first is that even in situations where an engineer is usually involved, for example in the design of a chemical plant, there is a systems problem to be solved before conventional engineering can begin. Thus the design of a plant must be seen as part of the Investment Planning System of the Company. The systems problem is concerned with answering such questions as "Why should we make this product anyway as compared with other products?" If the answer to this is in the affirmative, "What is the best way to manufacture it as cheaply as possible on a large scale?" "Does the venture still look economically viable, bearing in mind the activities of competitors?" "How big should the plant be and where should it be located?", and so on. Finally, a systems problem has to be solved first to determine the interactions which the various plant units must meet in order that they can play their designed role in optimizing an overall economic criterion. It follows that when the systems job is finished, there is still a major engineering job to be done to convert the system specifications for each individual unit into detailed engineering design and hardware. However, this detailed engineering design can turn out to be pointless and wasteful in resources unless the systems job has been properly executed.

The second and more important reason that there is a difference between engineering and Systems Engineering is that the same systems thinking which can be applied to the design of hardware systems, such as space rockets, plants or ships, can also be applied, for example, to parts of firms, or whole firms, or to local government. Conventionally a firm is not regarded as a system analogous to a space rocket. But the systems engineer claims that, by definition, a firm is a system of at least similar complexity and, moreover, that to look at a firm in this way provides an opportunity for improving its efficiency markedly. Thus the systems approach to design can be applied to these much wider Commercial, Management and "software" systems just as well as to the "hardware" systems with which the engineer is more familiar.

Operations (Operational) Research and Systems Engineering

Whereas Systems Engineering has its origins in the design of Engineering Systems, Operations Research is said to have been born in the Battle of Britain which led to the, by now, legendary studies into the optimal size of convoys. Operations Research is defined by the British Operational Research Society as "The application of the methods of science to complex problems arising in the direction and management of large systems of men, machines, materials and money in industry, business, government, and defence. The distinctive approach is to develop a scientific model of the system, incorporating measurements of factors such as chance and risk, with which to predict and compare the outcomes of alternative decisions, strategies or controls. The purpose is to help management determine its policy and actions scientifically." Hence, it is clear that Operations Research and Systems Engineering have much in common. In fact, during an article in a book entitled "Operations Research and Systems Engineering", Roy [13] states that "in a certain sense, operations research and systems engineering *are* the same". However, in the same breath he goes on to say that "The operations research team is more likely to be concerned with operations *in being* rather than with operations *in prospect*" and "systems engineers are more likely to be engaged in the design of systems yet to be, rather than in the operation of systems in being". This supposed difference has also been suggested by several other people, including Hall [19]. However, it is a false and artificial distinction since Systems Engineering is just as much concerned with redesigning existing systems as in designing new ones. In fact, designing a new system is usually the most satisfactory of situations since a fresh start can be made. However, the majority of systems have just evolved in a haphazard way without any clear objectives, or if there were clear objectives originally, these have now changed without any consequent change in the system. In such a situation, the application of systems engineering is as valuable, or even more valuable, as when a fresh start can be made.

To a certain extent it is true that, in its early days, Systems Engineering was more concerned with the design of what may be called Technical Systems, such as military and space systems, new plants, the improvement of existing plants and systems engineering studies leading to the installation of on-line computers on plants. (These are areas which involve the development of new technology and as such lie outside the field of practice of operational research workers.) However, in recent years, this activity has been extended considerably and the systems approach is now being applied, for example, to such wide ranging problems as the design of integrated production–distribution planning and control systems, corporate planning systems, transportation systems and to improving the efficiency of departments in a city administration.

The writer cannot speak for Operational Research but can only repeat that Systems Engineering is concerned with placing a big emphasis on the design of the total system and not individual sub-systems. It is difficult to escape the conclusion that if Operational Research is really concerned with the overall design of complex human-machine systems, the subject has a rather unfortunate title and that many of its exponents are nevertheless content to tinker with the sub-systems.

Systems Analysis and Systems Engineering

The word Systems Analysis has been used, for example by the RAND Corporation [30] and the Systems Development Corporation [31] to describe the application of the systems approach to the wider "non-hardware" systems mentioned earlier. It is unfortunate perhaps that, in their attempts to emphasize that Systems Analysis is concerned with the design of much wider systems than "engineering" or "hardware" systems, the users of this word have only told part of the story. This is because systems synthesis is an equally important step as systems analysis in the design of systems. However, the word "engineering" covers the processes of analysis and synthesis and it is for this reason that the Department of Systems Engineering at the University of Lancaster has been so named. The word "engineer" has been used in the general sense of the man who designs or engineers systems whether they be composed of hardware or software or just people.

Clearly, semantic discussions resolve nothing. However, since Operational Research, Systems Analysis and Systems Engineering have a great deal in common, then the systems approach would suggest that there is need for better communication between them.

B. Stages in the application of Systems Engineering

This section gives some general guide lines as to how a systems engineer would tackle any problem. The various stages to be described represent a breakdown and amplification of the four steps

1. Systems Analysis,
2. Systems Design,
3. Implementation,
4. Operation,

mentioned earlier and are summarized in the flow diagram of Figure 4.5. It is *not* suggested that all systems projects will proceed along the following lines. Rather, the stages are intended as rough guide lines to aid clear thinking and to emphasize above all that *Systems Engineering is an orderly and well-disciplined way of getting things done.*

1. SYSTEMS ANALYSIS

1.1. Formulation of the problem
1.2. Organization of the project
1.3. Definition of the system
1.4. Definition of the wider system
1.5. Objectives of the wider system
1.6. Objectives of the system
1.7. Definition of overall economic criterion
1.8. Information and data collection

↓

2. SYSTEMS DESIGN

2.1. Forecasting
2.2. Model building and simulation
2.3. Optimization
2.4. Control
2.5. Reliability

↓

3. IMPLEMENTATION

3.1. Documentation and Sanction Approval
3.2. Construction

↓

4. OPERATION

4.1. Initial operation
4.2. Retrospective appraisal
4.3. Improved operation

Figure 4.5 Detailed stages in a Systems Approach to problems.

B.1. Systems Analysis

The first step in Systems Engineering is Systems Analysis. This involves the following stages:

1.1. Recognition and formulation of the problem.
1.2. Organization of the systems project.
1.3. Definition of the system.
1.4. Definition of the wider system of which the system being studied forms part.
1.5. Definition of the objectives of the wider system.
1.6. Definition of the objectives of the system being studied.
1.7. Definition of the overall economic criterion.
1.8. Information and data collection.

1.1. *Recognition and formulation of the problem*

Firms and organizations do not exist to provide employment for systems engineers! Rather, they have problems which arise in the day-to-day running of their organizations, the solution of which would lead to improved efficiency and profitability. The job of the systems engineer is to provide effective solutions to those problems.

A problem arises because some manager needs *help* – he may have noticed that something is going wrong or he may need help to make a planning decision or to implement a planning decision made higher up. He may then decide to consult the systems engineer as an individual accustomed to taking an overall point of view towards solving problems. In these circumstances the systems engineer should interrogate the manager very thoroughly and also all other persons within the organization who are likely to be able to help. In particular he should ask

(1) How did the problem arise?
(2) Who are the people who believe it to be a problem?
(3) If it involves implementing a planning decision made higher up, what is the chain of argument leading to the making of the decision?
(4) Why is the solution important? How much money might it save?
(5) Is it the right problem anyway? Might it not be just a manifestation of a much deeper problem? Would greater benefits accrue if that problem were solved rather than the one posed by the manager?
(6) Inevitably, the resources available to the firm are limited. On the evidence available at this stage, does it seem that there would be a reasonable return on systems effort if applied to the project or would this effort be better employed in tackling a different problem?

As a result of this dialogue, a clearer picture should now begin to emerge about the scope of the problem and the likely benefits which would result from its solution.

1.2. *Organization of the project*

1. *Composition of the systems team.* Once the scope of the problem has been defined, the way in which it is to be tackled should be mapped out. Systems Engineering is a team activity not an individual activity and an *ad hoc* systems team should now be set up. Ideally this should be able to draw some of its members from a small central systems department within the company. In addition, it will be necessary to supplement these systems men by people, drawn from various departments within the company, whose specialist knowledge can be brought to bear upon the problem.

Many companies do not have a central systems organization nor a great deal of systems experience. This does not mean that a systems team cannot be formed. The best resources available within the organization for tackling the problem should be brought together and, with efficient leadership, they can develop into a systems team very quickly.

A typical systems team would contain some or all of the following:

(*a*) *Team leader* – ideally an experienced systems engineer, or alternatively, someone with a keen intellect and a great deal of knowledge of the problem which is to be tackled.
(*b*) *User* – a representative of the team which will operate the engineered system. For example, if the problem is to design and build a chemical plant, production department should be represented, preferably by the plant manager designate.
(*c*) *Model builders* – to take part in the model building itself and also to liaise with and stimulate those specialist functional departments (such as research, process development and sales departments) which will be able to provide information for the sub-systems models.
(*d*) *Designers* – if the system involves the building of hardware, representatives of the engineering team who will be responsible for the design of this hardware to meet the system specification. Similarly, if the system involves software, representatives of the team which will design this software, for example computer programmers, data processing and computer experts.
(*e*) *Computer Programmers/Mathematicians* – to programme the systems models and to help with the optimization of the design.
(*f*) *An economist or accountant* – to provide information on the general economic environment of the problem which will help in defining the overall economic criterion, and to assist in obtaining cost information for the model.
(*g*) *Systems engineers* – these may contribute to a greater or lesser extent in the model building, programming, optimization, economic evaluation, etc. In some situations, they may have to do most of this work themselves. Above all, however, they should be taking an overall view of the development of the project.

2. *Terms of reference.* The systems team should take steps to ensure that they are given the widest possible terms of reference and are given access to any information or

person. In other words, they should be given the maximum opportunity to stand back and take a fresh look at the problem.

3. *Scheduling the project*. The systems engineering team should apply the systems approach to the conduct of its own activities to ensure that the work is carried out logically and systematically and that the implementation of the systems study can take place by an allotted time. Thus a decision network should be constructed (for example a critical path schedule), targets set and duties allocated. In this way, maximum effort can be concentrated in areas which are most important. The systems team will then ensure, by its critical approach to its own method of working, that problems are tackled in their correct order of importance.

1.3. *Definition of the system*

The next task of the systems engineering team is to define in precise terms the system which is to be studied. This is a process of *analysis* in which the system has to be broken down into its important sub-systems and the interactions between these sub-systems indicated by drawing a *flow-block diagram*. The subsequent task of the systems engineer is one of *synthesis*, that is to design or engineer the individual sub-systems so that they work together towards achieving an overall objective. As indicated in Section A.1, the flows typically appearing in a flow-block diagram represent money, energy, materials, information and decisions. In constructing a flow diagram, it is sometimes helpful to use a different flow convention for these different types of flow, or alternatively, the flows should be clearly labelled.

The extent to which the system needs to be broken down into sub-systems may not be known initially. For example, certain processing units in a chemical plant may usefully be lumped together and considered as one unit for the purpose of building a model to establish the overall system requirements. To avoid making the system description too complex, it is usually better to work with a simple representation of the system first and then elaborate later if necessary. If the system already exists, as for example in a process improvement study of an existing plant, the simplicity of the system diagram, and its description, is justified if the resulting model is able to describe those areas which are sensitive to cost. Even when nothing very much is known about the system, as in the initial stages of the design of a new plant, there is a great deal to be said for making the system description simple to start with and then examining the consequences of gradual elaboration later.

A related question to the simplicity of the system description is that the systems design process must be sufficiently *flexible*, so that this description can be changed as further knowledge and experience is accumulated during the course of the project. This is especially important in the design of a new system when there may be very inadequate knowledge at the start of a project. As the project proceeds, the system description will become clearer as the process of *innovation* develops. Indeed, there may have to be several iterations of the design process before a satisfactory solution can be found.

1.4. *Definition of the wider system which contains the system being studied*

To define the objectives of the system, it is necessary to display very clearly the role which the system plays in the wider system of which it forms part. A separate flow-block diagram should be constructed to display this role very clearly. By contrast to the flow-block diagram of the system itself, the flow-block diagram of the system as part of a wider system should include as much detail as is available. Invariably, it happens that the relationship of the system to the wider system is hazy and unclear to most people at the beginning. Hence a great deal of clear thinking will be necessary to fill in sufficient detail on this flow-block diagram so that proper account is taken of the interactions between the system and the wider system when formulating objectives. The flow diagram of the wider system provides an excellent tool for clarifying this thinking.

1.5. *Definition of the objectives of the wider system*

The block diagrams of the system and of the wider system provide invaluable tools for analysing and then formulating objectives. Because systems form part of a hierarchy of systems, it is impossible to dissociate the objectives of the system being studied from those of the wider system of which it forms part. In fact, it is the objectives of the wider system which are the crucial ones since they determine the *environment* within which the system has to function. If this environment changes, then so will the objectives of the system change. To take a very simple example, the objectives of a single chemical plant must fit into the overall production plan of the company. At different times the company plan may stipulate one of several objectives.

For example,

(1) the plant must make a fixed tonnage of the right quality product at minimum cost per ton. This is commonly referred to as a *production limited* situation, or

(2) the plant must make as much product as possible while satisfying the same or possibly less stringent quality constraints than in (1), or
(3) the plant must make a fixed tonnage at minimum cost per ton but using a cheaper, less pure, raw material, and so on.

Therefore, in a process improvement exercise, there may not be one unique objective but rather a *catalogue* of possible objectives, each resulting in a different way of operating the plant. Thus the relevant objective at any given instant must be dictated by the needs of the wider system. Definition of the objectives of the wider system brings several advantages:

(1) It focuses attention on the fact that systems must be designed so that:

(*a*) junior systems in a hierarchy should play their designed role in achieving the objectives of more senior systems in the hierarchy,
(*b*) senior systems in the hierarchy should make a clear, unambiguous statement of what the junior systems are expected to contribute.

Without such clearly defined responsibilities a system may operate very inefficiently, or if the objectives are very vague, anarchy may result.
(2) We have seen that the objectives of systems at the same level in the hierarchy are usually in conflict – such systems are usefully described as *interlocking* or *competitive* systems. Thus, defining the objectives of the wider system is essential so that the objectives of the competitive systems can be formulated in such a way that they contribute effectively to the objectives of the wider system instead of pulling in different directions. For example, two plant managers can very easily operate their own plants in such a way that they optimize their performance individually but at the expense of not meeting the company's objectives efficiently. Such conflicts are of frequent occurrence in industry and in other walks of life and stem from a failure to define objectives clearly.
(3) Facing up to the objectives of the senior systems in the hierarchy will counteract the tendency to omit these systems altogether. There will always be a tendency to omit the systems at the top because they seem too vague, or more importantly, *too difficult* to formulate. In fact, the systems at the top should not be omitted until the consequence of doing so on the system being studied have been carefully ascertained and shown to be of no consequence. For example, it might be sensible to omit the influence of the British Economy in deciding how to operate a plant but to omit the influence of raw material availability within the company might be very short sighted.
(4) By defining the objectives of the senior systems in the hierarchy, the system under study can be designed so that it is capable of *adapting itself to change* quickly. For example, by knowing that the production system of a company may make changing demands on an individual plant, the operation of that plant can be planned so that it is able to react quickly to the changes when they occur. This can be achieved, for example, by tabulating the best operating conditions for different values of the throughput.
(5) The performance of junior systems in the hierarchy may improve, especially the performance of the people involved. By communicating the objectives of the higher systems to those involved in the junior systems, their sense of involvement is increased, with a consequent increase in efficiency. Sometimes immediate and worthwhile benefits can result by communicating objectives of the senior systems since it can pin-point areas where conflicts are leading to a deterioration in efficiency.

1.6. *Definition of the objectives of the system*

The end product of any definition of objectives is the formulation of the criterion, usually an economic one, which measures the efficiency with which the system is achieving its objective. The formulation of the economic criterion may require a careful study in its own right. However, in the early stages it is better to define the objectives in broad terms.

There will usually be conflicting objectives and at the start of a systems study it is essential to make a *comprehensive list* of all possible objectives in *their anticipated order of importance*. One, or possibly a few, objectives might then be singled out as being the most important ones. What weight to give these conflicting objectives in the formulation of the overall economic criterion will have to be considered later.

At this important stage of any systems study, much questioning will have to be done and many different points of view listened to. In the end the systems engineer must make up his own mind about the correct objectives, then get agreement with all concerned and finally communicate his findings to everybody so that their future cooperation can be relied upon.

Important points to be borne in mind at this stage are:

(1) The systems engineer may meet resistance when he tries to define objectives. People who have got along quite well in the past with vague objectives may object to the influx of new ideas. However, the systems engineer must persist because no system can be designed properly unless it is clearly known what it is trying to achieve.

(2) He may be frustrated because the objectives are not clear. If after persistent attempts, objectives are still not satisfactorily defined, this must be faced up to. This means that the resulting system will have to be acknowledged to be imperfect and improved later if more precise information becomes available. His motto should be "It ain't much but it's all I've got" rather than "I'm sorry mate but I don't know enough to start!"

(3) Where possible, objectives should be *simple and direct*. If simple quantitative objectives are not possible, they should be replaced by simple subjective objectives. Failure to quantify objectives should be recorded and brought to the attention of people so that attempts can be made at some future stage to improve precision. In other words, the systems engineer can make a big contribution by being objective about subjective matters!

1.7. *Definition of the overall economic criterion*

Once the objectives have been agreed, the next step is to define, in as precise terms as possible, a criterion which measures the efficiency with which the system can achieve its objective. Usually, but not invariably, this criterion will be an economic one, for example a company may measure performance by its rate of return on capital, or the performance of a chemical plant may be measured by the cost per ton of manufacturing the main product. The more precise the objectives, the easier it is to set up quantitative criteria. Conversely, if the overall objectives are not precise, then there will have to be some subjective criterion of performance.

An overall economic criterion should be:

(1) *Related to objectives* – care should be taken to avoid a precise mathematical criterion which does not embrace all the objectives.

(2) *Simple and direct* – if possible, both objectives and performance criterion should be expressed as concisely and simply as possible.

(3) *Clearly agreed and accepted, even if qualitative* – in practice confusion often arises because of the application of contradictory criteria. For example, one manager may decide that a certain course of action is desirable because it is based on a certain criterion, whereas another manager in the same company may reach the opposite decision because his criterion is different. It is very important that criteria of performance be agreed by everyone, notified to everyone and applied by everyone in the same way – otherwise confusion results and efficiency is impaired.

Resolution of conflicting objectives. To formulate an economic criterion, it is necessary to decide on the compromises which have to be achieved between the conflicting objectives of a system. There are two basic ways in practice in which conflicting objectives can be met.

(1) By *weighting* alternative objectives in the overall criterion.

(2) By imposing *constraints* (sometimes objective, sometimes subjective) on certain variables which enter into the economic criterion.

These two methods are now discussed briefly.

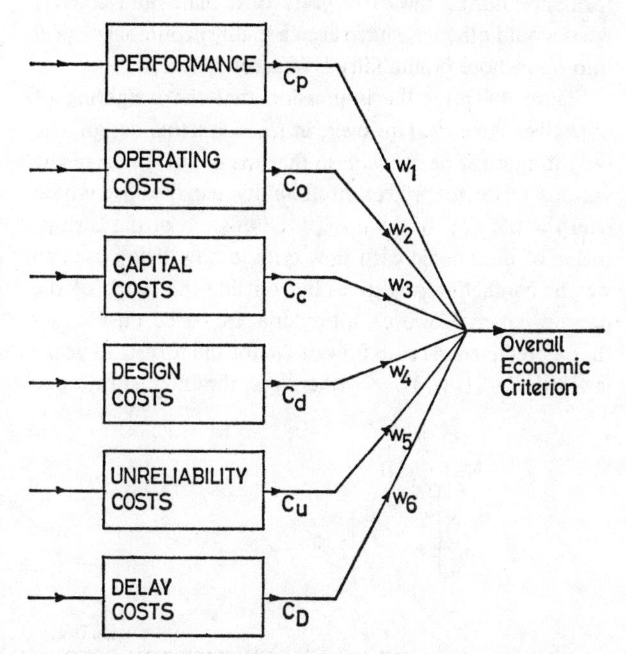

Figure 4.6 Conflicting objectives in the design of a system.

1. *Weighting of objectives.* Figure 4.6 summarizes typical conflicting objectives which have to be met by any system. This shows that the overall economic criterion has to be obtained by attaching weights $w_1, w_2, \ldots, w_6$ to:

1. The performance of the system.
2. Operating and production costs.

3. Capital costs.
4. Design costs.
5. Unreliability costs.
6. "Delay to build" costs.

Very often these weights will not be known exactly and it may be necessary to make *value judgements* of what the real weights are. This does not detract from the exercise but on the contrary forces designers to think more clearly about the subjective judgements that they have to make. For example, in the case of a plant design study, the influence of these costs on total cash flow as the project develops is illustrated in Figure 4.7. Thus the time and money spent on each of the four stages – research, development, construction and start-up – can have a considerable effect on the profitability of the project.

Most plant design studies nowadays try to balance capital against operating costs. Unfortunately, insufficient attention seems to be paid to unreliability and "delay to build" costs. Bad judgement of these costs can result in poor performance during the early years of a plant and convert what would otherwise have been a highly profitable project into one whose profitability is unacceptably low.

Figure 4.6 gives the impression that the weighting of objectives is *static*. However, in most systems design, the weighting must be *dynamic* so that the contribution of the various costs to project profitability can be discounted over the life time of the project. Discussion of the various forms of discounted cash flow criteria now being used to weight conflicting objectives lies outside the scope of the present paper. However, all criteria are based on the fact that a predicted net cash flow of £a_i for the project in year i is worth £$a_i/(1+r)^i$ *now*, where r is the interest rate one could reasonably expect to earn by investing one's money in other projects inside the company, or possibly outside. For example, the *net present worth* (N.P.W.) of a project is defined by

$$\text{N.P.W.} = \sum_{i=1}^{n} a_i/(1+r)^i,$$

where n is the expected life time of the project. Again, from the point of view of systems design, such criteria should not be used in a mechanical fashion. In practice, it is necessary to look at the individual predicted cash flows and their *uncertainty* in greater detail because two projects with widely different cash flows can give rise to the same net present worth. Therefore, it is usually necessary to compare the *risks* associated with two projects before a decision in favour of one or the other can be made.

2. *Constraints.* It is not always possible to achieve a compromise between conflicting objectives by weighting alone. In addition, the conflicts may have to be resolved by imposing certain constraints on the design of the system. These constraints are usually of two kinds:

(*a*) physical constraints,
(*b*) constraints which are a recognition of the fact that *it is sometimes difficult to quantify.*

Once again, this is illustrated by considering the design of a chemical process. Under (*a*) come constraints on quality and physical properties imposed by customers and by equipment performance, etc. Under (*b*) come, for example, factors associated with the safe running of the process at conditions as near as possible to the inflammable or explosive limits or other process constraints. If an explosion happens, there may be a loss of human life and the plant may be shut down for long periods.

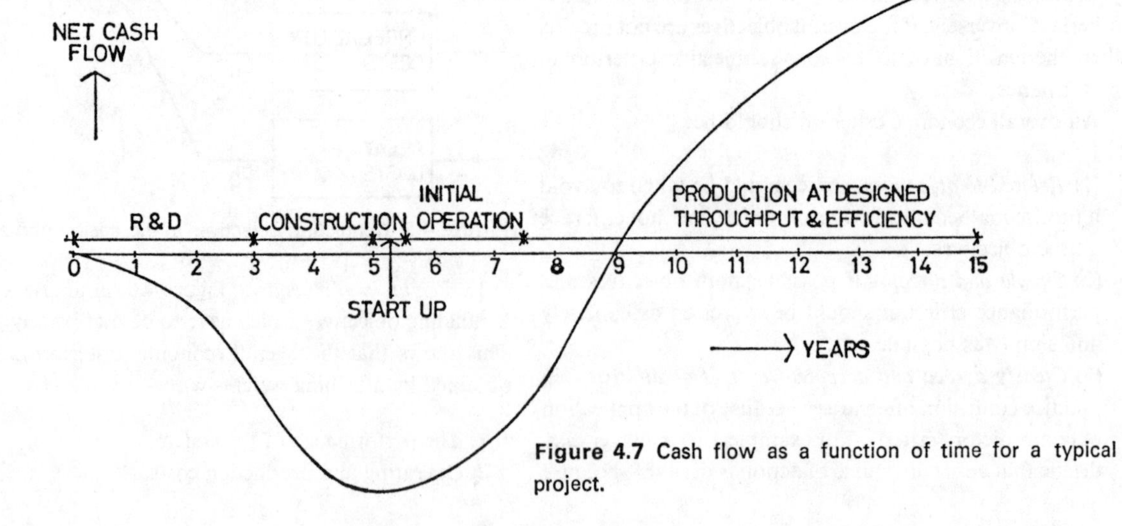

Figure 4.7 Cash flow as a function of time for a typical project.

To allow for such factors in a quantitative way would be too difficult or too complex. As mentioned earlier, much is to be gained by making the overall economic criterion simple. In other words, one should try to build in those main economic features which are well understood and then impose certain constraints on the freedom of choice of certain variables in the design to allow for factors which are much more difficult to quantify, such as the loss in customers due to bad quality and to shutdowns due to explosions, the loss of lives, etc.

This illustrates the general point made about objectives earlier, namely that a detailed analysis of the economics of a situation may reveal areas which are difficult to quantify. As with objectives, it is better to bring to the foreground these subjective elements rather than bury one's head in the sand and to pretend that one is making progress by using a criterion which is easy to quantify but *irrelevant* to the real-life situation. The fact that subjective elements are present does not mean that nothing can be done. On the contrary, it means that subjective economic judgements have to be made and these are usually best expressed in the form of constraints imposed on the design. Indeed, once this is done, the possibility arises of varying the constraints to see what effect they have on the economic performance of the system.

1.8. *Information and data collection*

The final, and probably most extensive, stage in Systems Analysis is the gathering of data and information which will form the basis of any future modelling of the system. If the system is not in existence, then the fact finding will have to be based on new research and development and on other similar systems if they exist. Data will be required not only to provide information about the operation of the system but also to make *forecasts* of the *environment* in which the system will have to operate in future.

Efficient information and data collection requires

(i) *Clear thinking* about the problem so that relevant sources of information can be tapped.

(ii) *Ability to communicate* in speech and in writing so that people are stimulated into parting with information and into volunteering new information.

(iii) *A grasp of statistical techniques* so that the significance of the data can be appreciated and used in subsequent model building and decision making.

B.2. Systems Design (or Synthesis)

The systems analysis stage should have opened up the problem, enabled objectives to be defined and initiated fact finding. Based on these foundations, the Systems Design stage can then be tackled with confidence. It is convenient to discuss the design stage under the five headings

2.1. Forecasting,
2.2. Model Building and Simulation,
2.3. Optimization,
2.4. Control,
2.5. Reliability.

2.1. *Forecasting*

Forecasting is the first important step in the design of any system. For example, in the design of a production control or production planning system, forecasts of demand may be needed for each fortnight or month for up to a year ahead (short term forecasting). Similarly, to design a plant, forecasts of demand will be needed for a period of several years ahead (medium term forecasting) and to design a Corporate Plan for a company, forecasts of the firm's activities and environment will be needed for up to 15 or even 20 years ahead (long term forecasting, technological forecasting).

Accurate forecasts are essential for the efficient design of any system. By contrast, if the forecasts are inaccurate, they can not be compensated for by sophisticated model building and optimization at a later stage. Indeed such sophistication at a later stage may then turn out to be a complete waste of time.

Forecasts of future values are not of much use on their own. In addition, it is essential to obtain estimates of their *accuracy*, so that this can be allowed for in assessing the *risk* associated with the design of the system.

2.2. *Model building and simulation*

To compute the costs associated with different ways of running a system, it is necessary to predict the performance of the system over a wide range of operational conditions. To do this, a *model* of the system needs to be built. By a model is meant a *quantitative* description of the behaviour of the system which can be used to predict performance over a relevant range of operating conditions and real life environments. In its crudest form, a model could consist of a set of tables or graphs; at a more sophisticated level it might be written in mathematical form, namely as a set of algebraic or differential equations.

Model building is not a straightforward, but rather, a highly creative activity. It is of necessity an *iterative* or *adaptive* process in which one moves from a state of little knowledge to one of greater knowledge – in other words, it is a process of hauling oneself up by one's bootstraps. To design a system, many different types of model may need to be developed. Experience and good judgement are needed to decide which type of model should be used for any particular situation, so that the system can then be designed as efficiently as possible with the minimum expenditure of time and money.

Quantitative models useful in Systems Engineering may be classified into four types:

(i) *Descriptive* models, which provide a qualitative description and insight, as compared with *predictive* models, which should be able to give predictions of the economic performance of a system to a required degree of accuracy.
(ii) *Mechanistic* models, which are based on a mechanism by the way that the system behaves, as compared with *empirical* or *statistical* models, obtained by fitting data obtained from the system.
(iii) *Steady state* models, which are based on average performance, as compared with *dynamic* models which allow for the way that performance fluctuates with time.
(iv) *Local* models, which describe the behaviour of sub-systems, as compared with *global* models which tie together the sub-system models into a model for the overall system.

Model building is the cornerstone of any scientific activity and, as such, plays a very important role in systems engineering – as mentioned in Section 1.2, a model replaces guesses by facts. However, there are important differences between the approach required for model building in systems engineering compared with that required for other scientific subjects. For example, in physics and chemistry, model building is almost an end in itself since the objective is to subsume as many facts as possible under the umbrella of a single model. In Systems Engineering, on the other hand, the end objective is to *optimize* the performance of a system and hence model building must be subservient to this objective. Thus a Systems Engineering team must:

(1) Ensure that model building is carried out *with a sense of purpose*, that is to enable a system to be designed as cheaply and efficiently as possible.
(2) *Tie together* the various specializations that may be needed for building models of the sub-systems. In particular, the systems team will always be emphasizing to the specialist model builders that the important thing is to model those areas to which the overall economic criterion is sensitive.
(3) *Ensure that work is concentrated where it is most needed.* Because it takes an overall view, the systems team will be in a much better position to pin-point those areas which are cost sensitive and, if necessary, to ask for more work to be done in those areas. Conversely, it will be in a better position to stop activity in those areas which are not cost sensitive. For example, in a plant design project an over-enthusiastic chemist may believe that it is necessary to predict a certain physical property to say within 1%, whereas from a systems viewpoint it may only be necessary to predict it to within 5% or even 10%. *As a general rule, models should be kept as simple as possible.*
(4) *Decide when the model is adequate* for the purposes for which it is needed, namely to predict overall system performance to a sufficient degree of accuracy. Model building is a fascinating activity and people, if left to their own initiative, tend to be carried away, tend to become emotionally involved and tend to waste time and money by over-elaborating the model. By contrast, the systems approach requires discipline to ensure that when the model is adequate, model building is terminated and the optimization stage begun as soon as possible.
(5) If the model is to be used for planning, see that an *effective dialogue* is conducted between the systems team and the managers who will use the model. Experience suggests that in such areas as production, investment and corporate planning, that this *dialogue must start whilst the model is being built* and not when the model is handed over. This is because models must be thoroughly understood by management if they are to be used effectively and moreover, because it is management that makes the decisions, their involvement at an early stage can often lead to suggestions resulting in substantial improvements and simplifications in the models themselves.

Simulation. Once a model of the overall system has been built, the model has to be converted from a *passive* device (a set of graphs or equations) into an *active* device which can then be used to *simulate* the behaviour of the system when subjected to:

(*a*) realistic inputs typical of those which the operational system will have to meet in practice,
(*b*) realistic disturbances, which will cause the behaviour of the system to fluctuate from its average or steady state performance.

Hence a simulation is a *working* or *operational* model of the system which should be able to reproduce the actual behaviour of the real system to an accepted degree of accuracy. If the simulation is accurate, data generated from the simulation should agree closely with data from the operational system. In other words, from the point of view of systems design, we can treat the simulation *as if it were* the actual system and make changes to the parameters in the simulation to see how to optimize the system.

The inputs or disturbances, in the simulation may be:

(i) deterministic, or
(ii) statistical,

and correspondingly, the simulation is said to be *deterministic* or *statistical*. An example of a deterministic simulation is provided by a model of a chemical plant which is set up in such a way that it can be fed with average values of process inputs and used to calculate the corresponding values of the average outputs. An example of a statistical simulation is provided by a model which describes the random times of arrival of tankers at a port with a view to optimizing the berthing facilities.

In a statistical simulation, it may be possible to collect enough data relating to the statistically fluctuating variables directly from an operating system. However, it often happens that insufficient data is available to simulate the behaviour of the system over a sufficiently long period for the results to have meaning. In these cases, a statistical analysis will be needed to analyse the data so that a statistical (probability) model can be set up to describe the data. Armed with a statistical model, unlimited amounts of hypothetical data can then be generated so that the system can be simulated over a sufficiently long period of time.

In recent years, powerful computer simulation languages have been developed to speed up the simulation stage of a systems design. As with modelling, the systems engineer will emphasize that a simulation should be made as simple as possible so that optimization can go ahead with minimum delay.

2.3. *Optimization*

The next step after simulation is to optimize the system. Armed with a model which can predict performance, it is then possible to compute the value of the economic criterion corresponding to different modes of operating the system. For example, when designing an industrial plant, it enables the rate of return on capital to be calculated for different sizes of equipment and operating conditions, and so on. Choosing the system which results in the most favourable value of the economic criterion is what is meant by optimization. It is at this point that the importance of defining the overall objectives should become apparent. If the system and its objectives are too narrow, then the most profitable mode of operation for this narrow system will almost certainly be in conflict with the most profitable conditions for the wider system which has been overlooked. This is usually referred to as *sub-optimization*.* One of the most important jobs of the systems engineer is to ensure that sub-optimization does not take place. He will constantly need to emphasize that the optimization of each sub-system independently will not in general lead to a system optimum. More strongly, he will have to emphasize that improvement to a particular sub-system *may actually worsen* the overall system.

During the last fifteen years, the need to optimize systems has stimulated the development of a number of mathematical techniques called *optimization techniques*.†

Although these mathematical techniques are very important, from a systems engineering point of view, optimization involves a great deal more than a mathematical problem. Assuming that the problem is one of maximization, optimization is equivalent to climbing a multi-dimensional hill. Very often one is debarred from climbing to the top because of the constraints on the design variables. However, in systems engineering work, one is not only interested in getting to the highest point possible on the hill and then planting the Union Jack at the summit! In the design of a system, three other much more important considerations have to be investigated, namely:

(1) The sensitivity of the economic criterion to changes in the parameters near the "optimum".

* It is stated [25] that the word "sub-optimize" was first used by C. J. Hitch [32]. In his analysis of the wartime problem where it was shown that convoys should be made as large as possible, he points out that although the final answer reached was approximately correct, it was arrived at for the wrong reasons. He argues that the reasons were wrong because too much emphasis was placed on a sub-system, namely a skirmish between a convoy and a submarine pack, rather than on the wider system, that is winning the Battle of the Atlantic or more importantly still, optimizing an even wider system, namely winning the war.

† From a mathematical point of view, the problem of optimization is one of maximizing or minimizing an overall economic criterion

$$f(x_1, x_2, \ldots, x_n)$$

which depends on various design parameters $x_1, x_2, \ldots, x_n$, (e.g. operating conditions, sizes of equipment, when designing a plant) and is subject to certain inequality constraints

$$g_i(x, x_{21}, \ldots, x_n) \leqslant c, \; i = 1, 2, \ldots, l$$

and certain equality constraints

$$h_j(x_1, x_2, \ldots, x_n) = d_j, \; j = 1, 2, \ldots, m.$$

(2) The sensitivity of the economic criterion to various assumptions made in the design.
(3) The effect of uncertainty in the forecasts of the environment in which the system will have to operate.

(1) In many optimization studies, there are usually more design variables to choose from than one usually can, or wants to, work with. As a first approach, one can choose those variables which appear to be the most important on intuitive grounds and optimize with respect to these. It is useful then to plot the economic criterion as a function of each variable, holding the others fixed at their best values – in this way one can identify the variables to which the value of the economic criteria is sensitive. Those variables to which it is insensitive can be left out and further optimization runs made with these insensitive variables replaced by new variables.

When all the sensitive variables have been discovered, the end point of any optimization study should be:

(*a*) to quote the best values of all variables.
(*b*) to give plots of the economic criterion with respect to each variable, holding all other variables at their best values.
(*c*) if possible, to give contour plots for certain pairs of variables, holding all other variables at their best values. This type of exercise can be very useful in discovering how one can "trade-off" one parameter against another in achieving a given cost or performance.

When (*a*), (*b*) and (*c*) are completed, it should be possible to obtain a clear picture of how the system is sensitive to cost.

Thus, if the overall economic criterion is smooth with respect to changes in the design parameters, then this information tells us that a wide range of systems is acceptable, each with a value of the economic criterion that is roughly the same as that of the "best" system.

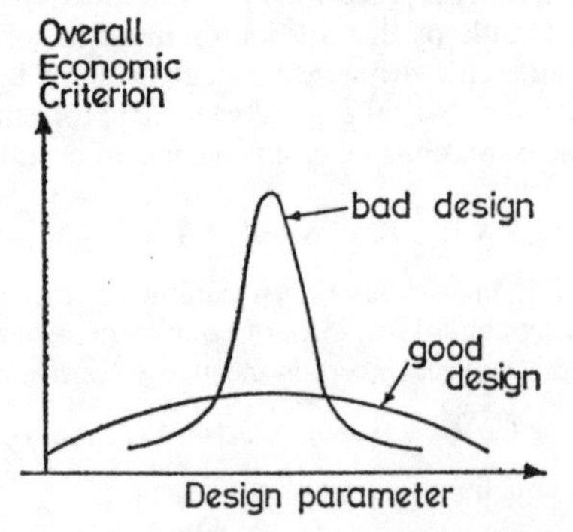

Figure 4.8 Sensitivity of a design to the design parameters.

This constitutes a good design, as indicated in Figure 4.8. On the other hand, if the economic criterion is steep and pointed, small changes in the design of the system will produce very marked changes in its profitability. Such sensitivity to small changes in the design parameters is undesirable and leads to a bad design since one cannot be absolutely certain that the assumptions, on which the design was based, are valid. If the assumptions are wrong, there might be a high probability that the realized economic performance will occur in the foothills of the surface rather than near the top.
(2) This brings us to a very important part of the optimization stage, namely that it is essential that changes should be made in the assumptions on which the design is based and an examination made of the effects which these changes have on the system. If the changes produce markedly different systems, then one will have to think again. On the other hand, if they produce systems having the same general features, then one can feel more assured about recommending the design. The system can then be said to be *robust* with respect to the underlying assumptions.
(3) Not only must sensitivity to assumptions about performance of sub-systems be investigated but also sensitivity to the uncertainty in the forecasts of the environment within which the system will have to operate. For example, in designing a plant, if the sales forecasts are very uncertain, it has to be faced up to that the most pessimistic forecasts might be the ones that actually occur, and hence that low profitability may result. However, one should clearly not design in the expectation that the most pessimistic forecasts will be realized. A *risk analysis* is then very important in helping to assess the most reasonable approach to the design problem. For example, this form of analysis can be especially helpful if a multi-product complex is being designed when the sales forecasts for each product are subject to different degrees of uncertainty. The size of these uncertainties can have an appreciable effect on the way in which the complex is designed.

To summarize, at the optimization stage the systems engineer should pay particular attention to the following:

(1) He should be aware of the dangers of sub-optimization and indeed of *leaving out altogether* certain important variables.
(2) After locating the optimum design conditions, he should examine very carefully the economic criterion surface to see which parameters are cost sensitive.

(3) He should guard against "very sharp optima" since a system which is very cost sensitive may be highly dependent on the assumption made in the design.

(4) He should make a *sensitivity analysis* to see whether changes in the assumptions lead to systems with the same general features. If not, he should think again.

(5) Finally, he should be very conscious of the fact that, when optimization is complete, a *very costly decision* may have to be made to allow the optimized design to be built. Therefore, at the optimization stage, he should be guided by certain techniques for making decisions in the presence of uncertainty, such as risk analysis.

2.4. *Control*

When a system has been optimized, a control system must be set up so that the most profitable design conditions can be realized when the system is operated in a real life environment. Control is necessary because unpredictable *disturbances* enter a system which cause its *actual* performance to deviate from its *predicted* performance. For example, local control devices will need to be installed on a chemical plant to regulate automatically the flow of materials, the levels of liquids in tanks and the temperatures and pressures required to achieve the most profitable operating conditions. Similarly, a management control system is needed to control a company production plan. This means that when a company production plan has been formulated, it must be backed up by an accounting system which feeds information to senior managers indicating whether the plan is being "kept on target". Management can then be helped to take corrective action to compensate for deviations from the plan. As argued elsewhere [4]:

(1) Control should be thought of as an integral part of the design of the system and not an afterthought, as often happens for example in the design of an industrial plant.

(2) A systems approach takes away the undue attention that is sometimes given to the mechanics of the control of local loops and focuses attention on the wider questions of where should control be exercised, how sophisticated should it be, what equipment is needed, whether an on-line computer is justifiable, and so on.

(3) A systems approach focuses attention on the economic benefits, both tangible and intangible, which result from control and demands that the cost should be justified as part of the design of the overall system.

(4) Most important of all, the advantages of local control within a firm can only be realized effectively when it plays its planned role within a hierarchy of technical and financial control systems within the company.

2.5. *Reliability*

The importance of system reliability has already been mentioned in the section dealing with the definition of the economic criterion (stage 1.7). A good control system will go part of the way towards ensuring good reliability. For example, if there are process constraints which must not be isolated in the operation of a plant, the control system will have to be both efficient and reliable to keep clear of these constraints. However, systems reliability involves looking at aspects other than control by considering the overall effect of *uncertainty* on the design of the system. Uncertainty in forecasting environmental factors has already been discussed in Section 2.1. Other sources of unreliability are, for example, breakdown of equipment, non-availability of resources in the form of raw materials and men to perform operating and maintenance tasks. To cope with such unpredictable events requires redundant or stand-by equipment, extra storage and capacity, extra manpower and so on. This inevitably leads to higher capital and operating costs so that reliability questions must be considered as an integral part of the overall optimization of the design. The singling out of systems reliability as a separate stage in this paper is a reflection of the fact that reliability questions are all too often regarded as an *after thought*, invariably resulting in disastrous effects on performance and profitability. It is hoped to prepare a paper for a later issue of this journal to illustrate how neglect of reliability problems had serious consequences on the profitability of a large petrochemical complex.

Another reason for separating out reliability from optimization is that extensive simulation and modelling may be needed after the optimization stage to check that system unreliability has been reduced to an acceptable level.

B.3. Implementation

No systems study, however well carried out, is of much use unless it leads to positive action and is properly implemented. The important steps in this stage are:

3.1. Documentation and sanction approval.
3.2. Construction.

3.1. *Documentation and sanction approval*

The end product of a systems engineering study is a report (or reports) which should highlight proposals for ACTION.

Failure of *communication* at this very important stage can sometimes ruin what is otherwise a perfectly good systems study. To avoid this happening,

(1) *before issue*, the form and content of the final reports should be discussed and agreed with those senior managers whose backing will be needed to implement the results of the study, and not planted on their desks as a "fait accompli" before any discussion has taken place.

(2) the report should be simple, direct and logical. A three tier report is recommended for work of this kind.

(i) (*a*) A brief statement of the problem,
(*b*) a summary of the main conclusions and benefits of the study,
(*c*) a summary of the recommendations, including an overall figure for the savings and the cost of implementing them.

Many senior managers will only have time to read this part of the report so that it should make the maximum impact in the smallest amount of space.

(ii) A middle section where the main substance of the recommendations and benefits are argued so that any intelligent manager can understand them.

(iii) A final, more detailed, section containing the technical details of the study.

(3) a separate document will usually be needed within the organization to highlight the recommendations and to draw up a time-table or critical path schedule for their implementation. Such a sanctioning document may need to go to higher management to obtain approval for the project to proceed.

This is the most critical stage of any systems study since a *decision* has now to be reached as to whether to go ahead and build the system. Clearly, this decision should be made in an objective way and no attempt should be made to "dress up" the sanction proposal. On the other hand, the systems engineer should avoid the all-too-common mistake of putting forward bad arguments for a good case. In this way, not only does he not do justice to himself and his team but will have turned out to be a bad systems engineer.

3.2. *Construction*

Some projects may require the building of special hardware and software before implementation is possible. For example, if a plant is to be built, site works must be constructed and equipment ordered, delivered and installed on site. Similarly, if an on-line computer is to be installed on a plant, hardware in the form of instrumentation, control equipment and the computer itself must be ordered, delivered and installed. In addition, software in the form of computer programmes must be designed to implement the control and optimization algorithms.

The majority of the systems team may have moved on to other work by this stage. However, it is important to realize that the construction stage is still a part of the overall systems design. Thus, bad planning at the construction stage can have a big effect on the profitability of a project by holding up its start-up date. A systems approach at this stage would ensure that

(i) the systems team has specified the system in a clear and unambiguous way down to the minutest detail of all subsidiary sub-systems.

(ii) by the continued involvement of the systems team, the builders of the system are kept constantly in touch with the philosophy underlying the design and the way that it is to be implemented when in operation,

(iii) the project is well planned, duties allocated and targets met on time. Critical path scheduling emerges as the important technique in this area. One of the big advantages of C.P.S. is that it forces planners to concentrate attention on those areas which are critical for completion of the project.

B.4. Operation

After the system has been designed, built and installed in its working environment, then follow the steps:

4.1. *Initial Operation* of the system.
4.2. *Retrospective Appraisal* of the system in the light of operational experience, and if necessary,
4.3. *Improved Operation* of the system.

4.1. *Initial operation*

Effective liaison between the systems team and the users of the system is essential if the full benefits of the systems study are to be realized. In the writer's experience, this is the part of the exercise which is least well done in practice. The process of handing over is helped if:

(1) adequate documentation and training have been provided in advance,

(2) one of the users of the operational system has been involved as a member of the systems team, so that he is fully conversant with the design philosophy,

(3) adequate communication is maintained between the systems team and the users so that any defects and misunderstandings can be cleared up without delay.

Difficulties can usually be expected at the *start-up* of the system. However, good systems thinking will ensure that these upsets are minimized and that the operation works smoothly according to plan.

4.2. *Retrospective appraisal of the project*

After the new system has been operating for a sufficiently long period, the systems team should collaborate with the users of the system in making a retrospective appraisal of system performance. If the system is shown to be working according to plan, or better, everyone will be happy and the original systems thinking will have been vindicated. On the other hand, if system performance is not up to expectation, a postmortem will be needed. Nothing but harm, both to himself and to the systems approach, will be achieved if the systems engineer regards his job as finished when the original systems study has been completed. He must be prepared to accept full responsibility for the successful operation of the system and to identify himself with its success or failure. The retrospective appraisal may show that

(1) the original system study has overlooked certain factors, or

(2) the system has had to operate in an environment which is different from that for which it was designed. For example, a chemical plant may have had to switch to feedstocks of different quality, or the plant may have to operate at throughput in excess of design, and so on.

In either of the situations (1) or (2), a certain amount of reoptimization will be inevitable, however painful this may turn out to be. In the retrospective appraisal, a splendid opportunity arises of checking the economic assessment that was made at the design stage and hence to check the systems thinking. For example, in a computer control scheme, a list should be made of all the benefits, both tangible and intangible, which have been achieved. This is usually a very worthwhile exercise despite the fact that some of the benefits may be difficult to assess.

4.3. *Improved operation*

Improved operation of the system may be needed

(i) if the retrospective appraisal of the system shows that actual performance is falling short of plan, or

(ii) in certain cases, because certain parameters can only be optimized when operational experience has been obtained.

The system must then be "tuned" to its environment, leading to improved operation. If the retrospective appraisal shows that reoptimization is necessary, the systems engineering work at the design stage may have to be re-examined and modified. If the original systems study was conducted in an orderly way and well documented, this should not prove difficult. However, if this was not well done and people have left the company in the meantime, chaos may result – Joe Bloggs will have disappeared into thin air taking away in his head all the details of the system!

Finally, if redesign or retuning of the system is necessary, a decision will have to be made as to whether the improved system performance is acceptable.

Summary of stages in the application of Systems Engineering

The various stages in a Systems Approach are summarized below in the form of certain key questions which need to be answered.

1. *Systems Analysis*

1.1. *Formulation of the problem.* What is the problem? How did it arise? Who believes it to be important? Why is it important? Is it the right problem anyway? Will it save money? Is there the prospect of a reasonable rate of return on systems effort to justify tackling it as compared with some other problem?

1.2. *Organization of project.* What is the best composition for the systems team? Are its terms of reference sufficiently wide? When is the project to be completed? Has a preliminary project schedule been constructed? Have duties been allocated so that a more detailed critical path schedule can be set up?

1.3. *Definition of the system.* What is the precise nature of the system being studied? How is it best broken down into convenient sub-systems? What are the interactions between these sub-systems? Has a flow-block diagram been constructed?

1.4. *Definition of the wider system.* What is the environment (commercial, social, political) in which the system will have to operate? How does the system fit into the flow-block diagram of the wider system?

1.5. *Definition of the objectives of the wider system.* Have *competitive* and *senior* systems been properly taken into account? How do they influence the objectives of the system being studied? Is there a danger of sub-optimization?

1.6. *Definition of the objectives of the system.* Has a list of objectives in order of importance been drawn up? Have the constraints been listed? Are the objectives simple and direct? Are there some subjective features which are difficult to quantify? Have the objectives been agreed? Have they been communicated to all concerned with the design of the system?

1.7. *Precise definition of economic criterion.* Have conflicting objectives been properly weighted? Are the constraints reasonable? Are both criterion and constraints simple and direct? Have they been clearly agreed and accepted, even if qualitative?

1.8. *Information and data collection.* Have all important persons and sources of data been interrogated? Has all relevant data been assembled and presented in the best way?

2. *Systems Design*

2.1. *Forecasting.* Has all relevant data been used to forecast system environments? How accurate are these forecasts?

2.2. *Model building and simulation.* What type of model is best suited to the purposes at hand? Is the model building concentrated where it is most needed? Does the simulation performance agree sufficiently well with actual or conjectured performance for optimization to proceed?

2.3. *Optimization.* What optimization technique should be used? Have the performance sensitive variables been isolated? Have plots of the economic criterion with respect to the main variables been prepared? Has the sensitivity of the design to assumptions been checked? Would a risk analysis help?

2.4. *Control.* How can the optimized conditions be realized most economically as part of the overall design by installing a control system? Where should control be exercised and what instrumentation is needed? What type of control is required? How sophisticated should it be?

2.5. *Reliability.* Has the effect of uncertainty on system reliability been properly taken into account? Has the unreliability been reduced to an acceptable level?

3. *Implementation*

3.1. *Documentation and sanction approval.* Have the conclusions been agreed? Is the report well written so as to make the maximum impact? Has a time-table been prepared for implementation? Have the users of the system been fully involved and briefed?

3.2. *Construction.* Is the hardware and/or software fully integrated into the system specification? Have the users of the system been kept in touch with the philosophy underlying the design? Has a critical path schedule been drawn up for building and installation?

4. *Operation*

4.1. *Initial operation.* Have adequate plans been made for start up? Is there effective liaison with the users so that the system can be handed over? Have all the operational snags been ironed out?

4.2. *Retrospective appraisal.* Is the actual performance as predicted at the design stage? If not, why not? Has the retrospective appraisal been adequately documented?

4.3. *Improved operation.* Does the system need reoptimizing either by retraining or redesign? How is this best achieved? Finally, is the resulting improved operation now adequate?

Role of the Systems Engineer

A great deal of the work described above will involve people other than systems engineers. In this interdisciplinary approach, it is useful to highlight the *key roles* played by the systems engineer.

(1) He tries to distinguish the wood from the trees – what's it all about?

(2) He stimulates discussion about objectives – obtains agreement about objectives.

(3) He communicates the finally agreed objectives to all concerned so that their co-operation can be relied upon.

(4) He always takes an overall view of the project and sees that techniques are used sensibly.

(5) By his overall approach, he ties together the various specializations needed for model building.

(6) He decides carefully when an activity stops.

(7) He asks for more work to be done in areas which are sensitive to cost.

(8) He challenges the assumptions on which the optimization is based.

(9) He sees that the project is planned to a schedule, that priorities are decided, tasks allocated, and above all that the project is finished on time.

(10) He takes great pains to explain carefully what the systems project has achieved, and presents a well-argued and well-documented case for implementation.

(11) He ensures that the users of the operational system are properly briefed and well trained.

(12) He makes a thorough retrospective analysis of system performance.

References

[1] Jenkins, G. M. (1967) "Systems and their Optimisation", Inaugural lecture, University of Lancaster publication.

[2] Andrew, S. M. (1969) "Computer Modelling and Optimisation in the Design of a Complete Chemical Process", *Trans. Inst. Chem. Eng.* **47** (4), T79–T84.

[3] Jenkins, G. M. and Youle, P. V. (1968) "A Systems Approach to Management", *Opl. Res. Quart.* **19**, 5.

[4] Jenkins, G. M. (1968) "Control as Part of a Wider Systems Philosophy", *Meas. and Contr.* **1**, 105.

[5] Stephenson, G. G. (1965) "Production Optimisation at I.C.I.'s Complex at Wilton", *Chem. Proc.* **11**, 48.

[6] Taylor, A. W. and Youle, P. V. (1969) "The Optimisation of a Large Petrochemical Complex", *Jour. Syst. Eng.* **1**, 55.

[7] Jenkins, G. M. (1969) "A Systems Study of a Petrochemical Plant", *Jour. Syst. Eng.* **1**, 99.

[8] Schlager, K. J. (1956) "Systems Engineering – Key to Modern Development", *IRE Trans., Prof. Group Eng. Man.* **3**, 64.

[9] Schlager, K. J. (1957) "Organisation for Effective Systems Engineering", *Problems and Practices in Engineering Management*, Amer. Man. Assoc. Special Report **24**, 115.

[10] Goode, H. H. and Machol, R. E. (1959) *Systems Engineering*, McGraw-Hill, New York.

[11] Engstrom, E. W. (1957) "Systems Engineering – A Growing Concept", *Elec. Eng.* **76**, 113.

[12] Morton, J. A. (1959) "Integration of Systems Engineering with Component Development", *Elec. Man.* **64**, 85.

[13] Flagle, C. D., Huggins, W. H. and Roy, R. H. (eds.) (1960) *Operations Research and Systems Engineering*, Johns Hopkins Press, Baltimore.

[14] Williams, T. J. (1961) *Systems Engineering in the Process Industries*, McGraw-Hill, New York.

[15] Williams, T. J. (1962) "Systems Engineering a Large Chemical Plant Complex", *Elec. Eng.*, **82**, 590, August.

[16] Levine, S. and Buegler, R. J. (1962) "Large-Scale Systems Engineering for Airline Reservations", *Elec. Eng.*, **82**, 604, August.

[17] Chadwick, W. L. (1962) "Systems Engineering for Automation of a Large Power Station", *Elec. Eng.*, **82**, 598, August.

[18] Dommasch, D. O. and Landeman, C. W. (1962) *Principles Underlying Systems Engineering*, Pitman Publish. Corp., New York.

[19] Hall, A. D. (1962) *A Methodology for Systems Engineering*, Van Nostrand, Princeton, N.J.

[20] Feigenbaum, D. S. (1963) "Systems Engineering – A Major New Technology", *Ind. Qual. Con.* **20**, 9.

[21] Johnson, R. A., Kast, F. E. and Rosenzweig, J. E. (1963) *The Theory and Management of Systems*, McGraw-Hill, New York.

[22] Affel, H. A. (1964) "Systems Engincering", *Intern. Sci. Technol.* **35**, 18.

[23] Rosenstein, A. B. (1964) "Systems Engineering and Modern Engineering Design", *Amer. Ceram. Soc. Symposium Proc.* (Chicago, Ill., April 19), Miscellaneous Publication 267, Nat. Bur. Stand., Washington, D.C.

[24] Chestnut, H. (1965) *Systems Engineering Tools*, John Wiley, New York.

[25] Machol, R. E. (1965) *Systems Engineering Handbook*, McGraw-Hill, New York.

[26] Savas, E. S. (1965) *Computer Control of Industrial Processes*, McGraw-Hill.

[27] Chestnut, H. (1967) *Systems Engineering Methods*, John Wiley, New York.

[28] Jeffreys, T. O. (1967) "Nuclear Power and the Systems Engineer", Inaugural Lecture, University of Swansea Publication.

[29] Kline, M. B. and Lifson, M. W. (1968) "Systems Engineering", Chapter 2 of *Cost Effectiveness*, J. M. English (ed.), John Wiley, New York.

[30] Kahn, H. and Mann, I. (1957) *Techniques of Systems Analysis*, R. M. – 1829 – 1, The Rand Corporation.

[31] Adelson, M. (1966) "The System Approach – a Perspective", *Systems Development Corporation Magazine*, **9** (10), 1.

[32] Hitch, C. (1953) "Sub-optimisation in Operations Problems", *Jour. Op. Res. Soc. Am.* **1**, 87.

[33] Anthony, R. N. (1965) *Planning and Control Systems: A Framework for Analysis*, Harvard Grad. School of Bus. Admin., Boston, Mass.

[34] Schoderbek P. (ed.) (1967) *Management Systems*, John Wiley, New York.

Appendix

Techniques useful in Systems Engineering

The following table lists, alongside the various stages in the development of a system project, the techniques which are most relevant at each stage. The final column in the table gives a selected list of references which describe these techniques – note that the numbering refers to a selected

Stage	Technique	Reference
1. *Systems Analysis*		
1.1. Formulation of the problem	Critical Examination	1, 2
1.2. Organization of the project	Critical path scheduling	3, 4, 5
1.3. Definition of the system		
1.4. Definition of the wider system		
1.5. Objectives of the wider system }	Corporate Planning	6, 7, 8
1.6. Objectives of the system }	Management by Objectives	9
1.7. Definition of overall economic criterion	Financial Practice and Economic Modelling	10
	Discounted cash flow techniques	11, 12
1.8. Information and data collection	Statistical techniques	13
2. *Systems Design*		
2.1. Forecasting	Short and medium term forecasting	14
	Technological forecasting	15, 16
2.2. Model Building	Empirical – steady state models	17, 18, 19
	Empirical – dynamic models	14
	Mechanistic – steady state models	20, 21
	Mechanistic – dynamic models	22
	Probabilistic models	23
	Economic Modelling	10
	Numerical Mathematics	24
Simulation	Statistical (Monte Carlo) Simulation	25, 26, 27, 28
	Deterministic Simulation	29
	Analogue Computer Simulation	30
2.3. Optimization	Linear Programming	31, 32
	Hill climbing	33, 34
	Variational methods (Dynamic programming)	35, 36
	Risk Analysis	37
	Decision Theory	38
2.4. Control	Technical Control	39
	Computer Control	40
	Statistical Control	14
	Production Control	41, 42
	Management Control	43, 44
2.5. Reliability	Reliability Techniques	45
3. *Implementation*		
3.1. Documentation and Sanction Approval	Report Writing	46
3.2. Construction	Critical path scheduling	3, 4, 5
4. *Operation*		
4.1. Initial operation }	Systems Analysis	
4.2. Retrospective appraisal }	Techniques	
4.3. Improved operation	Systems Design Techniques	

bibliography of techniques given at the end of this appendix and is quite distinct from the reference numbering used in the main body of the paper. Most of the references are to books, but in some cases, recent papers are included.

Finally, there is a brief reference to the literature on Operational Research, including a very valuable survey by P. G. Moore.

Selected Bibliography of Techniques

1 Currie, R. M. (1963) Chapter 7 of *Work Study*, Pitman, London (2nd ed.).

2 Whitmore, D. A. (1968) Chapter 5 of *Work Study and Related Management Sciences*, Heinemann, London.

3 Lambourne, S. (1967) *Network Analysis in Project Management*, Industrial and Commercial Techniques, Ltd. Handbook.

4 Battersby, A. (1964) *Network Analysis for Planning and Scheduling*, Macmillan, London.

5 Archibald, R. D. and Villoria, R. L. (1967) *Network based Management Systems (PERT/CPM)*, John Wiley, New York.

6 Argenti, J. (1968) *Corporate Planning*, George Allen and Unwin, London.

7 Jackson, A. S., Stephenson, G. G. and Townsend, E. C. (1968) "Financial Planning with a Corporate Financial Mode", *The Accountant*, Jan. 27th–Feb. 17th.

8 Spencer, R. S. (1966) "Modelling strategies for corporate growth. Paper presented at the Society for General Systems Research session at the conference of the American Association for the Advancement of Science, Washington, D.C., Dec. 27th.

9 Odiorne, G. S. (1965) *Management by Objectives*, Pitman, London.

10 Baumol, W. J. (1965) *Economic Theory and Operations Analysis*, Prentice-Hall, Englewood Cliffs, N.J. (2nd ed.).

11 Merrett, A. J. and Sykes, A. (1963) *The Finance and Analysis of Capital Projects*, Longmans, London.

12 English, J. M. (ed.) (1968) *Cost-Effectiveness, The Economic Evaluation of Engineered Systems*, John Wiley, New York.

13 Lowe, C. W. (1968) *Industrial Statistics, Vol. 1*, Business Books, London.

14 Box, G. E. P. and Jenkins, G. M. (in the press) *Time Series Analysis, Forecasting and Control*, Holden-Day, San Francisco.

15 Jantsch, E. (1967) *Technological Forecasting in Perspective*, O.E.C.D. Publication, Paris.

16 Quinn, J. B. (1967) "Technological Forecasting", *Harvard Bus. Rev.* **45**, 89.

17 Draper, N. R. and Smith H. (1966) *Applied Regression Analysis*, John Wiley, New York.

18 Davis, O. L. (ed.) (1963) *The Design and Analysis of Industrial Experiments*, Oliver and Boyd, Edinburgh (2nd ed.).

19 Box, G. E. P. and Draper, N. R. (1969) *Evolutionary Operation*, John Wiley.

20 Himmelblan D. M. and Bischoff, K. B. (1968) *Process Analysis and Simulation*, John Wiley, New York.

21 Hunter, W. G., Kittrell, J. R. and Mezaki, R. (1967) "Experimental Strategies for Mechanistic Model Building", *Trans. Inst. Chem. Eng.* **45**, 146.

22 Campbell, D. P. (1958) *Process Dynamics*, John Wiley, New York.

23 Feller, W. (1966) *Probability Theory and its Applications*, John Wiley, New York, Vol. 1, 1950, Vol. 2.

24 Conte, S. D. (1965) *Elementary Numerical Analysis*, McGraw-Hill, New York.

25 Tocher, K. D. (1964) *The Art of Simulation*, English Universities Press, London.

26 Naylor, T. H. *et al.* (1966) *Computer Simulation Techniques*, John Wiley, New York.

27 *Bibliography on Simulation*, I.B.M. publication (Form 320-0924-0), White Plains, New York.

28 Lee, A. M. (1966) *Applied Queuing Theory*, Macmillan, London.

29 Sargent, R. W. H. (1968) "Developments in Computer Aided Process Design", *Chem. Eng. Prog.* **64** (4), 39.

30 Rogers, A. E. and Connolly, T. W. (1960) *Analogue Computers in Engineering Design*, McGraw-Hill, New York.

31 Beale, E. M. L. (1968) *Mathematical Programming in Practice*, Pitman, London.

32 Gass, S. I. (1958) *Linear Programming*, McGraw-Hill, New York.

33 Box, M. J., Davies, D., Swann, W. H. (1969) *Non-Linear Optimisation Techniques*, I.C.I. Monographs in Mathematics and Statistics, No. 5, Oliver and Boyd, Edinburgh.

34 Wilde, D. and Beightler, C. (1966) *Foundations of Optimisation*, Prentice-Hall.

35 Jacobs, O. L. R. (1967) *An Introduction to Dynamic Programming*, Chapman and Hall, London.

36 Rosenbrock, H. H. and Storey, C. (1966) *Computational Techniques for Chemical Engineers*, Pergamon, London.

37 Adelson, R. M. (1965) "Criteria for Capital Investment: An Approach Through Decision Theory", *Opl. Res. Quart.*, **16**, 19 .

38 Morris, W. T. (1964) *The Analysis of Management Decisions*, Irwin, Homewood, Ill.

39 Kuo, B. C. (1962) *Automatic Control Systems*, Prentice-Hall, Englewood Cliffs.

40 Savas, E. S. (1965) *Computer Control of Industrial Processes*, McGraw-Hill, New York.

41 Magee, J. F. (1967) *Production Planning and Inventory Control*, McGraw-Hill, New York, 1958 2nd ed. J. F. Magee and D. Boodman, McGraw-Hill, New York.

42 Forrester, J. W. (1961) *Industrial Dynamics*, M.I.T. Press, Cambridge, Mass. (and see Article 12).

43 Boyce, R. O. (1967) *Integrated Managerial Control*, Longmans, London.

44 Anthony, R. N. (1965) *Planning and Control Systems: A Framework for Analysis*, Harvard Grad. School of Bus. Admin., Boston, Mass.

45 Sandler, G. H. (1963) *Systems Reliability Engineering*, Prentice-Hall, Englewood Cliffs, N.J.

46 Cooper, B. M. (1969) *Writing Technical Reports*, Penguin Books Ltd., Harmondsworth.

Bibliography of Operations Research

1 *Philosophy*

(*a*) Rivett, B. H. P. and Ackoff, R. L. (1963) *The Manager's Guide to Operational Research*, John Wiley, New York.

(*b*) Rivett, B. H. P. (1968) *Concepts of Operational Research*, C. A. Watts, London.

2 *Techniques*

Sasieni, M. Yaspan, A. and Friedman, L. (1959) *Operational Research – Methods and Problems*, John Wiley, New York.

3 *Survey*

Moore, P. G. (1966) "A Survey of Operational Research", *Jour. Roy. Stat. Soc. A*, **129**, 399.

Reprinted from Jenkins, G. M. (1969) "The Systems Approach", *Journal of Systems Engineering*, **1**, 1.

5 Towards a system of systems concepts

by Russell L. Ackoff

The concepts and terms commonly used to talk about systems have not themselves been organized into a system. An attempt to do so is made here. System *and the most important types of system are defined so that differences and similarities are made explicit. Particular attention is given to that type of system of most interest to management scientists:* organizations. *The relationship between a system and its parts is considered and a proposition is put forward that all systems are either variety-increasing or variety-decreasing relative to the behavior of its parts.*

Introduction

The concept *system* has come to play a critical role in contemporary science.[1,2,3] (Churchman, Emery.) This preoccupation of scientists in general is reflected among Management Scientists in particular for whom the *systems approach* to problems is fundamental and for whom *organizations*, a special type of system, are the principal subject of study.

The systems approach to problems focuses on systems taken as a whole, not on their parts taken separately. Such an approach is concerned with total-system performance even when a change in only one or a few of its parts is contemplated because there are some properties of systems that can only be treated adequately from a holistic point of view. These properties derive from the *relationships* between parts of systems: how the parts interact and fit together. In an imperfectly organized system even if every part performs as well as possible relative to its own objectives, the total system will often not perform as well as possible relative to its objectives.

Despite the importance of systems concepts and the attention that they have received and are receiving, we do not yet have a unified or integrated set (i.e., a system) of such concepts. Different terms are used to refer to the same thing and the same term is used to refer to different things. This state is aggravated by the fact that the literature of systems research is widely dispersed and is therefore difficult to track. Researchers in a wide variety of disciplines and interdisciplines are contributing to the conceptual development of the systems sciences but these contributions are not as interactive and additive as they might be. Fred Emery[3] has warned against too hasty an effort to remedy this situation:

> It is almost as if the pioneers [of systems thinking], while respectfully noting each other's existence, have felt it incumbent upon themselves to work out their intuitions in their own language, for fear of what might be lost in trying to work through the language of another. Whatever the reason, the results seem to justify the standoffishness. In a short space of time there has been a considerable accumulation of insights into system dynamics that are readily translatable into different languages and with, as yet, little sign of divisive schools of thought that for instance marred psychology during the 1920s and 1930s. Perhaps this might happen if some influential group of scholars prematurely decide that the time has come for a common conceptual framework (p. 12).

Although I sympathize with Emery's fear, a fear that is rooted in a research perspective, as a teacher I feel a great need to provide my students with a conceptual framework that will assist them in absorbing and synthesizing this large accumulation of insights to which Emery refers. My intent is not to preclude further conceptual exploration, but rather to encourage it and make it more interactive and additive. Despite Emery's warning I feel benefits will accrue to systems research from an evolutionary convergence of concepts into a generally accepted framework. At any rate, little harm is likely to come from my effort to provide the beginnings of such a framework since I can hardly claim to be, or to speak for, "an influential group of scholars".

The framework that follows does not include all concepts relevant to the systems sciences. I have made an effort, however, to include enough of the key concepts so that building on this framework will not be as difficult as construction of the framework itself has been.

One final word of introduction. I have not tried to identify the origin or trace the history of each conceptual idea that is presented in what follows. Hence few credits are provided. I can only compensate for this lack of bibliographic bird-dogging by claiming no credit for any of the elements in what follows, only for the resulting system into which they have been organized. I must, of course, accept responsibility for deficiencies in either the parts or the whole.

Systems

1. A *system* is a set of interrelated elements. Thus a system is an entity which is composed of at least two elements and a relation that holds between each of its elements and at least one other element in the set. Each of a system's elements is connected to every other element, directly or indirectly. Furthermore, no subset of elements is unrelated to any other subset.

2. An *abstract system* is one all of whose elements are concepts. Languages, philosophic systems, and number systems are examples. *Numbers* are concepts but the symbols that represent them, *numerals*, are physical things. Numerals, however, are not the elements of a number system. The use of different numerals to represent the same numbers does not change the nature of the system.

In an abstract system the elements are created by defining and the relationships between them are created by assumptions (e.g., axioms and postulates). Such systems, therefore, are the subject of the so-called "formal sciences".

3. A *concrete system* is one at least two of whose elements are objects. It is only with such systems that we are concerned here. Unless otherwise noted, "system" will always be used to mean "concrete system".

In concrete systems establishment of the existence and properties of elements and the nature of the relationships between them requires research with an empirical component in it. Such systems, therefore, are the subject of study of the so-called "non-formal sciences".

4. The *state of a system* at a moment of time is the set of relevant properties which that system has at that time. Any system has an unlimited number of properties. Only some of these are relevant to any particular research. Hence those which are relevant may change with changes in the purpose of the research. The values of the relevant properties constitute the state of the system. In some cases we may be interested in only two possible states (e.g., off and on, or awake and asleep). In other cases we may be interested in a large or unlimited number of possible states (e.g., a system's velocity or weight).

5. The *environment of a system* is a set of elements and their relevant properties, which elements are not part of the system but a change in any of which can produce* a change in the state of the system. Thus a system's environment consists of all variables which can affect its state. External elements which affect irrelevant properties of a system are not part of its environment.

6. The *state of a system's environment* at a moment of time is the set of its relevant properties at that time. The state of an element or subset of elements of a system or its environment may be similarly defined.

Although concrete systems and their environments are *objective* things, they are also *subjective* insofar as the particular configuration of elements that form both is dictated by the interests of the researcher. Different observers of the same phenomena may conceptualize them into different systems and environments. For example, an architect may consider a house together with its electrical, heating, and water systems as one large system. But a mechanical engineer may consider the heating system as a system and the house as its environment. To a social psychologist a house may be an environment of a family, the system with which he is concerned. To him the relationship between the heating and electrical systems may be irrelevant, but to the architect it may be very relevant.

The elements that form the environment of a system and the environment itself may be conceptualized as systems when they become the focus of attention. Every system can be conceptualized as part of another and larger system.

Even an abstract system can have an environment. For example, the metalanguage in which we describe a formal system is the environment of that formal system. Therefore logic is the environment of mathematics.

7. A *closed system* is one that has no environment. An *open system* is one that does. Thus a closed system is one which is conceptualized so that it has no interaction with any element not contained within it; it is completely self-contained. Because systems researchers have found such conceptualizations of relatively restricted use, their attention has increasingly focused on more complex and "realistic" open systems. "Openness" and "closedness" are simultaneously properties of systems and our conceptualizations of them.

Systems may or may not change over time.

8. A system (or environmental) *event* is a change in one or more structural properties of the system (or its environment) over a period of time of specified duration; that is, a change in the structural state of the system (or environment). For example, an event occurs to a house's lighting system when a fuse blows, and to its environment when night falls.

9. A *static* (*one-state*) *system* is one to which no events occur. A table, for example, can be conceptualized as a

* One thing (x) can be said to produce another (y) in a specified environment and time interval if x is a necessary but not a sufficient condition for y in that environment and time period. Thus a producer is a "probabilistic cause" of its product. Every producer, since it is not sufficient for its product, has a coproducer of that product (e.g., the producer's environment).

static concrete system consisting of four legs, top, screws, glue, and so on. Relative to most research purposes it displays no change of structural properties, no change of state. A compass may also be conceptualized as a static system because it virtually always points to the Magnetic North Pole.

10. A *dynamic* (*multi-state*) *system* is one to which events occur, whose state changes over time. An automobile which can move forward or backward and at different speeds is such a system, or a motor which can be either off or on. Such systems can be conceptualized as either open or closed; closed if its elements react or respond only to each other.

11. A *homeostatic system* is a static system whose elements and environment are dynamic. Thus a homeostatic system is one that retains its state in a changing environment by internal adjustments. A house that maintains a constant temperature during changing external temperatures is homeostatic. The behavior of its heating subsystem makes this possible.

Note that the same object may be conceptualized as either a static or dynamic system. For most of us a building would be thought of as static, but it might be taken as dynamic by a civil engineer who is interested in structural deformation.

System Changes

12. A *reaction* of a system is a system event for which another event that occurs to the same system or its environment is sufficient. Thus a reaction is a system event that is deterministically caused by another event. For example, if an operator's moving a motor's switch is sufficient to turn that motor off or on, then the change of state of the motor is a reaction to the movement of its switch. In this case, the turning of the switch may be necessary as well as sufficient for the state of the motor. But an event that is sufficient to bring about a change in a system's state may not be necessary for it. For example, sleep may be brought about by drugs administered to a person or it may be self-induced. Thus sleep may be determined by drugs but need not be.

13. A *response* of a system is a system event for which another event that occurs to the same system or to its environment is necessary but not sufficient; that is, a system event produced by another system or environmental event (the *stimulus*). Thus a response is an event of which the system itself is a coproducer. A system does not have to respond to a stimulus, but it does have to react to its cause. Therefore, a person's turning on a light when it gets dark is a response to darkness, but the light's going on when the switch is turned is a reaction.

14. An *act* of a system is a system event for the occurrence of which no change in the system's environment is either necessary or sufficient. Acts, therefore, are self-determined events, autonomous changes. Internal changes – in the states of the system's elements – are both necessary and sufficient to bring about action. Much of the behavior of human beings is of this type, but such behavior is not restricted to humans. A computer, for example, may have its state changed or change the state of its environment because of its own program.

Systems all of whose changes are reactive, responsive, or autonomous (active) can be called reactive, responsive, or autonomous (active), respectively. Most systems, however, display some combination of these types of change.

The classification of systems into reactive, responsive, and autonomous is based on consideration of what brings about changes in them. Now let us consider systems with respect to what kind of changes in themselves and their environments their reactions, responses, and actions bring about.

15. A system's *behavior* is a system event(s) which is either necessary or sufficient for another event in that system or its environment. Thus behavior is a system change which initiates other events. Note that reactions, responses, and actions may themselves constitute behavior. Reactions, responses, and actions are system events *whose antecedents are of interest*. Behavior consists of system events *whose consequences are of interest*. We may, of course, be interested in both the antecedents and consequences of system events.

Behavioral Classification of Systems

Understanding the nature of the classification that follows may be aided by Table 1 in which the basis for the classification is revealed.

16. A *state-maintaining system* is one that (1) can react in only one way to any one external or internal event but (2) it reacts differently to different external or internal events, and (3) these different reactions produce the same external or internal state (outcome). Such a system only reacts to changes; it cannot respond because what it does is completely determined by the causing event. Nevertheless it can be said to have the *function* of maintaining the state it produces because it can produce this state in different ways under different conditions.

Thus a heating system whose internal controller turns it

Table 1 Behavioral Classification of systems

TYPE OF SYSTEM	BEHAVIOR OF SYSTEM	OUTCOME OF BEHAVIOR
State-Maintaining	Variable but determined (reactive)	Fixed
Goal-Seeking	Variable and chosen (responsive)	Fixed
Multi-Goal-Seeking and Purposive	Variable and chosen	Variable but determined
Purposeful	Variable and chosen	Variable and chosen

on when the room temperature is below a desired level, and turns it off when the temperature is above this level, is state-maintaining. The state it maintains is a room temperature that falls within a small range around its setting. Note that the temperature of the room which affects the system's behavior can be conceptualized as either part of the system or part of its environment. Hence a state-maintaining system may react to either internal or external changes.

In general, most systems with "stats" (e.g. thermostats and humidistats) are state-maintaining. Any system with a regulated output (e.g., the voltage of the output of a generator) is also state-maintaining.

A compass is also state-maintaining because in many different environments it points to the Magnetic North Pole.

A state-maintaining system must be able to *discriminate* between different internal or external states to changes in which it reacts. Furthermore, as we shall see below, such systems are necessarily *adaptive*, but unlike goal-seeking systems they are not capable of learning because they cannot choose their behaviour. They cannot improve with experience.

17. A *goal-seeking system* is one that can respond differently to one or more different external or internal events in one or more different external or internal states and that can respond differently to a particular event in an unchanging environment until it produces a particular state (outcome). Production of this state is its goal. Thus such a system has a *choice* of behavior. A goal-seeking system's behavior is responsive, but not reactive. A state which is sufficient and thus deterministically causes a reaction cannot cause different reactions in the same environment.

Under constant conditions a goal-seeking system may be able to accomplish the same thing in different ways and it may be able to do so under different conditions. If it has *memory*, it can increase its efficiency over time in producing the outcome that is its goal.

For example, an electronic maze-solving rat is a goal-seeking system which, when it runs into a wall of a maze, turns right and if stopped again, goes in the opposite direction, and if stopped again, returns in the direction from which it came. In this way it can eventually solve any solvable maze. If, in addition, it has memory, it can take a "solution path" on subsequent trials in a familiar maze.

Systems with automatic "pilots" are goal-seeking. These and other goal-seeking systems may, of course, fail to attain their goals in some situations.

The sequence of behaviour which a goal-seeking system carries out in quest of its goal is an example of a process.

18. A *process* is a sequence of behavior that constitutes a system and has a goal-producing function. In some well-definable sense each unit of behavior in the process brings the actor closer to the goal which it seeks. The sequence of behavior that is performed by the electronic rat constitutes a maze-solving process. After each move the rat is closer (i.e., has reduced the number of moves required) to solve the maze. The metabolic process in living things is a similar type of sequence the goal of which is acquisition of energy or, more generally, survival. Production processes are a similar type of sequence whose goal is a particular type of product.

Process behavior displayed by a system may be either reactive, responsive, or active.

19. A *multi-goal-seeking* system is one that is goal-seeking in each of two or more different (initial) external or internal states, and which seeks different goals in at least two different states, the goal being determined by the initial state.

20. A *purposive system* is a multi-goal-seeking system the different goals of which have a common property. Production of that common property is the system's purpose. These types of system can pursue different goals but they do not select the goal to be pursued. The goal is determined by the initiating event. But such a system does choose the means by which to pursue its goals.

A computer which is programmed to play more than one game (e.g., tic-tac-toe and checkers) is multi-goal-seeking. What game it plays is not a matter of its choice, however; it is usually determined by an instruction from an external source. Such a system is also purposive because 'game winning' is a common property of the different goals it seeks.

21. A *purposeful system* is one which can produce the same outcome in different ways in the same (internal or external) state and can produce different outcomes in the same and different states. Thus a purposeful system is one which can change its goals under constant conditions; it selects ends as well as means and thus displays *will.* Human beings are the most familiar examples of such systems.

Ideal-seeking systems form an important subclass of purposeful systems. Before making their nature explicit we must consider the differences between goals, objectives, and ideals and some concepts related to them. The differences to be considered have relevance only to purposeful systems because only they can choose ends.

A system which can choose between different outcomes can place different values on different outcomes.

22. The *relative value of an outcome* that is a member of an exclusive and exhaustive set of outcomes, to a purposeful system, is the probability that the system will produce that outcome when each of the set of outcomes can be obtained with certainty. The relative value of an outcome can range from 0 to 1.0. That outcome with the highest relative value in a set can be said to be *preferred.*

23. The *goal* of a purposeful system in a particular situation is a preferred outcome that can be obtained within a specified time period.

24. The *objective* of a purposeful system in a particular situation is a preferred outcome that cannot be obtained within a specified period but which can be obtained over a longer time period. Consider a set of possible outcomes ordered along one or more scales (e.g., increasing speeds of travel). Then each outcome is closer to the final one than those which precede it. Each of these outcomes can be a goal in some time period after the "preceding" goal has been obtained, leading eventually to attainment of the last outcome, the objective. For example, a high-school freshman's goal in his first year is to be promoted to his second (sophomore) year. Passing his second year is a subsequent goal. And so on to graduation, which is his objective.

Pursuit of an objective requires an ability to change goals once a goal has been obtained. This is why such pursuit is possible only for a purposeful system.

25. An *ideal* is an objective which cannot be obtained in any time period but which can be approached without limit. Just as goals can be ordered with respect to objectives, objectives can be ordered with respect to ideals. But an ideal is an outcome which is unobtainable in practice, if not in principle. For example, an ideal of science is errorless observations. The amount of observer error can be reduced without limit but can never be reduced to zero. Omniscience is another such ideal.

26. An *ideal-seeking system* is a purposeful system which, on attainment of any of its goals or objectives, then seeks another goal and objective which more closely approximates its ideal. An ideal-seeking system is thus one which has a concept of "perfection" or the "ultimately desirable" and pursues it systematically; that is, in interrelated steps.

From the point of view of their output, six types of system have been identified: state-maintaining, goal-seeking, multi-goal-seeking, purposive, purposeful, and ideal-seeking. The elements of systems can be similarly classified. The relationship between (1) the behavior and type of a system and (2) the behavior and type of its elements is not apparent. We consider it next.

Relationships between systems and their elements

Some systems can display a greater variety and higher level of behavior than can any of their elements. These can be called *variety increasing.* For example, consider two state-maintaining elements, *A* and *B*. Say *A* reacts to a decrease in room temperature by closing any open windows. If a short time after *A* has reacted the room temperature is still below a specified level, *B* reacts to this by turning on the furnace. Then the system consisting of *A* and *B* is goal-seeking.

Clearly, by combining two or more goal-seeking elements we can construct a multi-goal-seeking (and hence a purposive) system. It is less apparent that such elements can also be combined to form a purposeful system. Suppose one element *A* can pursue goal G_1 in environment E_1 and goal G_2 in another environment E_2; and the other element *B* can pursue G_2 in E_1 and G_1 in E_2. Then the system would be capable of pursuing G_1 and G_2 in both E_1 and E_2 if it could select between the elements in these environments. Suppose we add a third (controlling) element which responds to E_1 by "turning on" either *A* or *B*, but not both. Suppose further that it turns on *A* with probability P_A where $0 < P_A < 1.0$ and turns on *B* with probability P_B where $0 < P_B < 1.0$. (The controller could be a computer that employs random numbers for this purpose.) The resulting system could choose both ends and means in two environments and hence would be purposeful.

A system can also show less variety of behavior and operate at a lower level than at least some of its elements. Such a system is *variety reducing.* For example, consider a simple system with two elements one of which turns lights on in a room whenever the illumination in that room drops below a certain level. The other element turns the lights off whenever the illumination exceeds a level that is lower than that provided by the lights in the room. Then the lights will go off and on continuously. The system would not be state-maintaining even though its elements are.

A more familiar example of a variety-reducing system can be found in those groups of purposeful people (e.g., committees) which are incapable of reaching agreement and hence of taking any collective action.

A system must be either variety-increasing or variety-decreasing. A set of elements which collectively neither increase nor decrease variety would have to consist of identical elements either only one of which can act at a time or in which similar action by multiple units is equivalent to action by only one. In the latter case the behavior is non-additive and the behavior is redundant. The relationships between the elements would therefore be irrelevant. For example, a set of similar automobiles owned by one person do not constitute a system because he can drive only one at a time and which he drives makes no difference. On the other hand a radio with two speakers can provide stereo sound; the speakers each do a different thing and together they do something that neither can do alone.

Adaptation and Learning

In order to deal with the concepts "adaptation" and "learning" it is necessary first to consider the concepts "function" and "efficiency".

27. The *function(s)* of a system is production of the outcomes that define its goal(s) and objective(s). Put another way, suppose a system can display at least two structurally different types of behavior in the same or different environments and that these types of behavior produce the same kind of outcome. Then the system can be said to have the function of producing that outcome. To function, therefore, is to be able to produce the same outcome in different ways.

Let C_i ($1 \leq i \leq m$) represent the different actions available to a system in a specific environment. Let P_i represent the probabilities that the system will select these courses of action in that environment. If the courses of action are exclusive and exhaustive, then $\sum_{i=1}^{m} P_i = 1.0$. Let E_{ij} represent the probability that course of action C_i will produce a particular outcome O_j in that environment. Then:

28. The *efficiency* of the system with respect to an outcome O_j which it has the function of producing is $\sum_{i=1}^{m} P_i E_{ij}$.

Now we can turn to "adaptation".

29. A system is *adaptive* if, when there is a change in its environmental and/or internal state which reduces its efficiency in pursuing one or more of its goals which define its function(s), it reacts or responds by changing its own state and/or that of its environment so as to increase its efficiency with respect to that goal or goals. Thus adaptiveness is the ability of a system to modify itself or its environment when either has changed to the system's disadvantage so as to regain at least some of its lost efficiency.

The definition of "adaptive" implies four types of adaptation:

29.1. *Other-other adaptation:* A system's reacting or responding to an external change by modifying the environment (e.g., when a person turns on an air conditioner in a room that has become too warm for him to continue to work in).

29.2. *Other-self adaptation:* A system's reacting or responding to an external change by modifying itself (e.g., when the person moves to another and cooler room).

29.3. *Self-other adaptation:* A system's reacting or responding to an internal change by modifying the environment (e.g., when a person who has chills due to a cold turns up the heat).

29.4. *Self-self adaptation:* a system's reacting or responding to an internal change by modifying itself (e.g., when that person takes medication to suppress the chills). Other-self adaptation is most commonly considered because it was this type with which Darwin was concerned in his studies of biological species as systems.

It should now be apparent why state-maintaining and higher systems are necessarily adaptive. Now let us consider why nothing lower than a goal-seeking system is capable of learning.

30. To *learn* is to increase one's efficiency in the pursuit of a goal under unchanging conditions. Thus if a person increases his ability to hit a target (his goal) with repeated shooting at it, he learns how to shoot better. Note that to do so requires an ability to modify one's behavior (i.e., to display choice) and memory.

Since learning can take place only when a system has a choice among alternative courses of action, only systems that are goal-seeking or higher can learn.

If a system is repeatedly subjected to the same environmental or internal change and increases its ability to maintain its efficiency under this type of change, then it *learns how to adapt*. Thus adaptation itself can be learned.

Organizations

Management Scientists are most concerned with that type of system called "organizations". Cyberneticians, on the other hand, are more concerned with that type of system called "organisms", but they frequently treat organizations as though they were organisms. Although these two types of system have much in common, there is an important difference between them. This difference can be identified once "organization" has been defined. I will work up to its definition by considering separately each of what I consider to be its four essential characteristics.

(1) An organization is a purposeful system that contains at least two purposeful elements which have a common purpose.

We sometimes characterize a purely mechanical system as being well organized, but we would not refer to it as an "organization". This results from the fact that we use "organize" to mean, "to make a system of", or, as one dictionary puts it, "to get into proper working order", and "to arrange or dispose systematically". Wires, poles, transformers, switchboards, and telephones may constitute a communication system, but they do not constitute an organization. The employees of a telephone company make up the organization that operates the telephone system. Organization of a system is an activity that can be carried out only by purposeful entities; to be an organization a system must contain such entities.

An aggregation of purposeful entities does not constitute an organization unless they have at least one common purpose: that is, unless there is some one or more things that they all want. An organization is always organized around this common purpose. It is the relationships between what the purposeful elements do and the pursuit of their common purpose that give unity and identity to their organization.

Without a common purpose the elements would not work together unless compelled to do so. A group of unwilling prisoners or slaves can be organized and forced to do something that they do not want to do, but if so they do not constitute an organization even though they may form a system. An organization consists of elements that have and can exercise their own wills.

(2) An organization has a functional division of labor in pursuit of the common purpose(s) of its elements that define it.

Each of two or more subsets of elements, each containing one or more purposeful elements, is responsible for choosing from among different courses of action. A choice from each subset is necessary for obtaining the common purpose. For example, if an automobile carrying two people stalls on a highway and one gets out and pushes while the other sits in the driver's seat trying to start it when it is in motion, then there is a functional division of labor and they constitute an organization. The car cannot be started (their common purpose) unless both functions are performed.

The classes of courses of action and (hence) the subsets of elements may be differentiated by a variety of types of characteristics; for example:

(*a*) by *function* (e.g., production, marketing, research, finance, and personnel, in the industrial context),
(*b*) by *space* (e.g., geography, as territories of sale offices),
(*c*) by *time* (e.g., waves of an invading force).

The classes of action may, of course, also be defined by combinations of these and other characteristics.

It should be noted that individuals or groups in an organization that *make* choices need not *take* them: that is, carry them out. The actions may be carried out by other persons, groups, or even machines that are controlled by the decision makers.

(3) The functionally distinct subsets (parts of the system) can respond to each other's behavior through observation or communication.*

In some laboratory experiments subjects are given interrelated tasks to perform but they are not permitted to observe or communicate with each other even though they are rewarded on the basis of an outcome determined by their collective choices. In such cases the subjects are *unorganized.* If they were allowed to observe each other or to communicate with each other they could become an organization. The choices made by elements or subsets of an organization must be capable of influencing each other, otherwise they would not even constitute a system.

(4) At least one subset of the system has a system-control function.

This subset (or subsystem) compares achieved outcomes with desired outcomes and makes adjustments in the behavior of the system which are directed toward reducing the observed deficiencies. It also determines what the desired outcomes are. The control function is normally exercised by an executive body which operates on a feed-back principle. "Control" requires elucidation.

31. An element or a system *controls* another element or system (or itself) if its behavior is either necessary or sufficient for subsequent behavior of the other element or system (or itself), and the subsequent behavior is necessary or sufficient for the attainment of one or more of its goals.

* In another place, Ackoff [1], I have given operational definitions of "observation" and "communication" that fit this conceptual system. Reproduction of these treatments would require more space than is available here.

Summarizing, then, an "organization" can be defined as follows:

32. An *organization* is a purposeful system that contains at least two purposeful elements which have a common purpose relative to which the system has a functional division of labor; its functionally distinct subsets can respond to each other's behavior through observation or communication; and at least one subset has a system-control function.

Now the critical difference between organisms and organizations can be made explicit. Whereas both are purposeful systems, organisms do not contain purposeful elements. The elements of an organism may be state-maintaining, goal-seeking, multi-goal-seeking, or purposive; but not purposeful. Thus an organism must be variety increasing. An organization, on the other hand, may be either variety increasing or decreasing (e.g., the ineffective committee). In an organism only the whole can display will; none of the parts can.

Because an organism is a system that has a functional division of labor it is also said to be "organized". Its functionally distinct parts are called "organs". Their functioning is necessary but not sufficient for accomplishment of the organism's purpose(s).

Conclusion

Defining concepts is frequently treated by scientists as an annoying necessity to be completed as quickly and thoughtlessly as possible. A consequence of this disinclination to define is often research carried out like surgery performed with dull instruments. The surgeon has to work harder, the patient has to suffer more, and the chances for success are decreased.

Like surgical instruments, definitions become dull with use and require frequent sharpening and, eventually, replacement. Those I have offered here are not exceptions.

Research can seldom be played with a single concept; a matched set is usually required. Matching different researches requires matching the sets of concepts used in them. A scientific field can arise only on the base of a system of concepts. Systems science is not an exception. Systems thinking, if anything, should be carried out systematically.

References

1 Ackoff, R. L. (1967) *Choice, Communication, and Conflict*, a report to the National Science Foundation under Grant GN-389, Management Science Center, University of Pennsylvania, Philadelphia.

2 Churchman, C. W. (1968) *The Systems Approach*, Delacorte Press, New York.

3 Emery, F. E. (1969) *Systems Thinking*, Penguin Books Ltd., Harmondsworth, Middlesex, England.

Reprinted from Ackoff, R. L. (1971) "Towards a system of systems concepts", *Management Science*, **17**, 11.

Section II

Introduction

The demands made upon men by society and technology are constantly increasing. Machines operate faster, organizations become large and more complex, simple manual control is no longer adequate to cope with industrial and military equipment. Since the Second World War there has been a steady growth in the study of the interactions and combinations of man and machine. Engineers and psychologists together with physiologists, anthropologists, and others have come to work together as human factors engineers, or engineering psychologists to tackle the problems which arise. For some time attention was directed to the smaller-scale problems, the design of easy-to-grip knobs and easy-to-read dials for example, but latterly it has been realized that there is a larger, systems aspect to the design of even the small parts of a system.

Kenyon De Greene's article sets out the development of systems thinking in psychology, particularly in relation to the problems of engineering psychology. Many of the terms and ideas presented in earlier articles are repeated but with different examples. We have left these in to provide alternative explanations which we feel will assist in making some of the more difficult concepts clearer. The second article by George Bekey deals more specifically with problems of control by human operators and the design of displays and controls in relation to the whole system.

Associated reading

Several general books on Human Factors Engineering, or Ergonomics as it is sometimes called in Britain, are available which deal with the much wider field.

The more systems oriented approach which is slowly spreading is well expressed by Singleton "Psychological aspects of man machine systems". In: Warr, P. B. (ed.) "Psychology at Work"; Penguin, 1972, pp. 97–120.

A case study of a man-machine system is dealt with in *Air traffic control* Units 3 and 4 of *Systems Behaviour* T241 (Open University Press). Many aspects of man-machine systems are dealt with in the Open University's course *Systems Performance: Human Factors and Systems Failures*, (described on page 5).

6 Systems and psychology

by Kenyon B. De Greene

Introduction

It's a typical day. The car starts OK, but you think with a flash of irritation that it really shouldn't have been necessary to remove the engine just to replace the starter motor. At the stop signal, it seems the flow of cross traffic will never end. You've got to present a briefing to military higher management at the other end of the country this afternoon and you're not satisfied. The pieces just don't seem to fit together. Perhaps that combination of tranquilizers and sleeping pills had left you unduly grouchy, but your wife shouldn't have bugged you about spending so much time on your job and neglecting the kids and all. An hour to catch your plane and cars piled up on the freeway as far as the eye can see. Start, move a few feet, and stop, start. . . . You reflect on the events in this morning's paper. Big jet pancakes down in the ocean eight miles short of the runway, killing 147. More student and minority group riots. Danger of imminent starvation in Africa and Asia. Strikes curtail services in still another city. You're still worried about your briefing. If you only had a better gauge that the men (how many *are* really necessary) could really do the job in that environment. You glance over your left shoulder. Traffic going the other way isn't much better. Car piled with skis and boats and people headed for the mountains and desert, even though it's a weekday. Airport's a couple of miles ahead now, but you can't see the tower through the smog. Surely there ought to be a way of getting an overall grasp of things, things that should fit nicely together but always seem to be operating at cross purposes. People pretty well manage to foul things up. People. Human nature again. If only the headshrinkers. . . . You'll have to walk about half a mile to your plane from the parking place you were just able to ace away from that other guy. But no sweat, there's still time for a quick cup of coffee, and you'll get points from your brisk walk in that new exercise system. System! You remember a book you leafed through at the company book stall on the quad. There was sort of a catchy quotation: "When we try to pick out anything by itself, we find it hitched to everything else in the universe." McGraw-Hill as you recall. Right! *Systems Psychology*. Have to look into it when you get a chance and see what it's about. . . .

This book* is about people and how they relate to complex technology and its consequences – how they relate to machines, buildings, communications, roads, and one another. Because human behavior varies from one situation to another, it is also about how people relate to environments. You may already be familiar with the *Three-M* – man–machine–medium – concept of interaction and its more recent extension to include a fourth and fifth *M*, mission and management. In this book, we determine whether apparently diverse and unrelated problems can be investigated in a *general yet systematic* manner, a manner that at once provides a basis for both definition and solution of these problems. Of basic concern is the *effectiveness* of people and the missions to which they contribute. We look at the way people behave and the effects of various conditions on their behavior.

At the same time, we examine the capabilities and limitations of machines. We thus are concerned with the man–machine *interface* – how man and machine can complement one another most effectively in accomplishing some end. Optimum design of the man–machine package alone does not guarantee the effectiveness we desire, however, and our success in man–machine design, even optimized to meet the constraints of a physicochemical environment, may introduce problems that seem less easy to handle. Our concept of effectiveness must be extended to include the variables of individual need, reward, expectation, and attitude. Not all problem situations structured through engineering will involve the same *psychological factors*. Sometimes it is necessary to single out specific factors – for example, alertness, vigilance, or decision making – for special study and consideration. However, under operating conditions of, say, piloting an aircraft, vigilance and decision making are functions of the pilot's needs at a given moment.

* [i.e., *Systems Psychology*, *Ed.*]

Clearly, *interrelatedness* and *interaction* among things are important themes in systems science. As we attempt to lend order and meaning to complicated situations, we apply the framework concepts and methods of *systems*, represented along the two dimensions of *time* and *depth*, which characterize, respectively, the sequence of activities by which a system accomplishes a given *purpose*, and the amount of knowledge available and applicable at given times. It is necessary first to postulate a hierarchy of subwholes,* leading eventually to the level of the individual human being, who can conveniently be viewed as a "*black-box* component" that, by some poorly understood *transfer function*, converts *inputs* into *outputs*. Unraveling the mechanisms involved in the human transfer function is a primary ongoing responsibility of psychology and physiology. You will notice the engineering language used here: Black-box component can be translated into the more psychological *stimulus–organism–response* (*S–O–R*) *paradigm*, or model.

It is sometimes convenient to single out constructs such as perception, motivation, learning, thinking, and intelligence. But these constructs are interrelated and are, in turn, reflections of other, perhaps more basic processes and of the environment at any instant. This difficulty has long been evident in the field of accident investigation. Suppose a pilot has had a quarrel with his wife and later, on a routine flight, in good weather, collides with another aircraft. Is the "cause" of this accident perception, attention, emotion, poor judgment, or what? Let's say we attribute it to degraded attention. We must then ask: Was attention degraded because of poor equipment design, conflicting task demands, boredom, or preoccupation with marital difficulties? The point we must never forget is this: While it is necessary to identify an abstraction such as "vigilance" for study in the laboratory, the vigilance of one situation may have little predictive value for the vigilance in another situation.

This book† attempts to integrate several established practices in psychology and in engineering. Part I covers the sequence of processes of analysis, synthesis, and evaluation applicable to the engineering of all systems. Because we are concerned primarily with human behavior, matters of particular interest to psychology applied within the context of systems engineering are covered in greater detail; specifically, this is the field of *human factors*‡ or engineering psychology.

As viewed by the U.S. Air Force, the management of human factors within systems engineering involves development of partially sequential, partially parallel, but interrelated specialties. These *personnel subsystem elements* (see the preceding footnote and the last section of this chapter) are considered in Parts I and IV of this book. The concept of man as a constituent of man–machine systems is expressed n Part II, where we examine input, throughput, and output in terms of *information*, *decision*, and *control theory*, respectively. In Parts II, III, and V, we discuss perception, attention, cognitive processes, perceptual-motor behavior, individual differences, and motivation, usually in the context of a particular system operational requirement. The modifying effects of physicochemical and psychosocial environmental factors on behavior are discussed in Part V.

Specific system problem areas of salient concern to the psychologist, and not treated widely elsewhere, are considered in detail in Parts III and VI. Throughout the book, attempts are made to integrate psychology with other disciplines, to determine a common language for the intercommunication of ideas, and to develop a body of systems psychological methods applicable to the study of any system problem. Particular emphasis is placed on the recognition, definition, measurement, and, where possible, quantification of basic psychological factors; the identification of human capabilities and limitations; the relating of psychological factors to systems factors; the prediction of effective and ineffective human behavior; and the highlighting of situations in which failure to follow these procedures leads to degraded human and system performance, human frustration or misery, danger, waste of resources, and other unsatisfactory results. Applications include those in which psychologists have had extensive experience and those in which we urge far greater participation.

* This process introduces epistemological difficulties, discussed later, that are associated with terminology and the meaningfulness of constituents and abstractions. (Note the use of *constituents* rather than *units*, *components*, or *elements*.)

† [i.e., *Systems Psychology*, *Ed.*]

‡ For the benefit of our human factors readers who are not psychologists, we acknowledge the important contributions of physiologists, physicians, anthropologists, sociologists, engineers, and mathematicians to the field. Choice of either term – *human factors* or *engineering psychology* – is a matter of personal preference, in part reflecting the professional organization addressed (Human Factors Society or Society of Engineering Psychologists, a division of the American Psychological Association).

Here, human factors is considered a broader term, which includes training, manpower determinations, analysis, evaluation, equipment design, and so forth. On the other hand, engineering psychology can be equated most readily to human engineering – equipment, facilities, and environments designed for compatibility with human capabilities and limitations.

In the remainder of this chapter, we examine, in some detail, different ways of looking conceptually at systems, at psychology, and at psychology within the systems context. A number of dramatic examples are provided of failure to see things as systems, particularly with regard to psychological factors in systems. The chapter concludes with an examination of the practical aspects of systems engineering and management.

Systems science and psychology

Every educated person recognizes that a "system" imposes an order or consistency on similar interrelated constituents (for example, solar system, nervous system, tax-evasion system); yet there is no integrated body of system knowledge acceptable to the educated public as a whole. We will use the term *systems science* here in the most inclusive sense to include conceptual, theoretical, and applied developments.* There usually is no close relationship between the theoretical, as represented most typically in the *Yearbook of the Society for General Systems Research*, and the applied, as represented by the burgeoning advances in systems engineering and the derived systems management. In fact, it is often stated that applications lack valid theoretical underpinnings. This is perhaps as it should be for a field in the stage of initial rapid growth, but we should caution against theorizing apart from insightful interpretation of our experiences building systems, and conversely, continuing to implement outside of a conceptual *system for systems*. (Following popular practice, we use the term *systems*, instead of *system*, as a noun modifier.)

In this section, we examine various conceptual ways of approaching the study of systems and then attempt to integrate psychology and systems. Systems engineering and management methodology are discussed later in this chapter.

What is a system?

A look into a dictionary reveals that definition of the word *system* entails consideration of a set or arrangement, of relationship or connection, and of unity or wholeness. Further, the term has had longstanding use both as general methodology for achieving order and as a specific modifier in sciences such as astronomy, mathematics, chemistry, geology, and biology. In the most general sense, then, *system* can be thought of as synonymous with *order*; the opposite of chaos.

Experience has led to modification of this simple definition, in most cases in terms of *level* or *hierarchy*, *purpose*, and *environment*. Hall and Fagen (1956) provide the following definition:

> A system is a set of objects together with relationships between their attributes.

As Hall and Fagen see it, *objects* are simply the parts, or components of a system – for example, stars, atoms, neurons, switches, and mathematical laws. *Attributes* are the properties of objects – for example, the temperature of a star. *Relationships* tie the system components together, and which relationships are meaningful at a given time is a matter of discretion by the investigator. It is important to determine interconnections and dependencies, as well as the static or dynamic nature of the relationship.

The *environment* of a system has been defined by Hall and Fagen as follows:

> For a given system, the environment is the set of all objects a change in whose attributes affects the system and also those objects whose attributes are changed by the behavior of the system.

Subdivision of a universe into system and environment is obviously often quite arbitrary. Yet to specify completely an environment, one must know all the factors affecting or affected by the system. This is easier in the physical sciences than in the life, behavioral, and social sciences. Differentiation between system and environment is an immensely complex problem in the last two.

Any system can be further subdivided hierarchically into subsystems, which can in turn by subdivided into sub-subsystems, components, units, parts, and so forth (Figure 6.1 and Table 6.2).†

Objects that are parts of one system or subsystem can be considered parts of the environment of another system or subsystem. Also, systems may unite as subsystems of a still larger system, and under some conditions subsystems can be considered systems. Often the behavior of subsystems is not completely analogous to that of the system itself.

Systems may be studied at macroscopic or microscopic levels, depending on one's training, specialization, and philosophy. Analytic, "atomistic", "elementaristic", or "molecular" approaches versus "holistic" or "molar" approaches are discussed later.

* For example, a definition of systems science given by the Institute of Electrical and Electronic Engineers professional group in Systems Science and Cybernetics (see Rowe, 1965) is: *The scientific theory and methodology that is common to all large collections of interacting functional units that together achieve a defined purpose.*

† We discuss problems of terminology involving these subdivisions later in this section.

Many workers do not consider as systems those natural organizations or structures that lack *purpose*, where purpose is construed to be the discharge of some function. For example, minerals can be classified into one of six systems (halite into the cubic system and calcite into the hexagonal system); however, some authors (Gérardin, 1968) argue that crystals cannot be said to form a system, because they perform no function, are end-products in themselves, and do not change except by application of external force. For man-made systems, purpose or mission is an important, integral feature. A general definition of such a system might be: an assemblage of constituents (people and/or hardware and/or software) that interact to fulfill a common purpose transcending the individual purposes of the constituents.

Systems properties and types

A survey of the systems literature reveals a plethora of definitions, but there is almost universal acceptance of the following:

1. Basic system constituents (components, elements, parts, objects) may or may not be similar and possess peculiar characteristics (attributes, behaviors).
2. What constitutes a *basic* constituent is an arbitrary decision within a hierarchical arrangement and a function of one's specialization and the exigencies of the moment.
3. Upon incorporation into the system, constituents are modified through *interactions* with other constituents.
4. The system's characteristics are usually quantitatively greater than, and qualitatively different from, the inferred sum of the characteristics of the constituents.
5. The system exists within an environment, defined as a function of the hierarchical level chosen, that modifies the behavior of the system and may be modified by it: The *boundary* between system and environment should be clearly recognizable.
6. Some systems have a recognized purpose.

Systems are often considered to possess other, often interrelated properties (see, for example, von Bertalanffy, 1956; Hall and Fagen, 1956). The properties outlined below represent a good approximation of *first-order* properties from which it is possible to make second-order derivations, third-order derivations, and so on. For example, out of the disruption of equilibrium, we can observe what might be called *drive* or goal-direction, which in turn can lead to *competition*. Recognition of secondary and tertiary properties is particularly important in biological and social sciences.

From certain dominant properties, we can designate *types* of systems – for example, feedback control systems,* adaptive control systems, self-organizing systems, or information systems.

Equilibrium. Equilibrium may be static but is usually dynamic and occurs in concepts of chemistry, geology, biology, and other sciences. A familiar example is *homeostasis*.†

Change over time. Changes that occur over time (especially *growth* and *decay*) are important in almost all sciences.

Dominance or centralization. On subsystem may play a dominant role in the behavior of the system. With caution, we can offer the nervous and endocrine systems in vertebrates as examples.

Independence. The hierarchical nature of systems has been noted, as well as the arbitrary nature of system designation. However, there is some virtue in excluding from consideration as systems those entities that cannot exist independently. Thus, the nervous system cannot operate outside the body; the propulsion system of a spacecraft has no ongoing function outside that spacecraft.

Feedback. The system output is sampled, measured, and fed back to the input with subsequent modification, if necessary, of the output. Feedback is especially important to the control mechanisms of organisms and machines, and it provides one of the major bases for cybernetics, discussed later in this section. Feedback systems are typically called *closed-loop* systems. Systems without feedback are called *open-loop* systems. These terms should not be confused with *open* and *closed systems*, respectively (see below).

The type and amount of feedback is important to system stability and equilibrium. The terms *positive and negative feedback* are commonly used. Familiar machine examples of feedback are servomechanisms such as antiaircraft fire-control mechanisms, ship-steering mechanisms, and target-seeking guided missiles. In mammalian psychophysiology, increased secretions of adrenal cortical, thyroid, and gonadal hormones inhibit the secretion of the relevant anterior pituitary hormones in a negative-feedback loop; the release of epinephrine by the adrenal medulla in stress helps enhance the action of the sympathetic nervous system by a positive-feedback mechanism. Integration of control mechanisms in animal and machine is, of course, the main role of *cybernetics* discussed later.

* [See Article 7. *Ed.*] † [See Article 13. *Ed.*]

Entropy and information. In the strictest sense, entropy indicates the theoretical amount of energy (as in steam) in a thermodynamic system that cannot be transformed into mechanical work. It is a function of the probabilities of the states of gas particles. In a closed system, entropy must increase to a maximum with eventual cessation of the physical process and equilibrium. The concept of entropy has also been applied to information systems, where one can speak of "source entropy", "channel entropy", and the like. Entropy can be considered a measure of probability in that the most probable distribution is one of randomness or disorder. It is thus the opposite of information, which can be used as a measure of order in a system.

Open and closed systems. Open systems exchange information, energy, or materials with their environments. Biological systems are the best examples of open systems. One of the most important jobs of the biologist and psychologist is understanding the transfer function whereby inputs (information, energy, or materials) are converted into outputs. Many "test-tube" physical–chemical reactions occur in closed systems (see entropy above), which are considered isolated from their environments.

Differentiation. Related to growth and especially characteristic of biological systems, differentiation refers to the formation, over time, of new constituents from old. *Reproduction* is closely related to differentiation.

Adaptation. Adaptation refers to the ability to modify a system in accord with environmental demands to assure continued function. Individual learning and biological evolution are examples. There is tremendous interest in developing artificial systems possessing qualities of biological systems. Such systems are referred to as *self-regulating*, *self-adapting*, or *self-organizing*. The self-organizing system is described by Yovits *et al.* (1962) as one "which changes its basic structure as a function of its experience and environment".

Predetermined control. Predetermined control of a sequence of structural or functional changes is a consequence of information coding. It is characteristic of living systems in which control over reproduction, growth, differentiation, and behavior reflects information coded in all genes, in organizer tissues in embryological development, and in nervous tissue. Predetermined control is an important feature also of computerized systems. In some man–machine systems, termed *procedural* systems, it is possible to specify beforehand the sequence of operator tasks. The alternative is a *contingency* system.

Naturalness or artificiality. This might appear to be a simple dichotomy, but the distinction fades with man's ability to synthesize proteins and to design systems on the basis of natural bionic analogs.

Compatibility with the environment. The biological world provides myriad examples of organisms marvelously adapted to an environment or, conversely, extinct because the changing environment has passed them by. In a related sense, artificial systems may not function if the environment for which they were designed changes (*cf.* general-purpose and special-purpose computers). Similarly, we may find that one system does not interface well with another when we attempt to build a new system on an old base (the "grow like Topsy" approach) or to effect retrofits. The job of system integration indeed may be formidable in such cases.

Randomness. If constituents are assembled at random, the situation is chaos and a system cannot be said to exist. Yet all physical, biological, and social systems have random properties or functions. Vacuum-tube noise is considered due to random emission of electrons from the cathode. Many people believe that, at first, connections among neurons in the retina and brain may be purely random. There is evidence that random errors and accidents occur in complex systems.

Ways of looking at systems

Systems are observed, studied, and evaluated primarily to: (1) improve the system or its successor; (2) determine general theories or methods for new system development; and (3) advance science. Study of man-made, natural, and semi-natural systems more or less fulfills all three objectives, depending on the purposive or adventitious human contribution to the original "design".

A growing body of systems science methods is beginning to reconcile the conflicting definitions of what a system is and differences among systems. These methods include: (1) generalization across systems; (2) analysis and synthesis; and (3) modeling and simulation. These general approaches must be modified in terms of system level and definition of environment and the degree of practical relationship to the "real world".

Generalization across systems. Examination of the examples given in the earlier discussion of system properties reveals that different systems may have much in common. In several cases, workers in different specialties have arrived independently at similar concepts. The term *isomorphism* (*an* isomorphy) refers to structural similarities among systems in different fields. The concept of isomorphism suggests that the various fields of science can be united at basic levels through underlying principles. An analog based on only two variables – input and output – has the lowest degree of isomorphism, and the underlying function may be vastly dissimilar. Such an oversimplified model is most useful as a representation of a subsystem, which is then linked to other simplified subsystems. *The lack of precision and detail should not transcend the subsystems.* Isomorphism can also refer to structural–functional relationships between a living prototype and a model. Practical attempts to determine isomorphic properties of different systems are particularly spectacular in bionics.

Analysis and synthesis. Complexity in all systems can be approached by breaking the whole into simpler constituents – that is, by analysis. Often, however, there is reason to suspect the significance of the abstracted constituents when we attempt to synthesize them as a means of explaining and predicting the whole. This has long been a major problem within psychology, a most recent example of which is provided by efforts to predict human error.

Modeling and simulation. Models and simulations are analogies ranging from physical operating devices with definite shapes to block diagrams, figures, and computer programs (*abstract* or *mathematical models*). They aid in explaining natural phenomena. Whether present mathematics can be applied is a function of the extent to which a given system can be analyzed. Realistic models can be constructed fairly easily for physical systems but not for complete biological and social systems, in which real system relationships are obscure, the actual number of variables may not be known, and quantification poses a formidable problem. Modeling and simulation have an advantage for these systems, however, in that they allow detachment of the observer from the system he is studying and therefore reduce personal bias.

Selection of level and environment. All systems possess hierarchical structure: A system at one level may be considered a subsystem at another. Similarly, how much of the environment is included in the system helps determine the system properties. Knowledge gleaned at one level may have limited applicability to another level.

Relationship to the "real world". Even if a particular level and environment are chosen for study, only some of the properties of systems may find application in the study of real-world engineering problems. The systems engineer and manager are particularly concerned with interactions, processing, feedback, environmental compatibility, evolution and change, and purpose (mission); the systems theorist is more concerned with such factors as entropy and differentiation. Accordingly, we can recognize a "systems approach" that may be partially qualitative, even intuitive, and lack the theoretical basis of systems science.

Examples of systems

Over the years, systems have been classified in a number of ways on the basis of properties such as those discussed earlier. It is interesting to combine the properties in different ways or to describe definite entities in terms of relevant properties. A more formidable problem arises, however, when we try to achieve unanimity of agreement as to what *is* and *is not* a system. Most people will agree that a man is a system and that a submarine is a system, and that each can be characterized in terms of certain properties; most also will agree that a single perceptual-motor action is not a system and that an electron is not a system. On the other hand, many will disagree as to whether a crystal or a city is a system, primarily because of differing conceptions of *purpose*, *normal independence*, and *clearly recognizable boundary*. Ultimate resolution of these disparities *is* a responsibility of systems science, but for purposes of this discussion, we will consider some nonequivocal examples of systems arranged in order of complexity.

Wilson (1965) illustrates a simple mechanical system that shows rudimentary features of negative and positive feedback and exemplifies that the whole is greater than sum of the parts. This system has only five constituents: a pipe, a valve, a supportive arm, a spring, and a container. Its mission is to fill the container to a certain level and shut off the valve automatically. When the constituents are *properly connected* – that is, the fulcrum is placed near the container – the increasing weight of the liquid results in greater tension on the spring, which closes the valve. The system mission is faithfully discharged. Conversely, an improper connection of the constituents – moving the fulcrum to the middle of the pipe – precludes start of the mission. If the system is started in the middle of its operating cycle, the valve opens progressively wider as the container fills. The miller's grain-feeder (Gérardin, 1968), known by at least the eighteenth century, illustrates an early practical design of a similar mechanical system with aspects of automatic control.

A telephone network illustrates an intermediate level of complexity. It can be viewed at the level of a local exchange, a nationwide network, or a worldwide network connected by radio, undersea cable, and communications satellites.

Each succeeding level can be viewed as a system or a subsystem of a larger system. At different levels, specialty subsystems, such as central switching and direct dialing, can be recognized. Despite its immense number of constituents, a telephone network is not the most complicated of systems. Its several basic functions are relatively simple, straightforward, and generalizable from one level to the next, and thus are amenable to considerable automation.

Nearly the most complicated systems are the large-scale, computerized information-acquisition, -processing, and -display, control and command-and-control systems. In the broadest sense, these systems acquire radar, sonar, microwave, ionizing radiation, system status, biomedical, voice, and other data from a variety of terrestrial and solar system environments. Conversions of information, digital to analog, parallel to serial, and vice versa, and data compressions are almost always necessary. Almost every aspect of computer and display technology is relevant. Large numbers of personnel are required, sometimes in a great variety of types and skills. The many environmental stresses are both acute and chronic. Vehicles are controlled directly or indirectly and are themselves complex systems that may or may not cooperate with the system. This large systems category includes the various Air Force "L" systems, the manned spaceflight systems, and the naval control systems. The Semi-Automatic Ground Environment (SAGE) air defense system, the Air Force Satellite Control Facility, and the Apollo manned spacecraft system are considered in some detail throughout this book, because they exemplify system problem areas particularly well and because several of the authors have had experience with them.

The mission of SAGE is to detect, track, identify, intercept, and destroy enemy bomber aircraft. It has no antiballistic missile destruction capability. Major inputs are the dynamic position and speed data from radar and flight plans. Large, duplexed, digital computers compute aircraft tracks, determine identifications, calculate intercept points, etc. Various data are displayed on computer-related cathode-ray tubes. Enemy bombers are intercepted by manned aircraft or by Nike or Bomarc missiles. At one time, four-storied, duplexed SAGE Direction Center blockhouses were distributed over most of the contiguous United States and Canada. Now being phased out, SAGE is of particular interest because: (1) it can be viewed as the "granddaddy" of the electronic command and control systems, a laboratory of what was done correctly and incorrectly; (2) it exemplifies the long lead times between system conceptualization and system implementation; (3) in a related sense, it dramatizes the possibility that by the time a system is operational, its mission may be quite incidental – SAGE went into operation just as the enemy threat changed from "air-breathing" bomber to ballistic missile, a still unsolved problem; (4) almost no attention was paid at high engineering and management levels to human factors and other psychological problems; and (5) it contributed a great deal to the design of computerized systems – for example, the presently important concepts of man–computer interaction, time-sharing, and display buffer design owe much to SAGE.

The Air Force Satellite Control Facility (SCF) has evolved considerably since its inception. The general aspects of the system and its mission can be gleaned from an article by White (1963), although the system has changed in detail since White's publication appeared. The mission of the SCF is to track, receive, and process telemetry data, test and check out, and control satellites. The SCF does not launch satellites, although it monitors prelaunch checkout and launch, which is a responsibility of other agencies at Cape Kennedy and at Vandenberg Air Force Base in California. Once launched, the satellite is tracked by, and telemetry data are received from, a worldwide network of tracking stations. Raw data received by these tracking stations are processed, compressed if necessary, converted, and transmitted mostly via digital data link to the Satellite Test Center in Sunnyvale, near San Francisco, California. After analysis of the tracking and telemetry data, especially the latter, voice commands are sent to the tracking station, which transmits them in nonvoice form to the satellite, correcting its attitude and so on. This is an oversimplification, and data are not always easily transmitted or easily analyzed. For our purposes, the major lessons learned from the SCF are: (1) it is an outstanding example of system development wherein operation requirements came too fast, too heavily, and from too many separate users, without central planning – what had started out as a fairly simple research and development (R&D) effort for testing the Discoverer unmanned satellite within several years became a superimposed mass of satellite-support equipment, methods, and personnel; (2) it provides an example of managerial debate as to whether a system is "operational" or "R&D", raising questions as to the applicable type of management control; (3) it has long demonstrated a challenging number of human factors problems connected with the allocation of functions between man and computer, automation of other functions, information availability, diagnosis and troubleshooting, problem solving, display and control design, personnel numbers and training, and formulation of operational procedures.

The mission of Apollo 11 was to bring three American men into a lunar orbit, land two of them on the moon, bring all three together again, and return them to Earth. Apollo involves a marvelous integration of test and checkout, launch, tracking and telemetry, data processing and display, control and recovery capabilities. It is perhaps *par excellence* the example of successful planning and the management of thousands of contractors and tens of thousands of specialist workers to bring about the successful implementation of a mission. For our purposes, it is of particular interest because: (1) it demonstrates that sophisticated management of complex processes can lead to the solution of quite formidable *technical* problems; (2) it can serve as a type example for application to the sociotechnical area; (3) in all systems, unforeseen interactions can result in costly waste and in tragedy; and (4) it exemplifies the concept of system hierarchy. The "system" can be considered the complete spacecraft-launch vehicle assemblage plus the worldwide network of tracking stations, the launch and recovery facilities, the Integrated Mission Control Center (IMCC) in Houston, Texas, and the simulation and training facilities; or we could consider the "system" as comprising only the spacecraft itself, consisting of an Escape Module (Launch Escape System jettisoned shortly after earth launch), Service Module, Command Module, and Lunar Module.* During descent to and ascent from the moon, the system could consist of either the Lunar Module (containing two astronauts) or Command Module (containing one astronaut). During the return cislunar voyage, and reentry and earth recovery, the system is the combined Service-Command Modules and the Command Module, respectively. Of course, the unmanned spacecraft or an individual astronaut can also be considered the system.

Systems such as those discussed above are *nearly* the most complicated with which we must deal. What then is more complex? The entire universe? Probably not, if complexity may be defined in terms of the overall problems with which we must live – and survive. The universe is complex and inspiring, but *in toto* has little effect upon our everyday lives. The most complicated system, or system environment, can be delineated as comprising the earth, Earth's moon, the sun and the five nearest planets to the sun. How is this a system? Consider our earlier properties and the concept of boundary. Boundary need not be a physical wall or even the effective force of the sun's gravitation. It can also be managerial, organizational, economic, psychological, conceptual. There is no question that space exploration, undersea exploration, poverty, automation and technological change, education, population growth, human happiness, democracy, and communism are today inextricably intertwined. This monstrous system, absolutely the most important for you and for us during our lifetimes and probably for a long time thereafter, can be characterized, however crudely, in terms of boundary; inputs, throughputs, and outputs; equilibrium; feedback loops; growth, differentiation, and decay; dynamic interactions; control and other properties. In this most macroscopic of macrosystems, the North American air defense network (NORAD) of which SAGE can be viewed a subsystem, the SCF, and Apollo, as well as individual cities and nations, can be recognized only as important subsystems (Figure 6.1). The interactions are quite evident. Important contributory factors of this system are discussed in the last three chapters of this book. We believe that one of the most useful discoveries of the twentieth century will be the application of systems know-how to the solution of problems in *sociotechnical* systems.

* At earth launch, the Lunar Module is physically separated from the Command and Service Modules by the S-IVB (third) stage of the three-stage Saturn V launch vehicle. A reconfiguration of the modules is required shortly after injection into the cislunar path.

The partially human-designed, semispontaneous, and fortuitous systems and organizations of mankind possess many of the systems properties defined earlier. Yet there are important differences: man-made equipment-oriented systems reflect the purposes of a few users and are designed largely to function *in spite of* environments. Seminatural systems reflect the vagaries of numerous economic, social, and political needs, and geographic and climatic environments. Organizations, as in business and industry, are examples of systems in which man–man interactions predominate over man–machine interactions.

System interactions occur in these less, as well as in the more, structured systems. The interactions and their results may be apparently *unpredictable*, *uncontrollable*, and as in our present technological society, *unmanageable*. Psychology has the chance of its lifetime to demonstrate its worth in dealing with these late twentieth-century problem areas, especially in relation to growth and decline, need and goal direction, stability, and internally and externally generated stress and change.

The search for universals

Three separate approaches to the identification of underlying principles of structure and behavior have developed in response to observations of isomorphies among systems and the development of similar concepts in different fields on the one hand, and the increasing specialization of knowledge on the other. To varying extents, general systems theory, cybernetics, and bionics seek universals that can relate the specifics of different sciences and technologies.

General systems theory. By the 1950s, it was evident that scientific specialization was leading to increasing difficulties of communication across disciplines. A number of philosophies, methods, and approaches, based on attempts to understand organization and the behavior of wholes, integrative mechanisms, dynamic interaction, and environmental effects, had evolved over the last 100 years in several sciences. However, this evolution took place in one discipline independently of developments in other disciplines. Examples include the field concept of physics, homeostasis and synergy in biology, servo theory in engineering, and Gestalt psychology.

Pressures to integrate similarities and relationships among the sciences, to enhance communication across disciplines, and to derive a theoretical basis for *general* scientific education had several philosophical roots, of which we will mention three (see von Bertalanffy, 1956; Boulding, 1964).*

1. Science in the late nineteenth and early twentieth centuries was largely analytic, with the whole being reduced to even smaller units, the study of which would allegedly result in understanding the whole. Eventually, many theorists hoped to achieve unity within science by reduction to the particles and mechanisms of physics. Thus, molecules were broken down into atoms; atoms into electrons, protons, and other particles; organisms into cells; behavior into reflexes; perception into sensations; the mind into ideas; and so forth. Simple additive and static cause–effect explanations were offered in describing the properties of the whole. Concepts of organization and of interaction were ignored. Countering this *reductionistic* approach was the increasing awareness of the importance of interaction and of dynamics that emerged in several fields of science during the first third of the twentieth century. Such terms as *field theory*, *Gestalt*, *holistic*, *organismic*, *adaptiveness*, and *goal-direction* reflect this newer tenor. The independent development of these concepts in different sciences can be considered a forerunner of the development of a general systems theory.
2. The second stimulus toward development of a general systems theory – and limitation of specific-system approaches – came from the so-called Heisenberg Principle of Indeterminacy. Thus, information cannot be applied to or withdrawn from a system without changing it, and the very process of observation or study distorts the system itself and hence the meaning of results. A wide variety of experiences, especially in the biological, behavioral, and social sciences, substantiates this objection. For example, in experimental psychology the experimenter's behavior itself or the design of his equipment may offer subtle cues to the human subjects; in opinion polling, respondents tend to give answers they believe will seem "right" to the pollster.
3. Many systems are probabilistic or stochastic rather than deterministic. A single observation can tell us little or nothing about the probability of occurrence of the event observed. Again this holds especially true in the biological, behavioral, and social sciences.

In 1956, a group of scientists established the Society for General Systems Research (originally called the Society for the Advancement of General Systems Theory). The Society issues a yearbook of articles on systems approaches from virtually all the sciences. Young (1964) has surveyed general systems theory after nearly a decade of its existence, summarizing the attempts of workers to apply general systems theory to their specific fields. *Emphasis on specific applications to enhance the general theory was found to be far greater than the applications of general systems theory to specific disciplines.* Work could be broken down into four categories:

1. *Systematic and descriptive factors.* This category dealt with classifications of types of systems, their data and internal organization, and system environments. Particular attention was given to openness and closedness, organismic or nonorganismic properties, centralization, independence, differentiation, interaction, and boundaries.
2. *Regulation and maintenance.* This category dealt with control and stabilization. Concepts of equilibrium, feed-back and communication, and control were important.
3. *Dynamics and change.* This category dealt with nondisruptive internal and external environmental changes. Of particular importance were adaptation, learning, growth, and goal-seeking.
4. *Decline and breakdown.* This category dealt with disruption and dissolution, and emphasized stress, overload, entropy, and decay.

* The term "general systems theory" was coined by von Bertalanffy.

Young states that the typical literature is strong on regulation and maintenance and weak on decline and breakdown. Social scientists are showing an increased, sometimes overriding, interest in the general systems field, perhaps because of training and interests at the given time. The usefulness of the literature is diminished by the tendency of some authors to cite general concepts without indicating how these concepts helped specific applications.

Material on general systems can be found in the *Yearbook of the Society for General Systems Research* and in the Institute of Electrical and Electronic Engineers *Transactions on Systems Science and Cybernetics*. Boulding (1956) is a good general reference.

Cybernetics. Since World War II, there have been several concerted, often highly mathematical attempts to determine universals applicable to the explanation of the behavior of both organisms and machines. Work has been directed to increasing understanding of organisms (and societies) and making machines more adaptive, more flexible, and more in tune with given environments. In this book, we will consider two main developments: *cybernetics* and *bionics*.* Cybernetics can be thought of as an attempt to understand organisms through making analogies to machines, and bionics as an attempt to develop better machines through understanding of biological design principles. Cybernetics traditionally has emphasized understanding of a given process *per se*, while bionics seeks understanding of a given process as a means of generalizing to another situation. Both cybernetics and bionics involve theory, model building, experimentation, and application; they have been compared with the two sides of a coin.

A few individuals have participated in developments in both cybernetics and bionics; similarly, there have been tangential developments, closely akin to these, but given other names such as *self-organizing systems* (Yovits *et al.*, 1962), *adaptive systems*, *learning machines*, *automata*, and *artificial intelligence*, which have yet to be interrelated. These approaches appear to rely more heavily on "armchair", rational, and intuitive methods, while cybernetics and bionics emphasize empirical and experimental methods.

As we attempt to deal with the increasing complexity of our world, those most complex of things, living organisms, can provide clues to better design for small and compact power supplies, for reliability, for greater adaptability, for more effective organization, and so forth. Of particular interest to both cybernetics and bionics are the following:

1. The reception ("sensation") and recognition ("perception") of information.
2. Integrative processes.
3. Storage and retrieval of information.
4. Self-regulatory ("homeostatic") processes.
5. Adaptive ("learning") processes.
6. Control processes.

The world first became widely aware of cybernetics in 1948 when Norbert Wiener published the first edition of *Cybernetics or Control and Communication in the Animal and the Machine* (the second edition appeared in 1961). However, Wiener had formulated his ideas earlier in World War II, when he was faced with problems of automatic aiming of antiaircraft guns. It was necessary to shoot the projectile not at the aircraft itself, but along a trajectory such that the two would intersect in space sometime in the future. Accordingly, it was necessary to predict the future position of the aircraft. Wiener was able to formulate equations describing a closed-loop system (the input to a computer was part of the output signal). Thus, the computer, utilizing a feedback loop, could calculate the time of the trajectory of a projectile and predict the point at which the gun should aim. Working with Arturo Rosenblueth, a biologist, and with other prominent engineers, mathematicians, biologists, and psychologists, Wiener formulated principles common to machines, animals, and societies. The term *cybernetics* itself comes from the Greek word for *steersman*, a tribute to the fact that a ship's steering engines provide one of the earliest types of feedback mechanism. Wiener's book was eclectic and contained discussions of normal and abnormal physiological, psychological, and sociotechnical processes, as well as of information, communications and feedback *per se*.

Wiener viewed cybernetics as encompassing the entire field of control and communication theory, whether in the animal or the machine. The study of *automata*, machine or animal, was regarded as a branch of communication engineering, and was concerned with the concepts of information amount, coding and message, with noise, and so on. Automata are related to the outside world through sensors/receptors and control mechanisms/effectors, which are interconnected by central integrating mechanisms.

Wiener recognized that the value of cybernetics would be shaped by the limitations of the data we can obtain. Yet he felt there were two areas in particular offering practical results: the development of prostheses and the development

* In some parts of Europe, bionics is considered to be synonymous with applied cybernetics.

of automatic computing machines. Subsequent developments have borne out Wiener's expectations, particularly in the second area.

The importance of cybernetics to the psychologist or physiologist interested in neuro-endocrine integrative action, self-organizing behavior, homeostasis, perception, learning, and so forth should be quite evident. In another area, Wiener was remarkably prophetic: he expressed concern over our abilities to construct machines of almost any degree of sophistication of performance, believing that we are confronted with "... *another social potentiality of unheard-of-importance for good and for evil* ..." (emphasis added). Just as the industrial revolution *devalued* the human arm through competition with machinery, so the present technological revolution is bound to devalue the human brain, at least in its more routine processes. Wiener believed, as we emphasize later, that the alternative is a society based on human values other than buying and selling – *a society that would require a great deal of planning and struggle.* He hoped that a better understanding of man and society, as "fall-out" of cybernetics efforts, would outweigh the concentration of power (in the hands of the most unscrupulous) incidental to the applications of cybernetics, but he concluded in 1947, "... that it is a very slight hope".

Cybernetics is an integrated body of concepts applicable to orderly study within physical, biological, and social sciences, and in the "crossroads" interdisciplinary sciences between. In each case, problems can be represented in terms of information content and flow and in terms of feedback and control. Yet, like some other concepts, cybernetics has proved no universal panacea. Its initial reception was lurid with the connotation, "The robots are here." Extensions were interesting and led to coining of new terms – *cyborg* for an organism with a machine built into it with consequent modification of function, *cybernation* for automation involving especially information and control systems – but cybernetics generally did not live up to expectations. The term itself remained an obscure one in the United States, although it became popular in Germany and in the Soviet Union, where theoretical cybernetics is considered to include information theory, automata theory, programming theory, and the theory of games. More recently, interest in cybernetics has renewed in the United States, as reflected in the establishment of the Professional Group in Systems Science and Cybernetics (1965) within the Institute of Electrical and Electronics Engineers and the American Society for Cybernetics (1968).

Cybernetics applications include adaptive teaching machines and pattern perception devices; the best examples are provided by automata and by prostheses. *Locomotion automata*, which may be bi- or quadrupedal, are of potential value for use in difficult terrain, such as mountains, polar regions, swamps, and the lunar surface. In a simple quadruped automaton, each leg has only two output states: on the ground pushing backward and in the air pushing forward. The sequence of motions for each leg and the gait are controlled by a binary-sequence generator using a different program for each gait and based on *finite-state* logic (the machine can have only a finite number of states) (Kalisch, 1968; Swanson, 1968). Other applications deal with powered prostheses. The most sophisticated concepts involve sensing and amplifying bioelectric potentials from muscles, or even better, nerves in the stump of the severed limb itself. Devices based on utilization of muscle potentials and including an electric motor enable the patient to perform fairly precise activities such as writing and to lift weights of about 10 pounds. Other cybernetic machines under study include those that amplify a normal operator's strength, enabling him to lift 1,500 pounds, or increase his locomotion speed to 35 mph over rough or dangerous terrain.

An automaton possessed of adaptive ("homeostatic") behavior was Ashby's (1960) *homeostat*, an electromechanical device, which always returned to equilibrium by means of switches, regardless of input. Another was Shannon's mechanical mouse, which was programmed to "learn" a checkerboard maze after one trial by "remembering" the direction in which it had left a given square for the last time (Lindgren, 1968). A Russian automaton, based on a hierarchy of heuristic computer programes, purportedly also possesses feeling and consciousness (Lindgren, 1968).

Bionics. Bionics is another of the important interdisciplinary areas that emerged toward the late 1950s and early 1960s. The term was coined by U.S. Air Force Major J. E. Steele in 1958, but first received widespread recognition at the first bionics symposium in 1960 (*Bionics Symposium: Living Prototypes – the Key to New Technology*, 1960). Since 1960, other bionics congresses have been held (e.g., *Bionics Symposium: Information Processing by Living Organisms and Machines*, 1964). The word "bionics" suggests a coalescence of biology and electronics, but bionics protagonists emphasize the integration of *analysis* (from biology) and *synthesis* (from engineering design). This is reflected in an official insignia: the scalpel representing analysis, an integral sign representing synthesis. Over the years biologists, psychologists, engineers, and mathematicians have participated in bionics efforts.

Bionics can be defined as the study of living systems to identify concepts applicable to the design of artificial systems; alternatively, it can be defined as the study of systems whose functions have been derived from the study of living systems.

The philosophical and rational basis for bionics rests on the time-based, dynamic organism–environment interactions that have characterized all living systems since the first appearance of life on earth some two to three billion years ago. The environment stresses the organism, which either adapts to fit a particular ecological niche at a given time or perishes. Hence, living systems can be thought of as being good, sometimes even the best, approximations of adjustments to the demands of given environments at a given time.

The next question concerns the appropriate degree of isomorphism between the natural and artificial system. Attempts to pattern design too rigidly after the living prototype often lead to dead ends, as shown by early (sometimes fatal) attempts to fly by avian methods. Modeling is widely used in bionics and provides a bridge between different specialists. Model building, however, always presents the possibility of too great abstraction and mathematical precision at the cost of minimum relation to the real world. Also, there has long been a tendency both in biology and in psychology, and now perhaps also in bionics, to concentrate on knowledge that may be incomplete, distorted out of context, incidental, or artifactual. Examples are the undue emphasis on the electrical activity of the nervous system and on reflex activity, and attempts to equate nervous system and computer functioning. Nevertheless, a rigid insistence on complete understanding of a biological process may retard useful serendipitous discovery. It seems desirable, therefore, to qualify the definition of bionics to include processes that directly and wholly explain a natural phenomenon, those that seem to explain *some* aspects of a natural phenomenon, those recognized as incidental, and those that clearly are only analogies.

Bionics thus can be seen to be the study of living organisms with the intention of deriving technological knowledge. As the flight of aircraft and of spacecraft demonstrate, the capabilities of the artificial systems – *along some dimensions* – may greatly exceed those of the original prototypes. Actual or potential system design applications based on *living prototypes* are summarized in Table 6.1.

Often the living prototype indicates only that a process *is* possible, but information as to how the process works is scant. Thus, in many bionics studies, limited biological or psychological knowledge is extended by simulation and modeling and by intuition on the part of the bionicist.

The question has been raised as to the usefulness of the bionics approach. That there are probably more workers in the area of artificial neurons or *neuromimes** than in any other derives from the hope that greater understanding of the nervous system will aid in the construction of smaller, more flexible computers. On the other hand, neural modeling should provide better understanding of the nervous system *per se*. What we know about the neuron has enabled us to build electronic analogs that simulate neuron behavior. Many different kinds of neuromimes have been built, depending on the interests of the designers. Some emphasize central processes such as memory; others emphasize peripheral processes such as excitation-inhibition, threshold, summation, and refractoriness. Van Bergéijk and Harman (1960) have attempted to produce as precise an analog of the peripheral nervous system as possible, and report that this approach has helped elucidate both anatomical and physiological features.

Reichardt (1961) and his interdisciplinary coworkers at the Max Planck Institut für Biologie, Tübingen, Germany, have studied visual processes in the beetle *Chlorophanus*. This beetle responds optokinetically (in terms of head or eye movements) to relative movements of light in its optical environment. The most elementary succession of light changes found capable of eliciting an optomotor response consisted of two stimuli in adjacent ommatidia (facets) of the compound eye. A stimulus received by one ommatidium can interact only with that received by the adjacent ommatidium or those adjacent to the latter. Transformation and interaction within the central nervous system were found to agree with known principles. The results were expressed in the language of control systems, suggesting that the beetle could derive velocity information from a moving, randomly shaded background. This finding led to the design of a ground-speed indicator for aircraft based on the function of two of the hundreds of facets comprising the compound eye (see Savely's article in Steele, 1960).

A very readable book that discusses most of the developments in bionics has been written by Gérardin (1968). Specific original papers of representative interest are those of Rosenblatt (1958), Lettvin *et al.* (1959), Newell and Simon (1961), and Simon (1961).

* There is a need for nomenclature to differentiate between the natural and analog entity. Following van Bergéijk (1960), we can consider the suffix *mime* to indicate the most general type of artificial cell or organ. Accordingly, a neuristor would be one type of neuromime.

Table 6.1 Bionic developments (Prepared partially from text in Gérardin, 1968)

PROTOTYPE	APPLICATION
Olfactory receptors of moths and butterflies; infrared receptors of pit vipers	Lightweight sensors
Compound eye of beetle *Chlorophanus*	Aircraft ground-speed indicator
Compound eye of king crab *Limulus*; retina of frog	Automatic recognition of pattern, movement
Eel and ray electrical-field generation, detection	Submarine detection
Bat and cetacean echo-location behavior and related physiology; bat–moth interactions	Radar and sonar with better antijamming and antievasive capabilities
Bat ear structure and echo location	Location aid to the blind
Neuronal (generally peripheral) electrophysiology	Artificial neuron or neuristor to propagate a "signal" without attenuating it*
Retina and brain of higher vertebrates	Pattern-perception and learning machines (perceptrons)
Animal short-term memory (apparently electro-chemical or synaptic) and long-term memory (apparently chemical and inter- and intracellular involving both neurons and neuroglia)	Computer memory
Human problem solving	Adaptive (heuristic) problem-solving computer programs
Dolphin swimming behavior and double (turbulence-reducing) skin	Streamlined torpedo
Migratory, orientation, homing behavior; related physiology of birds, turtles, fish, insects	Navigation devices
Bioluminescence	Cold (100% efficient) light

* Neuristors are capable of performing complex calculations, leading to attempts to build computers using them as basic constituents.

Systems psychology: a new field

Conceptually, systems theory and psychology have long had much in common. Concepts that have arisen independently include those of field and environment, dynamics, interaction, and evolution and change. Most significantly, both organisms and systems consist of wholes that transcend the sum of the dynamically interacting parts. In turn, each part affects the properties of the whole. The organism can be thought of in terms of a hierarchy expressed from most general and tenuous to most elementary and precise: the social grouping of organisms and the man–machine system; the total intact organism; the organ system such as the nervous system; the tissue such as nervous tissue; the individual cell such as the neuron; the cell nucleus; the complex molecule or colloid such as deoxyribonucleoprotein; the simpler molecules such as the nucleotides, nucleosides, purines, and pyrimidines; and finally the atom and subatomic particle. At the upper end of this hierarchy, psychology interrelates with sociology, cultural anthropology, economics, and political science; at the lower end with physiology, biochemistry, and biophysics. Systems concepts and methods are applicable at all levels, and psychological problems at each level can be couched in systems terms. Examples at each end of the continuum are provided by simulation studies of the industrial organization and by relating memory to nucleic acid and protein metabolism within the neuron cell body and associated neuroglia.

In a similar vein, systems are arranged as *macrosystems* such as the Apollo systems as defined earlier; as *systems* such as the Apollo spacecraft; and as *subsystems*, *subsubsystems*, or *modules*, *components* (individual subassemblies), *units*, and *parts*. Psychological factors or psychology-related problems can be defined at each level, for example, by the use of computers in military decision making involving national defense, at one extreme, and by training required for the assembly of printed-circuit boards at the other. A major problem derives from the lack of consistency in use of systems terminology. Such terms as *unit*, *part*, *component*, and *element* are used interchangeably. We have used the neutral term *constituent*, as appropriate, to indicate the most general case. The same semantic difficulties apply to such terms as *job*, *task*, *element*, and others in the behavioral hierarchy. In some chapters, the term *element* is used in a general sense, although there are objections to doing so. The nomenclature and definitions given in Table 6.2, and used throughout this book, are based generally on usage in the aerospace industry and specifically on practice at Northrop Corporation on the Skybolt air-to-ground missile project.

Table 6.2 Definitions of basic terms in systems hierarchy

GENERAL

Mission. A statement of *what* the system is to do to solve a given problem and *when* and *where* – an expression of purposes and objectives. It can be arbitrarily segmented in terms of identifiable beginning and end points. Mission determination involves many subjective or judgmental factors.

Requirement. A statement of an obligation the system must fulfill to effect the mission. Requirements are expressed first in qualitative terms and progressively in quantitative performance terms relative to some criterion(ia). They further delineate the system mission.

Function. A general means or action by which the system fulfills its requirements. Functions are usually expressed in verb form (monitor, control) or participial form (monitoring, controlling). They are the first expression of the *hows* of the system. They are expressed progressively more precisely. Ideally, functions are conceived apart from implementation by men and/or by machines; in practice, they are usually expressed along with machine design implications.

EQUIPMENT

Subsystem. At its *most basic level*, a single module, or combination of modules, plus independent components that contribute to modular functions, all interconnected and interrelated within a system and performing a specific function. Examples: guidance and control subsystem, propulsion subsystem.

Module (*sub-subsystem*). A combination of components contained in one package or so arranged that together they are common to one mounting, which provides a complete function(s) to the subsystem and/or systems in which they operate. Examples: guidance and control computer, astrotacker.

Component. A combination of units or parts independent of, or an independent entity within, a complete operating module or subsystem, providing a self-contained capability necessary for proper module, subsystem, and/or system operation. Can be replaced as a whole. Examples: DC power supply, digital display readout.

Unit. A combination of parts constituting a definable entity of a component, possessing a functional potential essential to the proper operation of that component. Example: chip.

Part. The smallest *practical* equipment subdivision of a system; an individual piece having an inherent functional capability, but unable to function without the interaction of other parts or forces; ordinarily not subject to further disassembly without destruction. Examples: transistor, diode, resistor, capacitor.

BEHAVIORAL

Job operation. A combination of duties and tasks necessary to accomplish a system function. A job operation may involve one or more positions or career specialties or fields.

Position. A grouping of duties and responsibilities constituting the principal work assignment of *one* person. The position may be that of operator, maintainer, controller, etc. Positions related in terms of ability, education, training, and experience can be grouped as career specialties and fields. Synonym: *job.*

Duty. A set of operationally related tasks within a given position. These may involve operating, maintaining, training, and supervising, etc.

Task. A composite of related (discriminatory–decision–motor) activities performed by an individual, and directed toward accomplishing a specific amount of work within a specific work context. Involves, for example, a group of associated operations or inspections.

Subtask. Actions fulfilling a limited purpose within a task – for example, making a series of related machine adjustments.

Task element. A basic S–O–R constituent of behavior comprising the smallest *logically* definable set of perceptions, decisions, and responses required to complete a task or sub-task. Involves, for example, identifying a specific signal on a specific display, deciding on a single action, actuating a specific control, and noting the feedback signal of response adequacy. Synonym: *behavior* or *job behavior.*

SOURCE: Modified from text in Headquarters Air Force Systems Command, *Personnel Subsystems*, (1969) and from practices at Northrop Corporation.

Figure 6.1 summarizes the above aspects of systems hierarchy. Macro-macrosystems can be subdivided into many other ways – for example, in terms of communications; transportation, or use of resources. An immensely complicated figure would be required to indicate *organizational* hierarchy and all subdivisions at the lower hierarchical levels. Identification and analysis of all segments of the mission profile and the constituent functions, tasks, and the like require detailed documentation, and constitute one of the main businesses of *human factors. Systems* science is concerned with determination of interrelationships among the various concepts, levels, and terms. This does not imply, however, simple linear, additive, multiplicative, or deterministic relationships, either laterally or hierarchically. For example, at the moment we can clearly relate system job performance neither to the biochemistry of the brain nor to task-element performance. Hopefully someday we will be able to do much of both.

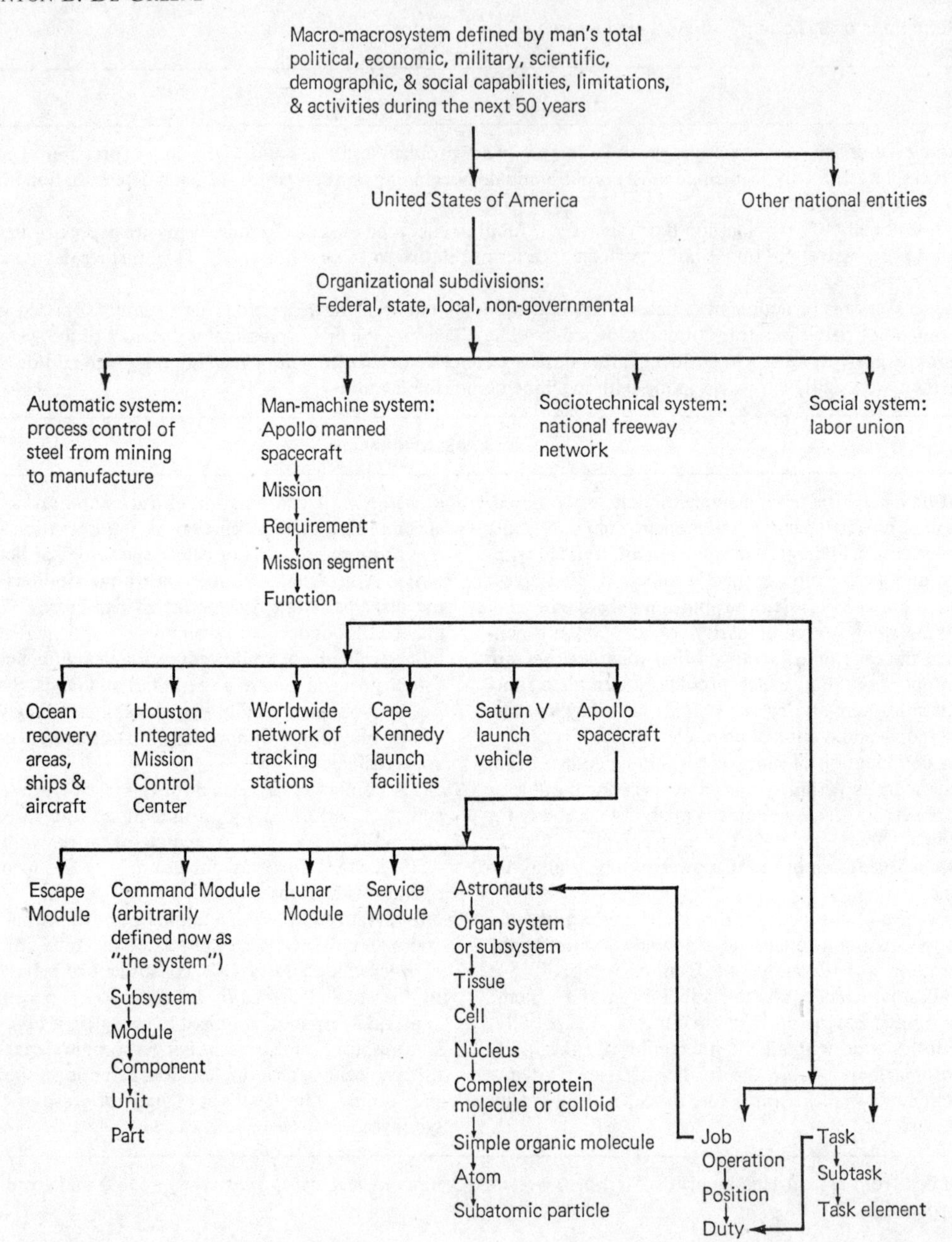

Figure 6.1 Examples of systems subdivision and organismic, equipment, and behavioral hierarchy (see Table 6.2 for definitions of basic terms).

Systems and organisms can also be studied and described in terms of *feedback control*. Independently of physical scientists, biologists and psychologists came up with the concepts of *milieu internale* and homeostasis, and extensions thereof, which describe the maintenance of constancy of physiological, behavioral, and social parameters, internal to the organism or group, despite wide variations in stimuli. However, the organism is immensely complex when compared to the machine: Control loops in the organism are superimposed upon one another, and its internal nonlinear feedback mechanisms may be dissimilar to those in the machine.

Conceptualization of organisms and systems in hierarchical terms like those indicated earlier has been associated with the development of philosophies concerned with methods of approach. Can the complex whole best be understood at that level, or by studying the individual constituents? Is analysis that defines the behavior of these constituents isolated from the system a more meaningful approach than synthesis, the attempt to deduce behavior of the system from knowledge of the constituent functions? Terms such as *holism*, *Gestalt*, *molar*, *molecular*, *atomism*, *elementarism*, *associationism*, *reductionism*, *stimulus-response unit*, *mechanism*, and *vitalism*, long used by psychologists or biologists, attest to the continuing lack of agreement.

Finally, systems techniques have the heuristic or epistemological benefit of providing rigor in the definition of psychological terms such as *intelligence*, *learning*, *thinking*, and *feeling*. This is especially evident when we try to answer such questions as: Do machines think? What is artificial intelligence? How does problem solving relate to decision making? In our attempts to provide answers to these questions, assist colleagues in other fields to understand how psychology can contribute to solving their problems, and evaluate the statements they so frequently vouchsafe, we are forced to consider even more far-reaching questions. For example: What *is* psychology? How good are its basic methods? Are the right problems being recognized, defined, and attacked? Does psychology have a body of theory and an approach amenable to the study of *real-world* problems? How good and how useful is psychological research? How can we better apply psychological research to the crushing problems of technology and society? Is psychology poorly understood by the layman and by other scientists and engineers so that its findings, while valid and generalizable, are poorly applied? Throughout this book, we attempt to provide answers to each of these questions, mostly within the context of specific subject areas.

The nature of psychology and psychological theory. A science must be defined in terms of the events of a given time in history; the efforts of its practitioners, the problems they recognize and identify, the tools they use, and interfaces with other sciences. It is not always clear just *what* psychology is. Certainly, the customary definitions do not provide a realistic framework for a science that encompasses a greater vertical range than any other, including at one extreme human behavior in groups and organizations, and at the other the biophysics and biochemistry of learning. More and more, psychologists find themselves associated with specialists in other fields – clinicians with psychiatrists, educational psychologists with teachers, engineering psychologists with engineers – cut off from their fellows who share the science of mind, of man, of behavior, and of experience.

We can expect a shifting of boundaries within science, the emergence of interdisciplinary "cross-roads sciences", which eventually achieve an intrinsic sufficiency of their own, and the absorption of subsciences that have not proved their worth. This may sound like an unduly pragmatic view, and there is indeed danger in compromising development of basic knowledge in the name of immediate returns on research grants. There is just as great a danger in retreating into the contented isolation of our laboratories while the world collapses without, secure in our grasp of an idea, or method, or shibboleth of questionable relevance.

The history of science provides us with many examples of intellectual *culs de sac*. There is always the risk of misunderstanding the problem, selecting the wrong level or the fortuitous artifactual, rather than the lasting and real, or of simply grasping the most convenient. It is always tempting to build elaborate theories on limited or premature data, only to become caught up in the excitement and momentum of the times and pushing applications that may be invalid at best and downright harmful at worst. Jones and Gray (1963) cite the selection of neuron pulse interval or pulse frequency, while ignoring pulse amplitude or width, as an example of grasping a phenomenon that is easier to deal with conceptually or mathematically in model building: the problem and unit of measure has been selected to fit available mathematics, rather than new mathematics developed to fit the problem. The traditional attempts of psychology to explain learning and memory in terms of simple conditioned reflexes or in terms of electrophysiological events (ignoring the chemical) probably represent premature theorizing based on limited fact. Goslin (1968) has reviewed the field of standardized ability tests and testing, and has pointed out the many questions of validity and predictability and the real danger of individual and social harm.

A considerably body of psychological theory and data has been based on experimentation with the albino variety of the brown rat, *Rattus norvegicus*. What if all this research is at best incidental and at worst artifactual and decidedly wrong? Some insight into this serious problem is offered by Lockard (1968), who presents considerable evidence that the albino rat is an *atypical* organism – a poor one, indeed, on which to base generalizations of behavior – that its very evolution is adventitious and artificial, and that results based thereon are bound to be distorted. Here is an excellent example of our misguided hope in finding a standard unit (as in physical science); there are biologically many types of white rats.

When the engineering psychologist turns to the experimental psychologist and asks for basic data on human performance, he is likely to find that there are no data or that the data are inapplicable to real-world problems of analysis, design, and operation. Again and again, workers have complained about the lack of application, even relevancy, of the results of psychological experimentation to pressing engineering and social requirements (Chapanis, 1967; Alluisi, 1967, and Meister, 1964; and Boulding, 1967, and Mackie and Christensen, 1967). At the same time engineers, computer programmers, chemists, mathematicians, and others are assuming, and – to an extent more than psychologists – mastering problems long considered within the domain of psychology. At present, problem solving, especially man–computer problem solving, is quite *in*. But where is the basic groundwork in psychology developed over the years as an aid to the psychologists and others now specializing in the field? Why were so many of us psychologists running rats through mazes over several decades and so few studying human thinking and problem solving? Simultaneously, in the streets throughout the world, social and sociotechnical problems cry out for solution – and the cry is becoming louder – in terms of skill definitions *vis-à-vis* automation and technological change, in terms of training, education, attitudes, emotions, mental disease, and so forth. Psychology's record of accomplishment in helping to ameliorate the world's woes, perhaps also in advancing basic science, has not been great, especially considering the number of psychologists. Demonstrable results often come from outside psychology. The major advance in the treatment of the mentally ill in the last decade or so has stemmed from developments in pharmacology, not from developments in clinical psychological analysis and therapy. Separate abstractions such as personality, mental illness, intelligence, learning, and memory may see unity through extension of Pauling's (1968) concept of *orthomolecular psychiatry* to include gene action and specified biophysical and biochemical processes involving membrane permeability, metabolism, waste product accumulation, and the like.

Psychology, as we have now seen, s a remarkably diverse science that often seems at odds with itself and with its neighbor sciences. Internecine battle has long raged within psychology: clinician against experimentalist, "brass-instrument" man against "field theorist", "rat man" against "head shrinker", pure scientist against applied worker. Many psychologists believe that this conflict has been for the better and will lead to a truly stable eclectic science. Actually, this is far from true, and at no time more evident than when we try to answer the question, just what *are* the psychological factors in systems? At first the answer seems deceptively simple. We could say, why they're perception, learning, memory, motivation, emotion, psychomotor behavior, and so on. Closer inspection, however, reveals that these entities themselves are interrelated, varied within themselves, and time- and context-dependent. There seem to be at least two types of memory, for example, short term and long term, differing at the cellular level. There are undoubtedly several levels of perception, learning, and emotion, which might be called "peripheral", "subcortical", and "cortical".

Further, in operational situations we find it necessary to deal with factors like judgment and intuition that have long been pariahs to objective psychology. Even when we reduce the psychological factors to basics like visual acuity, vigilance, and reaction time, we find that these are dependent on the temporal and environmental context. Clearly what is needed is a *general* and *flexible* approach, adaptable to different problems, levels, times and environments. Systems theory seems to provide this approach. Throughout psychology, it is meaningful to conceptualize a person – or a human group, or a brain, or a mitochondrion – as a constituent processing energy, materials, and information; interacting within a given environment, at a given time, and at a given state of equilibrium and internal consistency. Systems theory should provide a common framework for posing, studying, and solving systems problems that seem to involve as apparently disparate factors as pattern perception, altertness, decision making, language, fatigue and stress, individual skill and performance differences, morale, and interpersonal relations.

It is probably premature to try to develop a comprehensive theory encompassing the *continuum*: automatic – man–machine – sociotechnical – social systems. An intermediate step is to use various models: information processing, feedback control, probabilistic, input–throughput–output, and man–machine–environmental. Further, it is probably fair to state that *there is no comprehensive theory tying together what we know about human behavior in systems*. In view of the many problems discussed later in this book, any systems psychological theory clearly must account for factors long faced by psychological theory in general. What are these factors? One approach is suggested by Coan's (1968) recent study of basic trends in psychological theory over time and at any given time. Coan determined 34 variables (divided into emphasis on content, methodology, basic assumption or mode of conceptualization) related to 54 theorists by the ratings of a couple of hundred experts in the history and theory of psychology. The theories included those of personality, abnormal behavior, learning, brain mechanisms, homeostasis, peripheral nervous activity, mental abilities, individual differences, sensation, integrative activity of the nervous system, and so forth.* For each psychological theorist, the experts' ratings were averaged and a 54 × 34 matrix obtained. Factor analysis and multiple-regression analysis revealed six factors and a placement of each theorist along a continuum represented by that factor. The factors are summarized as follows:

1. Subjectivistic versus objectivistic.
2. Holistic versus elementaristic; these two factors emerged as the factors of greatest variance, but other factors were also necessary.
3. Transpersonal versus personal, or experimental versus clinical.
4. Quantitative versus qualitative.
5. Dynamic versus static.
6. Endogenist ("biological") versus exogenist ("social" or, we might add, "environmental").

The six factors were intercorrelated and further analysis revealed two more general factors:

1. Synthetic (subjectivistic, holistic, and qualitative) versus analytic (objectivistic, elementaristic, and quantitative).
2. Functional (dynamic, personal, and internal and biological) versus structural (static and transpersonal).

In turn, the two second-order factors had a correlation of 0·55, leading to isolation of a final general factor: fluid (relaxed) orientation versus restrictive (controlling, compartmentalized) orientation. These theoretical orientations were seen to be a function of the personality of the scientist, a point systems researchers could do well to bear in mind. Does the need for systematization or mathematical rigor in fact reflect inner insecurities? Coan's work is summarized in Figure 6.2.

From the trends revealed in his work, Coan observes that neither at the factor level nor at the variable level is there a basis for confidently extrapolating to future developments. The clearest trend was toward greater objectivism. The participation of psychologists in system theory and system development over the past two and one-half decades suggests an approach (remember there is not yet a theory) that can be characterized as objectivistic, elementaristic, transpersonal, quantitative, dynamic, and exogenist. This approach *could* serve as a launching point for a comprehensive theory of human behavior in both man–machine and sociotechnical systems (which should be viewed not as dichotomous but rather as operating along a continuum). This theory must incorporate the input–throughput–output paradigm, and it must be expressible in terms of the modification or processing of inputs, probably in terms of information processing. Most important, it must express interactions at different levels and the same level, and predict the effects of interactions on the disruption and maintenance of equilibrium. Other systems properties could be factored in to a lesser extent. From here there could be an infinite quest for universals, perhaps the "elements" or "tasks" or "behaviors". One factor of particular importance is *expectation*: when the system situation, operation, or environment does not accord with an individual's expectation, degraded behavior occurs: the pilot loses control of his aircraft, the maintenance man's morale decreases, the minority group member or college student riots.

* Unfortunately, Coan emphasized classical theorists to the detriment of pragmatic theorists such as organization theorists.

The definition, value, and directions of systems psychology. What we've said so far suggests the time has arrived to say, "Whoa, let's take stock of ourselves!" We must define systems psychology within its context of the other sciences, and carefully evaluate its future directions. Formally we define systems psychology as *the science and technology of understanding, describing and predicting; and generalizing the total effective performance and need-gratification behavior of intact organisms (usually human beings) under conditions of interaction with other intact organisms or machines within given environments.*

From preceding discussions, we may define *systems psychology* more generally as a new interdisciplinary specialty characterized by a *level*, a *breadth*, and a *method of approach.* At its most basic, the *level* is that of the individual intact organism (almost invariably, but not exclusively, a human being) interacting with at least one other entity, either man or machine, at about the same hierarchical level of complexity. Usually, at least ten of these basic constituents are involved. *Breadth* refers to similar consideration of man–man interactions within an environment, and man–machine interactions within an environment, and combinations thereof. The *method of approach* is that of systems

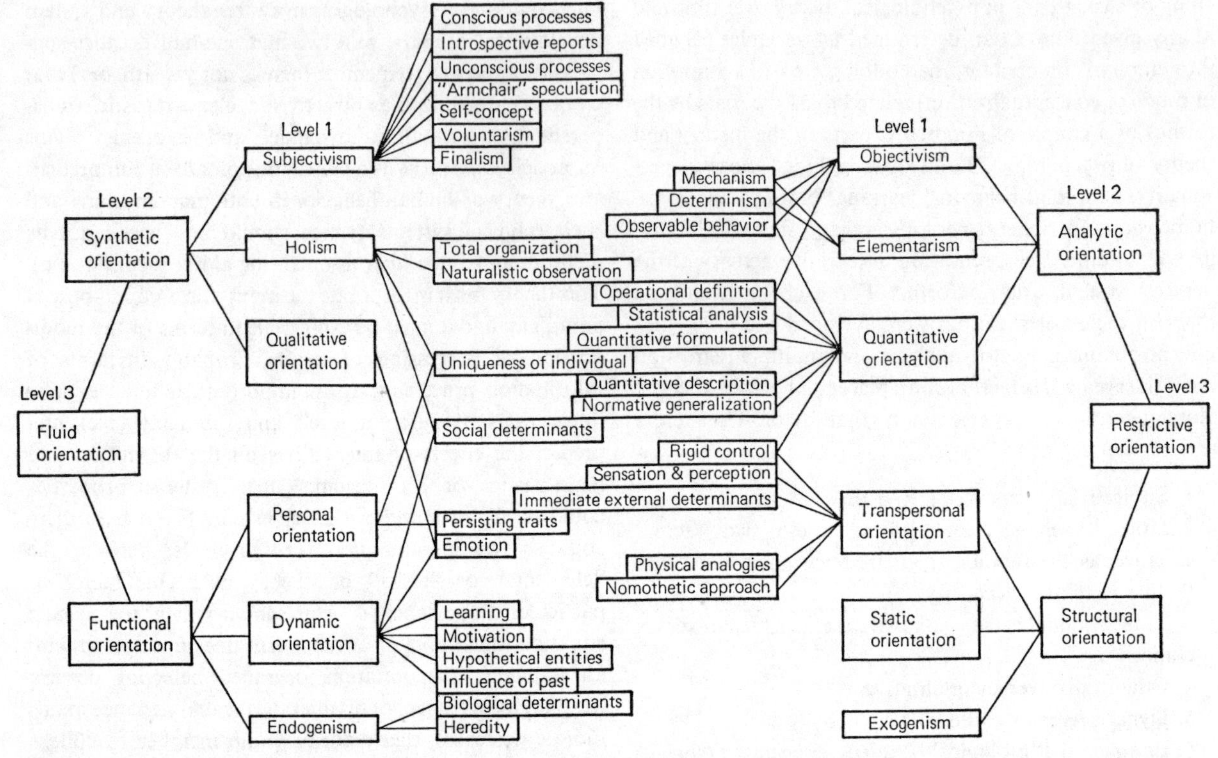

Figure 6.2 Three levels of factors common to psychological theories (From Coan, 1968).

theory *and* practice as discussed in detail throughout this chapter. It is general enough to permit qualitative, quantitative, or mixed techniques as appropriate.

Clearly, systems psychology cannot stand alone, and recognition of its relationships to other fields is important. If the same methods are applied at the level of the brain, the scientist is a physiological psychologist or physiologist. If the focus is on applying knowledge as to *specific* human capabilities and limitations, especially performance capabilities and limitations, to equipment design, the worker is an engineering psychologist; finding the requisite information is the work of an experimental psychologist. The man interested in the interaction of people in groups is a social psychologist.

We already have *specialists* capable of dealing with all sorts of system-related problems. Why establish still another science? One reason is that it is often difficult to relate problems to disciplines. Is self-regulated decrease in viable births, consequent to over-crowding, apparently involving sensory–neural–endocrine mechanisms and observed in mammals from rats to elephants, the province of physiological psychology or social psychology? Is the study of attitudes toward machines a concern of engineering psychology, industrial psychology, or social psychology? The choice of problems within our discipline is largely a function of how we define our jobs. A second reason is that, in practice, scientific generalists are few and far between. Industrial psychology, nominally broad, has proved to be rather static and has generated few, if any, exciting ideas. Human factors has made much progress in integrating human engineering, training, and man-power determination, but so far has largely ignored motivation, attitudes, and social factors in systems.

Systems psychology has roots in physiological, experimental, general, individual, and social psychology, as well as in physiology, the other social sciences, engineering, the computer and information sciences, and mathematics. It is both an applied and a theoretical/conceptual field, but at the present time, emphasis is on applications. Although systems psychology possesses methods appropriate to its level (for example, computer modeling and simulation), it is reciprocally dependent on other sciences and technologies: *One of the most healthful relationships entails recognizing and defining complex real-world problems in terms amenable to solution by the contributory sciences and referring these problems to the relevant specialty science.*

It is not sufficient to establish systems psychology as a well-defined and worthwhile science: We must insist on a careful evaluation of where we are going. Your response might now be, "Eureka! The millenium has arrived." A look to the past in psychology suggests caution, however. Uncertain of its own identity, psychology has all too often tried to emulate the rigor of the physical sciences. We have already mentioned the "standard rat". In the early days of experimental psychology, the human subject was known as the "reagent". The present author, whose initial university education was in chemistry and biology, is still uncomfortable with applications to psychology of such terms as *genotype* and *phenotype*, *molar* and *molecular*. Currently, many of the *in*-concepts and terms derive from information and control theory, but terms like *bit* and *coding* and *channel* should not be used to camouflage a poorly defined problem. Psychology must stand on its own two feet.

We must encourage the development of theory of man-related systems, but not at the expense of basic data collection. As psychologists, we should not be ashamed of perhaps emphasizing fieldwork over experiment. Meteorology, astronomy, geology, zoology, botany, anthropology, and sociology too must rely largely on field observations and measurements. Attempts to predict novae, earthquakes, or volcanic eruptions largely on the basis of laboratory abstractions would seem doomed to failure. Complex socio-technical systems in particular require that basic field data be collected before subaspects of the systems problem can be relegated to the laboratory for detailed study.

Finally, we must urge the expansion of interdisciplinary, systems-related education in a changing psychology. One example is Fein's (1961) discussion of education in the computer-based sciences in 1975; the present book is another.

Systems science is general, widely applicable; we believe that, when judiciously used, it will reshape psychology and psychology's effect upon the world. And you, the reader, should you care to launch yourself into a career as a systems psychologist, will find your world very exciting indeed – and pleasantly remunerative, if you plan carefully, discipline yourself, and don't mind some risk and uncertainty.

How systems come into being

In the last section, we examined the theoretical basis of systems and psychology, an area that we will help advance. At the same time, however, there is a multitude of immediate systems to implement, of "brush fires to put out". The remainder of this chapter and much of the rest of this book is devoted to the practical demands of systems analysis, engineering, evaluation, and management. The most relevant theoretical, experimental, and empirical findings that provide substance to human factors will be discussed in appropriate sequence. A first step in learning to work with systems involves understanding how they originate and develop. In this section, we offer four examples of levels of rigor, taken from the natural and artificial worlds.

The origin, development, and evolution of systems can be conceptualized in terms of spontaneity, type, and degree of control; prior planning and integration; and exigencies of the time. Thus, the development of biological systems is mainly a function of control coded in the genes by the sequence of the nitrogenous bases adenine, cytosine, thymine, and guanine within the deoxyribonucleic acid molecule; by organizers within certain tissues during embryonic development (as appropriate); and by environmental constraints introduced, especially by the presence of chemical and ionizing radiation agents. Such systems are effectively preprogrammed and integrated.

Next come systems based on the most thorough possible long-range analyses of mission, operational, and performance requirements (Ferguson, 1965), involving much prior planning and freedom from ephemeral exigencies. This category includes most modern engineering systems such as the Apollo manned spacecraft system.

In the past, however, and to some extent still, systems have "grown like Topsy". Long-range planning has been absent, often because of military pressures to "get the system operational on schedule". In other cases, subsystems have been superimposed on subsystems, as, for example, in the SCF, outlined earlier, which included specialized – that is, "program-peculiar" – test support facilities (antennas, computers, communications, recorders, conversion equipment, telemetry ground stations, and so on). This specialization reduced the effectiveness of the network of control, tracking, and communications facilities by increasing station "turnaround time", and by necessitating replicate and redundant equipment, additional specialized personnel, and computer programs, and so on. Subsequently, attempts have been made to simplify and integrate the network toward development of a real system. But many interesting human factors problems still remain unsolved in the areas of display, control, man–machine task allocation, and personnel numbers. Unpublished studies by the present author, for example, suggest that personnel reallocations could result in cost savings sufficient to pay for a new computer-generated cathode-ray-tube display system.

Most civil or social systems are just reaching this state. Hence, we can expect obsolescent or obsolete transportation or urban networks superimposed on one another before a true system is evolved. Similarly, buildings in cities tend to be designed apart from their function in the total urban ecology.

Human roles in systems: the degree of mannedness

Human capabilities in the *operation* of such equipment as aircraft and radar were recognized as long ago as World War II. This recognition gave rise to fields variously called aviation psychology, applied experimental psychology, engineering psychology, human engineering, and ergonomics. Even after the importance of human capabilities and limitations in the operation of equipment had been recognized, albeit at the subsystem or "knob and dial" level, equipment continued to be poorly designed for *maintainability*. Still later, it was recognized that *controllability* is an attribute that can be differentiated from operability by virtue of complexity and the dynamic behavior of at least some equipment or vehicles.

At the present time there is considerable interest in man's role as *decision maker* and *manager*. A related role, particularly in complex military and space systems, is that of *analyst*. For example, analysts help evaluate the military threat, and are indispensable to the evaluation and interpretation of satellite tracking, trajectory, and telemetry (both hardware and biomedical) data. The role of analysis in system development is self-evident.

Other roles of men in systems include those of *planner*, *designer*, *producer*, and *evaluator*. These roles can be conceptualized and studied in terms of the amount of human involvement, and real-time or non-real-time nature of human participation, and the types of psychological factors involved. Systems are sometimes dichotomized as manned, e.g., projects Mercury, Gemini, and Apollo; or unmanned, e.g., projects Ranger, Surveyor, and Mariner (all these involve space missions). However, it is more meaningful to speak of *degrees of mannedness*. By definition, all man-made systems involve the participation of human beings.

Human roles also vary with time. Most attention has been paid to the more dramatic real-time operations and control activities. There *is* justifiable concern with human capabilities and limitations involved in, say, piloting an aircraft or spacecraft and in ground-based air traffic control. Yet system success or failure may depend as much on planning, inspection, quality control, computer program design, maintenance, and similar factors.

Finally, behavioral and life scientists have long emphasized clearly identifiable perceptual-motor skills and related physiological (e.g., reaction to high g forces) and anthropometric (reach and dimension) parameters. Poorly understood concepts such as "judgment" were implied. Today, decision making, short- and long-term memory, problem solving, and creativity are receiving much attention. Motivational and emotional factors are still largely ignored in systems contexts, although they are of considerable importance to the field of psychology as a whole. The problem of interfacing these psychological factors with equipment and environmental parameters and with anticipation, analysis, planning, management, and real-time operations is a major one in systems psychology – and in this book.

What happens when the systems approach is not followed?

In conceptualizing and experimenting in the laboratory, if we make a mistake, leave out a variable, or neglect an interaction, we end up with an imprecise theory or limited or misleading results. The same type of mistake in the design or operation of a system can kill hundreds or potentially millions of people, can cost millions or billions of dollars, or can degrade the dignity of our lives.

For example, as this manuscript goes to press, there is considerable congressional, military, industrial, and public concern with military systems that have "gone wrong" in one way or another. In some cases, system hardware was never constructed. In other cases the systems were cancelled, prototype equipment and all, before they became operational. It is estimated that the cost of 68 such "historic" weapons systems over the past 15 years was about $10 billion.* Examples are the B-70 manned supersonic bomber, Skybolt B-52-launched air-to-ground missile, and a nuclear-powered aircraft. In addition to these historic examples, we must consider the contemporary F-111A and B fighter-bomber aircraft, C5A heavy transport aircraft, Cheyenne jet combat helicopter, and Air Force Manned Orbital Laboratory (MOL) spacecraft. We discuss some of these systems in more detail later. The point we emphasize is that we can learn much from our mistakes. We all make mistakes – whether government planner, military planner, industrial system developer, or urban planner. A critical look at what we have done, where we stand, and where we are going is in no way to be interpreted as "finger pointing".

The systems method provides a means for the orderly, integrated, and timely development of systems. Where

* For a list of these "systems that failed", see the *Congressional Record*, vol. 115, no. 59, April 15, 1969.

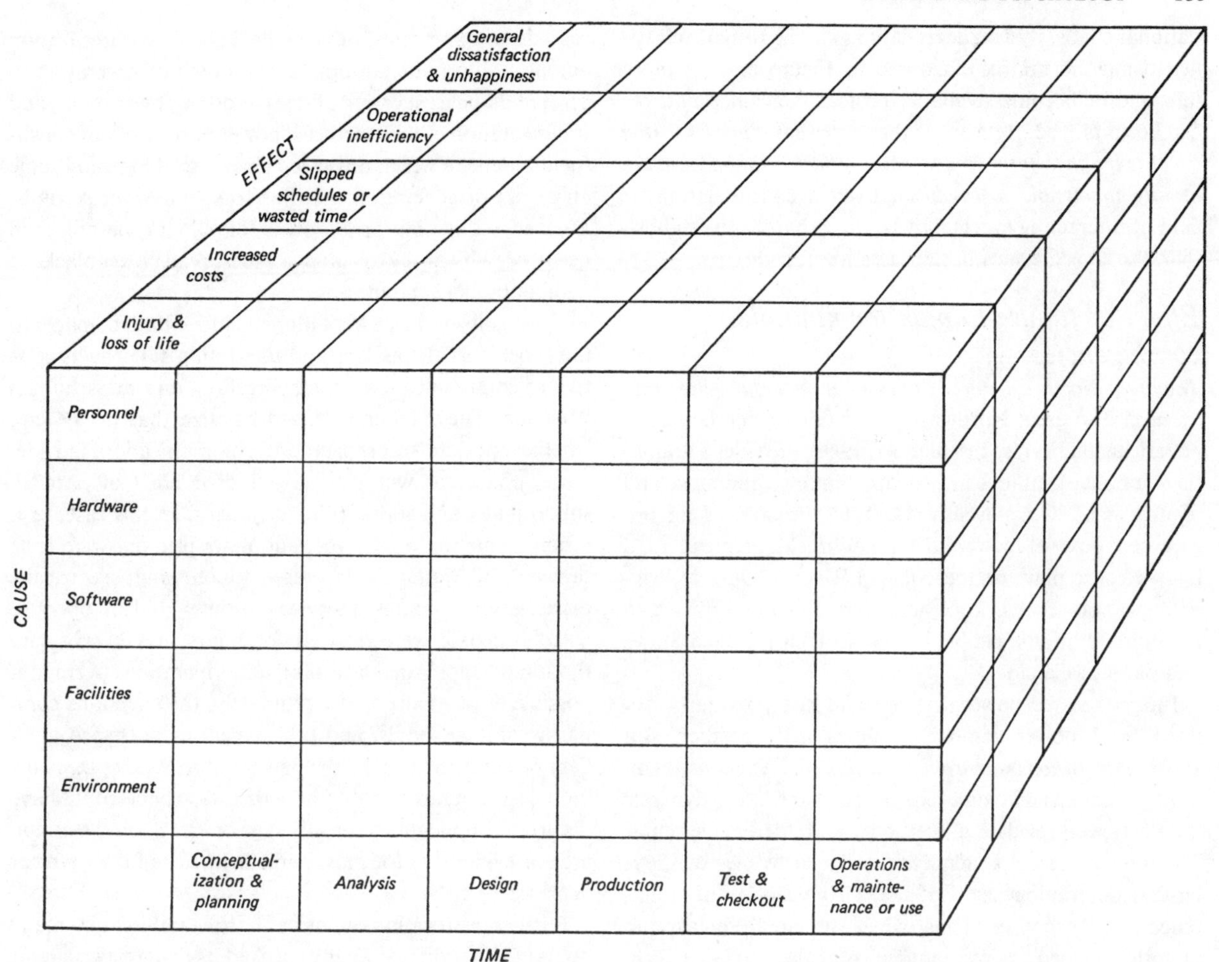

Figure 6.3 First-level cause-effect-time matrix for indicating system degradation. *Note:* Progressively more refined matrices may be constructed. Use extreme caution in establishing cause-effect relationships.

people are involved, careful consideration must be given to human capabilities and limitations. *When these guidelines are not followed, penalties must be paid in terms of increased costs, decreased performance, slipped schedules, accidents, and loss of life.*

A number of examples of failure to follow systems guidelines are outlined in this section. Whereas the *individual events* within each example are probably stochastic, the gross picture in each case has been qualitatively replicated and the cases are of *unequivocal* pedagogic value. Thus, there *have* been several Boeing 727 landing crashes under similar conditions, several midair aircraft collisions under similar conditions, several misinstallations of constituents in missile systems under similar conditions, and so on. All the examples demonstrate some aspect of "human error" (whatever the scientific limitations of that term).

The study of human errors and accidents – classification, prediction, investigation, and prevention – is, of course, an organized specialty in human factors. Accidents themselves provide a valuable source of data to systems psychological understanding of human capabilities and limitations and to design. Throughout this book, however, the theme recurs of human error and how to prevent it. The hardware- and software-related* examples that follow stress dollar losses, loss of lives, decreased operation efficiency, and near

* In this chapter, *hardware* means equipment, including communications equipment; *software* means computer programs and documentation; *facilities* means buildings, roads, airstrips, docks, etc. These three, plus *personnel*, are the four major constituent subdivisions of the systems with which we will deal most of the time. Interfaces are important both between and within major constituents.

national chaos. System degradations can be roughly classified using the matrix illustrated in Figure 6.3. As more information becomes available, progressively finer matrices can be employed. Remember that *every event occurs within a context; therefore, simple cause–effect relationships are usually impossible to establish.* Even a concatenation or field of interacting events may be much harder to establish than the causal events in the cases we now discuss.

Effect of failure to plan for multiple contingencies

The great Northeast power failure: national chaos narrowly avoided. The great Northeast power failure, the largest in American history (see Friedlander, 1966), provides a dramatic example of failures in systems thinking, planning, and management. On November 9, 1965, because of an unexpected flow of power, a circuit breaker tripped at a hydroelectric plant on the Niagara River in Ontario, triggering a power blackout that lasted many hours and involved 30 million people in the Northeast United States and parts of Canada.

Interconnected power systems comprise previously isolated local power company utilities tied together. But *system integration is often poorly advanced.* Each constituent utility company has different output capacities, dynamic inertia, speed-regulating methods, and loading systems. The interconnected systems are designed to meet a *single* large-scale contingency, but not a *combination* of simultaneous disturbances. Thus, when the circuit breaker in Ontario tripped, power shifted to other lines, which, because their own load-carrying capacities were now overloaded, tripped out in *cascade.* Significantly, the backup relay that triggered the disturbance had been set in 1963 at a capacity below that of the line, and perhaps unrelated to subsequent actual power flows. Thus, at about 5.16 P.M. on November 9, 1965, the power flow apparently reached the level at which the relay was set. The relay functioned properly, and its circuit breaker tripped out the line. *Operating personnel were not aware of the relay setting.*

Normally, in power systems, design provides for protection against loss of power generation, not against loss of a very large block of load. In this case, the outage of five transmission lines separated the Canadian power generation from its normal load, leading to a power reversal and superimposing additional power onto New York lines, exceeding the capacity of the New York system and leading to its breakup.

Thus, a simple power-limit pullout led to a chain reaction of power failures. This involved direction reversals of load, cascade tripping off of other units, loss of synchronization among utility units, automatic shutdown of generators to prevent damage, severing of lines to other power pools, and fragmentation of the Northeast power pool into four constituents, some of which possessed an excess of load and some an excess of generation. All this took only four seconds. Further sequential breakdowns of similar nature then occurred, with an end result of a widespread power blackout within about 12 minutes.

The results of the power failure included both damages to the generators themselves and interesting side effects relative to communications, transportation, and mass human behavior. These latter events dramatize that no system, whether power, transportation, communication, air defense, police, or what, exists out of context of parallel, subordinate, and supraordinate systems. In this case, television, some radio, teletype, and news printing were lost; railway and traffic lights ceased functioning; electrically driven gasoline pumps at service stations would not operate; 600,000 people were trapped for hours in subways, and thousands more in elevators; and operations were not possible at most airports. Fortunately, the telephone company, some hospitals, and other agencies possessed auxiliary power sources. Had the power failure taken place on a moonless, blizzardy night, a major catastrophe could have occurred. Implications for sabotage or attack on the nation are self-evident, as for mass panic behavior of the "War of Worlds" sort.

Further, restoration of power after the failure took many hours: Each utility company checked its own relays, circuit breakers, and switches; these companies' frequencies were synchronized with other parts of the system; city underground grid circuits were checked; the power generation sources, themselves to a greater (thermal) or lesser (hydro) extent dependent on electricity, were restored. At the time, the nature and location of the failure were unknown. This dramatizes again the need for sensing mechanisms, display of system status, and personnel trained in recognition of symptoms and contingency procedures.

Major lessons learned were recognition of the need for close human override of automatic equipment; planning for contingencies and improbable events; sensing, display, and monitoring of dynamic system status; better personnel training; and some backup, redundant, or emergency equipment. Further, the system failure dramatically demonstrated the integrated power system interface with other systems, and the resulting, even larger metropolitan or regional system. In avoiding a recurrence, guidance could perhaps be taken from such aerospace systems as SAGE (where track and other data must be maintained by Direction Centers adjacent to an "outage") or the SCF (where contact can be lost between the main control center and an outlying tracking station). An overall supervisory control system is also indicated.

Other power failures have occurred since the great November 9, 1965, power blackout.

Effects of improper analysis of operator requirements

SAGE: blue light and blinking displays. The SAGE air defense system is a salient example of increased computer programming and environmental costs, and reduced efficiency, resulting from the failure to analyze, in depth, operator and controller task information requirements. Too much information – and clutter – was displayed at each console. This increased the load on the display-generating computer program, which then was unable to refresh the (flickering) display on each cathode-ray tube rapidly enough to maintain a character brightness level legible under the conditions of tube and phosphor development of the time. To avoid this problem, the familiar broad-band blue lighting system was used, involving blue filters over ordinary fluorescent light sources and orange filters over the cathode-ray tube faces. The displays could be read and certain other tasks performed under the same level of blue illumination. However, the lighting system imparted an eerie atmosphere to SAGE Direction and Combat Centers, and personnel often complained of headaches after exposure to the blue light. To our knowledge, no studies were made of the behavioral or physiological consequences of exposure to this atypical environment. The system interactions among personnel information requirements, computer programming design, hardware availability, display design, and environmental design are evident. For a consideration of other problems in SAGE and similar systems, see Israel (1965).

Apollo Command Module: fire kills three astronauts. The launch pad fire in the Apollo Command Module in January 1967 exemplifies dramatically a failure of systems engineering and management. The interacting factors included the 100 percent oxygen breathing atmosphere (selected for all U.S. manned space programs), increased combustibility of normally noncombustible materials in a pure-oxygen atmosphere, failure to apply available knowledge of such fire danger, nonfireproof space suit worn by the astronauts, the length of time required to open the escape hatch, and the electric arcing permitted by the circuitry design. Three men died – yet prevention of this accident required no new advances in science or engineering. In systems work, little "knowns" seem often to add up to big "unknowns". The aftereffects of this accident included widespread managerial changes within both the National Aeronautics and Space Administration (NASA) and the contracting organizations, and a year's delay in implementation of the United States manned space program.

Effects of improper methods in design, production, and test and checkout

Nimbus B/Thorad/Agena: incorrect gyro installation leads to spacecraft destruction. On May 18, 1968, a Nimbus B/Thorad/Agena space vehicle was launched at the Western Test Range, Vandenberg Air Force Base, California. The space vehicle consisted of a Thorad (long-tank Thor) primary launch vehicle, a second-stage Agena orbital vehicle, and a Nimbus B spacecraft. The Nimbus was the largest, most heavily instrumented NASA meteorological (weather-mapping) satellite. Within two minutes after launch, during the Thorad-powered portion of the flight, direct visual, tracking, and telemetry data indicated that the vehicle was exhibiting increasing oscillations in yaw and deviating considerably to the left of the prescribed trajectory, finally crossing the range safety boundary. The range safety officer than destroyed the vehicle.

The accident investigation board* found that the malfunction of the Thorad launch vehicle was caused by a misinstallation of the yaw-rate gyro, which provided negative pitch-rate data to the autopilot yaw-control channel, rather than the correct yaw-rate feedback signals. Such a misinstallation required rotation of the gyro (looking forward) 90 degrees clockwise about its long axis. *It was a clear case of human error* associated with: (1) "incorrect installation of the gyro in a correctly fabricated and installed gyro-mounting bracket", or (2) "installation of the gyro in an improperly made gyro bracket", according to the accident investigation report. Specific deficiencies are outlined below.

Manufacturing. Dowel pin holes were drilled in a fixture permitting their mislocation, which resulted in a "proper appearing", but actually incorrect, installation.

Installation. The installation drawings did not show the dowel pin holes in place, and the projections used did not accord with the natural view of the worker. In addition,

* This discussion is based partially on the board's report; see also *Aviation Week*, May 27 and August 19, 1968.

it was possible to depress the guide pin inadvertently, thus losing an orientation to correct installation.

Installation of the long-tank Thor gyro in the vertical position was awkward because of the physical location of the mounting bracket. The installer had to reach 2 to 3 feet above his head to mount the gyro (a one-man operation) because the permanent step ladder was too low. A variety of rate gyro brackets, intended for use on different models of the Thor booster, could be interchanged, installed on the wrong vehicle, or misinstalled. *Test and Inspection Procedures.* Procedural inadequacies made it impossible to accurately determine correctness of installation. For example:

- It was possible to obtain valid-appearing data even from an improperly installed gyro, because it was nearly impossible physically for the operator to move the gyro precisely into its intended plane.
- Check lists of *critical items* to inspect when verifying an installation were not provided.
- Even when members of the accident investigation board were present (such observation was known to the personnel concerned), numerous attempts were made to misconnect electrical connectors, sensitive equipment was nearly damaged, etc.
- The acceptance testing procedures and specifications of the primary contractor permitted acceptance of an incorrectly installed gyro.
- Procedures for conducting gyro tests (not related to differences in test equipment) differed among launch sites, at the factory, and even on the same launch complex during different phases of prelaunch testing.
- During testing, operators did not wear their headsets, leading to miscommunication of information.

This case exemplifies the omission of long understood human factors principles associated with design, manufacture, testing, procedures, and supervision. We also can note again the importance of taking a systems overview that recognizes the reinforcing of interacting effects of a sequence of events. *The cost: estimated at $62 million!*

Apollo/Saturn: improper labeling leads to engine shutdown. An improper installation of signal wires led to premature shutdown of two of the five second-stage engines on the Apollo 6/Saturn 502 space vehicle launched April 4, 1968 (*Aviation Week*, April 15, 1968). This was the second unmanned flight of the Saturn 5, and the error might well have contributed to the requirement for an additional unmanned flight prior to the first manned flight and precluded hoped-for cost and time savings. NASA had hoped that a third unmanned flight, already programmed, could be cancelled. Postflight analysis revealed that a workman had followed the installation instructions properly, but that the wires had been mislabeled. The cost of another unmanned flight, if necessary to man-rate the launch vehicle, would be approximately $200 million. From the system standpoint, such an error could result from poor human engineering design for maintainability (including technical manuals), inadequate training or personnel selection, lack of motivation, inferior management, or some combination of these factors.

Mariner/Atlas/Agena Venus flyby: computer program error leads to vehicle destruction. In mid-1962, a Mariner/Atlas/Agena combination was launched from Cape Canaveral, Florida. The mission involved flight of the Mariner to the vicinity of the planet Venus. During operation of the second or sustainer stage of the Atlas, it was determined that the vehicle was off course and had to be destroyed. Subsequent investigation revealed a transcription error in coding the computer program, which resulted in the omission of *a hyphen* from the guidance equations before the equations were fed into the missile-guidance computer. The cost: over $35 million! This case exemplifies the importance of designing computer programs that do not assign rigidly prescribed meanings to easily overlooked or forgotten things like commas, spaces, and hyphens, and that ensure the identification of errors as they occur. It also illustrates that, although the actual perpetrator of an error or accident is usually an operator, factory worker, maintenance man, clerk, and so on, the error or accident situation may have received its essential structure long before, during problem formulation, analysis, or design.

Effects of failure to predict man–machine–environmental interactions in operations

DC-7B takeoff crash: evasive maneuver in response to optical illusion. That degraded perception can lead to fatal aircraft accidents is illustrated by the following two cases attributed to optical illusions. In the first case (*Aviation Week*, January 2, 1967),* an Eastern Airlines DC-7B climbing on departure from John F. Kennedy International Airport (New York City) on February 8, 1965, crashed in the Atlantic Ocean off Long Island about 6.26 P.M. (and hence in

* From time to time, the weekly *Aviation Week and Space Technology* presents detailed reports on the results of aircraft accident investigations. It is recommended that you keep a file of these if you wish to begin specializing in this field.

darkness), killing all 84 persons aboard. The Civil Aeronautics Board (CAB) determined the probable cause to be evasive action taken by the DC-7B to avoid an *apparent* collision with a Pan American Boeing 707 descending to land. This maneuver placed the DC-7B aircraft in an attitude from which recovery in time was not possible.

Detailed analysis indicated that one or more illusions, associated with a paucity of stimuli, misleading cues, or a conflict among sensory cues – separately amenable to laboratory study but difficult to isolate operationally – may have been involved. Other perceptual factors such as vertigo led to disorientation. Following recognition of disorientation, as much as 36 seconds would have been required to reestablish orientation using instruments. Under conditions of this particular flight, this period was too long.

The abrupt avoidance maneuver had originally been precipitated by the belated ability of the pilot of the DC-7B to detect the jet aircraft visually. The initial evasive maneuver of the DC-7B was paralleled by an evasive maneuver by the 707, which appeared to negate the former. A second evasive maneuver, a rapid pull up and roll to the right, led to spatial disorientation of the crew of the DC-7B. This vertical bank could be corrected only within a time period less than that required for instruments. Postaccident flight tests indicated five out of six pilots experienced similar illusions of collision under equivalent flight conditions.

Ground radar separation of the two aircraft had disappeared at the time of the initial turn of the DC-7B toward the 707, although in reality the aircraft were separated by 1,000 feet vertical elevation. The CAB recommends that pilots avoid vectoring aircraft on directly converging courses because of problems associated with spatial disorientation. Also, the Federal Aviation Administration (FAA) promulgated a procedure requiring that 2,000- rather than 1,000-foot vertical separation be maintained between inbound and outbound flights in areas where illusions have occurred. Development of automatic collision-sensing and display systems also show promise in preventing future accidents of this kind.

The above case exemplifies the interactions among visual angle and visual field, themselves a function of cockpit design, window size, and placement; depth and movement perception based on size, change, and rate-of-change cues; decision making; mission profile and maneuvers; lack of a horizon and night-time flying conditions; contingency phenomena and operational procedures in both aircraft and at ground control; psychophysiological cues and responses to visual, labyrinthine, and probably kinesthetic stimuli; reaction times and other aspects of task time stress; and cockpit display design. There is no alternative to our development of a total systems understanding.

Boeing 707-Lockheed 1049C midair collision: cloud forms distort perception. The midair collision of a Trans World Airlines (TWA) Boeing 707 and an Eastern Airlines Lockheed 1049C (*Aviation Week*, January 9, 1967) is another example of failure to consider all the variables in a dynamically operating system. This accident occurred on December 4, 1965, while the TWA and Eastern aircraft were enroute to John F. Kennedy International Airport and to Newark (New Jersey) Airport, respectively. The Boeing 707 was approaching the New York area, under an Instrument Flight Rules (IFR) flight plan at an assigned altitude of 11,000 feet. The Lockheed aircraft was approaching the area under an IFR flight plan at an assigned altitude of 10,000 feet. In close sequence, the crews of both aircraft perceived an apparent collision course and took evasive maneuvers, the Lockheed aircraft being pulled up and the Boeing then being rolled first to the right and then to the left. The aircraft then collided at approximately 11,000 feet. Both aircraft suffered structural damage; the Lockheed aircraft, forced to land in an open field, was destroyed by impact and fire. Four persons on the Lockheed aircraft suffered fatal injuries and 49, nonfatal injuries. There were no injuries aboard the Boeing 707. The CAB attributed the collision to misjudgment of altitude separation by the crew of the Lockheed aircraft because of an optical illusion created by the up-slope effect of cloud tops, followed by the previously described evasive measures. At the time of the collision, the area was overcast with cloud tops at just above 10,000 feet. Cloud tops tended to be higher to the north than to the south, resulting in an illusion of upward slope toward the north.

The above example dramatizes better than any laboratory abstraction the figure/ground relationships in perception. In this case, illusion appears to have been a function of the distances between the aircraft, their angular velocities, and the observed rate of change of the range-rate of each aircraft, all superimposed on the false horizon cues provided by the sloping clouds.

Boeing 727 landing crash: display design and inclement weather. A landing crash and burn of an American Airlines Boeing 727 aircraft occurred during darkness and inclement weather at Greater Cincinnati Airport. Fifty-eight of the 62 persons on board were fatally injured (*Aviation Week*, October 24, 1966). The CAB determined that the probable cause of the accident lay in the crew's failure to monitor properly the altimeters during a visual landing approach into environmental conditions leading to reduced visibility. However, careful scrutiny provides detail of a subtler nature involving psychological factors of attention, perception, motivation, and judgment interacting with task structure and load and with display design:

- Late departure from La Guardia (New York) Airport, together with increasing thunderstorm activity around Cincinnati, apparently prompted the crew to expedite their landing. A landing delay or instrument approach would have been a more prudent judgment. Much more psychological research should be performed on motivation, decision making, problem solving, and judgment under conditions of time- and load-stress and fatigue.
- Just before the crash, both pilots were preoccupied with maintaining visual reference to the runway, so that they neglected the altimeters.
- It is well known to human factors scientists that poor altimeter design has been associated with numerous aircraft accidents over the years. In this case, the three altimeters were Kollsman drum-pointer types; thousands of feet were indicated on a rotating drum visible through a slot, and hundreds of feet were indicated by a radial pointer. The range of the altimeters was from minus 1,500 feet to plus 50,000 feet. Crosshatching was printed on the drum adjacent to the critical values between plus 1,000 and minus 1,500 feet. Human factors and other developments leading to the incorporation of audible and additional visual alarms at low altitudes should prove valuable.
- System design should reflect contingency, low-probability, stress factor, and interaction events, particularly for critical segments of a mission profile, such as aircraft takeoff, approach, and landing. These factors should receive particular attention in information display design.

The systems approach as a problem-solving and decision-making methodology

Systems methods applied to practical engineering problems are now intrinsic facets of our technology. These are largely pragmatic, even intuitive, and have grown through trial and error in the face of complex, previously unmanageable problems. Thus, in many ways, the systems approach is one of conceptual problem solving, an attitude or ability to perceive wholes, different levels, and interrelationships, rather than a formal regimen! It cannot be said that practice usually follows the general systems theory discussed earlier in this chapter; on the contrary, theory often has been derived from observation of, and experience with, real-life problem areas and cases. The large-scale systems methods employed so far can thus be said to be more rational, empirical, and observational than experimental. Nevertheless, an appreciable engineering and management methodology has emerged, which has been associated with rather spectacular improvements in costs, efficiency, safety, and timeliness, and which is currently being applied to broader and broader vistas. For purposes of discussion, this methodology can be considered to have four iterative, usually overlapping and reinforcing segments: *analysis*, *engineering*, *management*, and *evaluation*. Collectively, these contribute to the *system-development process*, usually more realistically considered the *system-development cycle*.

Systems analysis

Systems analysis may have a future orientation in which the realism of requirements is degraded by increasing uncertainties (see, for example, Quade, 1966). Ferguson (1965) cautions against expecting formalized, unchanging requirements. Further, in dealing with social systems, most of the input derives from the poorly understood, poorly quantifiable, and the uncertain.

Two examples of the uses and limitations of systems analysis are provided by the B-70 high-altitude supersonic bomber and the F-111 (TFX) fighter-bomber, both reflecting decisions from high levels within the United States government.

In each case, lively political controversy was generated, involving the Department of Defense, the Congress, the Air Force, and other agencies; the Department of Defense's insistence on rigor in the analysis of weapons systems requirements prevailed over more partisan political pressures. The B-70 eventually cost about $1·5 billion; two experimental aircraft were built, one of which eventually crashed, with loss of the pilot's life, during a public relations filming (see Carter, 1968). As originally planned, the program was to lead to a $20 billion fleet of these aircraft. The program was, in Carter's words, "a classic example of failure to analyze rigorously the mission to be performed and the state of the technology the mission requires". One of the most severe limitations of the program was its dependence on the development of an extremely high-resolution radar, associated processing and display equipment, and (perhaps an impossible) *human interpretation capability*. The Department of Defense questioned that this equipment could be available by 1967, as planned, or, indeed, even by 1970. Further, even if the aircraft (later renamed the RS-70) could be developed, *cost/effectiveness* studies indicated that cheaper, less vulnerable, and more effective weapons systems were available. The RS-70 made a poor showing when compared to the Polaris missile-carrying submarine and even to the antiquated B-52. In retrospect, it appears that application of rigorous systems analysis saved the taxpayers some $8·5 billion.

The case of the F-111 aircraft is still unclear. Development of this aircraft was also clouded by political factors, in this case involving industrial contractors as well. Further consideration of these factors is beyond the scope of this book. Suffice it to say that the Department of Defense attempted to reduce the cost of developing separate weapons systems for the Air Force and Navy by developing two versions, the Air Force F-111A and Navy F-111B, of the same basic aircraft. Unfortunately, the program has been plagued with difficulties. Both versions have cost considerably more than originally programmed. The Navy version has been cancelled; among other reasons, the Navy maintains the aircraft is too heavy and cumbersome to operate off a carrier. Several F-111A aircraft have crashed, both in the United States and, under combat conditions, in Southeast Asia.

Systems engineering

Systems engineering connotes design of equipment, communications, computer programs, documentation, job or task assignments for personnel, or usually, some combination of all these factors. Where systems analysis leaves off and systems engineering begins is an arbitrary judgment: Systems analysis can be considered an integral part of systems engineering; alternatively, following Gilmore *et al.* (1967), for example, we might consider the dividing line to be the selection of performance requirements by the customer, user, or design team. Systems engineering, of course, usually has a hardware-design implication, as opposed to a pencil/paper or computer model and simulation implication. Systems engineering design starts when a system is fairly well defined. Traditionally, intuitions, judgments, economics, politics, and expediency have played as great a role in the selection of a design as has analysis, whether poor, good, or nonexistent.

The essence of systems engineering lies within recognizing and understanding constituent specialties – for example, structures, electronics, power and fuel, and crew – and allowing these to function only within the context of integrating these efforts into a whole. The systems engineer must be able to stimulate, direct, and utilize much work that is not "systems engineering", but only specialty or constituent engineering.

Systems engineering developments have taken place within industry; methodology has been generalized from case histories and is based more on inference derived from direct observation than on experimentation or model building. According to Hall (1966), the universities have been rather uncritical in accepting even extreme claims from industry; there is a need for the active development of better curricula, more scholarly analysis and less promotionalism in the present literature, and the formulation and testing of models against actual situations.

Systems management

We have seen that there are systems of all levels of complexity. The systems of primary concern in this book – weapons, aerospace, sociotechnical – are of *immense* complexity. The United States civilian space program, for example, has involved 20,000 contractors; some 300,000 engineers, technicians, and production workers; and expenditures of $35 billion over a decade, of which $24 billion alone has been spent on the manned space effort. This complexity of techniques, disciplines, specialists, contractors, concepts, resources, and building blocks must be brought together in an orderly manner. Engineering systems management or, more simply, systems management techniques have provided an impressive start in the right direction: that of *total system design and management.*

Systems management encompasses techniques developed since the early 1950s, primarily by the Air Force in connection with advanced aircraft, missile, and space systems. NASA's unmanned and manned space programs and the Navy's nuclear submarine projects have also contributed substantive ideas.* Systems management is a time-phased, monitorial, evaluative, and integrative activity involving the recognition of technical (for example, analytic) criteria; the assignment of organizational and contractual responsibilities; the definition of milestones; and the assignment of required documentation, hierarchies of contributing organizations, and feedback responses to initial documentation. In parallel with, and utilizing, systems analysis and engineering, it proceeds from the general to the specific, from the hypothetical to the real, and from the conceptual through research and development and production to the operational. The systems concepts of interface, evolution, and integration are of paramount importance.

* The Program Evaluation and Review Technique (**PERT**), a form of *network analysis* developed for the Navy, deserves special mention.

Before examining the main ideas of systems management in more detail, we will first review the situation before the early and mid-1960s. Neither the user nor the contractor was certain how the system would turn out until it appeared in the field. Often hardware was delivered, but no crews; if there were crews, they were untrained. Often, the operations and maintenance philosophy had not been thought out. Typically, the system was not available on schedule, was difficult to use, and required costly and time-consuming retrofit changes. To determine and correct deficiencies in the management of systems engineering, Air Force Systems Command (AFSC) of the U.S. Air Force conducted a series of management surveys (see, for example, AFSCP 375-2, 1963). Significant lessons were gained from 24 major contractors. Twenty-six major findings were cited. Those of particular relevance to this book include:

- Underestimation of costs.
- Limited standards of manpower utilization (the largest single element of costs) and the related, inefficient utilization of engineering manpower.
- Costly proliferation and duplication of reports.
- Insufficient systems analysis, systems design, and detailed design integration, leading to a cascading effect on production, logistics, and the like, and requiring costly design changes to correct early over-sights.
- Inadequate consideration of reliability in detailed design.
- Unrealistic engineering development and test schedules, leading to slipped schedules and increased costs.
- Late delivery of end items.
- Technical data late, costly, and inaccurate; related technical manuals not validated, late, and of poor quality.

As you might expect from the points emphasized so far, these deficiencies were seldom isolated problems but were *interrelated*, symptomatic of common management problems. As a result of these and other findings, the U.S. Air Force has developed a comprehensive body of systems management techniques, which have been extended to other branches of the Department of Defense. Much of applied systems management has also been derived from the cost/effectiveness systems analysis studies of the RAND school, which has received enthusiastic support in the Department of Defense. A descendant of the concept, the Planning–Programming–Budgeting System (PPBS), which attempts to provide a bridge between the already extant military planning and budgeting, is becoming rather widely accepted in state governments as well.

Systems management can be defined as *the process of planning, organizing, coordinating, controlling, and directing the combined efforts of contractors and other relevant organizations to accomplish system program objectives. It involves an integration, in a time-phased manner, of organizations, responsibilities, knowledge, and data and documentation.* The details of the Air Force program are given in the AFR 375-1 through 375-5 series and the AFSCM 375-1 through 375-5 series. Systems management concepts as applied to personnel subsystem (human factors) management are presented in the *Handbook of Instructions for Aerospace Personnel Subsystem Design* (HIAPSD), AFSCM 80-3.* Most important of the 375 series for our purposes is AFSCM 375-5, *Systems Engineering Management Procedures* (1964). A review of this document and its uses, available to the general reader, is provided by Gelbwaks (1967).

The management of the systems engineering process entails as early and accurate an identification as possible of *total system* requirements, control over the evolution of requirements and designs, integration of technical specialty efforts including human factors, and development of basic data and documentation. In the broadest sense, systems management is applied over the life cycle of the system, including conceptualization, design, development, test and evaluation, operations and maintenance, and senescence and replacement. In actuality, one could make a distinction between *systems development management* and *systems operational management*: The basic Air Force regulation AFR 375-1 states, for example, "systems management does not apply to *actual use* of a system during the operational phase". Thus, the degree and type of management control, the nature of the scientific or technical and engineering processes, and the organizations involved differ among say a conceptual system, an operational system, and a system about to be phased out. In the past, so-called "R&D systems" have been largely free of management control.

The heart of the management of complex systems lies in the identification and control over specific events in the life history of the system. The life cycle of *Air Force Systems* is divided into four formal phases which in turn may be further divided into subphases. The four phases and their salient events can be summarized as follows:

1. *Conceptual phase:* determination of a system concept that will satisfy the stated (mission) requirements and is

* The names and codes for military organizations frequently change, as does the nomenclature for documentation. HIAPSD has been revised and updated, and the design (but not the management) portions have been reissued as AFSC DH 1-3 (1969).

indicated by analysis as suitable, feasible, and acceptable in terms of *performance*, *cost*, and *schedule*; and specification of functions, system interfaces, and gross performance and design requirements.

2. *Definition phase:* definition in greater detail of the basic concept, personnel, computer programs, equipment, facilities, costs, and schedules leading to acquisition phase contractor selection.

3. *Acquisition phase:* design, production, and test (Categories I and II) of the system; the acquisition phase may overlap with the operational phase.

4. *Operational phase:* system test (Category III) and evaluation to determine accomplishment of the mission objectives; and actual operation of the system.

Within and, particularly, among the phases, specific documents are prepared and released by given organizations; these documents call for responses on the part of other organizations and are reflected in feedback modifications of the original documentation.

Such system development depends at each phase on the *previous* development of *basic building blocks*. A series of R&D categories has been established by the Department of Defense in the sequence: basic research, exploratory development, advanced development, and so on, eventually leading to the operational system. Psychologists have the opportunity to participate in basic research and all phases of the system life cycle. They have made contributions in universities, industry, and military and associated nonmilitary organizations.*

There is a burgeoning specialty area dealing with management and the manager, and psychologists have the opportunity to make major contributions both within the field of management, viewed as a system, and with regard to the interpersonal capabilities and limitations of the individual manager. The term *behavioral science* sometimes is used in a very specialized sense to describe the study of human behavior in organizations; in practice, it is more circumscribed with a small-group-dynamics emphasis. Fields of concentration include such areas as leadership, sensitivity training in which the manager hopes to develop a better awareness of himself and his effect on other people, and organizational change.

Management information and control systems are bona fide systems that present all the challenges of complex systems and provide the psychologist with all the opportunities to apply the techniques and information to which this book is devoted. Later, we discuss motivation, morale, job satisfaction, supervision, and group relations – all topics of particular interest to management psychology. For additional general reading on systems management see Johnson *et al.* (1967); on management psychology, see Leavitt (1964) and Leavitt and Pondy (1964); and on management systems, see Schoderbek (1967).

Experience with systems management programs has shown that effective documentation (including use of computer methods) is an absolute necessity for systems development. However, there is evidence of too strict adherence to *procedure*. Numerous industrial contractors have complained about stifled creativity and enforced conformity especially during the early conceptual phase, tight control over their design activities, and the time spent filling out and maintaining accountability forms that (the contractors maintain) could better be spent on design itself. Similarly, there is some evidence of misapplication of tight control to smaller, one-of-a-type or conceptual systems.

Also, it must be remembered that systems management approaches and techniques are evolving. We have a good thing, but it's not foolproof. Consider the C5A heavy-transport aircraft, developed under sophisticated systems management concept, which nevertheless has been estimated to cost twice as much per aircraft as originally budgeted. We must constantly refine our methods, adapt our techniques to fit experiences, and work toward the advancement of systems theory.

Systems evaluation

Systems evaluation begins with the first consideration of pencil–paper alternatives and continues through the use of nonwired, wired but static, and dynamic *mockups* and *simulators*, related in various degrees of fidelity to expected environmental conditions. Sophisticated simulations play a large role in such systems evaluations. As the operational date draws nearer, more and more realistic evaluations become possible. Thus, in the Air Force systems management terminology, *Category I testing* typically involves inplant (within the factory) subsystem testing and evaluation utilizing design engineer personnel; *Category II testing* is system testing in the field, usually with mixed design and user

* Systems analysis, engineering, and management skills, as opposed to specialty training, are not common. For this reason, the Air Force has used intermediary "think tank" organizations such as the RAND, Aerospace, and Mitre corporations, which perform advanced research, study, and planning, or act as intermediaries between the military user and the design, and frequently operations and maintenance, industrial contractor. Such organizations frequently have General Systems Engineering and Technical Direction (GSE/TD) responsibilities.

personnel; and *Category III testing* is system testing in the field under near-operational conditions. "Testing" cannot be separated from "evaluation" in these contexts. Personnel subsystem testing and evaluation is a specialized aspect of particular importance to this book.

The value of systems testing and evaluation – and of systems analysis – rests on successful *definition of criteria*, which presents a major, largely unresolved problem, especially in social systems.

Applied systems psychology: engineering psychology and human factors

In the practical, applied sense, systems psychology has meant engineering psychology and human factors. Engineering psychology, originally largely an outgrowth of experimental psychology,* is concerned mainly with the design of equipment, facilities, and environments to match the capabilities and limitations of people; and to a lesser extent with the selection and training of personnel. Engineering psychology originated in World War II (see Christensen, 1964; Grether, 1968), when it became clear that operational deficiencies in bombing, artillery targeting, submarine sonar detection, and aircraft were associated with poor equipment design and personnel selection and training.† Psychologists were asked to examine these problems, and when the war ended, several research laboratories and independent consulting companies were established. Interdisciplinary work with engineers began with the study of aircraft cockpits, radar and sonar displays, panel layouts, individual displays and controls, fire-control systems, and so on. Human factors groups were founded, grew, and proliferated in numerous military and industrial organizations. Many significant books were published and well received; some were revised and published in later editions (see especially Chapanis *et al.*, 1949; Morgan *et al.*, 1963‡ Gagné, 1963; Woodson and Conover, 1964; and Meister and Rabideau, 1965). The military released *design-criterion* documentation – guidance and specifications – which generally is evolving toward applicability across all departments of the Department of Defense.

Considerable progress in engineering psychology has been made, but the route has not always been clear, the road often rough, and much remains to be done by imaginative, aggressive people interested in specializing in the field. For example, there is considerable confusion in terminology. Engineering psychology and human factors are *not* synonymous: Some 40 percent of people who consider themselves to be human factors experts are *not* psychologists. Further, in practice it is easy to confuse engineering psychology with human engineering, which as we shall see is a single aspect of human factors. Also, in practice, human engineering has been much more subsystem-, even modular-oriented (we should say subsystem- or modular-limited) than should be the case. Then there is a whole group of people who variously call themselves training experts and training directors, who have their own society, the *American Society of Training Directors*, and who may have little association with human factors people. The same is true of the aerospace medicine specialists and the *Aerospace Medical Association*. Finally, we must consider our colleagues in Europe whose specialty, ergonomics, entails both the work of human engineers in the United States and things like work measurement (see, for example, Murrell, 1965). For these reasons, we will use the terms *human factors scientist* and *human factors psychologist* to indicate the expert, psychologist and otherwise, who is concerned with all the interacting elements of the personnel subsystem discussed in the next section. However, he is usually also a specialist in one area – for example, engineering psychology or anthropometry. With regard to psychology in the development of a *given* system, engineering psychology, training psychology, and personnel psychology might be thought of as specialties of human factors psychology. When the human factors psychologist begins to generalize across systems at all levels and to work toward development of basic theories and methods of application, he becomes a *systems psychologist*.

Even more crucial than semantic problems has been that of securing the rightful place of systems psychology among the other sciences and establishing the practical value of its findings. The early work cited above made good sense to most observers, including military managers, and after 1957, contractual requirements, particularly within the Air Force "family", forced the inclusion of human factors considerations – and the hiring of human factors specialists! Release of design-criterion documentation aided this effort immensely. Yet the human factors man, as a human engineer, usually found himself doing what became known as "knob-and-dial" work, minutiae such as indicating the

* The term *applied experimental psychology* was an early synonym.

† Even before World War II, designers were concerned with simplifying the aircraft pilots' job. Information from such instruments as the artificial horizon, rate-of-turn indicator, altimeter, and directional gyro was integrated at a single point of observation and displayed on a cathode-ray tube (Anonymous, 1938).

‡ This so-called "Joint Services (Army-Navy-Air Force) Guide" is scheduled for release in an updated version.

color coding of controls. Design was already frozen when he came into the picture, and isolated from the planning phase, he was powerless to alter situations that, according to his expertise, clearly violated sound principles relating to human capabilities and limitations. At the same time, in other parts of the organization, other human factors specialists were working, usually with great replication of effort, on training requirements, plans, and programs; manpower determinations; and test and evaluation. Appreciation of human factors contributions was often quite perfunctory. Many corporations, both profit and not-for-profit, employed thousands of engineers, but at most a few human factors workers.

Clearly, there was a need both to integrate human factors efforts and to bring this integrated activity at least a couple of years "upstream" into the problem-definition, conceptualization, and planning stages of the system. Coupled with this management shift was the preeminent need to convince high-level decision makers in industrial organizations, the military, and government what human factors and psychology can do.

The establishment of the personnel subsystem concept and program was a step of fundamental importance in meeting the first two needs. Meeting the last need is a matter of (1) diversification of effort; (2) taking a good look at ourselves and where we are going; (3) developing a basic body of theory and methodology, including, but not limited to, quantified methodology; and (4) education, training, and selling ourselves. Human factors scientists now participate at the several stages in the development of many different kinds of systems: weapons and nonweapons; aerospace, ground-based, and undersea; governmental and commercial. They study problems as varied as those involving communication with computers, automobile safety, telephone-user preferences, and aerospace pilot models. For a review of human factors activities up to the mid 1960s, see Lindgren (1966a, 1966b).

The personnel subsystem: a framework for systems management

The importance of the total systems approach, involving development of the appropriate subsystems in *parallel*, is now widely recognized. Thus, a human or crew or personnel subsystem can be recognized and defined, and planned and designed in relation to the structure, power, electronic, environmental control, and other subsystems. Gone are the days, hopefully forever, when aircraft are delivered with no (trained) crews; or costly retrofits are required to make the aircraft or ground system operable or safe.

Systems interactions *within* human factors must also be recognized. The relationships among oxygen partial pressure breathing requirements of Apollo astronauts, the flammability of materials and consequent degradation of safety, the design of the space suit, and the design of the escape hatch have been mentioned. Similarly, there is a clearly recognizable tradeoff among human engineering, training, and personnel selection. The better human engineered a system, the easier and less specialized the training and the more typical the crew member chosen. For example, flexibility in personnel assignment is reduced by a lack of standardization in type and layout of equipment performing similar or identical functions in satellite tracking stations. And placement of some controls in the navigator's position of certain models of the B-52 manned bomber appear to require the arm reach of an orangutan, an Air Force (personnel) Specialty Code hard to fill!

To manage the integrated development of human factors both internally and in relation to other efforts such as reliability engineering, civil engineering, and equipment design, and in the overall context of systems management, the Air Force (about 1960) defined the personnel subsystem (PS) program. PS is a time-phased program involving management, analysis, design, selection, training, and test and evaluation.* It emphasizes man's capabilities and limitations, and their effects on system performance, from the earliest phases of consideration of a (possible) system, through the conceptual, definition, acquisition, and operational phases. Even an obsolescent or senescent system can provide general concepts and specific performance data for incorporation into a new-generation system.

The personnel subsystem concept involves definition of a number of so-called *elements*, which can be interpreted roughly as subfields within human factors and aggregates of the subfields. The number of these elements changes with experience. The 14 defined in Table 6.3 are now in the process of being integrated into a more manageable number; this involves some change in nomenclature. For example, many of the PS elements deal with training and could be grouped together for purposes of generalization. However, it should be noted that the listing in Table 6.3 is to some extent sequential. Thus, for the most part, analysis precedes design and the determination of types and numbers of personnel or training requirements and plans. The personnel equipment data can be thought of as prerequisite to human engineering, QQPRI, and so forth. Similarly, training concepts are basic to the preparation of training

* The term *effectiveness engineering* has been utilized to describe an integrated approach to human factors, maintainability, reliability, safety, cost analysis, and value engineering.

Table 6.3 Personnel subsystem elements (Prepared from text in Headquarters Air Force Systems Command, 1967)

1. *Personnel equipment data (PED)*. Centrally controlled, multilevel, multisource, analytic data on personnel interactions with equipment, environments, etc.
2. *Human engineering*. Application of knowledge of man's capabilities and limitations relative to equipment, facilities, environments, jobs, procedures, computer programs, and performance aids, to achieve optimum safety, comfort, and effectiveness compatible with system requirements
3. *Life support*. Application of physiological, anthropometric, and psychological principles to ensure man's integrity, health, safety, and comfort
4. *Qualitative and quantitative personnel requirements information (QQPRI)*. Determination of the kinds and numbers of persons required in the system for operating, maintaining, controlling, etc.
5. *Trained personnel requirements*. List of personnel requiring system-specific training to support the system through the acquisition phase
6. *Training concepts*. Early planing based on estimates of Air Training Command and of the User Command
7. *Manpower authorizations*. Early allocations required for advanced planning
8. *System manning and trained personnel requirements*. Specialized planning produced for higher headquarters when special manpower problems are anticipated
9. *Training equipment planning information*. Recommendations on the types and quantities of training equipment required
10. *Training equipment development*. Procedures for developing and producing training equipment
11. *Training facilities*. Real estate and buildings used exclusively for training
12. *Technical publications*. Development of manuals to support operations, maintenance, and training
13. *Training plans*. Methods and schedules to effect training
14. *Personnel subsystem test and evaluation (PST&E)*. Coordinated subsystem and system testing under preoperational conditions

plans. Note, however, that system development is usually *iterative*: Test and evaluation data may modify the original requirements and initial design.

Summary and conclusions

This chapter has set the theme of the book.* The artificial systems with which we are primarily concerned lie along a continuum of different types, degrees, and relative directness of human participation. For convenience, we can designate four types of systems: *automatic*, *man–machine*, *sociotechnical*, and *social*. Man does participate as manager, planner, analyst, designer, production worker, operator, controller, maintainer, supervisor, evaluator, and user. In any given position, error, accident, or inefficiency may occur. A consideration of man's capabilities and limitations thus is critical to successful system operation.

It is tempting to seek guidance in the traditional fields of psychology: perception, intelligence, learning, motivation, and so on. To an extent we have done so, as reflected in some of the chapter titles, but this is an insufficient model. It can be embellished by viewing man as a system constituent, which at a given time and in a given environmental context acts on, changes, or processes something – at the level of psychology, usually information, but also energy or materials.

* [*Systems Psychology, Ed.*]

As a system constituent, man modifies and is modified by his fellow constituents – the machines, other men, and facilities; by the communicative connections among these constituents; and by the general ambiance. When the outside forces – physical, chemical, biological, and social – exceed a tolerance level set at the moment within the man, or these forces interact with internal psychological forces to reduce that tolerance, we can say that *stress* has occurred. Associated with stress can be severely degraded performance, both on specific tasks and in terms of general efficiency. In all cases, inputs to the system constituent are converted somehow into outputs. The search for the appropriate transfer functions is one of our ongoing responsibilities; therefore, the S–O–R paradigm can be added to the model for the book and is accordingly reflected in the chapter organization.

It is probably futile for psychology to proceed by itself apart from integration within the mainstream of science and technology. The body of a science can be an ephemeral thing. This does not mean aping the terminology of physics, chemistry, biology – or of information theory! Sciences can mutually reinforce themselves, each science both giving specifics and receiving generalities. General systems theory, cybernetics, and bionics are attempts toward this mutual support. Progress in psychology should follow from using the generally applicable systems concepts and methods. Otherwise, there is a formidable communications barrier in trying to translate psychological terms into, for example, engineering terms; there is a real danger in being left by the wayside. We have defined a new field, *systems psychology*, because there is no appropriate term within psychology, and because specialties like engineering psychology, industrial psychology, and human factors have been limited in practice. Systems psychology is necessary because of the shifting boundaries of knowledge; because all new science and technology *must* be interdisciplinary; and because of the hierarchical nature, complexity, and generalizability of our effort. Systems psychology, a branch of systems science, then, provides the theoretical framework of our effort.

Burgeoning almost completely apart from the somewhat erratic developments of systems theory, the pragmatic advances of systems analysis, systems engineering, and systems management have shaken our society to its roots. Psychology itself has found expression in human factors. Since World War II, great progress has been made, but major battles remain to be won. A major advance has been the development of the personnel subsystem concept and program within systems engineering and management. This concept also is reflected in the chapter structure of this book.

A major problem still derives from the self-identity of psychology and of human factors; their direction, orientation, and extension; and general education in their practical usefulness. When a systems approach incorporating psychological factors is *not* followed in design and management, literally *terrible* things can result at worst, frustrating and costly things at best. We can be of greatest service to our organization, our society, our nation, and our world by developing our theory, coalescing our methods, assuming a large-scale perspective of problems, and *aggressively* attacking these problems. It's an exciting world, offering adventure at each step. You, our young champion, are a far cry from the poor, jaded chap who entered this chapter.

References and bibliography

Anonymous (1938) "Simplifying the Pilot's Task", *Scientific American*, December, 159(6), 308.

Anonymous (1968a) "Use of Existing Spares Urged for Early Nimbus Replacement", *Aviation Week and Space Technology*, May 27, 88(22), 31.

Anonymous (1968b) "Nimbus Abort Laid to Gyro Misalignment", *Aviation Week and Space Technology*, Aug. 19, 89(8), 17.

Ashby, W. R. (1960) *Design for a Brain*, New York: Wiley.

Boulding, K. E. (1956) "General Systems Theory: The Skeleton of Science", *Management Science*, April 2, 197–208.

Boulding, K. E. (1964) "General Systems as a Point of View", in Mesarovic, M. D. (ed.), *Views on General Systems Theory*, Proceedings of Second Systems Symposium at Case Institute of Technology, New York: Wiley.

Butsch, L. M. and H. L. Oestreicher (Cochairmen) (1964) *1963 Bionics Symposium: Information Processing by Living Organisms and Machines*, Wright-Patterson AFB, Ohio: Aeronautical Systems Division, Technical Report ASD-TDR-63-946, March.

Carter, L. J. (1968) "The McNamara Legacy: A Revealing Case History – Death of the B-70", *Science*, Feb. 23, 159(3817), 859–63.

Christensen, J. M. (1964) *The Emerging Role of Engineering Psychology*, Wright-Patterson AFB, Ohio: Aerospace Medical Research Laboratories, Technical Report AMRL TR-64-88, September.

Chapanis, A., *et al.* (1949) *Applied Experimental Psychology: Human Factors in Engineering Design*, New York: Wiley.

Coan, R. W. (1968) "Dimensions of Psychological Theory", *American Psychologist*, 23(10), 715–22.

Fein, L. (1961) "The Computer-Related Sciences (Synnoetics) at a University in the Year 1975", *American Scientist*, June, 49(2), 149–68.

Ferguson, J. (1965) "Military Electronic Systems for Command and Control", *IEEE Transactions on Military Electronics*, April, MIL-9(2), 80–7.

Friedlander, G. D. (1966) "The Northeast Power Failure – A Blanket of Darkness", *IEEE Spectrum*, 3(2), 54–73.

Gagné, R. M. (ed.) (1963) *Psychological Principles in System Development*, New York: Holt, Rinehart and Winston.

Gelbwaks, N. L. (1967) "AFSCM 375-5 As a Methodology for System Engineering", *IEEE Transactions on Systems Science and Cybernetics*, June, SSC-3(1), 6–10.

Gérardin, L. (1968) *Bionics*, New York: McGraw-Hill.

Gilmore, J. S., *et al.* (1967) *Defense Systems Resources in the Civil Sector: An Evolving Approach, an Uncertain Market*, Washington, D.C.: U.S. Arms Control and Disarmament Agency, Paper E-103, July.

Goslin, D. A. (1968) "Standardized Ability Tests and Testing", *Science*, Feb. 23, 159(3817), 851–5.

Grether, W. F. (1968) "Engineering Psychology in the United States", *American Psychologist*, 23(10), 743–51.

Hall, A. D. and R. E. Fagen (1956) "Definition of System", in Bertalanffy, L. von and Rapoport, A. (eds.), *General Systems Yearbook of the Society for the Advancement of General Systems Theory*, I, 18–28.

Hall, A. D. (1966) "The Present Status and Trends in Systems Engineering", *IEEE Transactions on Systems Science and Cybernetics*, August, SSC-2(1), 1–2.

Headquarters Air Force Systems Command (1967) *Handbook of Instructions for Aerospace Personnel Subsystems Design*, Andrews AFB, Washington, D.C.: AFSCM 80–3.

Headquarters Air Force Systems Command (1969) *Personnel Subsystems*, Andrews AFB, Washington, D.C.: AFSC DH 1-3.

Israel, D. R. (1965) *System Design and Engineering for Real-Time Military Data Processing Systems*, Bedford, Mass.: Electronic Systems Division, Technical Report ESD-TDR 64-168, January (AD 610 392*).

Johnson, R. A., *et al.* (1967) *The Theory and Management of Systems*, New York: McGraw-Hill.

Jones, R. W. and J. S. Gray (1963) "System Theory and Physiological Processes", *Science*, May 3, 140(3566), 461–6.

Kalisch, R. B. (1968) "The Locomotion Quadruped", *OAR Research Review*, VII(8), 4–5.

Leavitt, H. J. (1964) *Managerial Psychology* (2nd ed.), Chicago: The University of Chicago Press.

Leavitt, H. J. and L. R. Pondy (eds.) (1964) *Readings in Managerial Psychology*, Chicago: The University of Chicago Press.

Lettvin, J. Y., *et al.* (1959) "What the Frog's Eye Tells the Frog's Brain", *Proceedings of the IRE*, 47(11), 1940–51.

Lindgren, N. (1966a) "Human Factors in Engineering: Part I – Man in the Man-Made Environment", *IEEE Spectrum*, 3(3), 132–9.

Lindgren, N. (1966b) "Human Factors in Engineering: Part II – Advanced Man–Machine Systems and Concepts", *IEEE Spectrum*, 3(4), 62–72.

Lindgren, N. (1968) "Purposive Systems: The Edge of Knowledge", *IEEE Spectrum*, 5(4), 89–100.

Lockard, R. B. (1968) "The Albino Rat: A Defensible Choice of a Bad Habit?", *American Psychologist*, 23(10), 734–42.

Meister, D. and G. F. Rabideau (1965) *Human Factors Evaluation in System Development*, New York: Wiley.

Morgan, C. T., *et al.* (eds.) (1963) *Human Engineering Guide to Equipment Design*, New York: McGraw-Hill.

Murphy, C. S. (Chairman) (1966) "Crew Monitoring Cited in 727 Crash (Civil Aeronautics Board Accident Investigation Report", *Aviation Week and Space Technology*, Oct. 24, 95–122.

* Defense Documentation Center (DDC) Acquisition Number.

Murphy, C. S. (Chairman) (1967a) "Collision-Course Illusion Cited in Accident (CAB Accident Investigation Report)", *Aviation Week and Space Technology*, Jan. 2, 86(1), 84–98.

Murphy, C. S. (Chairman) (1967b) "Optical Illusion Cited as Cause of Collision (CAB Accident Investigation Report)", *Aviation Week and Space Technology*, Jan. 9, 86(2), 81–97.

Murrell, K. F. H. (1965) *Human Performance in Industry*, New York: Reinhold.

Newell, A. and H. A. Simon (1961) *Computer Simulation of Human Thinking*, Santa Monica, Calif.: The RAND Corporation Paper, P-2276, April 20.

Normyle, W. J. (1968) "Nimbus B to Test New Weather Sensors", *Aviation Week and Space Technology*, May 6, 88(19), 71–9.

Pauling, L. (1968) "Orthomolecular Psychiatry", *Science*, April 19, 160(3825), 265–71.

Quade, E. S. (1966) *Some Problems Associated with Systems Analysis*, Santa Monica, Calif.: The RAND Corporation Paper, P-3391, June (AD 634 375).

Reichardt, W. (1961) "Autocorrelation, a Principle for the Evaluation of Sensory Information by the Central Nervous System", in Rosenblith, W. A. (ed.), *Sensory Communication*, 303–17, M.I.T. Press and Wiley.

Rosenblatt, F. (1958) "The Perceptron: A Probabilistic Model for Information Storage and Organization in the Brain", *Psychological Review*, November, 65(6), 386–408.

Rowe, W. D. (1965) "Why Systems Science and Cybernetics?", *IEEE Transactions on Systems Science and Cybernetics*, November, SSC-1(1), 2–3.

Schoderbek, P. P. (ed.) (1967) *Management Systems: A Book of Readings*, New York: Wiley.

Simon, H. A. (1961) *Modeling Human Mental Processes*, Santa Monica, Calif.: The RAND Corporation Paper, P-2221, February.

Steele, J. E. (Chairman) (1960) *Bionics Symposium: Living Prototypes – The Key to New Technology*. Wright-Patterson AFB, Ohio: Wright Air Development Division, Technical Report 60-600, December.

Swanson, R. W. (1968) "Automata in Motion", *OAR Research Review*, VII(10), 12–13.

Thomas, B. K., Jr (1968) "Apollo 6 Wiring Flaw Found as Next Mission Is Studied", *Aviation Week and Space Technology*, April 15, 88(16), 28–9.

van Bergéijk, W. A. (1960) "Nomenclature of Devices Which Simulate Biological Functions", *Science*, Oct. 28, 132(3435), 1248–9.

van Bergéijk, W. A. and L. D. Harman (1960) "What Good Are Artificial Neurons?", *Bionics Symposium: Living Prototypes – The Key to New Technology*, Wright-Patterson AFB, Ohio: Wright Air Development Division, Technical Report 60-600, December, 395–406.

von Bertalanffy, L. (1956) "General Systems Theory", *General Systems Yearbook of the Society for the Advancement of General Systems Theory*, I, 1–10.

White, V. (1963) "A Multiple Satellite Real-Time Control Network", *IEEE Transactions on Military Electronics*, October, MIL-7(4), 285–95.

Wiener, N. (1948) *Cybernetics, or Control and Communication in the Animal and the Machine*, New York: Wiley.

Wilson, W. E. (1965) *Concepts of Engineering System Design*, New York: McGraw-Hill.

Woodson, W. E. and D. W. Conover (1964) *Human Engineering Guide for Equipment Designers* (2nd ed.), Berkeley, Calif.: University of California Press.

Young, O. R. (1964) "A Survey of General Systems Theory", in von Bertalanffy, L. and Rapoport, A. (eds.), *General Systems Yearbook of the Society for General Systems Research*, IX, 61–80.

Yovits, M. C., *et al.* (eds.) (1962) *Self-Organizing Systems 1962*, Washington, D.C.: Spartan Books.

7 The human operator in control systems

by George A. Bekey

Introduction

The ultimate responsibility of the human element in a man–machine system is to take *action* that influences the system. The automobile driver receives inputs from the road ahead, from his instruments, from the sound of the engine, from the acceleration forces on his body, and so forth. He processes all this information to arrive at decisions regarding appropriate actions. Finally, he must act – by braking, turning the steering wheel, accelerating, or perhaps by some combination of these actions. This action is the man's *control input* to the machine. Thus, control may be viewed as *the end product of a chain of processes, which begin with information processing and continue through decisions.**

The *relation* between information theory, decision theory, and control theory may also be viewed from a mathematical point of view. Thus, a decision to act may be viewed from the standpoint of ultimate *utility* to the human controller. The relevant mathematical disciplines here are *value theory* or *utility theory*. On the other hand, a decision may be viewed from the standpoint of *maximizing* the information transfer through a system. The relevant dimension of decision theory here would be the *probability theory*. Finally, the decision to act, in the presence of uncertainty, may be constrained by system stability considerations, in which case the relevant discipline is *control theory*. In many practical situations (for example, the pilot's control of an aircraft following failure of a stability augmentation system) the maximization of information transfer or the evaluation of a rational basis for decisions may both have to be subordinated to a need to maintain stability or face complete disaster. Nevertheless, it should be clear that even in such extreme situations, the pilot receives and processes information, makes decisions and exerts control.

In the past, information theory, decision theory, and control theory have developed as *separate* disciplines. Only now is control theory beginning to use the available tools in the other disciplines. Thus, interest is beginning to focus on the adaptive and decision-making behavior of human controllers in complex systems. Although each discipline is treated separately in this book, as the reader reviews the classical problems of manual control at the beginning of this chapter, he should bear in mind the applicability of the concepts as discussed in earlier chapters. *It is quite possible that the most significant developments in the study of man–machine systems in the next decade will include those based on a synthesis of information theory, decision theory, and control theory.*

This chapter surveys the role of man as an element or constituent in a *control system*. Such systems include the steering of an automobile, manual attitude control of a spacecraft, the control of piloted aircraft, manual process control, air traffic control, and, in certain cases, man–computer systems. In all these systems, the human element provides certain inputs to a group of machines, devices, or other fixed elements (sometimes known collectively as "the plant"), and he receives feedback information regarding the state of the system. In general, *a control system involves the manipulation of certain variables to achieve desired or reference values.* Such a reference value may be *fixed*, as for instance the "set point" in the control of a furnace or chemical reactor, or it may be *variable*, as in the pursuit of an evasive target by means of an adjustable set of crosshairs. In general, the fundamental man–machine control system can be viewed as in the block diagram of Figure 7.1, where inputs to the plant are provided by means of a set of *controls* and feedback is obtained by means of *displays*. The man's receptors provide sensory inputs to the central nervous system from which a response R originates. Thus, *from a systems point of view, man can be viewed as an information-processing device. He converts sensory inputs into appropriately coded muscular outputs.* A complete analysis of man as an element in the system of Figure 7.1 requires an understanding of the characteristic of the receptors and effectors, the nature of the information processing in the central nervous system, the psycho-physical relationships existing between displays and receptors on the

* The application of information theory to the study of man's data processing is discussed in Chapters 6 and 7, and decision theory is covered in Chapter 8, of De Greene's book. *Ed.*

one hand and effectors and controls on the other, and the nature of the plant or controlled process. These will be reviewed briefly in the following pages.

Much basic study has gone into understanding the interaction between man and machine. Nevertheless, it is probably fair to say that, *except in certain simple cases, it is not possible at the present time to obtain a clear quantitative measure of the usefulness of man as a system constituent, in contrast to an automatic control device.* Man excels in environmental adaptability, versatility, and ability to discern signals in the presence of noise; his presence makes a control system adaptive and self-optimizing, within certain limits. However, the relative importance of these factors is hard to assess. The optimum selection of a control strategy for a proposed system involves a wide range of disciplines, including psychology, physiology, control systems theory, mechanics, and simulation techniques.

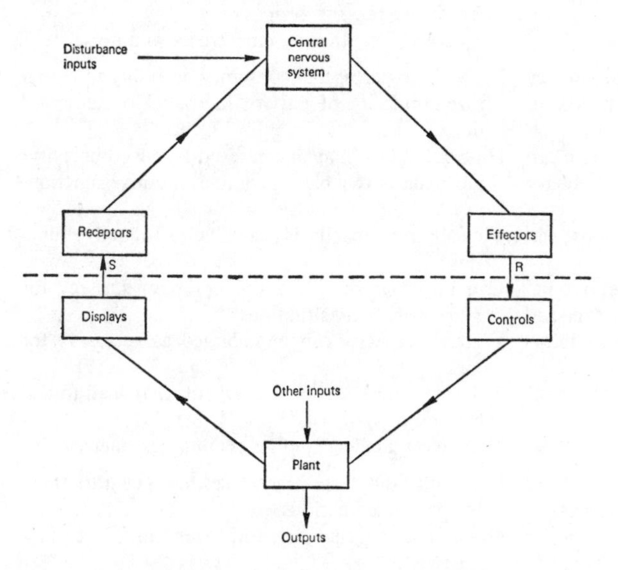

Figure 7.1 Structure of man–machine control system.

In this chapter, we first introduce some of man's input–output characteristics that are important to the design of man–machine systems. The psychological and engineering approaches to the description of man as a control element are then discussed. Display and control factors are reviewed briefly, with some examples of actual and proposed systems. The engineering approach to control systems is then indicated, and some mathematical models of the human operator's function are presented. Finally, simulation of manned systems is examined briefly in terms of stimuli, experimental design, and evaluation criteria.

Characteristics of human input and output channels

The design of a man–machine system, such as a manually controlled spacecraft, requires an understanding of man's characteristics. The effects of these characteristics, notably of the input channels (the senses) and the output channels (largely limb movements and speech) must be analyzed. Details on human characteristics may be found in Sinaiko (1961), Fogel (1963), Pew (1965), Kelley (1968), Lyman and Fogel (1961), and Gagné (1962). To aid in the allocation of sensing and operating functions in man–machine systems, it is convenient to refer to the apparent functional advantages and disadvantages of man and machine in typical system situations.* Nontechnical factors (e.g., government policies), a well as man or machine capabilities, must almost always be considered. A careful comparison of man and machine capabilities and limitations is presented in Table 7.1.

The major input channels useful in system operation are vision and audition, but other senses such as the kinesthetic sense and the perception of acceleration forces are extremely important in many cases. The major output channels are those requiring muscular movement by activation of hand controls, levers, pedals, and similar devices. The human voice is another important output channel.

Input channels

Vision. The major dimensions of *vision* are brightness discrimination, color discrimination, and spatial and time discrimination. Approximately 570 *relative* brightness and three to five *absolute* brightness levels can be distinguished. Spatial discrimination, usually considered excellent, involves visual acuity, depth, form, and movement sensitivity. Typically, the accuracy of spatial discrimination depends on exposure time, as do threshold levels. Temporal discrimination involves 0·04 to 0·4 second at the retina.

These features are important to man–machine systems design because:

- Displays can be coded by color and shape.
- Brightness sensitivity is used in display design.
- Vision is *the* major input sense in man–machine systems.

Audition. The major dimensions of *audition* are pitch, loudness, and duration. Aural spatial discrimination is poor compared with that of vision while time discrimination between sounds is one of audition's best features.

* [Limitations of this approach were discussed in Chapter 2 of De Greene's book. *Ed.*]

Table 7.1 Functional advantages and disadvantages of men and machines (Modified from Lyman and Fogel, 1961)

FUNCTIONAL AREA	MAN	MACHINE
Data sensing	Can monitor low probability events not feasible for automatic systems because of number of events possible	Limited program complexity and alternatives; unexpected events cannot be handled adequately
	Absolute thresholds of sensitivity are very low under favorable conditions	Generally not as low as human thresholds
	Can detect masked signals effectively in overlapping noise spectra	Poor signal detection when noise spectra overlap
	Able to acquire and report information incidental to primary activity	Discovery and selection of incidental intelligence not feasible in present designs
	Not subject to jamming by ordinary methods	Subject to disruption by interference and noise
Data processing	Able to recognize and use information, redundancy (pattern) of real world to simplify complex situations	Little or no perceptual constancy or ability to recognize similarity of pattern in spatial or temporal domain
	Reasonable reliability in which the same purpose can be accomplished by different approach (corollary of reprogramming ability)	High reliability may increase cost and complexity; particularly reliable for routine repetitive functioning
	Can make inductive decisions in new situations; can generalize from few data	Virtually no capacity for creative or inductive functions
	Computation weak and relatively inaccurate; optimal game theory strategy cannot be routinely expected	Can be programmed to use optimum strategy for high-probability situations
	Channel capacity limited to relatively small information throughput rates	Channel capacity can be enlarged as necessary for task
	Can handle variety of transient and some permanent overloads without disruption	Transient and permanent overloads may lead to disruption of system
	Short term memory relatively poor	Short term memory and access times excellent
Data	Can tolerate only relatively low imposed forces and generate relatively low forces for short periods	Can withstand very large forces and generate them for prolonged periods
	Generally poor at tracking though satisfactory where frequent reprogramming required; can change to meet situation. Is best at position tracking where changes are under 3 radians per second	Good tracking characteristics over limited requirements
	Performance may deteriorate with time; because of boredom, fatigue, or distraction; usually recovers with rest	Behavior decrement relatively small with time; wear maintenance and product quality control necessary
	Relatively high response latency	Arbitrarily low response latencies possible
Economic properties	Relatively inexpensive for available complexity and in good supply; must be trained	Complexity and supply limited by cost and time; performance built in
	Light in weight, small in size for function achieved; power requirement less than 100 watts	Equivalent complexity and function would require radically heavier elements, enormous power and cooling resources
	Maintenance may require life support system	Maintenance problem increases disproportionately with complexity
	Nonexpendable; interested in personal survival; emotional	Expendable; non-personal; will perform without distraction

Kinesthesis. An important sensory function in man–machine systems is that provided by the specialized receptors, known as *kinesthetic proprioceptors*, in muscles and joints; these provide feedback information on limb movement, its duration, and, to some extent, the applied force. Joint movements of 0·2 to 0·7 degree can be detected at a minimum rate of 10 degrees per minute.

Other senses. Smell and temperature senses are used mainly as alarm detectors, rather than for fine control. The sense of touch is incorporated in control systems mainly by the shape coding of knobs and other control devices. Note, however, that questions of tactile feedback also are important in connection with remote planetary exploration. In fact, it may be desirable to transmit back to earth signals that can be interpreted by earthbound observers as tactile stimuli regarding the nature of the surface of rocks or other planetary objects.

Angular and linear acceleration senses are of considerable importance in the design of aerospace vehicles. They are responsible in part for "seat of the pants" impressions regarding the movement of a vehicle. They also impose design limits on the acceleration rates of such vehicles to avoid vertigo and consequent disorientation and loss of control.

The ability of human controllers to be aware of the passage of time, and to detect the probability distribution of random events can also be considered as senses. Quantitative data regarding these "senses" are lacking, except under carefully controlled circumstances. It is also significant to note that the human being completely lacks sensors for ionizing and ultraviolet radiation. The detection of x-rays, certain radioactive particles, lethal but odorless gases, and so forth requires the use of specialized detection devices.

Problems with sensory inputs. In many cases there is "cross talk" between different dimensions of the same sense (such as brightness and color in vision) and between different sensory modalities. For example, the effect of strong auditory stimuli on pain thresholds is well known. From a systems viewpoint, such interaction makes it difficult to isolate particular stimulus–response relationships for mathematical analysis. This is particularly so in a complex system in which the human operator receives stimuli simultaneously through a number of sensory modalities, e.g., in a space vehicle where strong visual stimuli occur simultaneously with auditory alarms and violent pitching and rolling movements of the vehicle.

All the sensors are nonlinear. Nonlinearities are of particular importance:

1. The *threshold* phenomena present in all sensory modalities depend on a number of other variables such as vigilance and interaction from other senses.
2. There is a maximum signal that any particular sense is capable of receiving. Stimulation beyond this *saturation* level will produce organic damage or simply no additional change in the receptor output.
3. *Psychophysical* nonlinearities exist between a given stimulus and sensation. Even assuming that over the range between threshold and saturation stimuli a given sensor behaves as a linear transducer, this stimulus and the resulting subjective sensation are *not* linearly related. In some cases, the stimulus level P can be related to the sensation level S by means of approximate laws such as the Weber–Fechner law, $S = k_1 \log P$; or the Stevens power law, $S = k_2 P^n$, where k_1, k_2, and n are constants that depend on the sensory modality involved and the type of continuum being observed (Stevens, 1951).

Output channels

Muscular output. In most control systems, the human controller's input to the machine is obtained from the contraction of skeletal muscles. In manual control systems, the operator is dealing with such devices as toggle switches, buttons, knobs, levers, joysticks, cranks, and steering wheels. Footpedals are used as control devices in both aircraft and automobile applications. In extreme cases, other muscles have been used for control purposes. For example, tongue movement has been used as a control output by quadriplegics at the Rancho Los Amigos Hospital (Waring *et al.*, 1967). Ear movement has been used as a control output for the movement of artificial arms in experiments at the Case Institute of Technology (Bontrager, 1965). The accuracy of muscular movement depends on a number of factors, such as the muscles involved, the limb position and support, the amplitude and direction of motion, and the force required. Small movements tend to merge into involuntary tremors. Large movements tend to undershoot while small movements tend to overshoot the desired position. In general, muscular movement is of low accuracy unless monitored by appropriate feedback (usually visual) in both force and position.

Voice. The human voice is a control output of increasing importance. An aircraft "talked into a landing" by the control tower is evidently being controlled by a speech channel. Of growing significance is the availability of equipment that converts voice into a digital code, which is used directly as an input to a number of control devices. Voice control devices can be expected to assume a considerably larger share of man–machine system interaction in the next decade.

Other human outputs. Among other outputs available from a human operator are various electrophysical signals such as the electrocardiogram (ECG), the electroencephalogram (EEG), and the electromyogram (EMG); the galvanic skin response (GSR), eye movements, skin temperature, breathing rate, and blood pressure. Of these, only a few have been used for control purposes. EMG signals, which give an indication of muscle activity, can be detected and amplified and used as input to control devices. Eye movements can be detected by means of eye-movement cameras or simple biopotential electrodes mounted on the temples and forehead. Such electrodes can provide a useful signal, proportional to the position of the eyes, as an input to a control device.

Problems of output channels. From a systems point of view, two major problems of the output channels become readily apparent: *output rate limitations* and performance deterioration due to *fatigue*. The maximum rate of tapping with the fingers can be shown to be about 8 to 10 taps per second. Similarly, the maximum rate of repeating memorized syllables is about 8 per second. However, as accuracy requirements are imposed on movements, even these relatively low rates cannot be maintained. In fact, within certain limits, operators can trade speed for accuracy in a nearly linear ralationship, thus implying a fixed information-processing capacity. It is also important to note that accurate movements, especially movements requiring considerable amounts of force, cannot be maintained for long periods because of muscular fatigue.

The man–machine control loop: psychological and engineering approaches

The physiological and psychophysiological aspects of human input and output channels were introduced above. In a system design, however, the *overall* input–output transfer characteristics of the human element are of importance. *In many cases it is very difficult to isolate the specific physiological source for the human controller's behavior*.

The basic control system

The block diagram of Figure 7.2 may be considered a representation of a "tracking task", in which the human operator observes on a visual display the difference between a desired input quantity $i(t)$ and the feedback or system response $r(t)$, and adjusts a manipulator, joystick, handwheel, or similar output device in such a manner that the system response agrees with the input as closely as possible. *Tracking research*, involving an investigation of human behavior in systems of the type represented in Figure 7.2, has been performed by both psychologists and engineers for a number of years. It was initiated in connection with problems of tank-turret control and antiaircraft fire control during World War II. More recently, it has been applied to problems of aircraft control, spacecraft control, submarine control, and automobile control (Fogel, 1963; Kelley, 1968; Cooper, 1957; Brissenden *et al.*, 1961). Two overlapping classes of questions may be asked in connection with the block diagram of Figure 7.2. The first category, referred to loosely as the "psychological approach", is concerned with such factors as task difficulty, task loading, human operator vigilance, display–control compatibility, human operator training, learning effects, motivation, and stress. The second group of questions, which characterize the "engineering approach", includes such items as the effect of display gain on the stability of the feedback system, choice of forcing function frequency, the nature of the probability distribution of error, the relation between human operator performance and the performance of an appropriately defined "optimum controller", and the stability of the system with the human operator present. Both classes of questions are concerned with system performance; experience has shown that the degree of training of the human operator, for example, has a significant effect on loop stability margins.

Although the "psychological" and "engineering" approaches in the study of manual control systems are difficult to separate, there are differences of emphasis and motivation. Some psychologists (e.g., Adams and Webber, 1963) have found the engineering approach inadequate and overly confining for describing the details of human information processing. In many cases, psychologists have been concerned with variables such as training, motivation, and stress, while engineers have been concerned with variables such as spring loading and forcing function frequency. However, a more fundamental difference has arisen as a result of the variety of performance measures used in evaluating the quality or state of the complete tracking system. Engineers, as a result of their greater mathematical training, tend to specify the process in such a way as to permit the *deduction* of an appropriate measure. For example, much control system design is concerned with the use of mean-square performance criteria, since it is known that such criteria, when used as a basis of optimum design, lead to linear controllers. Tracking research in the psychological literature, on the other hand, has often been based on a convenient performance measure without a careful analysis of the limitations that may arise from its use. For example, "time-on-target" has been used as a performance measure for some time even though difficulties if interpretation of results have occasionally been demonstrated. An additional problem has arisen in connection with measures of task difficulty, which has been shown to be related in a complex and anomalous way to so many other system variables that it indicates little about the physical requirements of the task.

Types of tracking systems

Two basic types of tracking systems can be distinguished on the basis of the kind of display information presented to the operator:

1. *Pursuit tracking* refers to a situation where the target motion and response motion are separately displayed. The operator attempts to make his response output correspond to the target position, whether it be positioning an instrument needle to follow another one, or making a spot on a cathode-ray screen follow another.

2. *Compensatory tracking* refers to a situation where the display presents the error or difference between the target position and the controlled system response. Thus, in terms of Figure 7.2, the compensatory display presents only the difference between the input or forcing function $i(t)$ and the system output $r(t)$.

The two configurations are presented in more detail in Figure 7.3 (see also Licklider, 1960; McRuer and Krendel, 1957; Tustin, 1947).

Displays

Display design is an important part of manual control. As noted earlier, visual inputs are the most commonly used input channels. Most of the information used by automobile

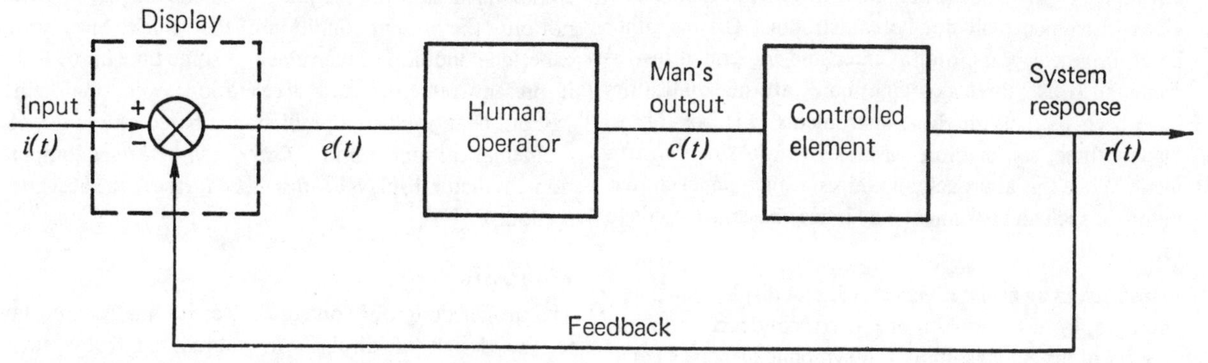

Figure 7.2 Block diagram of manual control systems.

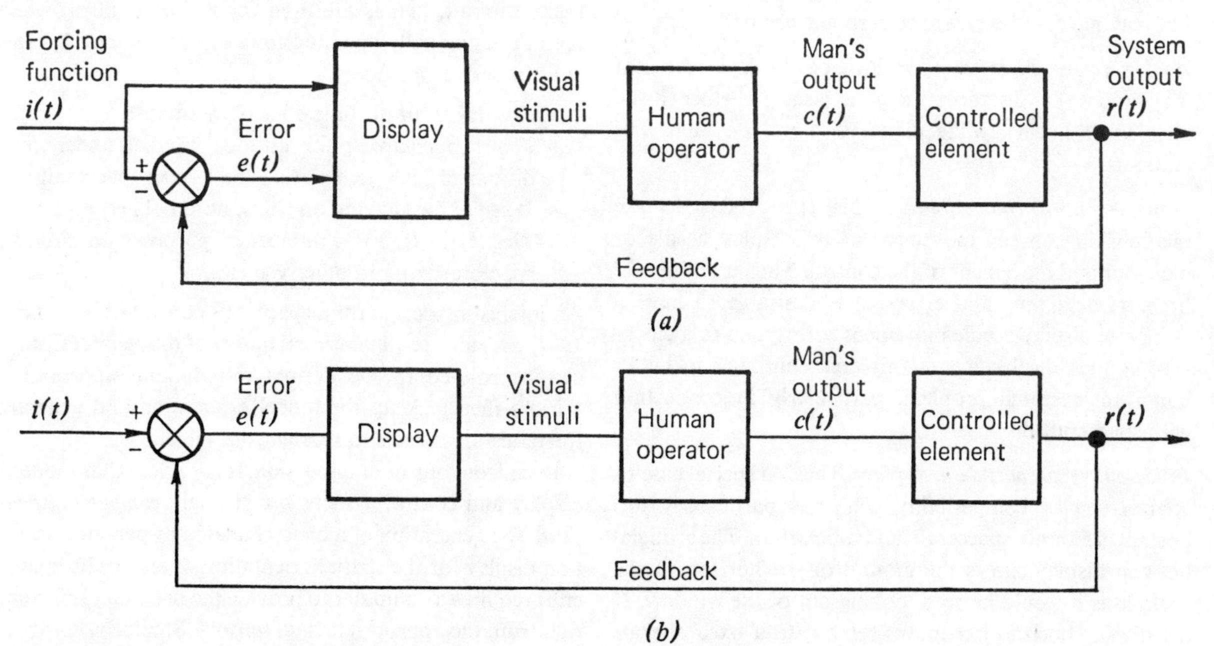

Figure 7.3 (*a*) Pursuit tracking and (*b*) compensatory tracking.

drivers or astronauts for control purposes comes by way of the visual channel, either by direct observation of the "outside world" or by reference to displays. Improved display design can substantially improve operator performance, ease the workload, and reduce skill requirements. A detailed discussion of the problems of display system design are beyond the scope of this chapter; the interested reader is urged to consult Poole (1966) and Luxenberg and Kuehn (1968). However, some aspects of display design are enumerated briefly here.

Separated versus integrated displays. The most common display concepts are based on the use of a separate indicator for each variable to be displayed, along with auditory alarms and warning lights for special purposes. The clear advantage of this approach is that, in general, failure of a given instrument will not be catastrophic. On the other hand, in recent years, multipointer and integrated instruments such as "three-axis eight-ball" attitude indicators have been used. With these instruments it is possible to display three, six, or more variables with a single instrument. While this approach minimizes display panel clutter, failure of such an instrument may indeed be catastrophic to the system.

Literal versus symbolic displays. A literal display, such as a photograph, has a one-to-one correspondence with the features of the actual situation. A symbolic display, such as a map, contains symbols that represent the actual objects but may have no necessary correspondence with them.

Analog versus digital displays. Analog displays represent magnitudes by distances along a scale (whether it be circular or linear), while digital displays use numerical readouts.

Display–control compatibility. This term refers to the relationship between movement of the display needle or indicator and movement of the control. Thus, to minimize both training time and errors, it is desirable to have a clockwise display-needle movement correspond to a clockwise control displacement. This area continues to be an important research problem, particularly in connection with spacecraft displays.

Inside-out versus outside-in displays. This is a special case of control–display compatibility, which is particularly important in aircraft, spacecraft, and submarines. The artificial horizon display shows the motion of the horizon in the cockpit as it would be seen looking out of the window. If the display horizon bar moves relative to a fixed aircraft symbol, it is an "inside-out" display. If an aircraft symbol moves relative to a fixed horizon, it is an "outside-in" display.

Types of displays. In addition to the commonly used dials and tapes, electroluminescent displays recently have been used, cathode-ray tubes are common in many modern display systems, three-dimensional displays are coming into use to provide the proper stimulus to spatial variables, and contact analogs have been in use since 1956. Predictor displays represent another important class. The *contact analog* display (Figure 7.4) is a computed pictorial display, which is an analog of the real-world and real-time situation presented in perspective to the observer. The pattern usually includes an artificial horizon, perspective information, and a textured ground plane. A flight-path generator produces a commanded path for the pilot. *Predictor* displays indicate not only the present condition of the vehicle, but also the expected condition of the vehicle at some time in the future if present velocities and accelerations were maintained without change. Such a prediction is based on the use of a mathematical model and a *faster-than-real-time* computation. Predictor displays are discussed further in a subsequent section.

Controls

The proper design of control devices is equally important to the design of displays in manual control systems. Controls may be hand- or foot-operated. For example, in many aircraft, pedals are used for rudder control, while levers are manually operated for elevator control. Typical control devices include:

- *Joysticks:* for attitude control in aircraft.
- *Fingertip controls:* for attitude control under conditions of high acceleration, such as in spacecraft.
- *Wheels:* for steering on ships, automobiles, etc.
- *Thumbwheels:* for a number of purposes on aircraft, spacecraft, and in other vehicles.

An interesting design for a spacecraft control (Besco *et al.*, 1964) uses a three-dimensional model of the spacecraft that can be rotated to the desired orientation. Appropriate sensors then pick up the model orientation and generate the necessary signals to reorient the vehicle.

Even from our brief discussion, it is evident that modern display and control systems are strongly *computer dependent*. The generation of a contact analog display or a situation display for the spatial orientation of an Apollo spacecraft requires a computer to provide the necessary information from the appropriate data sensors. Similarly, computers are necessary in the control systems of a modern passenger aircraft or a space vehicle, where the forces available to the human operator require augmentation by means of appropriate power assist devices, and where the integrated action of a number of controllers is required to maintain appropriate flight profile and stability. It is probably fair to state that advanced control systems involving human operators will continue to augment human capabilities by the use of computers to generate synthetic displays on which the operator can act and to process the relatively low levels of force under his limited degrees of freedom to obtain the desired vehicle performance.

Simulation of man–machine systems*

One of the most important applications of the simulation method is in the study of systems in which a human being participates, either as an element of the system such as the pilot of a vehicle or as a passenger whose tolerance to environmental characteristics is limited. Simulation techniques are so common in the design of piloted vehicles that in some quarters the term *simulator* is reserved for this type of activity.

Characteristics of manned simulation

In addition to all the characteristics of unmanned simulation, simulation involving man includes the following particular characteristics of human performance:

1. Human performance is inherently time varying. There is variation in successive trials of the same task by the same operator, and there is a variation in the responses of several operators trying the same task.
2. Human response includes elements that apparently are not determined by the input and can only be accounted for by statistical descriptions. Consequently, the description of systems involving human operators must make use of statistical methods, and the resulting descriptions will be in some sense statistical averages defined over particular populations.
3. The inherent variability of human performance implies that many repetitions of each particular experiment must be tried.
4. Simulation studies involving human operators must be run in real time, whereas studies involving nonhuman elements may be run in an accelerated time scale in many cases.
5. The simulation method and the experimental situation must be selected in such a way as to avoid any possible injury to the operators involved.

Simulation of manned systems takes on two primary forms: *environmental simulation* and *man-in-the-loop simulation.* Environmental simulation involves creation of an environment that reproduces one or more unusual system situations in which human beings may have to operate. Man-in-the-loop simulation involves an interaction between man and equipment.

Environmental simulation. Environmental simulators are needed because human beings are often subjected to environments drastically different from those of ordinary life. For example, man may be exposed to situations where high temperatures and high levels of pressure are involved, such as in certain types of mining or underground operations. Man may be asked to undergo long periods of weightlessness such as those occurring in interplanetary flight, or he may be asked to operate in atmospheres different from that of his normal habitat.

To test the adequacy of the proposed design techniques and to ensure human survival, it is absolutely necessary to simulate the characteristics of the particular environment before a vehicle is constructed. Generally, a characteristic problem in the design of such simulators is the selection of the particular quantities or variables to be investigated. For example, it may be decided to construct a simulated space cabin for an interplanetary voyage in which human passengers may be subjected to temperatures, radiation levels, and illumination levels similar to those encountered in the actual flight; it may be decided, however, that no attempt will be made to simulate the gravitational environment of free space. Other simulations may be designed for examining the ability of operators to perform certain tasks under conditions of reduced gravity; certain kinds of supporting harness structures have been used for this purpose. Note that decisions on what is to be simulated and what is to be omitted, what is important, and what is negligible must be made in connection with every simulation.

Environmental simulation has the following major characteristics:

* A portion of this section is based on a chapter entitled "Simulation", by Bekey, G. A. and Gerlough, D. L. in Machol, R. E. (ed.), *System Engineering Handbook*, McGraw-Hill (1965).

1. *Temperature simulation.* Variable climate chambers and hangars have been constructed, some with temperatures ranging from −300° to +1,000°F. Such a chamber may range in size from a cell barely adequate to accommodate one man to a chamber sufficient to accommodate an entire airplane or space vehicle.
2. *Acceleration.* The effect of acceleration and deceleration on human operators and passengers is usually measured using centrifuges and rocket sleds capable of imparting wide ranges of acceleration and deceleration. For example, the human centrifuge at Johnsville, Pennsylvania, has a cabin located at the end of a 50-foot arm. The centrifugal acceleration to which the operator is exposed may reach 40 and 50g. Rocket sleds, such as one located at Holloman Air Force Base, provide acceleration as high as 50 or 60g.
3. *Unusual atmospheric conditions.* Altitude chambers and environmental chambers have been constructed with a capability of generating ice and snow with atmospheric conditions ranging from sea level to 100,000 feet in altitude. Simulated desert sand and dust storms can be generated in certain simulators. Humidity ranges from 0 to 100 percent, salt spray, tropical rain storms, and similar unusual conditions have been produced in the laboratory.
4. *Vibration.* Simulators have been constructed that provide vibration and shock excitation ranging from 5 to 2,000 cycles, as well as random vibration sources with various spectral characteristics. Shock in the range of 0 to 100g and of various durations has also been simulated.
5. *Zero gravity.* Conditions of null or zero gravity have been simulated in airplane cabins while the airplane flies a particular type of trajectory known as a parabolic flight, during which gravitational and centrifugal accelerations exactly cancel, resulting in periods of weightlessness as long as 15 to 35 seconds. Zero g has also been simulated by spinning a man submerged in a fluid.
6. *Lack of atmosphere.* The lack of atmospheric friction and resistance in space for the performance of particular tasks has been simulated by means of minimum friction air-bearing tables.
7. *Complete cabin simulations.* A number of tests have been performed in simulated space cabins that include complete closed-cycle ecological systems. Human volunteers have stayed under simulated space cabin conditions for a number of days. In many cases, such simulated cabins have included temperature, atmospheric composition, and other aspects of the environment in simulated form.

Other environmental simulators have been constructed for the testing of equipment that does not involve human operators. Such simulators include methods for determining the effect of extreme levels of solar radiation, nuclear explosion effects, and so forth.

Since no simulator takes into account each and every effect encountered by a human operator or a human passenger in a particular situation, the addition or superposition of effects observed in various portions of the simulation must be handled with great care. In many cases a simple linear superposition of effects may not be valid. Performance decrement resulting from exposure to untoward environmental conditions is discussed in detail in Chapter 14;* see also Webb (1964).

Flight trainers and piloted simulators. Where a human pilot performs control or guidance functions in the operation of a system, some form of simulation is essential during the design phase. The simulation may be entirely an analog simulation, since in a control task the operator's input and output are generally continuous, or it may be a partially or entirely digital simulation, which may require some form of analog-to-digital and digital-to-analog conversion. In the design of flight-control systems, the simulation generally becomes some form of physical simulation in which there is an interrelationship between a human pilot, an actual or simulated portion of a vehicle control system (including manual controls, displays, dials, knobs, and so forth), and a general-purpose computer (analog or digital) that provides inputs, to the cockpit and operator, representing the variation of environmental characteristics during a particular flight mission. Where the pilot responds to simple dial movements, a general- or special-purpose computer may be adequate to provide the input signals. Where a more realistic simulation of the external environment is required, more elaborate equipment is also necessary.

Attempts to overcome various of the limitations of the fixed-base laboratory simulator of the type discussed above have resulted in a variety of more complex and generally considerably more costly simulators, some of which are outlined below.

1. *The moving-base simulator.* The simulated or actual

* [Of the original text. *Ed.*]

cockpit is mounted on gimbals, suspended on chains, mounted in a sled, or supported in other similar fashion and subjected to movement similar to that of actual mission conditions. All moving-base simulations involve limitations of dynamic range and consequently may provide faithful movement over only certain particular ranges of angular or linear displacement. Furthermore, motion cues may be misleading since a pilot in a simulated space mission will be subjected to the motion in space without the gravitational environment of space. Effects on both physiology and performance may be different from those in space.

2. *The variable-stability airplane.* In an attempt to provide a more realistic simulation of flight-control systems of vehicles under investigation, certain airplanes, helicopters, and other vehicles with adjustable handling characteristics have been developed. These vehicles include an airborne computer, analog or digital, which alters their handling characteristics to simulate the performance of the system under design. Many such variable-stability aircraft have been built and have proven an invaluable research and design tool in the aerospace industry. In fact, the simulation of certain phases of reentry from space has been accomplished using the variable-stability aircraft as a simulator.

3. *Increasing sophistication in physical simulation.* It is possible to include in the simulation a whole range of equipment from a simple simulator cockpit to a complete mockup of the actual vehicle. In airplane simulators, for example, it is common to include not only the cockpit itself, but also the servos, actuators, tail assemblies, hydraulic mechanisms, and similar devices as portions of the simulation to ensure that the performance of the pilot will not be distorted by a possibly inaccurate mathematical description included on a computer.

It is clear that simulation in one form or another is essential for the development of manned vehicles: It is important that man be subjected to simulated conditions before he is exposed to actual and possibly hazardous operating conditions. Thus manned simulation is both a *research tool* and a *design tool*. As a research tool, it enables us to determine conditions that will govern the design of future systems by providing envelopes of satisfactory performance. As an invaluable design tool, it proves the absolutely necessary verification by human subjects of a proposed system configuration. Note, however, that *simulation cannot and should not be a substitute for design.*

Computers used with manned simulators. Some form of computer is required to generate the input signals to the cockpit and process the pilot's output signals in accordance with a predetermined mission such as a particular flight trajectory, a landing on a carrier deck, or a reentry from space. Historically, *analog* computers have been used for flight-control simulators because of: (1) bandwidth requirements, since the mission characteristics as well as the input and output signals contained frequencies sufficiently high as to preclude real-time digital computation; and (2) accuracy compatibility, since in many cases the physical characteristics of the airframe and the atmosphere were known only to levels of accuracy compatible with those of analog elements. Recently, the picture has changed in two respects: first, the increasing speed of digital computers has made possible the real-time digital simulation of certain portions of aerospace missions; and second, airborne digital computers are being used to an increasing degree to handle the complex levels of data processing and computation characteristic of modern high-performance aerospace vehicles. Consequently, it is expected that an increasing use of *digital* computers in flight simulators will be seen in the future – in many cases, in the form of hybrid analog–digital equipment.

Bibliography

Adams, J. A. and Webber, C. E. (1963) "A Monte Carlo Method of Tracking Behavior", *Human Factors*, February, 5(1), 81–102.

Bates, J. A. V. (1947) "Some Characteristics of a Human Operator", *IEE Journal* (London), 94(IIA), 298–304.

Bekey, G. A. (1962) "The Human Operator as a Sampled-Data System", *IRE Transactions on Human Factors in Electronics*, September, HFE-3, 43–51.

Besco, R. O., *et al.* (1964) *Manual Attitude Control Systems: Parametric and Comparative Studies of Operating Modes of Control*, Washington, D.C.: National Aeronautics and Space Administration, Contractor Report 56, June.

Birmingham, H. P. and Taylor, F. V. (1954) "A Design Philosophy for Man–Machine Control Systems", *Proceedings of the IRE*, 42(12), 1748–58.

Birmingham, H. P. (1958) "The Optimization of Man–Machine Control Systems", *1958 IRE Wescon Convention Record*, 272–6, New York: Institute of Radio Engineers.

Bontrager, E. (1965) *The Application of Muscle Education Techniques in the Investigation of Electromyographic*

Control, Cleveland, Ohio: Case Institute, Technical Report 2DC-4-65-13.

Brissenden, R. F., *et al.* (1961) *Analog Simulation of a Pilot-Controlled Rendezvous*, Washington, D.C.: National Aeronautics and Space Administration, Technical Note D-747, April.

Cooper, G. E. (1957) "Understanding and Interpreting Pilot Opinion", *Aeronautical Engineering Review*, 16(3), 47–51, 56.

Elkind, J. I. and Green, D. M. (1961) *Measurement of Time Varying and Non-Linear Dynamic Characteristics of Human Pilots*, Wright-Patterson AFB, Ohio: Aeronautical Systems Division, Technical Report 61-225, December.

Elkind, J. I. and Miller, D. C. (1967) *Adaptive Characteristics of the Human Controller of Time-Varying Systems*, Edwards AFB, Calif.: Air Force Flight Dynamics Laboratory, Technical Report 66-60, December.

Fogel, L. J. (1957) "The Human Computer in Flight Control", *IRE Transactions on Electronic Computers*, September, EC-6, 195–201.

Fogel, L. J. (1963) *Biotechnology*, Englewood Cliffs, N.J.: Prentice-Hall.

Gagné, R. M. (1962) *Psychological Principles in System Design*, New York: Holt, Rinehart and Winston.

Gottsdanker, R. M. (1952) "The Accuracy of Prediction Motion", *Quarterly Journal of Experimental Psychology*, 43, 26–36.

Kelley, C. R. (1962) "Predictor Instruments Look into the Future", *Control Engineering*, 9(3), 86–90.

Kelley, C. R. (1968) *Manual and Automatic Control*, New York: Wiley.

Licklider, J. C. R. (1960) "Quasilinear Operator Models in the Study of Manual Tracking", in Luce, R. D. (ed.), *Developments in Mathematical Psychology*. Glencoe, Ill.: The Free Press.

Luxenberg, H. R. and Kuehn, R. L. (eds.) (1968) *Display Systems Engineering*, New York: McGraw-Hill.

Lyman, J. and Fogel, L. J. (1961) "The Human Component", in Grabbe, E. M. *et al.* (eds.), *Handbook of Automation, Computation and Control*, chap. 2, vol. 3. New York: Wiley.

McRuer, D. T. and Krendel, E. S. (1957) *Dynamic Response of Human Operators*, Wright-Patterson AFB, Ohio: Wright Air Development Division, Technical Report 56-524, October.

McRuer, D. T. and Graham, D. (1964) *Pilot–Vehicle Control System Analysis, Guidance and Control II, Progress in Astronautics and Aeronautics*, 13, 603–21, New York: Academic Press.

McRuer, D. T., *et al.* (1965) *Human Pilot Dynamics in Compensatory Systems*, Edwards AFB, Calif.: Air Force Flight Dynamics Laboratory, Technical Report 65-15, July.

McRuer, D. T., *et al.* (1968) *New Approaches to Human-Pilot/Vehicle Dynamic Analysis*, Edwards AFB, Calif.: Air Force Flight Dynamics Laboratory, Technical Report 67-150, February.

Murphy, G. J. (1962) *Basic Automatic Control Theory*, Princeton, N.J.: D. Van Nostrand.

Obermayer, R. W. and Muckler, F. A. (1963) "Performance Measures in Flight Studies", *Proceedings of the AIAA Simulation for Aerospace Flight Conference*, New York: American Institute of Aeronautics and Astronautics, August, 58–65.

Pew, R. (1965) "Human Information Processing Concepts for System Engineers", in Machol, R. E. (ed.), *System Engineering Handbook*. New York: McGraw-Hill.

Phatak, A. V. and Bekey, G. A. (1969) "Model of the Adaptive Behavior of the Human Operator in Response to a Sudden Change in the Control Situation", *IEEE Transactions on Man–Machine Systems*, September, MMS-10(3), 72–80.

Poole, H. H. (1966) *Fundamentals of Display Systems*, Washington, D.C.: Spartan Books.

Sinaiko, H. W. (1961) *Selected Papers on Human Factors in the Design and Use of Control Systems*, New York: Dover Publications.

Stevens, S. S. (ed.) (1951) *Handbook of Experimental Psychology*, New York: Wiley.

Todosiev, E. P., *et al.* (1966) *Human Tracking Performance in Uncoupled and Coupled Two-Axis Systems*, Washington, D.C.: National Aeronautics and Space Administration, Contract Report 532, August.

Tustin, A. (1947) "The Nature of the Operator's Response in Manual Control and Its Implications for Controller Design", *IEE Journal* (London), 94(IIA), 190–202.

Waring, W., *et al.* (1967) "Myoelectric Control for Quadriplegic Orthotics", *Prosthetics*, December, 21, 255–8.

Webb, P. (ed.) (1964) *Bioastronautics Data Book*, Washington, D.C.: National Aeronautics and Space Administration, Special Publication 3006.

F

Section III
Introduction

Systems ideas have had a considerable impact on sociological thinking. The concept of the organization as a system, of even nations and societies as systems, has been developed and used to try to explain their behaviour and the changes which occur in their structures. Buckley presents the system approach to the idea of society as an adaptive system and he uses the ideas of control and feedback to account for this adaptability. Buckley's article is not an easy one to read because he writes in the specialized "jargon" of the sociologist. He also assumes a familiarity with sociological work which will make the arguments difficult to follow for the non-sociologist. However, the article is powerful presentation of the impact of system thinking on sociology and as such will reward deeper study.

Easton uses a similar approach to model the political aspects of social systems and this short extract only gives the briefest flavour of his approach, which is developed in two books. The extract from a book by Hoos rightly draws attention to the question of value-judgements in systems treatments and makes some pertinent criticisms of the systems analysis technique.

Modelling of social and political systems may seem ambitious in itself but the next two articles go further still. Milsum, a noted control engineer, takes up the challenge of modelling virtually the total system around us, the geosphere, the biosphere, sociosphere, and the technosphere. He comes down in favour of experimenting on a small scale to predict the effects of changes on these systems. But Forrester, in the last article in this section, goes still further and presents a modelling method called system dynamics which, it is claimed, can project the outcomes of our whole world. Forrester's work has generated much controversy, but many of the objections to his work stem from disagreements about the validity of his starting data and assumptions rather than from the method.

Associated reading

Buckley's book *Sociology and Modern System Theory* (Prentice-Hall, 1967) presents his general ideas in more detail. An important work in this area is Katz and Kahn, *The Social Psychology of Organizations*, Wiley, 1966. Easton's two books are *A Framework for Political Analysis* and *A Systems Analysis of Political Life*, published by Prentice-Hall and Wiley respectively, in 1961 and 1965.

Forrester's work and that of his colleagues is presented in several books. The original work on the method appears in *Industrial Dynamics* and *Urban Dynamics*, both by Forrester (M.I.T. Press). Later work is in *World Dynamics*, by Forrester, and *Limits to Growth*, by Meadows, D. H. *et al.*, Earth Island, 1972.

Applications of systems ideas to social groups and organizations are dealt with in *Industrial Social Systems*, Units 5 and 6 of *Systems Behaviour* T241 (Open University Press) and in the complete course: *Systems Management* T242 (Open University Press) described on page 5. An evaluation of system dynamics and world modelling is available in *World Models – sense or nonsense:* Unit 16 of *Systems Modelling* T341 (Open University Press).

8 Systems approach in theoretical perspective
by I. R. Hoos

Said to have "roots as old as science and the management function",[1] and at the same time acclaimed as a product of the Space Age, the systems approach, in definition, theory, and practice, is fraught with paradox. Its dating is only the first of many inherent contradictions. For example, both strength and weakness lie in its myriad forms and manifestations, the very variety of which is encouraged by the latitude of interpretation as to what actually constitutes the systems approach. There is strength, because a concept so generously dimensioned and so encompassing in scope not only has widespread usefulness in many contexts but, through vagueness, maintains a kind of featherbed resilience against attack and, hence, a marked invulnerability to criticism. But lack of articulation conveys weakness, too, the more so because high among the attributes claimed for the systems approach is its precision, in the designation of parameters, identification of objectives, and measurement of inputs and outputs.

For a notion to have become the symbol of the "rational" and the "scientific" in management circles in business and government and yet to be so deficient as to clarity of meaning its truly anomalous. The phenomenon deserves inquiry here, not as a mere excursion into semantics but as a way to understanding the methodology that is so fundamentally influencing modes of thinking for the present and for the coming generation, at least.

Difficulties, contradictions, and complexities stem from three main sources: looseness of the word *system*; laxity as to usage of terms, with virtual interchange among *systems analysis*, *systems engineering*, and *systems management*, and even the occasional self-ascription of an honorific *systematic* to the analysis, engineering, management, or whatever the activity;[2] and convergence of a multiplicity of diverse disciplines and intellectual streams that have somehow been rendered congenial through semantic similitude.

Definitions of system

Countering the assumption of epistemological universality, Webster's *Dictionary* provides no less than fifteen different classes of meanings for *system*.[3] Number one is "an aggregation or assemblage of objects united by some form of regular interaction or interdependence; a group of diverse units so combined by nature or art as to form an integral whole, and to function, operate, or move in unison and, often, in obedience to some form of control; an organic or organized whole." Number two has several subparts, specifically: "(a) The universe; the entire known world;" "(b) The body considered as a functional unit;" and "(c) (colloquial) One's whole affective being, body, mind, or spirit." Number three shifts attention to the nonmaterial: "An organized or methodically arranged set of ideas; a complete exhibition of essential principles or facts, arranged in a rational dependence or connection." Also, "a complex of ideas, principles, doctrines, laws, etc., forming a coherent whole and recognized as the intellectual content of a particular philosophy, religion, form of government, or the like." Hence (number four): "(a) A hypothesis; a formulated theory. (b) Theory, as opposed to practice. (c) A systematic exposition of a subject; a treatise. *All now rare*" (their italics). Number five suggests structure: "A formal scheme or method governing organization, arrangement, etc., of objects or material, or a mode of procedure; a definite or set plan of ordering, operating, or proceeding; a method of classification, codification, etc." Number six carries the same notion further, *viz.*, "regular method or order; formal arrangement; orderliness." Among the meanings which precede and follow it, number seven stands out as worthy of sober contemplation: "The combination of a political machine with big financial or industrial interests for the purpose of corruptly influencing a government." Meanings numbered eight through fifteen are specialized, spanning the alphabet from *b* to *z* with a range of subjects from biological, through legal, to zoological.

Proponents of the systems approach, for all their claims to precision, have so far neglected to specify which of the above definitions they espouse. To judge by an almost universal predilection for the plural form, that is, the *systems* approach, one can only surmise that, in their ecumenism, they embrace all meanings, with the possible exception of number seven. Lack of firm definition leads persons engaged in systems analysis to indulge in a kind of solipsismal virtuosity which, contrary to the tenets of scientific method, generally yields irreproducible results. So reified and ratified, the system is what they say it is, what they conceive it to be. This they study; this they manipulate according to the rules they have set. In this way, other systems are perforce delimited, for they can only interface with, impinge upon, interact with, but never, therefore, be identical with or part of the first system.

Absence of sharp articulation creates another paradox: both arbitrary eclecticism and broad inclusiveness are possible within the rules of the game. For example, public welfare, frequently carved out as an area for system study, is approached as if it were an independent entity, apart from the economic condition, unrelated to the state of the job market, unaffected by cost and distribution of medical services, and divorced from history, geography, culture, and prevailing politics. In like fashion, its information system or, indeed, that of any enterprise, is treated as if it were a self-contained system having its own objectives and little congruence with those of the organization it is purported to serve. On the other hand, waste management – that is, pollutants in land, water, and air, and the even more comprehensive notion of resource management, or environmental quality – has been sketched as one system, with everything from farm, factory, transportation, and nuclear fission swept in.

Systems analysis, systems engineering, and systems management

While lexical laxity can, perhaps, account for the myriad interpretations, broad and narrow, of *system*, only casual usage can explain the virtual interchangeability among *systems analysis*, *systems engineering*, and *systems management*. E. S. Quade, for example, finds systems analysis closely akin to engineering because both are application rather than research oriented. He also includes *systems management* as one of a number of extensions of the body of knowledge originally called "operations analysis",[4] and there he lets the matter remain. Attempts at distinction among the three items have not necessarily led to clarification. One effort[5] views *systems analysis* as supplying the broad framework of the system and identifying the result wanted, *systems engineering* as creating a design that "incorporates the optimized technology," and *systems management* as the overall responsibility for control of the whole procedure.

But for the practical implications of the ready substitutions of these terms, one might dismiss assiduous examination of them as tedious pedantry or precious sophism. In real-life situations, we find that transference has borne with it a convenient incognito laden with vague promises of expertness at execution. Thus, "systems competence" or "systems capability" are often found to be ascribed without discrimination to persons experienced in any of the three poorly defined categories of activity. This has encouraged a display of the superficial appurtenances of each without the supporting or restraining discipline of any. Engineers who have worked on missiles or rocketry delivery systems are accredited with the capability of devising and delivering health care, education, and welfare programmes. In like fashion, persons who have been attached to certain of the research institutes known as "think tanks", with a roving talent for contracts at home and abroad, are accepted as qualified to set up or evaluate systems for everything from hospital administration to urban renewal, practically irrespective of the type of "systems work" they have previously performed. The notion of "systems capability," originated perhaps through lexicographical laziness, has been strengthened by a number of factors and accidents of history, economics, and politics, to be discussed later. The outcome has been a calculated avoidance of specificity, with easy slipover from one area to another accomplished largely by manipulation of the superficial platitudes common to all and a studied neglect of the particulars that often comprise the essential nature of each. Current usage suggests that he who has "systems capability" can analyze, engineer, and manage any system.

Actually, even in the field of engineering, where the term *system* is most at home, this is not the case. There are electrical, mechanical, fluid, and thermal systems, in each of which knowledge is quite specialized. Each classification has its own body of theory, its own developmental history, as well as its analytical descriptions and idealized models. Even though the engineering community itself, in search of a new image, is trying to broaden its range to include all kinds of systems, the principles and modes of thought within it remain substantially unchanged and do not reflect the widened embrace.

Several professors of electrical engineering set forth in

an authoritative textbook what *system* means to them.[6] Deriving a general definition of system as "a collection of objects united by some form of interaction or dependence" from Webster's *Dictionary*, they refine it by specifying that their concern is only with the quantitative aspects of system behaviour. They use mathematics extensively in order to "attain a high level of precision and clarity",[7] and they specify that "each object which is part of a system is characterized by a finite number of measurable attributes and that the interaction between such objects as well as the interdependence between the attributes of each object can be expressed in some well-defined mathematical form".[8] The authors make no claims as to the universal applicability of their tenets. On the contrary, they are meticulous in designating the conditions, which, it may be noted, rarely if ever obtain in situations and systems where social factors and human behaviour play an important part.

They point out that three fundamental questions arise in the analysis of most mechanical systems: "(1) What attributes of the objects of which the system is comprised need be considered? (2) What are the mathematical relations between the relevant attributes of each object in the system? (3) What are the mathematical relations between the attributes of different objects in the system; in other words, what are the relations representing the interactions of objects in the system?"[9] Asserting that learning how to arrive at the answers to these questions constitutes a considerable portion of the education of the engineer or physicist, the authors maintain that mathematical sophistication is essential. Also prerequisite for system design and engineering are rigorous and extensive training in circuit theory, information theory, control theory, optimization techniques, and computer programming.[10] Theorems, hypotheses, and proofs about the relations are expressed in specific mathematical formulas and equations and all of the terms and concepts have special meaning in the context in which they are used.

The recurrence in other milieus of the language and terms used in engineering tasks is noteworthy with reference to our view of systems analysis as a conglomerate of disciplines. Although carefully set forth under strictly circumscribed conditions in its original habitat, technical terminology often degenerates into convenient jargon when transplanted. And in the transfer there is the predilection to concentrate on the quantitative aspect of processes and problems as if these were most important or as if somehow mathematical techniques could help understand and balance all equations – human, social, economic, political – as they do those in engineering textbooks.

Engineers define system in more general terms as "a device, procedure, or scheme, which behaves according to some description, its function being to operate on information and/or energy and/or matter in a time reference to yield information and/or energy and/or matter".[11] Portrayed schematically, the system has been represented this way:[12]

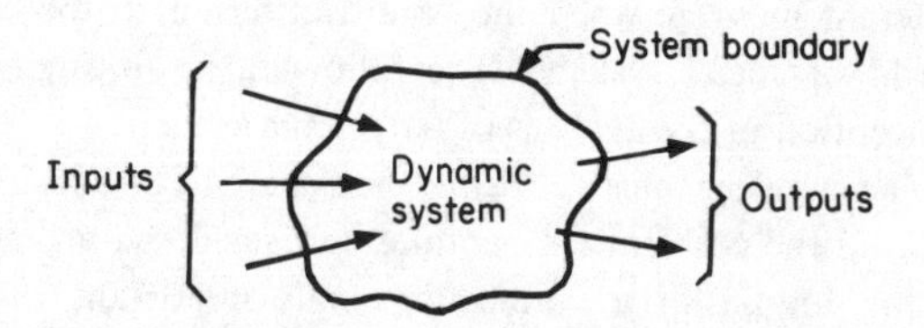

Figure 8.1. Schema of engineering system.

It is described as follows: "a collection of matter, parts or components which are included inside a specified, often arbitrary boundary".[13] The *modus operandi* for dealing with this conception of a system often takes a sort of generalized and simplistic handbook form:[14]

(1) define the system and its components;
(2) formulate the mathematical model;
(3) determine the system equations;
(4) solve the desired output;
(5) check the solution;
(6) analyze or design.

Another example of the procedures for engineering a large, complex system is somewhat similar:[15]

(1) understanding the problem;
(2) considering the alternative solutions;
(3) choosing the optimum system;
(4) synthesis of the system;
(5) updating equipment characteristics and data;
(6) testing the system;
(7) refining the design based on a correlation of test data and requirements.

This ready-made structure provides a convenient framework for analysis of almost any kind of system and is to be found frequently in the preambles to many proposals for a great variety – urban, social, educational, medical, and the like.

Closer examination of the system engineer's definitions as well as his task reveals that his professional preoccupations are much narrower and more precise than the above commonly used think-and-do outlines suggest. The mathematical definition of a system indicating the specific situation as he sees it is presented in Figure 8.2.[16]

When the systems engineering textbook refers to "analysis of a dynamic system", therefore, its use of such terms as *system definitions*, *inputs*, and *energetic interactions* relates to a specific frame of reference. The words have a special meaning in this context and convey quite a different message when they are transferred to the wider world of social systems. The following is an engineer's conception of the analysis of a dynamic system:

The engineer must ... *define* the system to be considered (should the ambient-temperature be considered an input? is the inertia of a connecting shaft important in this situation? etc.). He then may describe the system by means of the various dynamic system elements ... His next step is to investigate the energetic interactions between these two elements when they are interconnected and then excited by some signal. The engineer's most important (but frequently overlooked) job is to establish a *mathematical model* of the system to be analyzed. This involves the identification and idealization of their interconnection. ...

The mathematical statement of the governing relationships between systems variables is called the formation problem. Interconnection of the elements imposes constraints on the variation of system variables, and the convenient way of specifying these constraints is by a mathematical statement of the way in which the various *through-variables* are related and the way in which the various *across-variables* are related. The elemental equations then relate the through- and across-variables for each individual element. This package of equations is a complete mathematical description of the system.

To investigate the *dynamic behaviour* of the system, we must solve this set of equations. The input and outputs are selected and a single differential equation, called the *system equation*, relating each output and input must be determined. The initial state of the system must also be specified by a set of initial conditions. The system differential equation must be solved for the output response under the specified input signal and initial conditions. There are numerous approaches to the *solution* of the system equation, and the choice of a method will depend on the problem at hand. Briefly, these methods are (1) graphical; (2) numerical, with the possible use of a digital computer; (3) operational block diagram of the system with the eventual use of an analog computer; and (4) a purely mathematical solution.

DEFINITION: A system — is an object

$$\{T, \tau, \Gamma, \Sigma, \Omega, \{\Gamma_R\}, \{\Sigma_R\}, \{\omega_{\gamma\sigma}\}, \tilde{\omega}_{\gamma\sigma}\}\}$$

subject to postulates 1 to 5 below:

POSTULATE 1. Γ, Σ, and Ω are sets.

POSTULATE 2. T is a directed set, $(T, \leq)$, and τ is a set of directed subsets of T.

CONVENTION: Names are assigned as follows:

Symbol	Name of Set	Name of Element
T	Chronology	Time
τ	Staging space	Run
Γ	Input space	Input argument
Σ	Phase space	State
Ω	Output space	Output argument

CONVENTION: If $R \in \tau$, certain nets over R are named as follows:

Net	Name
$\gamma : R \to \Gamma$	Input
$\sigma : R \to \Sigma$	Staging
$\omega : R \to \Omega$	Output

POSTULATE 3. For each $R \in \tau$,

$$\Gamma_R \subset \Gamma^R \text{ and } \Sigma_R \subset \Sigma^R.$$

CONVENTION: The sets Γ_R and Σ_R are called the spaces of R-admissible inputs and R-admissible stagings, respectively.

CONVENTION: If $R \in \tau$ and $t \in R$, we denote by R_t the set $\{s \in R \mid s \leq t\}$.

CONVENTION: If $R \in \tau$, $\gamma \in \Gamma_R$, the sets R_γ and R_σ are understood to bear the quasi-orderings induced from R by γ and σ, respectively. Subsets of R_γ, for example, are understood to inherit this quasi-ordering.

POSTULATE 4. If $R \in \tau$, and $\gamma \in \Gamma_R$, and $\sigma \in \Sigma_R$, there is a mapping

$$\tilde{\omega}_{\gamma\sigma} : \Gamma \otimes \Sigma \to \Omega$$

defined and called the $\gamma\sigma$ correlatant.

CONVENTION: If $R \in \tau$, and $\gamma \in \Gamma_R$, and $\sigma \in \Sigma_R$, the mapping

$$\tilde{\omega}_{\gamma\sigma} : R \to \Omega$$

defined by

$$\omega_{\gamma\sigma} = \left(\frac{\gamma}{R_t}, \frac{\sigma}{R_t}\right) \tilde{\omega}_{\gamma\sigma},$$

is called the $\gamma\sigma$ resultant.

POSTULATE 5: If $\in \tau$, $S \in \tau$, $f : R \to S$ is an order isomorphism, $\gamma \in \Gamma_R$, $\sigma \in \Sigma_R$, $\hat{\gamma} \in \Gamma_s$, $\hat{\sigma} \in \Sigma_s$, $\hat{\gamma} = f^{-1}\gamma$ and $\sigma = f^{-1}{}_\sigma$, then

$$\omega_{\hat{\gamma}\hat{\sigma}} = f^{-1} \omega_{\gamma\sigma}.$$

Figure 8.2. Mathematical definition of a system.

After obtaining a solution for the response, the engineer has another major job. The tasks of initial modelling and final analyzing or designing are functions which *clearly* distinguish the engineer from the mathematician. The engineer must *check* his solution (is it correct dimensionally? does it correspond to physical reality? does it check for simplified situations which can be easily analyzed? etc.). His interest in the whole matter of dynamic systems is to eventually analyze to determine whether a certain performance is obtained and is satisfactory, or more generally, he must design the system so that it will meet certain performance specifications.

The process of designing is usually an iterative analysis. The results of one analysis point toward a change in the system which may improve the performance, etc. The design aspects distinguish the engineer from the scientist. The engineer is ultimately interested in building a system which will perform a useful function for the benefit of mankind.[17]

The notion of doing the public good, as expressed in the final sentence above, reflects the engineering community's quest for a social role. One observes it emerging on college campuses as well as in the corporate-industrial milieu. The provost of the Polytechnic Institute of Brooklyn, for example, reported "an astonishing increase in student dedication to the idea of utilizing technology for social and individual benefit".[18] The dean of the College of Engineering, University of California at Berkeley, deplored the "diffuse public image of the engineer as the man peering through a transit, slouched over a drawing board, or wiring resistors on an electronic breadboard". Instead, he cited as a range of possibilities for new social tasks a list presented by the dean of Engineering and Applied Physics at Harvard: "the technology of education, the technology of the delivery of medical care, the technology of urban planning; new transportation technologies; a new information technology evolving towards what one might term a social or collective brain; new technologies of environmental management such as weather modification or waste disposal; oceanic engineering".[19]

Protestations of benevolence to mankind as an underlying philosophy for teaching and learning systems engineering stand in juxtaposition, however, to the pedagogical content of educational materials and exercises in standard curricula at the present time. One finds such tasks as instrumentation tracking radar and command-and-control systems in antisubmarine warfare offered as practical uses for the techniques. A book prefaced by such generally irreproachable sentiments as the desire to construct "a system which will perform a useful function for the benefit of mankind"[20] concentrates on such matters as ascertaining the performance of missile systems. This, the students are taught, may be measured in terms of two probabilistic considerations: the "miss distance" of the missile and the "probability of kill" for a given miss distance.[21]

In contradistinction to precise formulas and in-depth exposition of procedures for calculating the design of mechanical systems, engineering texts and teachers lump social systems together in a happy hodge-podge and treat them cavalierly in "Technology and Society" curricula, the new look in engineering course planning. Organized as a response to what have been interpreted to be social exigencies, such courses are offered as "interdisciplinary" and are usually a parade of social scientists, with an anthropologist, a sociologist, a psychologist, an economist, and an urban planner, each giving a two-hour lecture. This kaleidoscopic gallimaufry of discrete offerings is supposed to convey the fundamentals of accumulated wisdom and experience in each field in sufficient measure to serve as input to the engineer's socialization process. Armed with this exposure, he goes forth to do battle with society's major problems: education, delivery of medical care, urban planning, and social welfare. Among the lessons which he has still to learn, however, is that many of society's pressing problems, which may have been generated or aggravated by technological change and development, are basically social in nature. Calling upon an engineer to cure them is much like asking an economist to treat a heart ailment because the patient became ill over money matters!

Like the curricula, current textbooks in systems engineering contain no specific methods for assessing the performance of the social, biomedical, educational, and other non-space and non-military arenas. Nor is there instruction as to how engineers might include in their cost-estimating the expenses involved in protecting the environment from the effects of the process or product they have devised. Moreover, we soon find that, belying the aura of precision lent by a plethora of formulas, charts, and diagrams, even in the engineer's territory of the calculable and ponderable, exactitude is not always attainable. Experience and judgement in the given field are vital. Many approximations and estimations must be made. With the engineered system the outcome of compromise among such factors as performance, reliability, cost, schedule, maintainability, power consumption, weight, and life expectancy, to mention only a few of the mechanically but not necessarily socially important matters, the process is a good deal less foolproof than the naive are led and inclined to believe.

In the Panglossian* glow surrounding systems engineering as the optimum method for approaching life's complexities, all but overlooked is the possibility that, even on its home ground, the technique cannot claim unqualified success. Some of the very systems cited as exemplars are, in fact, prime examples of miscalculation and mismanagement. In the development of many of them, "costs and times tend to be grossly underestimated and performance tends to be mercifully unmeasured".[22] SAGE, the Semi-Automatic Ground Environment system for air defence, is a case in point. Here, the number of man-hours of required programming was underestimated by six thousand, at a time when the total number of programmers in the world was hardly more than one thousand. Actual installation-wide tests found the programme on schedule at first but it soon slipped one year and then another year. Contrary to assumptions of uniformity, each location was discovered to behave idiosyncratically and, therefore, to require its own custom-tailored, lengthy, and costly programme. Obsolete before it was completed and long before it was paid for, SAGE was successful only because our enemies failed to attack.

BMEWS, the Ballistic Missile Early Warning System, was engineered to detect incoming ballistic missiles through the electronic sensing of the energy they reflected. But its designers apparently forgot that large, distant objects, such as the moon, can reflect as much energy as do lesser, nearer ones. When, early in its operational life, BMEWS detected "incoming ballistic missiles", only a lack of confidence in the system blocked the reflex of counterstrike, which would have precipitated one of the greatest tragedies in history – all because an untested electronic system had been relied on to launch nuclear missiles as a reaction to moonbeams.[23] After study of these and a number of other well-publicized systems, Dr. J. C. R. Licklider, Professor of Electrical Engineering at the Massachusetts Institute of Technology, issued some impressive caveats about the dangers of dependence on a anti-ballistic missile system in particular and technologically contrived and controlled systems in general. Even in tasks which may, in comparison to the DEW Line (Distant Early Warning system) and SACCS (Strategic Air Command Control System), seem modest, conventional, and undebatable, predominance of systems engineering as the accepted procedure may have obscured other, perhaps more promising, approaches. There may be more advantageous ways, with respect to cost, quality, and manpower utilization, to mention only a few of the standard checkpoints, to accomplish even the customary engineering tasks.

The unique contributions of the engineering community to the mythology and confusion surrounding the systems approach have been dealt with at length for several reasons. First, much of what now comprises the systems approach, whether practised by economists, political scientists, or sociologists, is rooted in the heritage from engineering and is permeated by the same basic philosophy. Second, the ascription of universal transferability and applicability has been iterated and reiterated by engineers until it has taken on a kind of reality, as it emerges in concrete form in the world of contract-getting. This is evident in the preamble to – and also generously lards the body of – almost every proposal crafted in response to government requests, practically irrespective of subject area. To judge from statements emanating from national meetings, the professional electrical engineering society, heady with their command and control over computers and mesmerized as to their capability of handling large masses of data, no matter what their substance, conceive that they have a mandate not only to *solve* social problems but actually to *formulate them.*[24] Such hubris is not universal among engineers, but there is widespread evidence of methodological arrogance, which has potential danger. As our research corroborates, a technological conception of a problem limits the focus to those aspects which can be expressed quantitatively and which fit certain models. The technological solution which results may be satisfactory from the engineering point of view but, because it has encompassed only selected facets, vital dimensions may have been neglected. Such violation of the essence of problems may, in the long run, exacerbate rather than ameliorate the troublesome condition.

General system theory

The fact that the systems approach is closely identified with engineering is not mere happenstance; many basic tools and procedures stem directly from that context: the centrality of measurement and mathematics, the structured organization of materials and ideas, the methodological conceptualization of the task. Basic to engineering science, these factors have found ready acceptance in, and are especially compatible with, many other disciplines affected by computer availability. Devotees of

* Pangloss was the character in Voltaire's *Candide* whose philosophy was "all is for the best in this best of all possible worlds."

technology's promise are persuaded that quantification will render their methods rigorous, too, and thus enhance their claim to being scientific.

Although current procedures in the systems approach strongly reflect the influence of engineering science, its intellectual heritage has deeper and wider dimensions. In fact, the approach, as we now encounter it, resembles the geological phenomenon known as "Roxbury pudding-stone" in both history and constitution. This formation, located in a suburb of Boston, Massachusetts, resulted from glacial movement, which over the miles and the centuries dragged with it, accumulated, and then incorporated a vast heterogeny of types of rock, all set in a matrix and solidified in an agglomerate mass. Many fragments still retain their original identity and character; some have undergone metamorphosis in varying degrees. In like manner, the systems approach is a kind of mosaic, made up of bits and pieces of ideas, theories, and methodology from a number of disciplines, discernible among which are – in addition to engineering – sociology, biology, philosophy, psychology, and economics.

Each discipline has its own intrinsic and fundamental conception of system, along with its own definitions, principles, assumptions, and hypotheses. But there is a dynamic which pulls them together, makes them *gemütlich*, and provides them with a mutually supportive kinship. This consists of their orientation to and emphasis on the *totality* of the experience, entity, or phenomenon under consideration. Common emphasis on wholes rather than parts has encouraged a sort of methodological superstructure to be built on a "Cottleston Pie" analogy[25] that serves to sustain a superficial but spurious impression of epistemological universality and consensus. From there, the next step is to the development of quantified measures, computerized computations, and the mass assembling and manipulation of data. While not necessarily germane to the philosophy underlying the total systems concept, these procedures have become practically *de rigueur*.

In sociology

When we review the history of sociology, we find that the conception of society as a total social system has gone in and out of style several times. In its early forms, the approach was bio-organismic. Spencer,[26] a pioneer proponent of this viewpoint, saw society as an organism, similar to biological organisms in a number of essential ways. Both, he claimed, experience growth and, in the process, undergo differentiation in structure and function. Both are composed of units, the one having cells, the other, individuals. Both have special sustaining systems, vasculatory and circulatory in the organism, arteries of commerce for society. Both have special regulatory systems, nervous in the organism, governmental in society. Perceiving in both a mutual dependence of parts, he based the case for his analogy on "the unquestionable community" between them.[27] Many theorists following Spencer espoused his conception of the human society as a homologue of the natural organism and carried it even further. Their social morphology went so far as to equate the epidermal tissue of animals with the protective network of army and police, and to ascribe sex to various social organisms, masculinity to the state, femininity to the church.

To the extent that the hypothesizing was based on a recognition that human society represents a kind of living unity different from a mere sum of the isolated individuals, this approach early demonstrated its usefulness and has continued to do so through its intermittent revivals and recurrences in the history of ideas. To be sure, the logical adequacy of many of the analogies was severely questioned and undermined even before the turn of the century.[28] But in general the resilience of the approach is demonstrated in the fact that it has managed to survive devastating criticism of its basic premises. P. A. Sorokin[29] long ago pointed up the speciousness of the syllogism underlying the fundamental inferences.

Just because human society may be considered a kind of unity in which the members are interdependent, and an organism is a unity of interrelated parts, it does not follow, he argued, that society is an organism. By the same token, the solar system, an automobile, a plant, an animal, a river, or a man represent a kind of unity with interdependent parts, but this does not mean that they are identical. Nor does establishment of their truismatic relationship imply that their intrinsic substance, or the rules governing their behaviour or performance, or the methods best applicable to an understanding of the dynamics of one are proper with respect to another. Criticizing the bio-organismic analogical methods as a prime example of a fallacy in analogical reasoning, Sorokin also pointed up the practical inferences made by various theorists of this persuasion. Some used their bio-organismic premises as an argument in favour of monarchy, administrative centralization, absolutism, or socialism as a form of the greatest integration of social organism.[30] Any pedagogical value likely to be gained from bio-organismic analogies which supply concrete images to help visualize the abstract and complex structure of society was, in Sorokin's view, greatly overweighted by their misuse and their scientific fallacies.[31]

Interpreting Sorokin's observations in light of the current definitions and use of systems concepts, we realize that his criticism is as appropriate and applicable to modern manifestations of the systems approach as it was a generation ago to bio-organismic theories. In the swing of the pendulum between the generalized and the specialized orientations, it is not surprising that Sorokin's criticisms, however cogent, have been all but ignored or forgotten. It is noteworthy that the systems approach, periodically dismissed and rejected as unscientific, has now been revived and is revered on the ground that it is rational and scientific. Its present quantitative underpinnings and technical trappings make it the all-purpose scientific methodology. It is acclaimed a Space Age phenomenon.

Modern sociologists, anxious to render their discipline more respectable and scientific, are tending toward quantification. Many have made the computer the keystone of their research activity and are more preoccupied with amassing and manipulating information than with conducting meaningful research. In response to what has been perceived as a need for rapprochement among theory and method, college curricula are now training the new generation of sociologists in the use of more sophisticated analytical tools; course offerings include symbolic logic, set theory, probability, and other methods derived from mathematics and statistics. The professional journals reflect this orientation, with a preponderance of articles concerned with technical niceties. If one were to rely on superficial review of respective tables of contents, sociology and econometrics journals would be almost undistinguishable. What is important is that beneath the apparent similarity, the common use of symbolic language, and even the technical calculations, there are deep-seated ideological differences.

In quest of more sophisticated analytical tools, modern sociologists have also embraced the systems approach as they construe it. A supporting frame of reference comes from Talcott Parsons' "systematic general theory",[32] which has been sketched as follows: "'System' is the concept that refers both to a complex of interdependencies between parts, components, and processes that involves discernible regularities of relationship, and to a similar type of interdependency between such a complex and its surrounding environment".[33] Furthermore, "the social system is . . . a very complex entity. As an organization of human interests, activities and commitments, it must be viewed as a system and in functional perspective".[34] An attempt at translating the total approach into operational terms was made by Walter Buckley,[35] who finds a striking similarity between society and an organism. He defines a system as "a whole which functions as a whole by virtue of the interdependence of its parts".[36] General system theory, for him, is a method intended to ascertain how the relationships are brought about, organized, and maintained. Developing the thesis that overemphasis on the scientific in various disciplines has led to an avalanche of findings but little knowledge, he argues a case for a "modern system point of view" as a necessary counteragent.[37] Such an approach will, he assumes, embrace and combine a number of divergent disciplines. His earnest assertions that the modern system point of view shows promise of re-establishing holistic approaches without abandoning scientific rigor and his apologia that analogies are not mere metaphors[38] have served more as a booster of the current trend toward indiscriminate application of something called systems analysis to all manner of social phenomena than to development of much needed analytic procedures. Despite the common-sense logic and philosophic congeniality of Buckley's criticism of splinter-minded psychologists and sociologists, his conception of the "modern system point of view" neither stems the avalanche of unrelated findings nor provides an insightful framework for the interpretation of the output of the disciplines and subdisciplines.

In psychology

In psychology, the total approach emerged in the early 1920s as a reaction against the tendency – then fashionable and subsequently never eclipsed – to achieve a scientific method of analyzing its subject matter into constituent elements. The formulations of *Gestalt-theorie* were first developed and tested in perceptual organization, a field of systematic psychology where atomistic reductionism had been entrenched. Wolfgang Köhler, one of the founders of Gestalt psychology,[39] observed striking similarities between certain aspects of field physics and facts of perceptual grouping and coherence. He argued that, just as there are in physics certain instances of functional wholes that cannot be compounded from the action of their separate parts, so also are there counterparts in such human experiences as, for example, memory and understanding.[40] In contrast to the neuron and dendrite counters who were wont to reduce psychology to its least biological denominator, proponents of Gestalt, sometimes translated as the "configurational" theory, chose to recognize form, sense, and value as striking characteristics of mental life.

Gestalt-theorie might have been eclipsed by the increasing force of the very techniques it sought to counteract. Because of its vulnerability to attack for being intuitive, the theory as developed by Köhler and his followers could have been short-lived but for the contributions of Kurt Lewin,[41] who reconciled it with the emergent trend toward symbolic mathematical conceptualization and scientific method generally. For Lewin, the total situation was paramount, with the psychological field or life space his fundamental construct. He created topological psychology or "field theory", interpreting all psychological events as functions of the life span, which consists of the individual and his environment in dynamic interaction. According to his theory, living systems tend to maintain labile equilibrium with their environments; need-tensions, level of aspiration, goal-directed action, and release of tension are the motivational processes in the restoration of equilibrium.

Lewin maintained that with the trend toward progressive narrowing of attention to a limited number of variables, the complexity of real-life situations could only be represented and interpreted by the broadening and continual crossing of the traditional boundaries between the social sciences. His field theory, an extension of the Gestalt approach, had considerable impact because of its focus on the inter-relationships between psychological events, between individual and group, and between groups and the wider social environment. Rarely acknowledged, Lewin's legacy is especially important because he was the spiritual godfather of much of the current activity in group dynamics, from T-groups in industrial relations, to encounter groups in race relations, to *ersatz* primary groups engaged in behaviour modification and attitude manipulation. In 1943, far in advance of the vogue of the 1970s, he put forward the notion of "psychological ecology", which embraced the social channels and gatekeepers that account for the way in which technological, cultural, and economic factors combine to influence choices and decisions.

With the possible exception of Lewin's topological concepts, which lend themselves to manipulation by modern technological tools, Gestalt theory eludes counting and measurement and is, consequently, more useful to psychiatrists and psychologists with a total environment orientation than to those whose methods reflect the current quantitative emphasis. Lack of popularity should not, however, be allowed to obscure the place of Gestalt theory in the historical development of the systems approach, which, by 1955, had already been revived by psychologists. The then accepted definition was:[42] "Systems are bounded regions in space-time, involving energy interchange among their parts, which are associated in functional relationships and with their environments."

Like Gestalt theory a generation earlier, systems theory in psychology was an antidote to the idea of simple stimulus-and-response behaviour patterns, in which are posited "penny-in-the-slot reactions by virtual automatons".[43] To the extent that the systems approach offered a framework for less mechanistic theories, it was welcomed. Less well received, however, were its organismic overtones. Entirely rejected was the attempt, probably in the name of methodological refinement and control, to make a case for the operation of closed systems. Parsons asserted that a social system, like all living systems, was inherently an open system, "engaged in processes of interchange (or 'input-output relations') with its environment, as well as consisting of interchanges among its internal units".[44] Having distinguished four levels of openness in psychological systems, Gordon W. Allport advised that closed systems be left to the realm of physics where they may be more appropriate, although even there, he maintained, "a question exists as to whether Einstein's formula for the release of matter into energy does not finally demonstrate the futility of positing a closed system".[45]

In economics

System concepts in various aspects and forms occupy an important place in the history and development of economic thought. A significant starting point appears to be the Physiocrats, a group known in their day as *Les économistes*, who enjoyed considerable influence in France from 1760 to 1770 and whose long-range imprint is indelible. François Quesnay and his disciples developed an analytic scheme with agriculture at its fulcrum and with *l'ordre naturel* the ideal dictate of human nature as revealed by human reason. Reorganizing the general interdependence of all sectors and all elements of the economic process, Quesnay developed an overall description, known as the *tableau économique*, which displayed the universal compatibility and even complementarity of individual interests in competitive society. The *tableau* is credited as the first method ever devised to convey an explicit conception of the nature of economic equilibrium.[46]

Recognition of an all-pervading interdependence, though a meaningful first step, did not supply answers to the question whether analysis of the interdependence would yield accurate and specific relations, so that prices and quantities of products and productive services that constitute the economic system could be calculated. A significant move forward was made by Leon Walras, who, using theoretical physics as his guide, formulated a mathematical mode of analysis to formulate a system of equations about economic relationships. So great was this achievement that Schumpeter called it "the Magna Carta of economic theory".[47] Convinced that quantitative, mathematical techniques were needed to assure the application of scientific method to economic matters, Walras set economics on a path from which it has not yet departed.[48] His theory of general equilibrium encompassed the fields of exchange, production, capital, and money in a unified formulation, and he assumed a closed system which, although ideal (in Weber's sense[49]), described the normal state toward which the economy spontaneously moves under free competition. With the logic of the system based on the logic of simultaneous equations, there was the implicit suggestion that a set of given conditions could inevitably determine its consequences.

Walras' monumental contributions to economic thought include several which are crucial to modern systems approaches. These are his painstaking application of quantitative methods and use of mathematical equations and his refinement of Quesnay's general equilibrium concept of a century earlier into a heroic abstraction encompassing the entire economic system. While in no way disputing the importance of Walras' work, it is necessary to take note of professional economists' evaluations of these two contributions. On the subject of his use of mathematics, Seligman makes the following observation:

> In Walras' system, for example, each good and productive service was to have its own equation, so that with 10,000 goods and 1,000 factors there would have to be 21,999 equations. For the whole economy the number of equations would be immense: empirical research, consequently, would have to group data in order to make economic investigation manageable. . . . Even more important was the fact that the analysis had little to say about joint and multiple products, *an almost fatal defect*. It may very well be that the institutionalists are right: perhaps more psychology and history and sociology are necessary if we are to grasp the true nature of the human animal. *Mathematics, queen of the sciences, had been shown to need some additional workers.*[50] (My italics.)

The purity of his model of the economic system was a simulated abstraction from its institutional setting. "It was a system devoid of human beings functioning in a complex social matrix. The factors of production remained what they had been throughout all such systems – abstract categories unmoved by the forces that give an economy its characteristic motion. . . . Economics thus became a science of exchange, buttressed by a mechanical technique of maximizing satisfactions. As Milton Friedman has said, the Walrasian model was a form of analysis without much substance".[51]

Nonetheless, economists continue to "walk with Walras",[52] the field of economic theory continues to become more and more the province of the mathematically anointed,[53] and refinements of the system equilibrium hypothesis have rendered it highly respectable. Nothing less than an encyclopedic work like Schumpeter's[54] would adequately trace the growth and development of mathematical techniques in conceptualizing economic relationships. Walras had worthy followers too numerous to list in this context. Perhaps proof of the prevailing and continuing trend toward recognition of this orientation has been persuasively established by the Swedish Academy of Science's selections in 1969 for the first two Nobel Prizes in Economic Science. Honoured were Dr. Ragnar Frisch of Norway and Dr. Jan Tinbergen of the Netherlands for their development of mathematical models for the analysis of economic processes. The high level recognition of the econometric aspect has been interpreted as endorsement of the emphasis on quantitative methodology, possibly even at a sacrifice of theoretical and philosophical considerations. A manifestation of the quest for more precise tools in all of the social sciences, new and improved quantitative techniques in econometrics have also been readily accepted and adopted by other disciplines. Among them are game theory, linear programming, and input-output analysis. With roots in Walras' mathematical conceptions and theory of mutual independence and with branches reaching into the very core of operations research and systems analysis, these closely related techniques are being widely used and should be included in this historical review.

Game theory, linear programming, and input-output analysis, although related in orgin, orientation, and application, came into being separately and merged gradually in economics, where they were developed and widely used.[55] Game theory, developed by John von Neumann[56] in 1928, rests on the hypothesis that there is a similarity between parlor games of kill and conflict situations in economic, political, and military life. Under certain assumptions, each participant can act so as to be guaranteed at least a certain minimum gain or maximum loss. By each participant's acting to secure his minimum guaranteed return, opponents are prevented from attaining anything more than their minimum guaranteeable gains. As a result, the minimum gains become the actual gains, and the actions and returns for all participants are determinate. Game theory has provided a framework in which to devise strategy under competitive circumstances. It has had important implications for economics, military decisions, and statistical theory.[57]

Input-output analysis, developed by Wassily W. Leontief in 1936,[58] is based on the notion that a large part of the dynamics of a modern economy is devoted to the production of intermediate goods, and the output of intermediate goods is closely associated with output of final products. In this closed system of economic activity, there is an inter-relationship linking all sectors of the economy. Equilibrium exists when the outputs of the various products are such that just enough of each is produced to meet the input requirements of all others. Leontief's primary objective is specification of this balance through aggregation, analysis, and interpretation of vast amounts of empirical data.[59] An input-output table, laying out the flows of goods and services, can portray the structure of an industry or an entire economy at any desired level of detail and, by revealing their dynamic inner characteristics, can serve underdeveloped countries in their planning for industrialization.[60]

Linear programming, like calculus, is primarily a mathematical tool, and not, strictly speaking, the province of the economist. Nonetheless, because a substantial class of economic problems fulfils its postulates, and because the technique has been considerably developed and refined by economists, it may appropriately be included here. Defined as "a deterministic model which assumes linear behavioral relationships and in which an optimal solution is sought (maximizing or minimizing) subject to one or more limiting restraints", this method is used to determine the best or optimum use of resources to achieve a desired result when the limitations on available resources can be expressed by simultaneous linear equations.[61] In other words, linear programming concerns itself with the allocation of limited resources in such fashion as to maximize a product or minimize its cost.[62] The method, developed by George B. Dantzig in 1947 for planning the diversified activities of the U.S. Air Force, stimulated two subsequent streams of activity – applications to managerial planning in industry and exploration of the implications for economic theory in general.[63]

Inclusion of game theory, input-output analysis, and linear programming under the rubric of economics is not to be construed as a suggestion that these methods are the private preserve of economists and that all other users are poachers. Perhaps it is because an equally persuasive case could have been made for placing them in the earlier discussion of engineering that we perceive the extent to which operations research and systems analysis, of which they form the essential core, represent a confluence and convergence of ideas. It is noteworthy that in fields as widely separated as engineering and economics there exists such commonality of tools, a phenomenon which merely underscores the ubiquity and widening reliance on mathematics. In many disciplines, the use of mathematical methods, logic, and symbols exemplifies the zenith of technical advance. This is a manifestation of the prevailing *Zeitgeist*, in which "mathematical precision" is accepted as a term that brooks no internal division – though mathematicians do not profess such faith in the infallibility of their methods. These methods, then, as an intrinsic part of systems analysis and applied by practitioners of widely disparate background to problems of public concern, signify the coincidence of the engineers' desire to apply their tools to peaceful, socially useful pursuits with the behavioural and social scientists' quest for more rigorous techniques.

Practice in the use of mathematical tools has not attained perfection nor even consensus that quantitative approaches have helped adjust the economy, improve public planning at home, or devise a livable strategy abroad. If, in any of these areas, refinements can be claimed, they are largely in the nature of nicer abstraction, always more comfortably remote from concrete reality. The more mathematics has been invoked in a particular problem, the greater the emphasis on technical aspects and the less accessible to scrutiny and understanding by persons outside the fraternal order. Interest has centered on building and rebuilding the toolboxes, described in the virtually incomprehensible language of symbols and formulas.[64] The longer and more widespread the experience with the quantitative approach, the longer grows the list of cautions and caveats.

On this subject, Leontief's views are apropos and telling. In his review of *Perspectives on Economic Growth*,[65] he wrote:

> Construction of abstract "models" intended to describe in mathematical terms the complex interrelationships governing the process of economic growth has become one of the favourite occupations of economic

> theorists. *Unfortunately, the lack of factual knowledge of conditions existing in the real world forces the model builder to base many if not all of his general conclusions on all kinds of a priori assumptions, chosen for their convenience rather than for their correspondence to observed facts.* (My italics.)

Commenting more recently on the fundamental imbalance of the present state of economics, Leontief scored the emphasis on hypothetical situations rather than with observable reality, with formal mathematical reasoning rating higher among his colleagues and their disciples than empirical analysis. "Uncritical enthusiasm for mathematical formulation tends often to conceal the ephemeral substantive content of the argument behind the formidable front of algebraic signs".[66]

These remarkable criticisms come from Leontief, whose work for many years has been regarded as the crucial link between Walras and modern users of input-output models. The point he makes is reminiscent of the message conveyed in the now classic review of Tinbergen's econometric accomplishments.[67] A commentary on a volume of papers in appreciation of Tinbergen reflects the criticisms: "While some might admire the technical ingenuity and elegance of the contributions, others might reply that if something is not worth doing, it is not worth doing well".[68]

Now, in many milieus, preoccupation with technicalities has been observed to deflect attention from assumptions and issues. Expressing trends and influences in figures tends to introduce a bias in favour of the calculable; that which cannot be weighed or measured is not taken into account. Crucial intangibles are neglected.[69] Concern for purity of model and rules of the game may impose so many constraints that the result is little more than a tautological exercise, satisfying only to the precious. Models which have gained acceptance because they complied with rules and postulates have been shown to have incorporated unsophisticated assumptions and hidden biases. Work on the same problem, with the same tools, can yield a model which is the exact opposite of one in general use.[70] Sober examination of some methodological disagreements now appearing as articles and book reviews in the professional journals causes one to reflect that it may not be an altogether unmixed blessing that the technique still remains to a considerable extent at an abstract level and has not been embraced as the planning tool to quite the total extent recommended by its creators.

The hazards of dependence on this kind of approach to public affairs have been outlined by Don K. Price. His main objection was that they made no allowance for humane sentiments or moral judgements. Kenneth E. Boulding, within the framework of a discussion of the impossibility of applying rational standards to public decision making, articulated the dangers of quantification as the way of assigning a rank order to goals in order to establish priorities.

> The quantification of value functions into value indices, whether this is money or whether it is more subtle and complicated measures of pay-off, introduces elements of ethical danager into the decision-making process, *simply because the clarity and apparent objectivity of quantiatively measureable subordinate goals can easily lead to failure to bear in mind that they are in fact subordinate.*[71] (My italics.)

General system theory as synergy

Identification and analysis of the individual pebbles trapped in the puddingstone have provided clues to the intellectual precedents of the systems approach. Through review of them, we have been able to ascertain their status, strengths, and weaknesses within their home disciplines. Brought together, they exemplify Boulding's observation about the ease with which the interdisciplinary deteriorates into the undisciplined.[72] How the synthetic agglomeration has been rationalized into a general theory must properly be considered against this backdrop. Arguments put forward by Ludwig von Bertalanffy, who claims to be among the first to introduce the notion, are based on a number of interrelated and nebulous postulates: a general tendency, in the various sciences, physical and social, toward integration; convergence of this tendency into a general theory of systems; possibility that such theory might be a likely means for approaching exact theory in nonphysical science fields; promise of a unity of science through development of unifying principles running through the individual disciplines.[73] By way of defense for a science of wholeness, he juxtaposes "molecular" against organismic biology, stimulus-response behaviorism against Gestalt psychology, social atoms against social systems. Observation of many phenomena in biology and in the social and behavioural sciences have convinced von Bertalanffy that not only are mathematical expressions and models applicable to them all but that these display remarkable structural similarity and isomorphism. On the argument that probability theory is applicable to such diverse fields as thermodynamics, biological and medical experimentation, and life insurance statistics, von Bertalanffy proposes a logico-mathematical methodology which uses computerization, simulation, and cybernetics. Information theory, theory of automata, games, decisions, and queuing are utilized.

Objections to the approach as perceived or imagined by Bertalanffy are threefold. One, the so-called isomorphisms are said to be nothing more than tired truisms about the universal applicability of mathematics; thus, $2 + 2 = 4$ prevails whether chicks, cheese, soap, or the solar system are under consideration. Two, superficial analogies may mislead, for they camouflage crucial differences and can lead to erroneous conclusions. Three, adherence to an alleged "irreducibility" doctrine renders the approach philosophically and methodologically unsound in that it could impede analytic advances, a dubious loss in Bertalanffy's estimation because they have effected the reduction of chemistry to physical principles and life phenomena to molecular biology. His counter-arguments depend more on iteration than on evidence, however. He is wont to repeat that his isomorphisms are not mere analogy and that his new organismic biology is not a revival of the long-discredited bio-organismic theories of old. Using the current jargon, he claims that his is a new conception of organized complexity, in which the general model, applicable to physiological, neurophysiological, behavioural, and social phenomena, is of the feedback system, elaborated mathematically or as a quantitative scheme, however the case allows.[74]

Bertalanffy has tempered his endorsement of the systems approach by a few thoughtful caveats. He warns against "incautious expansion to fields for which its concepts are not made"; dangers of distortion of reality through "forcible imposition" of mathematical models; detriment to scientific progress of a return to "vague analogies". But the warnings are too general to serve as useful guidelines and too weak to stem the flood of enthusiasm for quantitative methods and the systems approach. By its attribute of being all things to all people, the methodology has provided a vehicle for all who would ride. Perhaps not to so complete a *reductio ad ultimum* as the *Report from Iron Mountain*[75] but far along the road to hastening such negative utopianism as Huxley's *Brave New World*[76] and Orwell's *1984*,[77] von Bertalanffy saw the system orientation as serving and hastening "the process of mechanization, automation, and devaluation of man".[78]

References

1 Churchman, C. W., Ackoff R. L. and Arnoff E. L., *Introduction to Operations Research*, New York: John Wiley, 1957, p. 3.

2 One example: Rivlin, A. M., *Systematic Thinking for Social Action*, Washington, D.C., The Brookings Institution, 1971.

3 Webster's *New International Dictionary*, Second Edition Unabridged.

4 Quade, E. S. and Boucher, W. I., eds., *Systems Analysis and Policy Planning*, New York: American Elsevier, 1968, p. 3.

5 Black, G., *The Application of Systems Analysis to Government Operations*, New York: Frederick A. Praeger, 1968, pp. 6–10.

6 Zadeh, L. A. and Desoer, C. A., *Linear System Theory*, New York: McGraw-Hill, 1963, p. 2.

7 *Ibid.*, p. vii.

8 *Ibid.*, p. 2.

9 *Ibid.*, p. 2.

10 *Ibid.*, p. vii.

11 Ellis, D. O. and Ludwig, F. J., *Systems Philosophy*, Englewood Cliffs, New Jersey: Prentice-Hall, 1962, p. 3.

12 Shearer, J. L., Murphy, A. T. and Richardson, H. H., *Introduction to System Dynamics*, Reading, Massachusetts: Addison-Wesley, 1967, p. 3.

13 *Ibid.*, p. 103.

14 *Ibid.*, p. 103.

15 Shinners, S. M., *Technique for System Engineering*, New York: McGraw-Hill, 1967, pp. 16–17.

16 Ellis and Ludwig, *op. cit.*, pp. 129–131.

17 Shearer, Murphy, and Richardson, *op. cit.*, pp. 102–103.

18 Shinners, *op. cit.*, Foreword by Truxal, J. G., p. v.

19 Maslach, G. J., "In Search of a New Image", *The California Engineer*, Vol. 46, No. 2, December, 1967, p. 7.

20 Shearer, Murphy, and Richardson, *op. cit.*

21 Shinners, *op. cit.*, p. 38.

22 Licklider, J. C. R., "Underestimates and Overestimations", *Computers and Automation*, August, 1969, p. 48. From Chayes, A. and Wiesner, J. B., eds., *ABM: An Evaluation of the Decision to Deploy an Antiballistic Missile System*, New York, Harper & Row, 1969.

23 *Ibid.*, p. 50.

24 Assistant Secretary of Commerce Myron Tribus, Speech at IEEE Meeting, Washington, D.C., as quoted in *Modern Data*, Vol. 3, No. 1, January, 1970, p. 40.

25 "Cottleston Pie", a song sung by Winnie-the-Pooh, makes unlikes analogous through simple linkages:

A fish can't whistle and neither can I.
Ask me a riddle and I reply:
"Cottleston, Cottleston, Cottleston Pie".

Milne, A. A., *Winnie-the-Pooh*, 1926, in *The World of Pooh*, New York: E. P. Dutton, 1957, p. 72.

26 Spencer, H., *The Principles of Sociology*, Vol. 1, Part II, New York: Appleton, 1910.

27 *The Works of Herbert Spencer*, Germany: Osnabrück, 1871, in *Specialized Administration*, Vol. 15, p. 411.

28 See especially Gabriel Tarde, "La théorie organique des sociétés", *Annals Institut International de Sociologie*, Vol. IV, pp. 238–239, or Gabriel Tarde, *La Logique Sociale*, Paris: Alcan., 1895.

29 Sorokin, P. A., *Contemporary Sociological Theories*, New York: Harper & Brothers, 1928, especially chapter IV, pp. 195 ff.

30 *Ibid.*, p. 210.

31 *Ibid.*, p. 211.

32 Parson, T., *The Social System*, Glencoe: Free Press, 1951.

33 Parson, T., "Systems Analysis: Social Systems", *International Encyclopedia of the Social Sciences*, Vol. 15, 1968, p. 458.

34 *Ibid.*, p. 472.

35 Buckley, W., ed., Foreword to *Modern Systems Research for the Behavioral Scientist*, Chicago: Aldine, 1968, p. xxi.

36 *Ibid.*, p. xv.

37 *Ibid.*, p. 493.

38 *Ibid.*, p. xxi.

39 The other two men identified with the development of Gestalt psychology are Max Wertheimer (1880–1943) and Kurt Koffka (1886–1941).

40 Köhler, W., "Perception: An Introduction to the Gestalt-theorie", *Psychological Bulletin*, Vol. 19, 1922, pp. 531–585, and *Principles of Gestalt Psychology*, New York: Harcourt and Brace, 1935.

41 Lewin, K., *Field Theory in Social Science: Selected Theoretical Papers*, Dorwin Cartwright, ed., London; Tavistock, 1963.

42 Miller, J. G., "Toward a General Theory for the Behavioral Sciences", *American Psychologist*, Vol. 10, 1955, pp. 513–531.

43 Allport, G. W., "The Open System in Personality Theory", *Journal of Abnormal and Social Psychology*, Vol. 61, 1960, pp. 301–311.

44 Parsons, T., "Systems Analysis: Social Systems", *International Encyclopedia of the Social Sciences*, *op. cit.* p. 460.

45 Allport, G. W., *op. cit.*

46 Schumpeter, J. A., *History of Economic Analysis*, edited from manuscript by Schumpete, E. B., New York: Oxford University Press, 1968 (seventh printing), p. 242.

47 *Ibid.*, p. 242.

48 Seligman, B. B., *Main Currents in Modern Economics*, Glencoe: Free Press, 1962, p. 367. Friedman, M. is quoted in the article on Walras (in the *International Encyclopedia of the Social Sciences*, Vol. 16, p. 452): "We curtsy to Marshall, but we walk with Walras".

49 *Max Weber on the Methodology of the Social Sciences*, translated and edited by Shils, E. and Finch, H. A., Glencoe: Free Press, 1949.

50 Seligman, B. B., *op. cit.*, p. 385.

51 *Ibid.*, p. 385.

52 Cf. footnote 48.

53 Samuelson, P. A., "Economic Theory and Mathematics – An Appraisal", *American Economic Review*, Vol. XLII, No. 2, May, 1952, pp. 56–66.

54 Schumpeter, *op. cit.*

55 The concepts are listed in the chronology suggested by Dorfman, R., Samuelson, P. A., and Solow, R. M., *Linear Programming and Economic Analysis*, New York: McGraw-Hill, 1958. There may be some disagreement over the actual order of appearance. Hicks, J. R. apparently has some other time relationship in mind in his essay, "Linear Theory", *Surveys of Economic Theory*. Prepared for the American Economic Association and The Royal Economic Society, Vol. III, Surveys IX–XIII, pp. 75–114.

56 Neumann, J. von, "Zur Theorie der Gesellschaftsspiele", *Mathematische Annalen*, Vol. 100, 1928, pp. 295–320, and Neumann, J. von and Morgenstern, O., *Theory of Games and Economic Behavior*, Princeton University Press, 1944.

57 Dorfman, Samuelson, and Solow, *op. cit.*, p. 2.

58 Leontief, W. W., *The Structure of American Economy 1919–1929*, Cambridge: Harvard University Press, 1941.

59 Dorfman, Samuelson, and Solow, *op. cit.*, p. 3.

60 Leontief, W. W., "The Structure of Development", *The Scientific American*, Vol. 209, No. 3, September, 1963, pp. 148–166.

61 U.S. General Accounting Office, *Glossary for Systems Analysis and Planning-Programming-Budgeting*, Washington, D. C.: U.S. Government Printing Office, 1969, p. 32.
62 Seligman, B. B., *op. cit.*, p. 781.
63 Dorfman, Samuelson, and Solow, *op. cit.*, p. 4.
64 Seligman, *op. cit.*, p. 790.
65 Leontief, W., Review of *Perspectives on Economic Growth*, Heller, W., editor, New York: Random House, 1968, in *The New York Review*, October 10, 1968, p. 32.
66 Leontief, W., "Theoretical Assumptions and Non-observed Facts", *The American Economic Review*, Vol. LXI, No. 1, March, 1971, pp. 1–2.
67 Keynes, J. M., "Professor Tinbergen's Method", Review in *The Economic Journal*, Vol. XLIX, No. 195, September, 1939, pp. 558–569.
68 Streeten, P., Review of *Towards Balanced International Growth, Essays Presented to Tinbergen, J.*, Bos, H. C., ed., Amsterdam and London: North-Holland, 1969, *The Economic Journal*, Vol. LXXX, No. 319, September, 1970, pp. 679–681.
69 Brunner, J., "The New Idolatry", in *Rebirth of Britain*, London: Pan Books, in association with The Institute of Economic Affairs, 1964, p. 38.
70 Leibenstein, H., "Pitfalls in Benefit-Cost Analysis of Birth Prevention", *Population Studies*, Vol. XXIII, No. 2, July, 1969, pp. 161–171.
71 Boulding, K. E., "The Ethics of Rational Decision", *Management Science*, Vol. 12, No. 6, February, 1966, p. B-165.
72 Boulding, K. E., "General Systems Theory – The Skeleton of Science", *Management Science*, Vol. 2, No. 3, April, 1956, pp. 197–209.
73 Bertalanffy, L. von, *General System Theory*, New York: Braziller, G., 1968, p. 37.
74 Bertalanffy, L. von, "General Systems Theory and a New View of the Nature of Man", paper given at American Psychiatric Association Annual Meeting, 1968, pp. 5 and 6.
75 *Report from Iron Mountain on the Possibility and Desirability of Peace*, with Introductory Material by Leonard C. Lewin, New York: Dial Press, 1967.
76 Huxley, A., *Brave New World*, New York: Harper and Brothers, 1946.
77 Orwell, G., *1984*, London: Secker and Warburg, 1949.
78 Bertalanffy, von, "General Systems Theory and a New View of the Nature of Man", *op. cit.*, p. 5.

Reprinted from *Systems Analysis in Public Policy: A Critique* by Ida R. Hoos.

9 Society as a complex adaptive system

by Walter Buckley

We have argued at some length in another place[1] that the mechanical equilibrium model and the organismic homeostasis models of society that have underlain most modern sociological theory have outlived their usefulness. A more viable model, one much more faithful to the *kind* of system that society is more and more recognized to be, is in process of developing out of, or is in keeping with, the modern systems perspective (which we use loosely here to refer to general systems research, cybernetics, information and communication theory, and related fields). Society, or the sociocultural system, is not, then, principally an equilibrium system or a homeostatic system, but what we shall simply refer to as a complex adaptive system.

To summarize the argument in overly simplified form: Equilibrial systems are relatively *closed* and *entropic*. In going to equilibrium they typically *lose structure* and have a *minimum of free energy*; they are affected only by external "disturbances" and have *no internal or endogenous sources of change*; their component elements are *relatively simple* and *linked directly via energy exchange* (rather than information interchange); and since they are relatively closed they have no feedback or other systematic self-regulating or adaptive capabilities. The homeostatic system (for example, the organism, apart from higher cortical functioning) is open and negentropic, maintaining a moderate energy level within controlled limits. But for our purposes here, the system's main characteristic is its functioning to *maintain the given structure of the system* within pre-established limits. It involves feedback loops with its environment, and possibly information as well as pure energy interchanges, but these are geared principally to *self-regulation* (structure maintenance) rather than adaptation (*change* of system structure). The complex adaptive systems (species, psychological and sociocultural systems) are also open and negentropic. But they are *open "internally" as well as externally* in that the interchanges among their components may result in *significant changes in the nature of the components themselves* with important consequences for the system as a whole. And the energy level that may be mobilized by the system is subject to relatively wide fluctuation. Internal as well as external interchanges are mediated characteristically by *information flows* (via chemical, cortical, or cultural encoding and decoding), although pure energy interchange occurs also. True feedback control loops make possible not only self-regulation, but self-direction or at least adaptation to a changing environment, such that the system may *change or elaborate its structure* as a condition of survival or viability.

We argue, then, that the sociocultural system is fundamentally of the latter type, and requires for analysis a theoretical model or perspective built on the kinds of characteristics mentioned. In what follows we draw on many of the concepts and principles presented throughout this source-book to sketch out aspects of a complex adaptive system model or analytical framework for the sociocultural system. It is further argued that a number of recent sociological and social psychological theories and theoretical orientations articulate well with this modern systems perspective, and we outline some of these to suggest in addition that modern systems research is not as remote from the social scientists' interests and endeavors as many appear to believe.

Complex adaptive systems: a paradigm

A feature of current general systems research is the gradual development of a general paradigm of the basic mechanisms underlying the evolution of complex adaptive systems. The terminology of this paradigm derives particularly from information theory and cybernetics. We shall review these concepts briefly. The *environment*, however else it may be characterized, can be seen at bottom as a set or ensemble of more or less distinguishable elements, states, or events, whether the discriminations are made in terms of spatial or temporal relations, or properties. Such distinguishable differences in an ensemble may be most generally referred to as "*variety*". The relatively stable "causal", spatial and/or temporal relations between these distinguishable elements or events may be generally referred to as "*constraint*". If the elements are so "loosely" related that there is equal probability of any element or state being associated with any other, we speak of "chaos" or complete randomness, and hence, lack of "constraint". But our more typical natural environment is characterized by a relatively high degree of constraint, without which the development and elaboration of adaptive systems (as well as "science") would not have been possible. When the internal organization of an adaptive system acquires features that permit it to discriminate, act upon, and respond to aspects of the environmental variety and its constraints, we might generally say that the system has "*mapped*" parts of the environmental variety and constraints into its organization as structure and/or "information". Thus, a subset of the ensemble of constrained variety in the environment is coded and transmitted in some way via various channels to result in a change in the structure of the receiving system which is isomorphic in certain respects to the original variety. The system thus becomes selectively matched to its environment both physiologically and psychologically. It should be added that two or more adaptive systems, as well as an adaptive system and its natural environment, may be said to be selectively interrelated by a mapping process in the same terms. This becomes especially important for the evolution of social systems.

In these terms, then, the paradigm underlying the evolution of more and more complex adaptive systems begins with the fact of a potentially changing environment characterized by variety with constraints, and an existing adaptive system or organization whose persistence and elaboration to higher levels depends upon a successful mapping of some of the environmental variety and constraints into its own organization on at least a semi-permanent basis. This means that our adaptive system – whether on the biological, psychological, or sociocultural level – must manifest (1) some degree of "*plasticity*" and "*irritability*" vis-à-vis its environment such that it carries on a constant interchange with environmental events, acting on and reacting to it; (2) some source or mechanism for *variety*, to act as a potential pool of adaptive variability to meet the problem of mapping new or more detailed variety and constraints in a changeable environment; (3) a set of *selective* criteria or mechanisms against which the "variety pool" may be sifted into those variations in the organization or system that more closely map the environment and those that do not; and (4) an arrangement for *preserving and/or propagating* these "successful" mappings.[2]

It should be noted, as suggested above, that this is a *relational* perspective, and the question of "substance" is quite secondary here. (We might also note that it is this kind of thinking that gives such great significance to the rapidly developing relational logic that is becoming more and more important as a technical tool of analysis.) Also, as suggested, this formulation corresponds closely with the current conception of "information" viewed as the process of selection – from an ensemble of variety – of a subset which, to have "meaning", must match another subset taken from a similar ensemble.[3] Communication is the process by which this constrained variety is transmitted in one form or another between such ensembles, and involves coding and decoding such that the original variety and its constraints remains relatively invariant at the receiving end. If the source of the "communication" is the causally constrained variety of the natural environment, and the destination is the biological adaptive system, we refer to the Darwinian process of natural selection whereby the information encoded in the chromosomal material (for example the DNA) reflects or is a mapping of the environmental variety, and makes possible a continuous and more or less successful adaptation of the former system to the latter. If the adaptive system in question is a (relatively high-level) psychological or cortical system, we refer to "learning", whereby the significant environmental variety is transmitted via sensory and perceptual channels and decodings to the cortical centers where, by selective criteria (for example, "reward" and "punishment") related to physiological and/or other "needs" or "drives", relevant parts of it are encoded and preserved as "experience" for varying periods of time and may promote adaptation. Or, on the level of the symbol-based sociocultural adaptive system, where the more or less patterned actions of persons and groups are as crucial a part of the environment of other persons and groups as the non-social environment, the gestural variety and its more or less normatively defined constraints is encoded, transmitted, and decoded at the receiving end by way of the various familiar channels with varying degrees of fidelity. Over time, and again by a selective process – now much more complex, tentative, and less easily specified – there is a selective elaboration and more or less temporary preservation of some of this complex social as well as non-social constrained variety in the form of "culture", "social organization", and "personality structure".

On the basis of such a continuum of evolving, elaborating levels of adaptive system (and we have only pointed to three points along this continuum), we could add to and refine our typology of systems. Thus, we note that as adaptive systems develop from the lower biological levels through the higher psychological and sociocultural levels we can distinguish: (1) the *varying time span* required for exemplars of the adaptive system to map or encode within themselves changes in the variety and constraints of the environment; phylogenetic time scales for organic systems and for tropistic or instinctual neural systems; ontogenetic time scales for higher psychological or cortical systems; and, in the sociocultural case, the time span may be very short – days – or very long, but complicated by the fact that the relevant environment includes both intra- and inter-societal variety and constraints as well as natural environment variety (the latter becoming progressively less determinant); (2) the greatly *varying degrees of fidelity of mapping* of the environment into the adaptive system, from the lower unicellular organisms with a very simple repertoire of actions on and reactions to the environment, through the complex of instinctual and learned repertoire, to the ever-proliferating more refined and veridical accumulations of a sociocultural system; (3) the progressively greater separation and independence of the more refined "stored information" from purely biological processes as genetic information is gradually augmented by cortically imprinted information, and finally by entirely extrasomatic cultural depositories. The implications of these shifts, and others that could be included, are obviously far-reaching.

One point that will require more discussion may be briefly mentioned here. This is the *relative* discontinuity we note in the transition from the non-human adaptive system to the sociocultural system. (The insect society and the rudimentary higher animal society make for much less than a complete discontinuity.) As we progress from lower to higher biological adaptive systems we note, as a general rule, the gradually increasing role of other biological units of the same as well as different species making up part of the significant environment. The variety and constraints represented by the behavior of these units must be mapped along with that of the physical environment. With the transition represented by the higher primate social organization through to full-blown human, symbolically mediated, sociocultural adaptive systems, the mapping of the variety and constraints characterizing the subtle behaviors, gestures and intentions of the individuals and groups making up the effective social organization become increasingly central, and eventually equal if not overshadow the requirements for mapping the physical environment.[4]

It was these newly demanding requirements of coordination, anticipation, expectation and the like within a more and more complex *social* environment of interacting and interdependent others – where genetic mappings were absent or inadequate – that prompted the fairly rapid elaboration of relatively new system features. These included, of course: the ever-greater conventionalizing of gestures into true symbols; the resulting development of a "self", self-awareness, or self-consciousness out of the symbolically mediated, continuous mirroring and mapping of each unit's behaviors and gesturings in those of ever-present others (a process well described by Dewey, Mead, Cooley, and others); and the resulting ability to deal in the present with future as well as past mappings and hence to manifest goal-seeking, evaluating, self-other relating, norm-referring behavior. In cybernetic terminology, this higher level sociocultural system became possible through the development of higher order feedbacks such that the component individual subsystems became able to map, store, and selectively act toward, not only the external variety and constraints of the social and non-social environment, but also their own internal states. To speak of self-consciousness, internalization, expectations, choice, certainty and uncertainty, and the like, is to elaborate this basic point. This transition, then, gave rise to the newest adaptive system level we refer to as socio-cultural. As we argued earlier, this higher level adaptive organization thus manifests features that warrant its scientific study in terms as distinct from a purely biological system as the analytical terms of the latter are from physical systems.

The sociocultural adaptive system

From the perspective sketched above, the following principles underlying the sociocultural adaptive system can be derived:

1. The principle of the "irritability of protoplasm" carries through to all the higher level adaptive systems. "Tension" in the broad sense – in which stress and strain are manifestations under conditions of felt blockage – is ever-present in one form or another throughout the sociocultural system – how as diffuse, socially unstructured strivings, frustrations, enthusiasms, aggressions, neurotic or psychotic or normative deviation; sometimes as clustered and minimally structured crowd or quasi-group processes, normatively supportive as well as destructive; and now as socioculturally structured creativity and production, conflict and competition, or upheaval and destruction. As Thelen and colleagues put it:

> 1. Man is always trying to live beyond his means. Life is a sequence of reactions to stress: Man is continually meeting situations with which he cannot quite cope.
> 2. In stress situations, energy is mobilized and a state of tension is produced.

3. The state of tensions tends to be disturbing, and Man seeks to reduce the tension.

4. He has direct impulses to take action. . . .[5]

2. Only closed systems running down to their most probable states, that is, losing organization and available energy, can be profitably treated in equilibrium terms. Outside this context the concept of equilibrium would seem quite inappropriate and only deceptively helpful. On the other side, only open, tensionful, adaptive systems can elaborate and proliferate organization. Cannon coined the term "homeostasis" for biological systems to avoid the connotations of equilibrium, and to bring out the dynamic, processual, potential-maintaining properties of *basically unstable* physiological systems.[6] In dealing with the sociocultural system, however, we need yet a new concept to express not only the *structure-maintaining* feature, but also the *structure-elaborating and changing* feature of the inherently unstable system. The notion of "steady state", now often used, approaches the meaning we seek if it is understood that the "state" that tends to remain "steady" is *not to be identified with the particular structure* of the system. That is, as we shall argue in a moment, in order to maintain a steady state the system may change its particular structure. For this reason, the term "morphogenesis" is more descriptive.[7] C. A. Mace recognizes this distinction in arguing for an extension of the concept of homeostasis:

> The first extension would cover the case in which what is maintained or restored is not so much an internal state of the organism as some relation of the organism to its environment. This would take care of the facts of adaptation and adjustment, including adjustment to the social environment . . . the second extension would cover the case in which the goal and/or norm is some state or relation which has never previously been experienced. There is clearly no reason to suppose that every process of the homeostatic type consists in the maintenance or restoration of a norm.[8]

3. We define a system in general as a complex of elements or components directly or indirectly related in a causal network, such that at least some of the components are related to some others in a more or less stable way *at any one time*. The interrelations may be mutual or unidirectional, linear, non-linear or intermittent, and varying in degrees of causal efficacy or priority. The particular kinds of more or less stable interrelationships of components that become established at any time constitute the particular *structure* of the system at that time.

Thus, the complex, adaptive system as a continuing entity is not to be confused with the structure which that system may manifest at any time. Making this distinction allows us to state a fundamental principle of open, adaptive systems: *Persistence or continuity of an adaptive system may require, as a necessary condition, change in its structure*, the degree of change being a complex function of the internal state of the system, the state of its relevant environment, and the nature of the interchange between the two. Thus, animal species develop and persist or are continuously transformed (or become extinct) in terms of a change (or failure of change) of structure – sometimes extremely slow, sometimes very rapid. The higher individual organism capable of learning by experience maintains itself as a viable system vis-à-vis its environment by a change of structure – in this case the neural structure of the cortex. It is through this principle that we can say that the "higher" organism represents a "higher" level of adaptive system capable, ontogenetically, of *mapping the environment more rapidly and extensively* and with *greater refinement and fidelity*, as compared to the tropistic or instinct-based adaptive system which can change its structure only phylogenetically. The highest level adaptive system – the sociocultural – is capable of an even more rapid and refined mapping of the environment (including the social and non-social environment, as well as at least some aspects of its own internal state) since sociocultural structures are partially independent of both ontogenetic and phylogenetic structures, and the mappings of many individuals are selectively pooled and stored extrasomatically and made available to the system units as they enter and develop within the system.

Such a perspective suggests that, instead of saying, as some do, that a prime requisite for persistence of a social system is "pattern maintenance", we can say, after Sommerhof and Ashby,[9] that persistence of an adaptive system requires as a necessary condition the maintenance of the system's "essential variables" within certain limits. Such essential variables and their limits may perhaps be specified in terms of what some have referred to as the "functional prerequisites" of any social system (for example, a minimal level organismal sustenance, of reproduction, of patterned interactive relations, etc.). But the maintenance of the system's essential variables, we are emphasizing, may hinge on (as history and ethnography clearly show) *pattern reorganization or change*. It is true, but hardly helpful, to say that *some* minimal patterning or stability of relations, or integration of components, is necessary – by the very definition of "system" or adaptive organization. Nor can we be be satisfied, with the statement that persistence, continuity, or social "order" is promoted by the "institutionalization" of interactive relations via norms and values, simply because we can say with equal validity that discontinuity or social "disorder" is *also* promoted by certain kinds of "institutionalization".

To avoid the many difficulties of a one-sided perspective it would seem essential to keep before us as a basic principle that the persistence and/or development of the complex sociocultural system depends upon structuring, destructuring, and restructuring – processes occurring at widely varying rates and degrees as a function of the external social and non-social environment. Jules Henry, among others, has made this point:

> . . . the lack of specificity of man's genetic mechanisms has placed him in the situation of constantly having to revise his social structures because of their frequent failure to guide inter-personal relations without tensions felt as burdensome even in the society in which they originate . . . thus man has been presented with a unique evolutionary task: because his mechanisms for determining inter-personal relations lack specificity, he must attempt to maximize social adaptation through constant conscious and unconscious revision and experimentation, searching constantly for social structures, patterns of inter-personal relations, that will be more adaptive, as he feels them. Man's evolutionary path is thus set for him by his constant tendency to alter his modes of social adaptation.[10]

More generally, we recall that Karl W. Deutsch has seen restructuring as a basic feature distinguishing society from an organism or machine. Speaking of "the critical property which makes a given learning net into a *society*", he says:

> A learning net functions as a society, in this view, *to the extent* that its constituent physical parts are capable of regrouping themselves into new patterns of activity in response to changes in the net's surroundings, or in response to the internally accumulating results of their own or the net's past.
>
> The twin tests by machine we can tell a society from an organism or a machine, on this showing, would be the freedom of its parts to regroup themselves; and the nature of the regroupings which must imply new coherent patterns of activity – in contrast to the mere wearing out of a machine or the aging of an organism, which are marked by relatively few degrees of freedom and by the gradual disappearance of coherent patterns of activity. . . .
>
> This in turn may rest on specific properties of their members: *their capacity for readjustment to new configurations, with renewed complementarity and sustained or renewed communication.*[11]

4. The cybernetic perspective of control or self-regulation of adaptive systems emphasizes the crucial role of "deviation", seen in both negative and positive aspects. On the negative side, certain kinds of deviations of aspects of the system from its given structural state may be seen as "mismatch" or "negative feedback" signals *interpreted by certain organizing centers* as a failure of the system's operating processes or structures relative to a goal state sought, permitting – under certain conditions of adaptive structuring – a change of those operating processes or structures toward goal-optimization. (Thus, one facet of the "political" process of sociocultural systems may be interpreted in this light, with the more "democratic" type of social organization providing the more extended and accurate assessment of the mismatch between goal-attainment on the one hand, and current policy and existing social structuring on the other.)

On the positive side, the cybernetic perspective brings out the absolute necessity of deviation – or, more generally, "variety" – in providing a pool of potential new transformations of process or structure that the adaptive systems might adopt in responding to goal-mismatch. On the lower, biological levels we recognize here the principle of genetic variety and the role of gene pools in the process of adaptive response to organismic mismatch with a changed environment. (And in regard to the other major facet of the "political" process, the more democratic type of social organization makes available a broader range of variety, or "deviation", from which to select new orientations.) Ashby, in developing his very general theory of the adaptive or self-regulating system, suggests the "law of requisite variety", which states that the variety within a system must be at least as great as the environmental variety against which it is attempting to regulate itself. Put more succinctly, only variety can regulate variety. Although such a general principle is a long way from informing more concrete analysis of particular cases, it should help provide a needed corrective to balance (not replace) the current emphasis in social science on conformity, the "control" (as against the cultivation) of "deviants", and "re-equilibration" of a given structure.

Thus, the concept of requisite deviation needs to be proferred as a high-level principle that can lead us to theorize: A requisite of sociocultural systems is the development and maintenance of a significant level of non-pathological deviance manifest as a pool of alternate ideas and behaviors with respect to the traditional, institutionalized ideologies and role behaviors. Rigidification of any given institutional structure must eventually lead to disruption or dissolution of the society by way of internal upheaval or ineffectiveness against external challenge. The student of society must thus pose the question – What "mechanisms" of non-pathological deviance production and maintenance can be found in any society, and what "mechanisms" of conformity operate to counteract these and possibly lessen the viability of the system?

Attempts to analyze a society from such a perspective make possible a more balanced analysis of such processes as socialization, education, mass communication, and economic and political conflict and debate. We are then encouraged to build squarely into our theory and research designs the full sociological significance of such informally well-recognized conceptions as socialization for "self-reliance" and relative "autonomy", education for "creativity", ideational flexibility and the "open mind", communications presenting the "full spectrum" of viewpoints, etc., instead of smuggling them in unsystematically as if they were only residual considerations or ill-concealed value judgments.

5. Given the necessary presence of variety or deviance in an adaptive system, the general systems model then poses the problem of the *selection* and more or less permanent *preservation* or systemic structuring of some of this variety. On the biological level, we have the process of "natural selection" of some of the genetic variety existing within the interfertile species and sub-species gene pool, and the preservation for various lengths of time of this variety through the reproductive process. On the level of higher order psychological adaptive systems, we have trial-and-error selection, by way of the so-called "law of effect", from the variety of environmental events and the potential behavioral repertoire to form learned and remembered experience and motor skills more or less permanently preserved by way of cortical structuring.[12] As symbolic mapping or decoding and encoding of the environment and one's self becomes possible,[13] the selection criteria lean less heavily on direct and simple physiological reward and more heavily on "meanings" or "significance" as manifested in existing self-group structural relations. In the process, selection from the full range of available variety becomes more and more refined and often more restricted, and emerges as one or another kind of "personality" system or "group character" structure. On the sociocultural level, social selection and relative stabilization or institutionalization of normatively interpreted role relations and value patterns occurs through the variety of processes usually studied under the headings of conflict, competition, accommodation, and such; power, authority and compliance; and "collective behavior", from mob behavior through opinion formation processes and social movements to organized war. More strictly "rational" processes are of course involved, but often seem to play a relatively minor role as far as larger total outcomes are concerned.

It is clearly in the area of "social selection" that we meet the knottiest problems. For the sociocultural system, as for the biological adaptive system, analysis must focus on both the potentialities of the system's structure at a given time, and the environmental changes that might occur and put particular demands on whatever structure has evolved. In both areas the complexities are compounded for the sociocultural system. In developing a typology of systems and their internal linkages we have noted that, as we proceed from the mechanical or physical through the biological, psychic and sociocultural, the system becomes "looser", the interrelations among parts more tenuous, less rigid, and especially less directly tied to physical events as energy relations and transformations are overshadowed by symbolic relations and information transfers. Feedback loops between operating sociocultural structures and the surrounding reality are often long and tortuous, so much so that knowledge of results or goal-mismatch, when forthcoming at all, may easily be interpreted in non-veridical ways (as the history of magic, superstition, and ideologies from primitive to present amply indicate). The higher adaptive systems have not been attained without paying their price, as the widespread existence of illusion and delusions on the personality and cultural levels attest. On the biological level, the component parts have relatively few degrees of freedom, and changes in the environment are relatively directly and inexorably reacted to by selective structural changes in the species.

Sociocultural systems are capable of persisting within a wide range of degrees of freedom of the components, and are often able to "muddle through" environmental changes that are not too demanding. But of course this is part of the genius of this level of adaptive system: it is capable of temporary shifts in structure to meet exigencies. The matter is greatly complicated for the social scientist, however, by this system's outstanding ability to act on and partially control the environment of which a major determining part is made up of other equally loose-knit, more or less flexible, illusion-ridden, sociocultural adaptive systems. Thus, although the minimal integration required for a viable system does set limits on the kinds of structures that can persist, these limits seem relatively broad compared to a biological system.[14] And given the relatively greater degrees of freedom of internal structuring (structural alternatives, as some call them) and the *potentially* great speed with which restructuring may occur under certain conditions, it becomes difficult to predict the reactions of such a system to environmental changes or internal elaboration. Considering the full complexities of the problem we must wonder at the facility with which the functionalist sociologist has pronounced upon the ultimate functions of social structures, especially when – as seems so often the case – very little consideration is given either to the often feedback-starved social selective processes that have led to the given structures, or to the environmental conditions under which they may be presumed to be functional.

Although the problem is difficult, something can be said about more ultimate adaptive criteria against which sociocultural structures can be assessed. Consideration of the grand trends of evolution provides clues to very general criteria. These trends point in the direction of: (1) greater and greater flexibility of structure, as error-controlled mechanisms (cybernetic processes of control) replace more rigid, traditionalistic means of meeting problems and seeking goals; (2) ever more refined, accurate, and systematic mapping, decoding and encoding of the external environment and the system's own internal milieu (via science), along with greater independence from the physical environment; (3) and thereby a greater elaboration of self-regulating substructures in order – not merely to restore a given equilibrium or homeostatic level – but to purposefully restructure the system without tearing up the lawn in the process.[15]

With these and perhaps other general criteria, we might then drop to lower levels of generality by asking what restrictions these place on a sociocultural adaptive system if it is to remain optimally viable in these terms. It is possible that this might provide a value-free basis for discussing the important roles, for example, of a vigorous and independent science in all fields; the broad and deep dissemination of its codified findings; the absence of significant or long-lasting subcultural cleavages, power centers and vested interests, whether on a class or ethnic basis, to break or hinder the flow of information or feedback concerning the internal states of the system; and the promotion of a large "variety pool" by maintaining a certain number of degrees of freedom in the relations of the component parts – for example, providing a number of real choices of behaviors and goals. Thus we can at least entertain the feasibility of developing an objective rationale for the sociocultural "democracy" we shy from discussing in value terms.

6. Further discussion of the intricacies of the problem of *sociocultural selection processes* leading to more or less stable system *structures* may best be incorporated into the frame of discussion of the problem of "*structure versus process*". This is another of those perennial issues of the social (and other) sciences, which the modern systems perspective may illuminate.

Our argument may be outlined as follows:

– Much of modern sociology has analyzed society in terms of largely structural concepts: institutions, culture, norms, roles, groups, etc. These are often reified, and make for a rather static, overly deterministic, and elliptical view of societal workings.

– But for the sociocultural system, "structure" is only a relative stability of underlying, ongoing micro-processes. Only when we focus on these can we begin to get at the selection process whereby certain interactive relationships become relatively and temporarily stabilized into social and cultural structures.

– The unit of dynamic analysis thus becomes the systemic *matrix* of interacting, goal-seeking, deciding individuals and subgroups – whether this matrix is part of a formal organization or only a loose collectivity. Seen in this light, society becomes a continuous morphogenic process, through which we may come to understand in a unified conceptual manner the development of structures, their maintenance, and their change. And it is important to recognize that out of this matrix is generated, not only *social* structure, but also *personality* structure, and *meaning* structure. All, of course, are intimately interrelated in the morphogenic process, and are only analytically separable.

Structure, process, and decision theory

Though the problem calls for a lengthy methodological discussion, we shall here simply recall the viewpoint that sees the sociocultural system in comparative perspective against lower-level mechanical, organic and other types of systems. As we proceed upward along such a typology we noted that the ties linking components become less and less rigid and concrete, less direct, simple and stable within themselves. Translation of energy along unchanging and physically continuous links gives way in importance to transmission of information via internally varying, discontinuous components with many more degrees of freedom. Thus for mechanical systems, and parts of organic systems, the "structure" has a representation that is concrete and directly observable – such that when the system ceases to operate much of the structure remains directly observable for a time. For the sociocultural system, "structure" becomes a theoretical construct whose referent is only indirectly observable (or only inferable) by way of series of events along a time dimension; when the system ceases to operate, the links maintaining the sociocultural structure are no longer observable.[16] "Process", then, points to the actions and interactions of the components of an ongoing system, in which varying degrees of structuring arise, persist, dissolve, or change. (Thus "process" should not be made synonymous simply with "change", as it tended to be for many earlier sociologists.)

More than a half century ago, Albion W. Small argued that, "The central line in the path of methodological progress in sociology is marked by the gradual shifting of effort from analogical representation of social structures to real analysis of social processes."[17] This was an important viewpoint for many social thinkers earlier in this century, possibly as part of the trend in physical science and philosophy toward a process view of reality developing from the work of such people as Whitehead, Einstein, Dewey, and Bentley. Such views have apparently had little impact on those of recent decades who had developed the more dominant structure-oriented models of current sociology, but it seems clear that – with or without the aid of the essentially process-conscious general systems approach – a more even balance of process with structure in the analysis of sociocultural systems is gradually regaining lost ground.

C. H. Cooley, in his *Social Process*, focused on the "tentative process", involving inherent energy and growth as the dynamic agents, with ongoing "selective development" set in motion by the interaction of "active tendencies" and surrounding "conditions". He argued that for the social process, "that grows which works" is a better phrase than "natural selection" or "survival of the fittest", since "it is not so likely to let us rest in mechanical or biological conceptions".[18] R. E. Park, with his recognition of the central importance of communication, kept the notion of process in the foreground whether developing the forms of interaction or the foundations of social ecology. We should also recall the leaders of the so-called "formal" school: Whereas Simmel focused on "forms of interaction", the emphasis was always on the "interaction" as process rather than simply on the "forms"; and though the Wiese–Becker systematics developed in great detail a classification of action *patterns*, it gave equal attention to *action* patterns. For W. I. Thomas, all social becoming is viewed as a product of continual interaction of individual consciousness and objective social reality. (F. Znaniecki more recently reinforced this point of view.[19]) And at least one unbroken thread in this vein continuing from the early part of the century is the Dewey–Mead perspective referred to as social interactionism (which, we have noted, has established a strong base especially congenial to the modern cybernetic approach).[20] A reviewer of a recent collection of social interactionist essays was "reminded throughout of the continuous character of socialization, of the complexity and fluidity of interaction when it is viewed as a process rather than as the mere enactment of social forms . . .".[21]

We can take only brief note of a few of the more recent arguments for the process viewpoint. The anthropologists, for example, have become acutely concerned in the last few years with this issue. G. P. Murdock seems to be echoing Small when he says, 'All in all, the static view of social structure which seeks explanations exclusively within the existing framework of a social system on the highly dubious assumption of cultural stability and nearly perfect functional integration seems clearly to be giving way, in this country at least, to a dynamic orientation which focuses attention on the processes by which such systems come into being and succeed one another over time."[22] At about the same time, Raymond Firth was stating: "The air of enchantment which for the last two decades has surrounded the 'structuralist' point of view has now begun to be dispelled. Now that this is so, the basic value of the concept of social structure as an heuristic tool rather than a substantial social entity has come to be more clearly recognized."[23]

Soon after appeared the late S. F. Nadel's penetrating work, *The Theory of Social Structure*, which was preceded by his article on "Social Control and Self Regulation". This perspective is used effectively in *The Theory of Social Structure* as a basis for a critique of the current rather one-sided equilibrium model emphasizing the "complementarity of expectations" to the relative neglect of the several other crucial types of associative *and* dissociative social interrelationships considered equally important in earlier sociology.

Parsons' model has to do with "the conditions of relatively stable interaction in social systems", implying defined value "standards" and "institutionalized role expectations": any willful disagreement with them simply falls outside the stipulated stability and the model based on it.

I would argue that this is not necessarily so and that our model must allow for such disagreements. Even "relatively stable" social systems do not exclude them, or include them only in the form of purely fortuitous contingencies. Far from being fortuitous or idiosyncratic, the rejection of the sanctioning potentialities of other roles may itself be anchored in the existing institutions, reflecting the presence of diverse but equally legitimate "value patterns", ideologies or schools of thought, that is, that plurality of norms we spoke of before.[24]

Nadel's book as a whole explores the thesis that structural analysis is not, and should not be treated as, static analysis: "Social structure as Fortes once put it, must be 'visualized' as 'a sum of processes in time'. As I would phrase it, social structure is implicitly an event-structure. . . . "[25] And in concluding he reiterates his argument that

> . . . it seems impossible to speak of social structure in the singular. Analysis in terms of structure is incapable of presenting whole societies; nor, which means the same, can any society be said to exhibit an embracing, coherent structure as we understand the term. There are always cleavages, dissociations, enclaves, so that any description alleged to present a single structure will in fact present only a fragmentary or one-sided picture.[26]

As a final example in anthropology, we should mention the cogent argument of Evon Z. Vogt that the two concepts of structure and process must be integrated into a general theoretical model. As with Nadel, structure is seen as falsely conceived as static, with change pathological. Rather, Vogt feels, must we pose the primacy of change, considering structure the way in which moving reality is translated, for the observer, into an instantaneous and artificial observation: social and cultural structures are only the intersections in time and space of process in course of change and development.[27]

Among sociologists, a perennial critic of the overly-structural conception of the group is Herbert Blumer. Blumer has argued that it is from the process of ongoing interaction itself that group life gets its main features, which cannot be adequately analyzed in terms of fixed attitudes, "culture", or social structure – nor can it be conceptualized in terms of mechanical structure, the functioning of an organism, or a system seeking equilibrium, ". . . in view of the formative and explorative character of interaction as the participants *judge* each other and *guide* their own acts by that judgment."

> The human being is not swept along as a neutral and indifferent unit by the operation of a system. As an organism capable of self-interaction he forges his actions out of a process of definition involving *choice*, *appraisal*, and *decision*. . . . Cultural norms, status positions and role relationships are only *frameworks* inside of which that process [of formative transaction] goes on.[28]

Highly structured human association is relatively infrequent and cannot be taken as a prototype of a human group life. In sum, institutionalized patterns constitute only one conceptual aspect of society, and they point to only a part of the ongoing process (and, we might add, they must be seen to include deviant and disfunctional patterns: for conceptual clarity and empirical relevance, "institutionalization" cannot be taken to imply only "legitimacy", "consent", and ultimately adaptive values).

Finally, it should be noted that Gordon Allport, viewing personality as an open-system, stresses a very similar point concerning the organization of personality:

> . . . the best hope for discovering coherence would seem to lie in approaching personality as a total functioning structure, i.e., as a *system*. To be sure, it is an incomplete system, manifesting varying degrees of order and disorder. It has structure but also unstructure, function but also malfunction. As Murphy says, "all normal people have many loose ends". And yet personality is well enough knit to qualify as a system – which is defined merely as *a complex of elements in mutual interaction*.[29]

In the light of such views, we need only recall the many critiques pointing to the incapacity or awkwardness of the conventional type of framework before the facts of process, "becoming", and the great range of "collective behavior".[30]

Statements such as Blumer's, a continuation of the perspective of many neglected earlier sociologists and social psychologists, would seem to constitute a perspective that is now pursued by many under new rubrics such as "decision theory". For earlier antecedents it should be enough to mention W. I. Thomas's "definition of the situation", Znaniecki's "humanistic coefficient", Weber's "verstehen", Becker's "interpretation", and MacIver's "dynamic assessment".[31] Much of the current structural, consensus theory represents a break from this focus. As Philip Selznick has argued,

A true theory of social action would say something about goal-oriented or problem-solving behavior, isolating some of its distinctive attributes, stating the likely outcomes of determinate transformations. . . . In Parsons' writing there is no true embrace of the idea that structure is being continuously opened up and reconstructed by the problem-solving behavior of individuals responding to concrete situations. This is a point of view we associate with John Dewey and G. H. Mead, for whom, indeed, it had significant intellectual consequences. For them and for their intellectual heirs, social structure is something to be taken account of in action; cognition is not merely an empty category but a natural process involving dynamic assessments of the self and the other.[32]

It can be argued, then, that a refocusing is occurring via "decision theory", whether elaborated in terms of "role-strain" theory; theories of cognitive dissonance, congruence, balance, or concept formation; exchange, bargaining, or conflict theories, or the mathematical theory of games. The basic problem is the same: How do interacting personalities and groups define, assess, interpret, "verstehen", and act on the situation? Or, from the broader perspective of our earlier discussion, how do the processes of "social selection" operate in the "struggle" for socio-cultural structure? Instead of asking how structure affects, determines, channels actions and interactions, we ask how structure is created, maintained, and recreated.

Thus we move down from structure to social interrelations and from social relations to social actions and interaction processes – to a matrix of "dynamic assessments" and intercommunication of meanings, to evaluating emoting, deciding and choosing. To avoid anthropomorphism and gain the advantages of a broader and more rigorously specified conceptual system, we arrive at the language of modern systems theory.

Basic ingredients of the decision-making focus include, then: (1) a *process* approach; (2) a conception of *tension* as inherent in the process; and (3) a renewed concern with the role and workings of man's enlarged cortex seen as a complex adaptive subsystem operating within an *interaction matrix* characterized by *uncertainty*, *conflict*, and other dissociative (as well as associative) processes *underlying the structuring and restructuring of the larger psycho-social system.*

Process focus

The process focus points to information-processing individuals and groups linked by different types of communication nets to form varying types of interaction matrices that may be characterized by "competition", "cooperation", "conflict", and the like. Newer analytical tools being explored to handle such processes include treatment of the interaction matrix over time as a succession of states described in terms of transition probabilities, Markoff chains, or stochastic processes in general. The Dewey–Mead "transactions" are now discussed in terms of information and codings and decodings, with the essential "reflexivity" of behavior now treated in terms of negative and positive feedback loops linking via the communication process the intrapersonal, interpersonal and intergroup subsystems and making possible varying degrees of matching and mismatching of Mead's "self and others", the elaboration of Boulding's "Image",[33] and the execution of Miller's "Plans". And herein we find the great significance for sociology of many of the conceptual tools (though not, at least as yet, the mathematics) of information and communication theory, cybernetics, or general systems research, along with the rapidly developing techniques of *relational* mathematics such as the several branches of set theory – topology, group theory, graphy theory, symbolic logic, etc.

Conception of tension

Tension is seen as an inherent and essential feature of complex adaptive systems; it provides the "go" of the system, the "force" behind the elaboration and maintenance of structure. There is no "law of social inertia" operating here, nor can we count on "automatic" re-equilibrating forces counteracting system "disturbances" or "deviance", for, whereas we do find deviance-reducing negative feedback loops in operation we *also* find deviance-maintaining and deviance-amplifying *positive* feedback processes often referred to as the vicious circle or spiral, or "escalation".[34] It is not at all certain whether the resultant will maintain, change, or destroy the given system or its particular structure. The concepts of "stress" or "strain" we take to refer only to the greater mobilization of normal tension under conditions of more than usual blockage. And instead of a system's seeking to manage *tension*, it would seem more apt to speak of a system's seeking to manage *situations* interpreted as responsible for the production of greater than normal tension.

The "role strain" theory of William J. Goode is an illustrative attack on assumptions of the widely current structural approach, using a process and tension emphasis and contributing to the decision-theory approach. Goode analyzes social structure or institutions into role relations, and role relations into role transactions. "Role relations are seen as a sequence of 'role bargains' and as a continuing process of selection among alternative role behaviors, in which each individual seeks to reduce his role strain."[35] Contrary to the current stability view, which sees social system continuity as based primarily on normative consensus and normative integration, Goode thus sees "dissensus, nonconformity, and conflicts among norms and roles as the usual state of affairs. . . . The individual cannot satisfy fully all demands, and must move through a continuous sequence of role decision and bargains . . . in which he seeks to reduce his role strain, his felt difficulty in carrying out his obligations."[36] Goode also recognizes that there is no "law of social inertia" automatically maintaining a given structure.

> Like any structure or organized pattern, the role pattern is held in place by both internal and external forces – in this case, the role pressures from other individuals. Therefore, not only is role strain a normal experience for the individual, but since the individual processes of reducing role strain determine the total allocation of role performances to the social institutions, the total balances and imbalances of role strains create whatever stability the social structure possesses.[37]

It should be noted, however, that Goode accepts unnecessarily a vestige of the equilibrium or stability model when he states, "The total role structure functions so as to reduce role strain."[38] He is thus led to reiterate a proposition that – when matched against our knowledge of the empirical world – is patently false. Or, more precisely, not false, but a half-truth: it recognizes deviance-reducing negative feedback processes, but not deviance-amplifying positive feedback processes. Such a proposition appears reasonable only if we "hold everything else constant", that is, take it as a closed system. However, the proposition is unnecessary to his argument and, in fact, clashes with the rest of his formulation: ". . . though the sum of role performances ordinarily maintains a society it may also change the society or fail to keep it going. There is no necessary harmony among all role performances. . . . But whether the resulting societal pattern is 'harmonious' or integrated or whether it is even effective in maintaining that society, are separate empirical questions."[39]

Study of cognitive processes

A more concerted study of cognitive processes, especially under conditions of *uncertainty* and *conflict*, goes hand in hand, of course, with a focus on decision-making and role transactions. Despite the evolutionary implications of man's enlarged cortex, much social (and psychological) theory seems predicted on the assumption that men are decorticated. Cognitive processes, as they are coming to be viewed today, are not to be simply equated with the traditional, ill-defined, concept of the "rational". That the data-processing system – whether socio-psychological or electro-mechanical – is seen as inherently "rational" tells us little about its outputs in concrete cases. Depending on the adequacy and accuracy of the effectively available information, the total internal organization or "Image", the character of the "Plans" or program, and the nature of the significant environment, the output of either "machine" may be sense or nonsense, symbolic logic or psychologic, goal-attainment or oscillation.

Beyond giving us a deeper perspective on the concept of the "rational", current theories of cognitive processes give promise of transcending the hoary trichotomy of the cognitive, the conative, and the moral as analytical tools. Whether this amounts to a rejection of the distinction, or simply an insistence that what was analytically rent asunder must now be reunified, the ferment appears significant. We refer here, not only to the many neurological and schematic studies of the brain, or the processes by which it solves problems and attains concepts, but especially to the several theories of cognitive "dissonance" or "congruence" or "balance" represented in the works of Heider, Cartwright and Harary, Osgood and Tannenbaum, Festinger, and others, as well as the symbol-processing and interpersonal communication perspectives represented by the "psycholinguistics" of Osgood, the "communicative acts" of Newcomb, and the "two factor" theory of Mowrer.

The intricate meeting of the cognitive, the affective and evaluative (or attitudinal), and the semantic or symbolic in such theories is well illustrated in Osgood's treatment of "cognitive dynamics". Equating "cognitive elements" with the *meanings* of signs, Osgood proposes that "congruity exists when the evaluative meanings of interacting signs are equally polarized or intense – either in the same or opposite evaluative directions . . .".[40] In contrast to the theories of Heider and Festinger, this theory "*assigns affective or attitudinal values to the cognitive elements themselves*, and not to their relations . . .".[41] And in discussing the "process of inference through psycho-logic", Osgood says:

> Much of what is communicated attitudinally by messages and by behavior is based on such inferences; . . . The syntax of language and of behavior provides a structural framework within which meaningful contents

are put; the structure indicates what is related to what, and how, but only when the meaningful values are added does the combination of structure and content determine psycho-logical congruence or incongruence.[42]

Despite the incorporation of aspects of these several elements into their theories, however, the psychologically oriented theorist usually leaves the sociologist something to be desired, namely, something that transcends "the individual" and "his" attempts to minimize inconsistency or dissonance and maintain stability, and which views the group situation as inadequately characterized in terms of "myriad decisions in individual nervous systems". Thus Osgood hypothesizes that

> laws governing the thinking and behaving of individuals also govern the "thinking" and "behaving" of groups . . . with nothing but communication to bind us together, it is clear that "decisions" and "behaviors" of nations must come down to myriad decisions in individual nervous systems and myriad behaviors of individual human organisms.[43]

We are reminded here of Robert R. Sears' complaint that "psychologists think monadically. That is, they choose the behavior of one person as their scientific subject matter. For them, the universe is composed of individuals . . . the universal laws sought by the psychologist almost always relate to a single body."[44] Arguing for the desirability of combining individual and social behavior into a common framework, Sears noted that, "Whether the group's behavior is dealt with as antecedent and the individual's as consequent, or vice versa, the two kinds of event are so commonly mixed in causal relationships that it is impractical to conceptualize them separately."[45]

Fortunately, however, there are recent statements that rally to the side of the sociological interactionist theorists, whose perspective continues to be ignored or little understood by so many personality theorists who are nevertheless gradually rediscovering and duplicating its basic principles. A good beginning to a truly interpersonal approach to personality theory and the problem of stability and change in behavior is the statement of Paul F. Secord and Carl W. Backman, which remarkably parallels Goode's theory of stability and change in social systems discussed earlier. Pointing to the assumptions of several personality theorists that when stability of behavior occurs it is solely a function of stability in personality structure, and that this latter structure has, inherently, a strong resistance to change except when special change-inducing forces occur, Secord and Backman see as consequences the same kinds of theoretical inadequacies we found for the stability view of social systems:

> The first is that continuity in individual behavior is not a problem to be solved; it is simply a natural outcome of the formation of stable structure. The second is that either behavioral change is not given systematic attention, or change is explained independently of stability. Whereas behavioral stability is explained by constancy of structure, change tends to be explained by environmental forces and fortuitous circumstances.[46]

Their own theoretical view abandons these assumptions and "places the locus of stability and change in the interaction process rather than in intrapersonal structures." Recognizing the traditional two classes of behavioral determinants, the cultural-normative and the intrapersonal, their conceptualization

> attempts to identify a third class of determinants, which have their locus neither in the individual nor the culture, but in the interaction process itself. In a general sense this third class may be characterized as the tendencies of the individual and the persons with whom he interacts to shape the interaction process according to certain requirements, i.e., they strive to produce certain patterned relations. As will be seen, the principles governing this activity are truly interpersonal; they require as much attention to the behavior of the other as they do to the behavior of the individual, and it cannot be said that one or the other is the sole locus of cause.[47]

They go on to analyze the "interpersonal matrix" into three components: an aspect of the self-concept of a person, his interpretation of those elements of his behavior related to that aspect, and his perception of related aspects of the other with whom he is interacting. "An interpersonal matrix is a recurring functional relation between these three components."

In these terms, Secord and Backman attempt to specify the conditions and forces leading to or threatening congruency or incongruency, and hence stability or change, in the matrix. Thus, four types of incongruency, and two general classes of resolution of incongruency, are discussed. One of these latter classes

> results in restoration of the original matrix, leaving self and behavior unchanged (although cognitive distortions may occur), and the other leads to a new matrix in which self or behavior are changed.[48]

In sum, contrary to previous approaches, theirs emphasizes

that "the individual strives to maintain interpersonal relations characterized by congruent matrices, rather than to maintain a self, habits, or traits".

> Maintenance of intrapersonal structure occurs only when such maintenance is consistent with an ongoing interaction process which is in a state of congruency. That most individuals do maintain intrapersonal structure is a function of the fact that the behavior of others toward the individuals in question is normally overwhelmingly consistent with such maintenance.[49]

And this conception also, as most approaches do not (or do inadequately), predict or accounts for the fact that, should the interpersonal environment cease to be stable and familiar, undergoing great change such that others behave uniformly toward the individual in new ways, the individual "would rapidly modify his own behavior and internal structure to produce a new set of congruent matrices. As a result, he would be a radically changed person."[50]

As we have said, the Secord and Backman theory and Goode's role-strain theory may be seen as closely complementary views. The former argues that *personality* structure is generated in, and continues to have its seat in, the social interactive matrix; the latter argues that *social* structure is generated in, and continues to have its seat in, the social interactive matrix. Since it is the latter that we are focusing on here, we shall conclude with additional examples of current theory and research that explore further the mechanisms underlying the genesis or elaboration of social structure out of the dynamics, especially the role dynamics, of the symbolic interaction process.

Further examples

Ralph Turner has addressed himself to the elaboration of this perspective in that conceptual area fundamental to the analysis of institutions – roles and role-taking.[51] The many valid criticisms of the more static and overdetermining conception of roles is due, he believes, to the dominance of the Linton view of role and the use of an over-simplified model of role functioning. Viewing role-playing and role-taking, however, as a process (as implied in Meadian theory), Turner shows that there is more to it than just "an extension of normative or cultural deterministic theory" and that a process view of role adds novel elements to the notion of social interaction.

The morphogenic nature of role behavior is emphasized at the start in the concept of "*role-making*". Instead of postulating the initial existence of distinct, identifiable roles, Turner posits "a tendency to create and modify conceptions of self- and other-roles" as the interactive orienting process. Since actors behave *as if* there were roles, although the latter actually exist only in varying degrees of definitiveness and consistency, the actors attempt to define them and make them explicit – thereby in effect creating and modifying them as they proceed. The key to role-taking, then, is the morphogenic propensity "to shape the phenomenal world into roles"; formal organizational regulation restricting this process is not to be taken as the prototype, but rather as a "distorted instance" of the wider class of role-taking phenomena. To the extent that the bureaucratic setting blocks the role-making process, organization is maximal, "variety" or alternatives of action minimal, actors are cogs in a rigid machine, and the morphogenic process underlying the viability of complex adaptive systems is frustrated.

Role interaction is a tentative process of reciprocal responding of self and other, challenging or reinforcing one's conception of the role of the other, and consequently stabilizing or modifying one's own role as a product of this essentially feedback-testing transaction. The conventional view of role emphasizing a prescribed complementarity of expectations thus gives way to a view of role-taking as a process of "devising a performance on the basis of an imputed other-role", with an important part being played by cognitive processes of inference testing. In a manner consistent with models of the basic interaction process suggested by Goode and by Secord and Backman, Turner views as a central feature of role-taking "the process of discovering and creating 'consistent' wholes out of behavior", of "devising a pattern" that will both cope effectively with various types of relevant others and meet some recognizable criteria of consistency. Such a conception generates empirically testable hypotheses of relevance of our concern here with institutional morphogenesis, such as: "Whenever the social structure is such that many individuals characteristically act from the perspective of two given roles simultaneously, there tends to emerge a single role which encompasses the action."[52]

Turning directly to the implications for formal, institutional role-playing, Turner argues that the formal role is primarily a "skeleton" of rules which evoke and set into motion the fuller roles built-up and more or less consensually validated in the above ways. Role behavior becomes relatively fixed only while it provides a perceived consistency and stable framework for interaction, but it undergoes cumulative revision in the role-taking process of accommodation and compromise with the simple conformity demanded by formal prescriptions.

The purposes and sentiments of actors constitute a unifying element in role genesis and maintenance, and hence role-taking must be seen to involve a great deal of selective perception of other-behavior and relative emphasis in the elaboration of the role pattern. This selection process operates on the great variety of elements in the situation of relevant objects and other-behaviors which could become recognized components in a consistent role pattern. Not all combinations of behavior and object relations can be classed into a single role; there must be criteria by which actors come to "verify" or "validate" the construction of a number of elements into a consistent role. This verification stems from two sources: "internal validation" of the interaction itself, and "external validation" deriving from "the generalized other" of Mead. The former hinges on successful prediction or anticipation of relevant other-behavior in the total role-set, and hence on the existence of role patterns whereby coherent selection of behaviors judged to constitute a consistent role can be made. But the notion of fixed role prescriptions is not thereby implied, since, first, roles – like norms – often or usually provide a *range of alternative* ways of dealing with any other-role, or, as is most common, the small segment of it activated at any one time, and secondly, the coherence and predictability of a role must be assessed and seen as "validated", not in terms of any one other-role, but in terms of the Gestalt of all the accommodative and adjusted requirements set by the number of other-roles in the actor's role-set and generated in the ongoing role-making process.

An example is provided by the study by Gross *et al.* of the school superintendent role. It is found that incumbency in this role (1) actually involved a great deal of choice behavior in selecting among the alternative interpretations and behaviors deemed possible and appropriate, and that (2) consistency and coherence of an incumbent's behavior could be seen only in terms of the total role as an accommodation with correlative other-roles of school board member, teacher, and parent, with which the superintendent was required to interact simultaneously. As Gross puts it, a "system model" as against a "position-centric" model involves an important addition by including the interrelations among the counter positions. "A position can be completely described only by describing the total system of positions and relationships of which it is a part. In other words, in a system of interdependent parts, a change in any relationship will have an effect on all other relationships, and the positions can be described only by the relationships."[53]

Thus Turner sees the internal criterion of role validation as insuring a constant modification, creation, or rejection of the content of specific roles occurring in the interplay between the always somewhat vague and incomplete ideal role conceptions and the experience of their concrete implications by the interpreting, purposive, selectively evaluating and testing self and others.

The basis of "external validation" of a role is the judgment of the behavior to constitute a role by others felt to have a claim to correctness of legitimacy. Criteria here include: discovery of a name in common use for the role, support of major norms or values, anchorage in the membership of recognized groups, occupancy of formalized positions, and experience of key individuals as role models acting out customary attitudes, goals and specific actions.

Under the "normal loose operation of society" these various internal and external criteria of validation are at best only partially conveyant and consistent in identifying the same units and content as roles. The resulting inevitable discrepancies between formal, institutional rules and roles, and the goals, sentiments and selective interpretations arising from the experience of actually trying to play them out, make role conceptions "creative compromises", and insure "that the framework of roles will operate as a hazily conceived ideal framework for behavior rather than as an unequivocal set of formulas".[54]

In sum, "institutions" may provide a normative framework prescribing roles to be played and thus assuring the required division of labor and minimizing the costs of general exploratory role-setting behavior, but the actual role transactions that occur generate a more or less coherent and stable working compromise between ideal set prescriptions and a flexible role-making process, between the structured demands of others and the requirements of one's own purposes and sentiments. This conception of role relations as "fully interactive", rather than merely conforming, contributes to the recent trends "to subordinate normative to functional processes in accounting for societal integration"[55] by emphasizing the complex adaptive interdependence of actors and actions in what we see as an essentially morphogenic process – as against a merely equilibrial or homeostatic process.

Organization as a negotiated order

Next we shall look at a recently reported empirical study of a formal organization that concretely illustrates many facets of the above conceptualization of Turner and contributes further to our thesis. In their study of the hospital and its interactive order, Anselm Strauss and colleagues develop a model of organizational process that bears directly on the basic sociological problem of "how a measure of order is maintained in the face of inevitable changes (derivable from sources both external and internal to the organization)".[56] Rejecting an overly structural view, it is assumed that social order is not simply normatively specified and automatically maintained but is something that must be "worked at", continually reconstituted. Shared agreements, underlying orderliness, are not binding and shared indefinitely but involve a temporal dimension implying eventual review, and consequent renewal or rejection. On the basis of such considerations, Strauss and colleagues develop their conception of organizational order as a "negotiated order".

The hospital, like any organization, can be visualized as a hierarchy of status and power, of rules, roles and organizational goals. But it is also a locale for an ongoing complex of transactions among differentiated types of actors: professionals such as psychiatrists, residents, nurses and nursing students, psychologists, occupational therapists and social workers; and non-professionals such as various levels of staff, the patients themselves, and their families. The individuals involved are at various stages in their careers, have their own particular goals, sentiments, reference groups, and ideologies, command various degrees of prestige, esteem and power, and invest the hospital situation with differential significance.

The rules supposed to govern the actions of the professionals were found to be far from extensive, clearly stated, or binding; hardly anyone knew all the extant rules or the applicable situations and sanctions. Some rules previously administered would fall into disuse, receive administrative reiteration, or be created anew in a crisis situation. As in any organization, rules were selectively evoked, broken, and/or ignored to suit the defined needs of personnel. Upper administrative levels especially avoided periodic attempts to have the rules codified and formalized, for fear of restricting the innovation and improvisation believed necessary to the care of patients. Also, the multiplicity of professional ideologies, theories and purposes would never tolerate such rigidification.

In sum, the area of action covered by clearly defined rules was very small, constituting a few general "house rules" based on long-standing shared understandings. The basis of organizational order was the generalized mandate, the single ambiguous goal, of returning patients to the outside world in better condition. Beyond this, the rules ordering actions to this end were the subject of continual negotiations – being argued, stretched, ignored, or lowered as the occasion seemed to demand. As elsewhere, rules failed to act as universal prescriptions, but required judgment as to their applicability to the specific case.

The ambiguities and disagreements necessitating negotiation are seen by the researchers to be patterned. The various grounds leading to negotiation include: disagreement and tension over the proper ward placement of a patient to maximize his chances of improvement; the mode of treatment selected by the physician, which is closely related to his own psychiatric ideology and training; the multiplicity of purposes and temporal ends of each of the professional groups as they maneuver to elicit the required cooperation of their fellow workers; the element of medical uncertainty involved in treating the patient as a unique, "individual case", and the consequent large area of contingency lying of necessity beyond specific role prescription; and, finally, the inevitable changes forced upon the hospital and its staff by external forces and the unforeseen consequences of internal policies and the round of negotiations themselves. What is concretely observed, then, in researching the organizational order of the hospital, is negotiation between the neurologically trained and the psychotherapeutically oriented physician, between the nurses and the administrative staff, between the nonprofessional floor staff and the physician, between the patient and each of the others.

The negotiation process itself was found to have patterned and temporal features. Thus, different physicians institute their own particular programs of treatment and patient care and in the process develop fairly stable understandings with certain nurses or other institutional gatekeepers such as to effectuate an efficient order of behaviors with a minimum of communication and special instructions. Such arrangements are not called for by any organizational role prescriptions; nevertheless, they represent a concrete part of the actual organization generated in the morphogenic process of negotiation (or role-making and -taking, in Turner's terms). Thus, agreements do not occur by chance but are patterned in terms of "who contracts with whom, about what, as well as when . . .".[57] There is an important temporal aspect, also, such as the specification of a termination period often written into an agreement – as when a physician bargains with a head nurse to leave his patient in the specific ward for "two more days" to see if things will work themselves out satisfactorily.

In a final section of their paper, Strauss and his colleagues bring out the full implications of their negotiation model in dealing with genuine organizational change. The model presents a picture of the hospital – and perhaps most other institutionalized spheres of social life – as a transactional milieu where numerous agreements are "continually being established, renewed, reviewed, revoked, revised". But this raises the question of the relation between this process and the more stable structure of norms, statuses, and the like. The authors see a close systemic relation between the two. The daily negotiations periodically call for a reappraisal and reconstitution of the organizational order into a "new order, not the reestablishment of an old, as reinstituting of a previous equilibrium". And, we would add, it contributes nothing to refer to this as a "moving equilibrium" in the scientifically established sense of the term. The daily negotiative process not only allows the day-by-day work to get done, but feeds back upon the more formalized, stable structure of rules and policies by way of "a periodic appraisal process" to modify it – sometimes slowly and crescively, sometimes rapidly and convulsively. And, as a reading of history suggests, virtually every formal structure extant can be traced, at least in principle, from its beginnings to its present apparently timeless state through just such a morphogenic process – a process characteristic of what we have called the complex adaptive system.

The school superintendent and his role

We turn to the study by Gross and his associates of the role system of the school superintendent and his counter-role partners, the school board member, the teacher, and the parent. A major burden of this empirical study is to demonstrate the research sterility of the Lintonian conception of role, and structural theories built on it, due principally to the postulate of consensus on role definition. The study showed a majority of significant differences in the definitions of their own roles by a sample of incumbents of the same social position and by incumbents of different but interrelated counter positions. This fact led Gross and his associates to the demonstration of a number of important theoretical consequences derived from rejection of the postulate of role consensus. It is often assumed, for example, that the socialization process by which roles are "acquired" provides for a set of clearly defined and agreed-upon expectations associated with any particular position. But the empirically discovered fact of differential *degrees of consensus* seriously challenged this assumption. From our systems model viewpoint, recognition of degrees of consensus is tantamount to the recognition of a continuous source of "variety" in the role system, as defined earlier, which leads us to seek the various *selective*, choice processes occurring in the role transactions. At least for the occupational positions studied, it was found that the assumption of socialization on the basis of prior consensus on role definitions was untenable, and deserved "to be challenged in most formulations of role acquisition, including even those concerned with the socialization of the child".[58]

Secondly, the research showed that, instead of assuming role consensus and explaining variations of behavior of incumbents of the same position in terms of personality variables, one would better explain them in terms of the varying role expectations and definitions – which may be unrelated to psychological differences.

The implications are also great for a theory of social control. Instead of a model assuming that the application or threat of negative sanctions leads to conformity to agreed-upon norms, the research pointed to the numerous situations in which, due to variant or ambiguous role definitions, the same behavior resulted in negative sanctions by some role partners and positive sanctions by others, or failure to apply sanctions because of perceived ambiguity – or nonconformity to perceived expectations of another despite negative sanctions because other expectations were defined as more legitimate.

Another Lintonian postulate challenged by this research is that though an actor may occupy many positions, even simultaneously, he activates each role singly with the others remaining "latent". It is found, however, that individuals often perceive and act toward role partners as if simultaneous multiple roles were being activated. For example, one may hold different expectations regarding a teacher who is male, young and unmarried as against one who is female, older and married. In other words, standards and expectations are applied to the whole person as a result, in part, of the complex of positions the person is perceived as occupying at that time. A related consideration involves the time dimension over which two or more individuals interact; other positions they occupy enter progressively into their perception of each other and consequently modify evaluations and expectations. Thus the authors generalize their point to a broader theory of social interaction by suggesting that evaluative standards shift over time from those applied as appropriate to the incumbent of a particular position to those applied to a total person with particular personality features and capacities as the incumbent of multiple positions.

Finally, their rejection of the consensus model led these researchers to find a process of role-strain or role-conflict generation and resolution similar in principle to that conceptualized by others discussed above. Having defined the role set they were studying as a true *complex system* of interrelated components, and having then uncovered and analyzed the *variety* continuously introduced into the system by way of variant, ambiguous or changing role definitions, they then focused on the *selection process* whereby this variety was sifted and sorted in the give and take of role transactions. Thus, given the situation in which a role incumbent was faced with incompatible expectations on the part of two of his counter-role partners, a theory was constructed to answer the question of how the actor may choose from among four alternatives in resolving the role conflict. From our present perspective, the theoretical scheme suggested constitutes another important contribution to the forging of a conceptual link between the dynamics of the role transaction and the more stable surrounding social structure – a link that is too often skipped over by the consensus theorist's identification of social structure and consensual role playing.

This linkage is made in terms of the concepts of perceived *legitimacy* of the conflicting expectations, an assessment of the *sanctions* that might be applied, and predispositions to give primacy to a *moral* orientation, an *expedient* orientation, or a balance of the two. We face once again the reciprocal question of how role transactions are conditioned by the surrounding social structure and how that structure is generated and regenerated as a product of the complex of role transactions.

The four alternatives that Gross and colleagues see open to an actor to choose in attempting to resolve a role conflict between incompatible expectations A and B are: (1) conformity to expectation A; (2) conformity to expectation B; (3) compromise in an attempt to conform in part to both expectations; or (4) attempt to avoid conforming to either expectation. The first criterion that the theory postulates to underlie the particular alternative chosen is the actors' definition of the legitimacy of the expectations. Thus the prediction of behavior on this criterion is that, when only one expectation is perceived as legitimate the actor will conform to that one; when both are defined as legitimate he will compromise; and when neither is seen as legitimate he will engage in avoidance behavior. The second criterion is the actor's perception of the sanctions that would be applied for nonconformity, which would create pressures to conform if strong negative sanctions are foreseen otherwise. This predicts for three of the four combinations of two sets of expectations, but not for the case of both expectations being perceived as leading to weak or no negative sanctions.

It is assumed that for any role conflict situation an actor would perceive both of these dimensions and make his decision accordingly. Predictions on the basis of the theory so far provide for determinate resolutions of conflict in seven of the sixteen combinations of the four types of legitimacy and the four types of sanctions situations, but the other nine are left indeterminate with only the two criteria. This is because the criteria predispose in different directions, and at least a third criterion is needed to determine the outcome. The authors thus appeal to the actor's predisposition to give primacy to either the legitimacy or to the sanctions dimension, or to balance the two, thus leading to the postulation of three types of predisposing orientations to expectations as listed above – the *moral*, the *expedient*, and the balanced *moral-expedient*. All the combinations of situations now become predictive.

The accuracy of the predictions was tested empirically with the data from the superintendent-role study for four "incompatible expectation situations", and the evidence supported the theory, though with some incorrect predictions.

The implications of this conceptualization and empirical analysis are far-reaching, as already suggested, for general sociological theory. The study is concerned with what must be considered "institutional" organization and process, and supports a model of that structure and process that is quite different from the more traditional models. As the authors point out, one strong advantage of the theory is its conceptualization of institutional role behavior in terms of "expectations", whether legitimate or illegitimate, rather than in terms of "obligations" (legitimate expectations) as is assumed in consensus theory. The theory thus allows for the possibility that illegitimate expectations constitute a significant part of institutional role behavior, and underlie much of the conflict occurring – as we feel intuitively to be the case – within the institutional process. It follows, further, that deviance – nonconformity to expectations – is a more intimate and normal element in institutional behavior than conformity theory would permit. And it also permits theoretical recognition of the possibility that, as Etzioni has suggested,[59] a great deal of organizational behavior is based, not on internalized norms and values, but on an expedient calculation of self-interests and of possible rewards and punishments. This, in turn, leaves open the theoretical possibility that non-legitimized power, as well as legitimized authority, may often be a controlling factor in institutional behavior.

Role conflict and change among the Kanuri

The final empirical study we shall sketch is explicitly based on an understanding of the modern systems approach, focusing as it does on a theory of "self-generating internal change". Ronald Cohen, an anthropologist, reports a theoretically well-organized analysis of his field study of role conflict and change among the Kanuri of Nigeria.[60] The study focuses on "goal ambiguity" and "conflicting standards" within a facet of the joint native–colonial political administrative hierarchy, particularly on the pivotal position of native "district head" which had come to combine the quite diverse cultural orientations of the colonial British and the Kanuri. This diversity between, as well as within, the two cultures made for inconsistencies, ambiguity, and conflict in political goals and in role standards and performances, which were continuously exacerbated by the variety of pressures put on district heads by the central native administration, the colonial administration, and the colonial technical departments.

The consequences of this situation for the political system are analyzed in terms of A. G. Frank's theory of organizational process and change.[61] Given the conditions of ambiguity and conflict of standards and goals, it is postulated that a process of *selective performance* and *selective enforcement* of standards will occur, with subordinates being forced to decide on which expectations to meet, and superiors required to selectively evaluate performances and hence selectively enforce some standards over others. This postulate leads to a number of predictions that Cohen proceeded to test. In essence, a continuous process is set up that appears, though in more exaggerated form, much like the "role strain", "role-making", "negotiated order" situations we met earlier. Role players fail to meet, or feign meeting some standards, and differentially select those they will meet. As a result, the role system is postulated to exhibit a strain toward substantive rationality (in Weber's sense), shifting standards for members, widespread role innovation or "deviance", ready adaptation to environmental changes, and an active and widespread circulation of information about standards and goals by "intermediary dealers in information" and by members seeking to reduce the ambiguity and conflict concerning these standards and goals.

The process is thus a circular, feedback loop whereby superiors continuously modify their standards or expectations as definitions of political objectives change, and subordinates adapt their decisions and performances to these changing expectations and surrounding circumstances, which in turn changes the states of the situation toward which superiors are acting. The role system, then, is seen as continuously receptive and responsive to external and internal pressures which demand some kind of workable "mapping" of the abundantly available situational "variety", which in turn makes possible – though does not guarantee – the evolution of more or less adaptive, institutionalized internal system procedures.

Applying this theory to the Kanuri, Cohen found the predictions to be borne out to a substantial degree. We leave the detailed description of these phenomena to the original study, which drew the general practical conclusion that – in spite of its apparent conservative, anti-progressive traditionalism – the Kanuri political role system showed greater compliance to the varied pressures of superiors and situational exigencies than to the tenets of tradition and thereby proved to be a self-generating system containing mechanisms for its own transformation. The implications of this for policy relating to "developing countries" are of obvious importance.

On the theoretical side, Cohen clearly recognizes the implications of his mode of analysis for a genetic model of sociocultural evolution.

> This model depends basically on two conditions. First, the evolving phenomenon must be shown to be *variable* in terms of its constituent units, and second, there must be analytically distinct *selective factors* which operate on the variation within the phenomenon to produce a constantly adapting and thus an evolving history of development. Although there are more or less stable orientations of tradition present in Bornu, conflicts in the political organization produce a variability of response by the actors upon which selective pressures exerted by superiors in the political hierarchy may operate to bring about innovations and changes that are incremental in their nature, i.e., evolutionary rather than revolutionary.[62]

We opened our discussion of the decision-making, process approach to complex adaptive systems with a turn-of-the-century prognosis of Albion Small. We might remind ourselves further of important ties with the past by closing with the early fundamental insight of Edward Sapir:

> While we often speak of society as though it were a static structure defined by tradition, it is, in the more intimate sense, nothing of the kind, but a highly intricate network of partial or complete understandings between the members of organizational units of every degree of size and complexity. . . . It is only apparently a static sum of social institutions; actually it is being reanimated or creatively reaffirmed from day to day by particular acts of a communicative nature which obtain among individuals participating in it.[63]

Conclusion

We have suggested that much current thinking represents the coming to fruition of earlier conceptions of which Sapir's and Small's statements are harbingers. Although a science should not hesitate to forget its founders, it would do well to remain aware of their basic thought.

We have argued that a promising general framework for organizing these valuable insights of the past and present may be derived from the recent general systems perspective, embracing a holistic conception of complex adaptive systems viewed centrally in terms of information and communication process and the significance of the way these are structured for self-regulation and self-direction. We have clearly arrived at a point in the development of the "behavioral" sciences at which synthesis or conceptual unification of subdisciplines concerned with social life is challenging simple analysis or categorization. Not only is there growing demand that the "cognitive", "affective" and "evaluative" be conceptually integrated, but that the free-handed parceling out of aspects of the sociocultural adaptive system among the various disciplines (e.g., "culture" to anthropology, the "social system" to sociology, and "personality" to psychology) be reneged, or at least ignored. The potential of the newer system theory is especially strong in this regard.[64] By way of conclusion we recapitulate the main arguments.

1. The advance of science has driven it away from concern with "substance" and toward a focus on *relations* between components of whatever kind. Hence the concern with complex organization or systems, generally defined in terms of the transactions, often mutual and usually intricate, among a number of components such that some kind of more or less stable structure – often tenuous and only statistically delineated – arises (that is, *some* of the relations between components show *some* degree of stability or repetitiveness *some* of the time). Extremely fruitful advances have been taking place, especially since the rapid scientific progress made during World War II, in specifying basic features common to substantively different kinds of complex adaptive systems, as well as delineating their differences. In contrast to some of the general systems theorists themselves as well as their critics, we have argued that this is not simply analogizing but generalizing or abstracting as well (although the former is important, and scientifically legitimate also, when performed with due caution). To say that physiological, psychological, and sociocultural processes of control all involve the basic cybernetic principles of information flow along feedback loops is no more a mere analogy than to say that the trajectories of a falling apple, an artificial satellite, or a planet all involve the basic principle of gravitational attraction.

2. Complex adaptive systems are open systems in intimate interchange with an environment characterized by a great deal of shifting variety ("booming, buzzing confusion") and its constraints (its structure of causal interrelations). The concept of equilibrium developed for closed physical systems is quite inappropriate and usually inapplicable to such a dynamic situation. Rather, a characteristic resultant is the elaboration of organization in the direction of the less probable and the less inherently stable.

Features common to substantively different complex adaptive systems can be conceptualized in terms of the perspective of information and control theory. "Information" in its most general sense is seen, not as a thing that can be transported, but as a selective interrelation or mapping between two or more subsets of constrained variety selected from larger ensembles. Information is thus transmitted or communicated as invariant constraint or structure in some kind of variety, such that subsystems with the appropriate matched internal ensembles, reacting to and acting upon the information, do so in a situation of decreased uncertainty and potentially more effective adaptation to the variety that is mapped. Unless mapping (encoding, decoding, correlating, understanding, etc.) occurs between two or more ensembles we do not have "information", only raw variety or noise.

In these terms, adaptive systems, by a continuous selective feedback interchange with the variety of the environment, come to make and preserve mappings on various substantive bases, which may be transmitted generationally or contemporaneously to other similar units. By means of such mappings (for example, via genes, instincts, learned events, culture patterns) the adaptive system may, if the mappings are adequate, continue to remain viable before a shifting environment. The transmission and accumulation of such information among contemporaneous adaptive systems (individuals) becomes more and more important at higher levels until it becomes the prime basis of linkage of components for the highest level sociocultural system.

Some of the more important differences between complex adaptive systems include the substantive nature of the components, the types of linkage of the components, the kinds and levels of feedback between system and environment, the degree of internal feedback of a system's own state (for example, "self-awareness"), the methods of transmission of information between subsystems and along generations, the degree of refinement and fidelity of mapping and information transfer, the degree and rapidity with which the system can restructure itself or the environmental variety, etc.

3. Such a perspective provides a general framework which meets the major criticisms leveled against much of current sociological theory: lack of time and process perspective, overemphasis on stability and maintenance of given structure, and on consensus and cooperative relations, to the relative neglect – or unsystematic treatment – of deviance, conflict and other dissociative relations underlying system destructuring and restructuring.

4. Thus, the concept of the system itself cannot be identified with the more or less stable structure it may take on at any particular time. As a fundamental principle, it can be stated that a condition for maintenance of a viable adaptive system may be a change in its particular structure. Both stability and change are a function of the same set of variables, which must include both the internal state of the system and the state of its significant environment, along with the nature of the interchange between the two.

5. A time perspective is inherent in this kind of analysis – not merely historical but evolutionary. (It can probably be said that the time was ripe by 1959 for a Darwinian centennial ramifying well beyond the purely biological.) This perspective calls for a balance and integration of structural and processual analysis. As others have pointed out, the Linnean system of classification of structures became alive only after Darwin and others discovered the processes of variation, selection and recombination that gave them theoretical significance, though these discoveries leaned heavily in turn on the classification of systematically varying structures.

And among the important processes for the sociocultural system are not only cooperation and conformity to norms, but conflict, competition and deviation which may help create (or destroy) the essential variety pool, and which constitute part of the process of selection from it, such that a more or less viable system structure may be created and maintained (or destroyed).

6. In sociological terms, the "complementarity of expectations" model is an ideal type constituting only one pole of a continuum of equally basic associative and dissociative processes characterizing real societies – although the particular "mix" and intensities of the various types may differ widely with different structural arrangements. Further, the systemic analysis of a sociocultural system is not exhausted by analysis of its institutionalized patterns. By focusing on process, we are more prepared to include all facets of system operation – from the minimally structured end of the collective behavior continuum through the various degrees and kinds of structuring to the institutional pole. The particular characteristics of the process, especially the degrees and kinds of mappings and mismatchings of the interacting units, tell us whether we are in fact dealing with certain degrees of structuring and the dynamics underlying this structuring: de facto patterning may be anchored in coercive, normative, or utilitarian compliance, making for very different kinds of system.

7. "Institutionalized" patterns are not to be construed as thereby "legitimized" or as embracing only "conformity" patterns – at least for the sake of conceptual clarity and empirical adequacy. Processes of all degrees and kinds of structuring may be seen in terms of deviant as well as conformity patterns – relative to the point of reference selected by the observer. One may select certain institutional patterns and values (to be clearly specified) as an arbitrary reference point to match against *other* institutional patterns and values, along with less structured behaviors. The concept of *the* institutionalized common value system smuggles in an empirically dubious, or unverified, proposition – at least for complex modern societies.

8. The complex adaptive system's organization *is* the "control", the characteristics of which will change as the organization changes. The problem is complicated by the fact that we are dealing directly with two levels of adaptive system and thus two levels of structure, the higher level (sociocultural) structure being largely a shifting statistical or probability structure (or ensemble of constraints) expressing over time the transactional processes occurring among the lower level (personality) structures. We do not have a sociocultural system *and* personality systems, but only a sociocultural system *of* constrained interactions among personality systems.

We can only speak elliptically of "ideas" or "information" or "meanings" in the head of a particular individual: all we have is an ensemble of constrained variety embodied in a neurological net. "Meaning" or "information" is generated only in the process of interaction with other ensembles of similarly mapped or constrained variety (whether embodied in other neurological nets or as the ensemble of causally constrained variety of the physical environment), whereby ensemble is mapped or matched against ensemble via communication links, and action is carried out, the patterning of which is a resultant of the degree of successful mapping that occurred. (Of course, "meaning" on the symbolic level can be regenerated over a long period by the isolated individual through an internal interchange or "conversation" of the person with his "self", made possible by previous socially induced mappings of one's own internal state that we call "self-awareness". But in some respects, part of the world literally loses its meaning for such a person.)

If the ensembles of variety of two interacting units, or one unit and its physical environment, have no or little isomorphic structuring, little or no meaning can be generated to channel ongoing mutual activity; or in more common terms, there is no "common ground", no "meeting of minds" and thus no meaning or information exchange – only raw variety, uncertainty, lack of "order" or organization.

Unless "social control" is taken as simply the more or less intentional techniques for maintaining a given institutional structure by groupings with vested interests, it must refer to the above transactional processes as they operate – now to develop new sociocultural structures, now to reinforce existing ones, now to destructure or restructure older ones. Thus, we cannot hope to develop our understanding much further by speaking of one "structure" determining, "affecting", or acting upon another "structure". We shall have to get down to the difficult but essential task of (*a*) specifying much more adequately the distribution of essential features of the component subsystems' internal mappings, including both self-mappings and their mappings of their effective environment, (*b*) specifying more extensively the structure of the transactions among these units at a given time, the degree and stability of the given structuring seen as varying with the degree and depth of common meanings that are generated in the transaction process, and (*c*) assessing, with the help of techniques now developing, the ongoing process of transitions from a given state of the system to the next in terms of the deviation-reducing and deviation-generating feedback loops relating the tensionful, goal-seeking, decision-making subunits via the communication nets partly specified by (*b*). Some behavior patterns will be found to be anchored in a close matching of component psychic structures (for example, legitimized authority or normative compliance); others, in threats of goal-blockage, where there is minimal matching (for example, power or coercive compliance); still others, anchored in a partial matching, primarily in terms of environmental mappings of autonomous subunits and minimally in terms of collective mappings (for example, opportunism or utilitarian compliance). As the distribution of mappings shifts in the system (which normally occurs for a number of reasons), so will the transaction processes and communication nets, and thus will the sociocultural structure tend to shift as gradients of misunderstanding, goal-blockage, and tensions develop.

9. Finally, we have tried to show how this perspective bears on, and may help to integrate conceptually, the currently developing area of "decision theory" which recognizes individual components as creative nodes in an interactive matrix. In the complex process of transactions occurring within a matrix of information flows, the resulting cognitive mappings and mismappings undergo various stresses and strains as component units assess and reassess with varying degrees of fidelity and refinement their internal states and the shifting and partially uncertain, and often goal-blocking environment. Out of this process, as more or less temporary adjustments, arises the more certain, more expected, more codified sequences of events that we call sociocultural structure. In the words of Norbert Wiener, "By its ability to make decisions" the system "can produce around it a local zone of organization in a world whose general tendency is to run down".[65] Whether that structure proves viable or adaptive for the total system is the kind of question that cannot be reliably answered in the present state of our discipline. It most certainly demands the kind of predictive power that comes with the later rather than the earlier stages of development of a science. And later stages can arrive only at some sacrifice of ideas of earlier stages.

References

1 *Sociology and Modern Systems Theory* (Englewood Cliffs, N.J.: Prentice-Hall, 1967).

2 See Campbell, Donald T. (1959) "Methodological Suggestions from a Comparative Psychology of Knowledge Processes", *Inquiry*, **2**, 152–67.

3 See, for example, Rapoport and MacKay, Chapters 16 and 24 of *Modern Systems Research for the Behavioral Scientist*.

4 For an excellent recent overview of this transition, see Hallowell, A. Irving, "Personality, Culture, and Society in Behavioral Evolution", in Koch, Sigmund (ed.), *Psychology: A Study of a Science*, Volume 6: Investigations of Man as Socius (New York: McGraw-Hill, 1963), 429–509.

5 Thelen, Herbert A. (1956) "Emotionality and Work in Groups", in Leonard D. White (ed.), *The State of the Social Sciences* (Chicago: University of Chicago Press), pp. 184–6.

6 See Cannon, Article 13, p. 219.

7 Or perhaps we might take Cadwallader's suggestion and use Ashby's term "ultrastability". I dislike, how-

ever, the connotative overemphasis on "stability", which is sure to be misunderstood by many.

I prefer the term "morphogenesis" as best expressing the characteristic feature of the adaptive system. Thus, we might say that physical systems are typically equilibrial, physiological systems are typically homeostatic, and psychological, sociocultural, or ecological systems are typically morphogenic. From this view, our paradigm of the mechanisms underlying the complex system becomes a basic paradigm of the morphogenic process, perhaps embracing as special cases even the structuring process below the complex adaptive system level.

8 Mace, C. A. (1953) "Homeostasis, Needs and Values", *British Journal of Psychology*, **44**, 204–5. Gordon Allport reinforces this view for personality (but note his terminology): "Some theories correctly emphasize the tendency of human personality to go beyond steady states and to elaborate their internal order, even at the cost of disequilibrium. Theories of changing energies . . . and of functional autonomy . . . do so. These conceptions allow for a continual increase of men's purposes in life and for their morphogenic effect upon the system as a whole. Although homeostasis is a useful conception for short-run 'target orientation', it is totally inadequate to account for the integrating tonus involved in 'goal orientation'. . . . Although these formulations differ among themselves, they all find the 'go' of personality in some dynamic thrust that exceeds the pale function of homeostatic balance. They recognize increasing order over time, and view change within personality as a recentering, but not as abatement, of tension." – Allport, Gordon W. (1961) *Pattern and Growth in Personality* (New York: Holt, Rinehart & Winston), p. 569.

9 See Chapters 34 and 35 of *Modern Systems Research for the Behavioral Scientist*.

10 Henry, Jules (1959) "Culture, Personality, and Evolution", *American Anthropologist*, **61**, 221–2.

11 Deutsch, Karl W. (1948–9) "Some Notes on Research on The Role of Models in Natural and Social Science", *Synthese*, pp. 532–3.

12 See Campbell, *op. cit.*

13 See Osgood, Charles E., "Psycholinguistics", in Koch, S. (ed.), *Psychology, loc. cit.*, pp. 244–316; Mowrer, O. Hobart (1960) *Learning Theory and the Symbolic Processes* (New York: John Wiley), esp. Chapter 7: "Learning Theory, Cybernetics, and the Concept of Consciousness." For less behavioristic and more genetic and emergent views, see, for example, Mead, George H. (1934) *Mind, Self and Society* (Chicago: University of Chicago Press), and more recently, Werner, Heinz and Kaplan, Bernard (1963) *Symbol Formation* (New York: John Wiley).

14 See, for example, Sahlins, Marshall D. (1964) "Culture and Environment: The Study of Cultural Ecology", in Sol Tax (ed.), *Horizons of Anthropology* (Chicago: Aldine), pp. 132–47.

15 See the selections from Nett (Chapter 48), Deutsch (Chapter 46), Hardin (Chapter 55), and Vickers (Chapter 56) in *Modern Systems Research for the Behavioral Scientist*.

16 However, we should not deemphasize the important structuring role of concrete artifacts, for example, the structure of physical communication nets, road nets, cities, interior layouts of buildings, etc., as limiting and channeling factors for sociocultural action and interaction.

17 Small, Albion W. (1905) *General Sociology* (Chicago: University of Chicago Press), p. ix.

18 Cooley, Charles H. (1918) *Social Process* (New York: Scribner's).

19 Znaniecki, Florian (1952) *Cultural Sciences* (Urbana: University of Illinois Press).

20 Consider the explicit "feedback" and "self-regulation" conceptions in the following statements of G. H. Mead in *Mind, Self and Society*: ". . . the central nervous system has an almost infinite number of elements in it, and they can be organized not only in spatial connection with each other, but also from a temporal standpoint. In virtue of this last fact, our conduct is made up of a series of steps which follow each other, and the later steps may be already started and influence the earlier ones. The thing we are going to do is playing back on what we are doing now" (p. 71). "As we advance from one set of responses to another we find ourselves picking out the environment which answers to this next set of responses. To finish one response is to put ourselves in a position where we see other things. . . . Our world is definitely mapped out for us by the responses which are going to take place. . . . The structure of the environment is a mapping out of organic responses to nature; any environment, whether social or individual, is a mapping out of the logical structure of the act to which it answers, an act seeking overt expression" (pp. 128–9, and footnote 32, p. 129). "It

is through taking this role of the other that [the person] is able to come back on himself and so direct his own process of communication. This taking the role of the other, an expression I have so often used, is not simply of passing importance. It is not something that just happens as an incidental result of the gesture, but it is of importance in the development of cooperative activity. The immediate effect of such role-taking lies in the control which the individual is able to exercise over his own response. . . . From the standpoint of social evolution, it is this bringing of any given social act, or of the total social process in which that act is a constituent, directly and as an organized whole into the experience of each of the individual organisms implicated in that act, with reference to which he may consequently regulate and govern his individual conduct, that constitutes the peculiar value and significance of self-consciousness in those individual organisms" (p. 254, including part of footnote 7).

See also the extended discussion based on Mead's essentially cybernetic perspective in Shibutani, Chapter 39.

21 Seeman, Melvin, review of Rose, Arnold (ed.) *Human Behavior and Social Processes*, in *American Sociological Review*, **27** (August 1962), 557.

22 Murdock, George P. (1955) "Changing Emphasis in Social Structure", *Southwestern Journal of Anthropology*, **11**, 366.

23 Firth, Raymond (1955) "Some Principles of Social Organization", *Journal of the Royal Anthropological Institute*, **85**, 1.

24 Nadel, S. F. (1957) *The Theory of Social Structure* (New York: The Free Press of Glencoe), pp. 54–5.

25 *Ibid.*, p. 128.

26 *Ibid.*, p. 153.

27 Vogt, Evon Z. (1960) "On the Concept of Structure and Process in Cultural Anthropology", *American Anthropologist*, **62**, 18–33.

28 Blumer, Herbert (1953) "Psychological Import of the Human Group", in Sherif, Muzafer and Wilson, M. O. (eds.) *Group Relations at the Crossroads* (New York: Harper), pp. 199–201. Emphasis added.

29 Allport, Gordon W. (1961) *Pattern and Growth in Personality* (New York: Holt, Rinehart, & Winston), p. 567.

30 For example, see Gouldner, Alvin W. (1956) "Some Observations on Systematic Theory, 1945–55", in Zetterberg, Hans L. (ed.) *Sociology in the United States of America* (Paris: UNESCO), pp. 39–40. See also Moore, Jr., Barrington (1955) "Sociological Theory and Contemporary Politics", *American Journal of Sociology*, **61** (September), 111–15.

31 Recall, for example, the excellent treatment of "decision theory" in MacIver, Robert M. (1942) *Social Causation* (New York: Ginn & Co.), esp. pp. 291 ff.

32 Selznick, Philip, (1961) "Review Article: The Social Theories of Talcott Parsons", *American Sociological Review*, **26** (December), 934.

33 Boulding, Kenneth E. (1956) *The Image* (Ann Arbor: University of Michigan Press).

34 Recall Maruyama's discussion in Chapter 36 of *Modern Systems Research for the Behavioral Scientist*.

35 Goode, William J. (1960) "A Theory of Role Strain", *American Sociological Review*, **25** (August), 483.

36 *Ibid.*, 495.

37 *Ibid.*

38 *Ibid.*, 487.

39 *Ibid.*, 494.

40 Osgood, Charles E. (1960) "Cognitive Dynamics in the Conduct of Human Affairs", *Public Opinion Quarterly*, **24**, 347.

41 *Ibid.*, 347–8.

42 *Ibid.*, 351.

43 *Ibid.*, 363.

44 Sears, Robert R. (1951) "A Theoretical Framework for Personality and Social Behavior", *American Psychologist*, **6**, 478–9.

45 *Ibid.*, 478.

46 Secord, Paul F. and Backman, Carl W. (1961) "Personality Theory and the Problem of Stability and Change in Individual Behavior: An Interpersonal Approach", *Psychological Review*, **68**, 22.

47 *Ibid.*

48 *Ibid.*, 26.

49 *Ibid.*, 28.

50 *Ibid.*

51 Turner, Ralph H. (1962) "Role-Taking: Process Versus Conformity", in Rose, Arnold M. (ed.) *Human Behavior and Social Processes* (Boston: Houghton Mifflin Co.), Chapter 2.

52 *Ibid.*, 26.

53 Gross, Neal, *et al.* (1958) *Explorations in Role Analysis* (New York: John Wiley), p. 53.

54 Turner, *loc. cit.*, p. 32.

55 *Ibid.*, p. 38.

56 Strauss, Anselm *et al.* (1963) "The Hospital and Its Negotiated Order", in Freidson, Eliot (ed.) *The*

Hospital in Modern Society (New York: The Free Press of Glencoe), p. 148.

57 *Ibid.*, p. 162.

58 Gross, Neal *et al.*, *op. cit.*, p. 321. Also see Kahn, Robert L. *et al.* (1964) *Organizational Stress: Studies in Role Conflict and Ambiguity* (New York: John Wiley).

59 Etzioni, Amitai (1961) *A Comparative Analysis of Complex Organizations* (New York: The Free Press of Glencoe).

60 Cohen, Ronald (1964) "Conflict and Change in a Northern Nigerian Emirate", in Zollschan, George K. and Hirsch, Walter (eds.) *Explorations in Social Change* (Boston: Houghton Mifflin Co.), Chapter 19.

61 Frank, A. G. (1959) "Goal Ambiguity and Conflicting Standards: An Approach to the Study of Organization", *Human Organization*, **17**, 8–13.

62 *Op. cit.*, 519. Emphasis supplied.

63 Sapir, Edward (1931) "Social Communication", *Encyclopedia of the Social Sciences* (New York: Macmillan), Vol. 4, p. 78.

64 This still remains primarily a potential, however; perusal of the general systems literature shows treatment of the sociocultural level systems to be sparse compared to that of biological, psychological and other systems. Part of the reason for this is the failure of sociologists to participate and to make what could be significant contributions to a field rapidly leaving us behind.

65 Wiener, Norbert (1954) *The Human Use of Human Beings* (Garden City, N.Y.: Doubleday Anchor, 2nd ed., rev.), p. 34.

Reprinted from *Modern Systems Research for the Behavioral Scientist: A Sourcebook* edited by Walter Buckley (1968). Published by permission of Aldine Publishing Company, Chicago.

10 A systems analysis of political life

by David Easton

In *A Framework for Political Analysis*[1] I spelled out in considerable detail the assumptions and commitments that would be required in any attempt to utilize the concept "system" in a rigorous fashion. It would lead to the adoption of what I there described as a systems analysis of political life. Although it would certainly be redundant to retrace the same ground here, it is nonetheless necessary to review the kinds of basic conceptions and orientations imposed by this mode of analysis. In doing so, I shall be able to lay out the pattern of analysis that will inform and guide the present work.

Political life as an open and adaptive system

. . . The question that gives coherence and purpose to a rigorous analysis of political life as a system of behavior is as follows. How do any and all political systems manage to persist in a world of both stability and change? Ultimately the search for an answer will reveal what I have called the life processes of political systems – those fundamental functions without which no system could endure – together with the typical modes of response through which systems manage to sustain them. The analysis of these processes, and of the nature and conditions of the responses, I posit as a central problem of political theory.

Although I shall end by arguing that it is useful to interpret political life as a complex set of processes through which certain kinds of inputs are converted into the type of outputs we may call authoritative policies, decisions and implementing actions, at the outset it is useful to take a somewhat simpler approach. We may begin by viewing political life as a system of behavior imbedded in an environment to the influences of which the political system itself is exposed and in turn reacts. Several vital considerations are implicit in this interpretation and it is essential that we become aware of them.

First, such a point of departure for theoretical analysis assumes without further inquiry that political interactions in a society constitute a *system* of behavior. This proposition is, however, deceptive in its simplicity. The truth is that if the idea "system" is employed with the rigor it permits and with the implications currently inherent in it, it provides a starting point that is already heavily freighted with consequences for a whole pattern of analysis.

Second, to the degree that we are successful in analytically isolating political life as a system, it is clear that it cannot usefully be interpreted as existing in a void. It must be seen as surrounded by physical, biological, social and psychological *environments*. Here again, the empirical transparency of the statement ought not to be allowed to distract us from its crucial theoretical significance. If we were to neglect what seems so obvious once it is asserted, it would be impossible to lay the groundwork for an analysis of how political systems manage to persist in a world of stability or change.

This brings us to a third point. What makes the identification of the environments useful and necessary is the further presupposition that political life forms an *open* system. By its very nature as a social system that has been analytically separated from other social systems, it must be interpreted as lying exposed to influences deriving from the other systems in which empirically it is imbedded. From them there flows a constant stream of events and influences that shape the conditions under which the members of the system must act.

Finally, the fact that some systems do survive, whatever the buffetings from the environments, awakens us to the fact that they must have the capacity to *respond* to disturbances and thereby to adapt to the conditions under which they find themselves. Once we are willing to assume that political systems may be adaptive and need not just react in a passive or sponge-like way to their environmental influences, we shall be able to break a new path through the complexities of theoretical analysis.

As I have elsewhere demonstrated, in its internal organization, a critical property that a political system shares with all other social systems in this extraordinarily variable capacity to respond to the conditions under which it functions. Indeed, we shall find that political systems accumulate large repertoires of mechanisms through which they may seek to cope with their environments. Through these they may regulate their own behavior, transform their internal structure, and even go so far as to remodel their fundamental goals. Few systems, other than social systems, have this potentiality. In practice, students of political life could not help but take this into account; no analysis could even begin to appeal to common sense if it did not do so. Nevertheless it is seldom built into a theoretical structure as a central component; certainly its implications for the internal behavior of political systems have never been set forth and explored.

Equilibrium analysis and its shortcomings

It is a major shortcoming of the one form of inquiry latent but prevalent in political research – equilibrium analysis – that it neglects such variable capacities for systems to cope with influences from their environment. The equilibrium approach is seldom explicitly elaborated, yet it infuses a good part of political research, especially group politics and international relations. Of necessity an analysis that conceives of a political system as seeking to maintain a state of equilibrium must assume the presence of environmental influences. It is these that displace the power relationships in a political system – such as a balance of power – from their presumed stable state. It is then customary, if only implicitly so, to analyze the system in terms of a tendency to return to a presumed pre-existing point of stability. If the system should fail to do so, it would be interpreted as moving on to a new state of equilibrium and this would need to be identified and described. A careful scrutiny of the language used reveals that equilibrium and stability are usually assumed to mean the same thing.

Numerous conceptual and empirical difficulties stand in the way of an effective use of the equilibrium idea for the analysis of political life. But among these there are two that are particularly relevant for my present purposes.

In the first place, the equilibrium approach leaves the impression that the members of a system are seized with only one basic goal as they seek to cope with change or disturbance, namely, to re-establish the old point of equilibrium or, at most, to move on to some new one. This is usually phrased, at least implicitly, as the search for stability as though this were sought above all else. In the second place, little if any attention is explicitly given to formulating the problems relating to the path that the system takes insofar as it does seek to return to this presumed point of equilibrium or to attain a fresh one. It is as though the pathways taken to manage the displacements were an incidental rather than a central theoretical consideration.

But it would be impossible to understand the processes underlying the capacity of some kind of political life to sustain itself in a society if either the objectives or the form of the responses are taken for granted. A system may well seek goals other than those of reaching one or another point of equilibrium. Even though this state were to be used only as a theoretical norm that is never achieved, it would offer a less useful theoretical approximation of reality than one that takes into account other possibilities. We would find it more helpful to devise a conceptual approach that recognized that at times members in a system may wish to take positive actions to destroy a previous equilibrium or even to achieve some new point of continuing disequilibrium. This is typically the case where the authorities may seek to keep themselves in power by fostering internal turmoil or external dangers.

Furthermore, with respect to these variable goals, it is a primary characteristic of all systems that they are able to adopt a wide range of actions of a positive, constructive and innovative sort for warding off or absorbing any forces of displacement. A system need not just react to a disturbance by oscillating in the neighborhood of a prior point of equilibrium or by shifting to a new one. It may cope with the disturbance by seeking to change the environment so that the exchanges between the environment and itself are no longer stressful; it may seek to insulate itself against any further influences from the environment; or the members of the system may even transform their own relationships fundamentally and modify their own goals and practices so as to improve their chances of handling the inputs from the environment. In these and other ways a system has the capacity for creative and constructive regulation of disturbances as we shall later see in detail.

It is clear that the adoption of equilibrium analysis, however latent it may be, obscures the presence of system goals that cannot be described as a state of equilibrium. It also virtually conceals the existence of varying pathways for attaining these alternative ends. For any social system, including the political, adaptation represents more than simple adjustments to the events in its life. It is made up of efforts, limited only by the variety of human skills, resources, and ingenuity, to control, modify or fundamentally change either the environment or the system itself, or both together. In the outcome the system may succeed in fending off or incorporating successfully any influences stressful for it.

Minimal concepts for a systems analysis

A systems analysis promises a more expansive, more inclusive, and more flexible theoretical structure than is available even in a thoroughly self-conscious and well-developed equilibrium approach. To do so successfully, however, it must establish its own theoretical imperatives. Although these were explored in detail in *A Framework for Political Analysis*, we may re-examine them briefly here, assuming, however, that where the present brevity leaves unavoidable ambiguities, the reader may wish to become more familiar with the underlying structure of ideas by consulting this earlier volume. In it, at the outset, a system was defined as any set of variables regardless of the degree of interrelationship among them. The reason for preferring this definition is that it frees us from the need to argue about whether a political system is or is not really a system. The only question of importance about a set selected as a system to be analyzed is whether this set constitutes an interesting one. Does it help us to understand and explain some aspect of human behavior of concern to us?

To be of maximum utility, I have argued, a *political* system can be designated as those interactions through which values are authoritatively allocated for a society; this is what distinguishes a political system from other systems that may be interpreted as lying in its environment. This environment itself may be divided into two parts, the intra-societal and the extra-societal. The first consists of those systems in the same society as the political system but excluded from the latter by our definition of the nature of political interactions. Intra-societal systems would include such sets of behavior, attitudes and ideas as we might call the economy, culture, social structure or personalities; they are functional segments of the society with respect to which the political system at the focus of attention is itself a component. In a given society the systems other than the political system constitute a source of many influences that create and shape the conditions under which the political system itself must operate. In a world of newly emerging political systems we do not need to pause to illustrate the impact that a changing economy, culture, or social structure may have upon political life.

The second part of the environment, the extra-societal, includes all those systems that lie outside the given society itself. They are functional components of an international society or what we might describe as the supra-society, a supra-system of which any single society is part. The international political systems, the international economy, or the international cultural system would fall into the category of extra-societal systems.

Together, these two classes of systems, the intra- and extra-societal, that are conceived to lie outside of a political system may be designated as its total environment. From these sources arise influences that are of consequence for possible stress on the political system. The total environment is presented in Table 10.1 as reproduced from *A Framework for Political Analysis*, and the reader should turn to that volume for a full discussion of the various components of the environment as indicated on this table.

Disturbances is a concept that may be used to identify those influences from the total environment of a system that act upon it so that it is different after the stimulus from what it was before. Not all disturbances need strain the

Table 10.1 Components of the Total Environment of a Political System

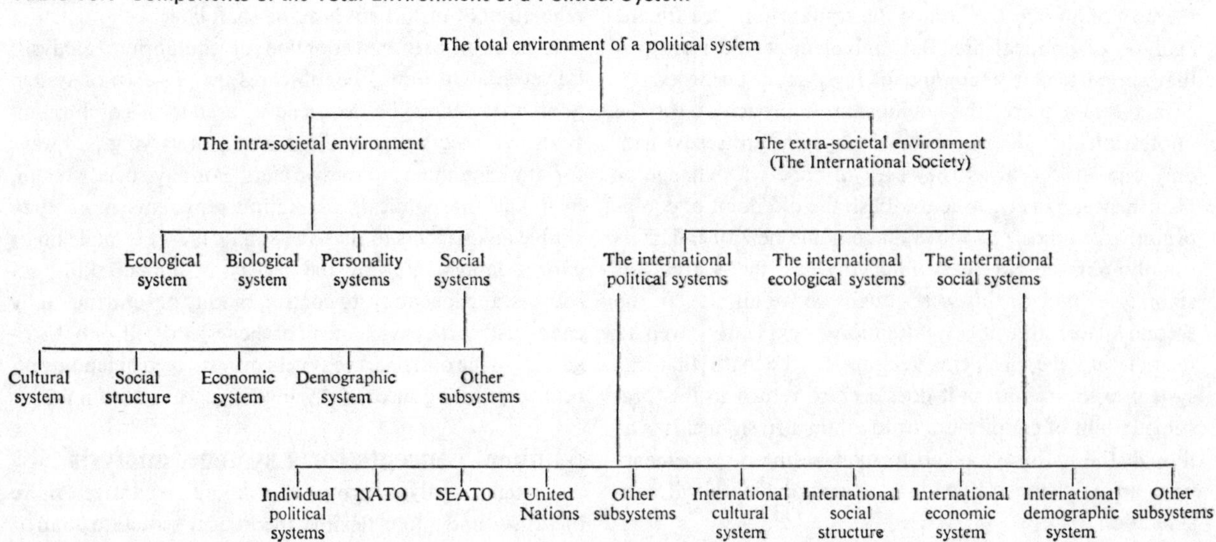

system. Some may be favorable with respect to the persistence of the system; others may be entirely neutral with respect to possible stress. But many can be expected to lead in the direction of stress.

When may we say that *stress* occurs? This involves us in a rather complex idea, one that has been treated at length. But since it does stand as a major pillar underpinning the analysis to be elaborated in the succeeding chapters, I must at least broadly sketch out its implications. It embodies several subsidiary notions. All political systems as such are distinguished by the fact that if we are to be able to describe them as persisting, we must attribute to them the successful fulfillment of two functions. They must be able to allocate values for a society; they must also manage to induce most members to accept these allocations as binding, at least most of the time. These are the two properties that help us to distinguish most succinctly political systems from other kinds of social systems.

By virtue of this very fact these two distinctive features – the allocations of values for a society and the relative frequency of compliance with them – are the *essential variables* of political life. But for their presence, we would not be able to say that a society has any political life. And we may here take it for granted that no society could exist without some kind of political system; elsewhere I have sought to demonstrate this in detail.

One of the important reasons for identifying these essential variables is that they give us a way of establishing when and how the disturbances acting upon a system threaten to stress it. Stress will be said to occur when there is a danger that the essential variables will be pushed beyond what we may designate as their *critical range*. What this means is that something may be happening in the environment – the system suffers total defeat at the hands of an enemy, or widespread disorganization in and disaffection from the system is aroused by a severe economic crisis. Let us say that as a result, the authorities are consistently unable to make decisions or if they strive to do so, the decisions are no longer regularly accepted as binding. Under these conditions, authoritative allocations of values are no longer possible and the society would collapse for want of a system of behavior to fulfill one of its vital functions.

Here we could not help but accept the interpretation that the political system had come under stress, so severe that any and every possibility for the persistence of a system for that society had disappeared. But frequently the disruption of a political system is not that complete; the stress is present even though the system continues to persist in some form. Severe as a crisis may be, it still may be possible for the authorities to be able to make some kinds of decisions and to get them accepted with at least minimal frequency so that some of the problems typically subjected to political settlements can be handled.

That is to say, it is not always a matter as to whether the essential variables are operating or have ceased to do so. It is possible that they may only be displaced to some extent as when the authorities are partially incapacitated for making decisions or from getting them accepted with complete regularity. Under these circumstances the essential variables will remain within some normal range of operation; they may be stressed but not in a sufficient degree to displace them beyond a determinable critical point. As long as the system does keep its essential variables operating within what I shall call their critical range, some kind of system can be said to persist.

As we have seen, one of the characteristic properties of every system is the fact that it has the capacity to cope with stress on its essential variables. Not that a system need take such action; it may collapse precisely because it has failed to take measures appropriate for handling the impending stress. But it is the existence of a capacity to respond to stress that is of paramount importance. The kind of response actually undertaken, if any, will help us to evaluate the probabilities of the system's being able to ward off the stress. In thus raising the question of the nature of the response to stress, it will become apparent, in due course, that the special objective and merit of a systems analysis of political life is that it permits us to interpret the behavior of the members in a system in the light of the consequences it has for alleviating or aggravating stress upon the essential variables.

The linkage variables between systems

But a fundamental problem remains. We could not begin the task of applying this kind of conceptualization if we did not first pose the following question. How do the potentially stressful conditions from the environment communicate themselves to a political system? After all, common sense alone tells us that there is an enormous variety of environmental influences at work on a system. Do we have to treat each change in the environment as a separate and unique disturbance, the specific effects of which for the political system have to be independently worked out?

If this were indeed the case, as I have shown in detail before, the problems of systematic analysis would be virtually insurmountable. But if we can devise a way for generalizing our method for handling the impact of the environment on the system, there would be some hope of reducing the enormous variety of influences into a relatively few, and therefore into a relatively manageable number of indicators. This is precisely what I have sought to effect through the use of the concepts "inputs" and "outputs".

How are we to describe these inputs and outputs? Because of the analytic distinction that I have been making between a political system and its parametric or environmental systems, it is useful to interpret the influences associated with the behavior of persons in the environment or from other conditions there as *exchanges* or *transactions* that cross the *boundaries* of the political system. Exchanges can be used when we wish to refer to the mutuality of the relationships, to the fact that the political system and those systems in the environmental have reciprocal effects on each other. Transactions may be employed when we wish to emphasize the movement of an effect in one direction, from an environmental system to the political system, or the reverse, without being concerned at the time about the reactive behavior of the other system.

To this point, there is little to dispute. Unless systems were coupled together in some way, all analytically identifiable aspects of behavior in society would stand independent of each other, a patently unlikely condition. What carries recognition of this coupling beyond a mere truism, however, is the proposal of a way to trace out the complex exchanges so that we can readily reduce their immense variety to theoretically and empirically manageable proportions.

To accomplish this, I have proposed that we condense the major and significant environmental influences into a few indicators. Through the examination of these we should be able to appraise and follow through the potential impact of environmental events on the system. With this objective in mind, I have designated the effects that are transmitted across the boundary of a system toward some other system as the *outputs* of the first system and hence, symmetrically, as the *inputs* of the second system, the one they influence. A transaction or an exchange between systems will therefore be viewed as a linkage between them in the form of an input–output relationship.

Demands and supports as input indicators

The value of inputs as a concept is that through their use we shall find it possible to capture the effect of the vast variety of events and conditions in the environment as they pertain to the persistence of a political system. Without the inputs it would be difficult to delineate the precise operational way in which the behavior in the various sectors of society affects what happens in the political sphere. Inputs will serve as *summary variables* that concentrate and mirror everything in the environment that is relevant to political stress. Thereby this concept serves as a powerful tool.

The extent to which inputs can be used as summary variables will depend, however, upon how we define them. We might conceive of them in their broadest sense. In that case, we would interpret them as including any event external to the system that alters, modifies or affects the system in any and every possible way. But if we seriously considered using the concept in so broad a fashion, we would never be able to exhaust the list of inputs acting upon a system. Virtually every parametric event and condition would have some significance for the operations of a political system at the focus of attention; a concept so inclusive that it does not help us to organize and simplify reality would defeat its own purposes. We would be no better off than we are without it.

But as I have already intimated, we can greatly simplify the task of analyzing the impact of the environment if we restrict our attention to certain kinds of inputs that can be used as indicators to sum up the most important effects, in terms of their contributions to stress, that cross the boundary from the parametric to the political systems. In this way we would free ourselves from the need to deal with and trace out separately the consequences of every different type of environmental event.

As the theoretical tool for this purpose, it is helpful to view the major environmental influences as coming to a focus in two major inputs: demands and support. Through them a wide range of activities in the environment may be channeled, mirrored, and summarized and brought to bear upon political life, as I shall show in detail in the succeeding chapters. In this sense they are key indicators of the way in which environmental influences and conditions modify and shape the operations of the political system. If we wish, we may say that it is through fluctuations in the inputs of demands and support that we shall find the effects of the environmental systems transmitted to the political system.

Outputs and feedbacks

In a comparable way, the idea of outputs help us to organize the consequences flowing from the behavior of the members of the system rather than from actions in the environment. Our primary concern is, to be sure, with the functioning of the political system. In and of themselves, at least for understanding political phenomena, we would have no need to be concerned with the consequences that political actions have for the environmental system. This is a problem that can or should be handled better by theories seeking to explore the operations of the economy, culture, or any of the other parametric systems.

But the fact is that the activities of the members of the system may well have some importance with respect to their own subsequent actions or conditions. To the extent that this is so, we cannot entirely neglect those actions that do flow out of a system into the environment. As in the case of inputs, however, there is an immense amount of activities that take place within a political system. How are we to sort out the portion that has relevance for an understanding of the way in which systems manage to persist?

Later we shall see that a useful way of simplifying and organizing our perceptions of the behavior of the members of the system, as reflected in their demands and support, is in terms of the consequences of these inputs for what I shall call the political outputs. These are the decisions and actions of the authorities. Not that the complex political processes internal to a system, and that have been the subject of inquiry for so many decades in political science, will be considered in any way irrelevant. Who controls whom in the various decision-making processes will continue to be a vital concern since the pattern of power relationships helps to determine the nature of the outputs. But the formulation of a conceptual structure for this aspect of a political system would draw us into a different level of analysis. Here I am only seeking economical ways of summarizing the outcomes of these internal political processes – not of investigating them – and I am suggesting that they can be usefully conceptualized as the outputs of the authorities. Through them we shall be able to trace out the consequences of behavior within a political system for its environment.

There would be little point in taking the trouble to conceptualize the results of the internal behavior of the members in a system in this way unless we could so something with it. As we shall see, the significance of outputs is not only that they help to influence events in the broader society of which the system is a part; in doing so, they help to determine each succeeding round of inputs that finds its way into the political system. As we shall phrase it later, there is a *feedback loop* the identification of which will help us to explain the processes through which the authorities may cope with stress. This loop has a number of parts. It consists of the production of outputs by the authorities, a response on the part of the members of the society with respect to them, the communication of information about this response to the authorities, and finally, possible succeeding actions on the part of the authorities. Thereby a new round of outputs, response information feedback, and reaction on the part of the authorities is set in motion and is part of a continuous never-ending flow. What happens in this feedback loop will turn out to have the deepest significance for the capacity of a system to cope with stress.

A flow model of the political system

It is clear from what has been said that this mode of analysis enables and indeed compels us to analyze a political system in dynamic terms. Not only do we see that it gets something done through its outputs but we are also sensitized to the fact that what it does may influence each successive stage of behavior. We appreciate the urgent need to interpret political processes as a continuous and interlinked flow of behavior.

If we apply this conceptualization in the construction of a rudimentary model of the relationships between a political system and its environment, we would have a figure of the kind illustrated in Figure 10.1. . . . In effect it conveys the idea that the political system looks like a vast and perpetual conversion process. It takes in demands and support as they are shaped in the environment and produces something out of them called outputs. But it does not let our interest in the outputs terminate at this point. We are alerted to the fact that the outputs influence the supportive sentiments that the members express toward the system and the kinds of demands they put in. In this way the outputs return to haunt the system, as it were. As depicted on the diagram, all this is still at a very crude level of formulation. It will be our task to refine these relationships as we proceed in our analysis.

But let us examine the model a little more closely since in effect this volume will do little more than to flesh out the skeleton presented there. In interpreting the diagram, we begin with the fact that it shows a political system surrounded by the two classes of environments that together form its total environment. The communications of the many events that occur here are represented by the solid lines connecting the environments with the political system. The arrowheads on the lines show the direction of flow into the system. But rather than attempting to discuss each disturbance in the environment uniquely or even in selected groups or classes of types, I use as an indicator of the impact that they have on the system, the way in which they shape two special kinds of inputs into the system, demands and support. This is why the effects from the environment are shown to flow into the box labelled "inputs". We must remember, however, that even though the desire for simplicity in presentation does not permit us to show it on

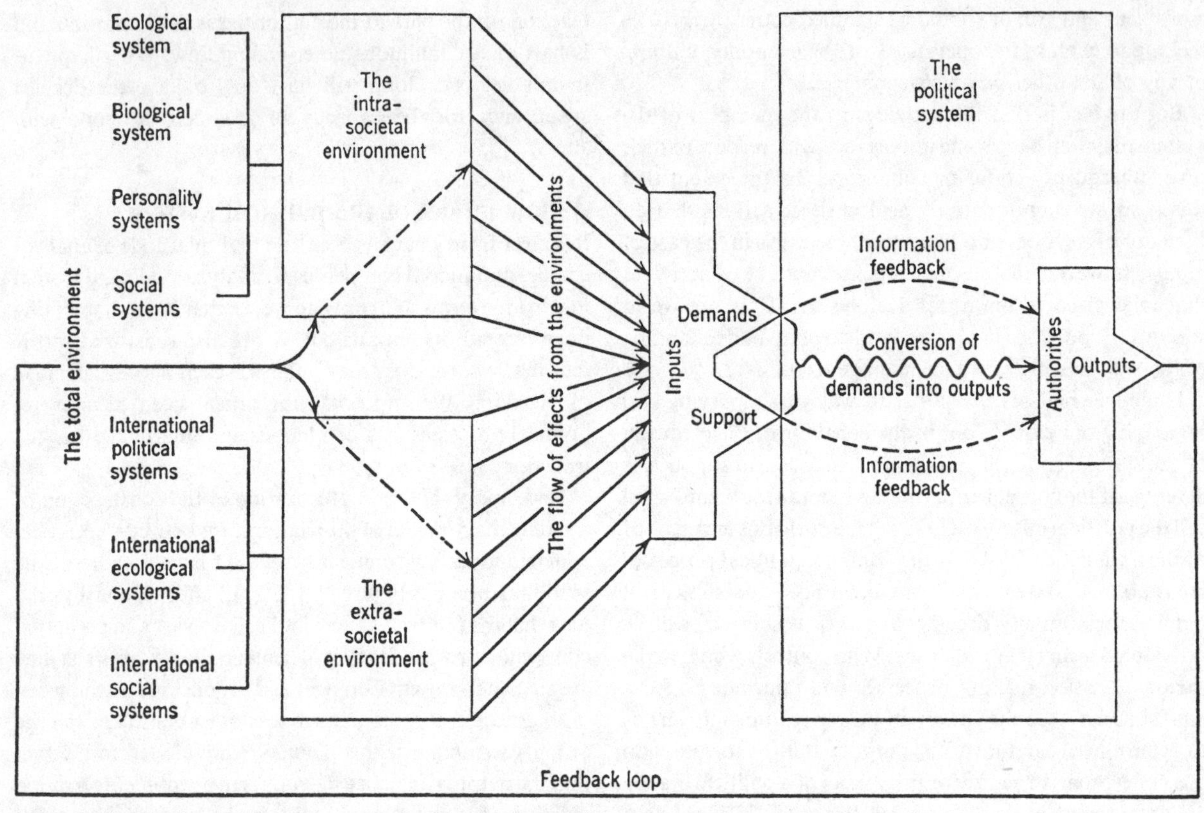

Figure 10.1 A Dynamic Response Model of a Political System.

the diagram, events occurring within a system may also have some share in influencing the nature of the inputs.

As is apparent, the inputs provide what we may call the raw materials on which the system acts so to produce something we are calling outputs. The way in which this is done will be described as a massive conversion process cavalierly represented on the diagram by the serpentine line within the political system. The conversion processes move toward the authorities since it is toward them that the demands are initially directed. As we shall see, demands spark the basic activities of a political system. By virtue of their status in all systems, authorities have special responsibilities for converting demands into outputs.

If we were to be content with what is basically a static picture of a political system, we might be inclined to stop at this point. Indeed much political research in effect does just this. It is concerned with exploring all those intricate subsidiary processes through which decisions are made and put into effect. This constitutes the vast corpus of political research today. Therefore, insofar as we were concerned with how influence is used in formulating and putting into effect various kinds of policies or decisions, the model to this point would be an adequate if minimal first approximation.

But the critical question that confronts political theory is not just the development of a conceptual apparatus for understanding the factors that contribute to the kinds of decisions a system makes, that is, for formulating a theory of political allocations. As I have indicated, theory needs to know how it comes about that any kind of system can persist long enough to continue to make such decisions. We need a theory of systems persistence as well. How does a system manage to deal with the stress to which it may be subjected at any time? It is for this reason that we cannot accept outputs as the terminal point either of the political processes or of our interest in them. Thus it is important to note on the diagram, that the outputs of the conversion process have the characteristic of feeding back upon the system and shaping its subsequent behavior. Much later I shall seek to demonstrate that it is this feature together with the capacity of a system to take constructive actions that makes it possible for a system to seek to adapt or to cope with possible stress.

In the figure, this feedback is depicted by the line that

shows the effects of the outputs moving directly back to the environments. As the broken lines within the environmental boxes indicate, the effects may reshape the environment in some way; that is to say, they influence conditions and behavior there. In this way the outputs are able to modify the influences that continue to operate on the inputs and thereby the next round of inputs themselves.

But if the authorities are to be able to take the past effect of outputs into account for their own future behavior, they must in some way be apprised of what has taken place along the feedback loop. The broken lines in the box labeled "The political system" suggest that, through the return flow of demands and support, the authorities obtain information about these possible consequences of their previous behavior. This puts the authorities in a position to take advantage of the information that has been fed back and to correct or adjust their behavior for the achievement of their goals.

It is the fact that there can be such a continuous flow of effects and information between system and environment, we shall see, that ultimately accounts for the capacity of a political system to persist in a world even of violently fluctuating changes. Without feedback and the capacity to respond to it, no system could survive for long, except by accident.

In this brief overview, I have summarized the essential features of the analytic structure. . . . If we condensed the figure still further, we would have the diagram shown on Figure 10.2. It reduces to its bare essentials the fundamental processes at work in all systems and starkly reveals the source of a system's capacity to persist. It may well stand temporarily as the simplest image, to carry in our minds, of the processes we are about to discuss in detail.

To summarize the conceptualization being reviewed here, our analysis will rest on the idea of a system imbedded in an environment and subject to possible influences from it that threaten to drive the essential variables of the system beyond their critical range. To persist, the system must be capable of responding with measures that are successful in alleviating the stress so created. To respond, the authorities

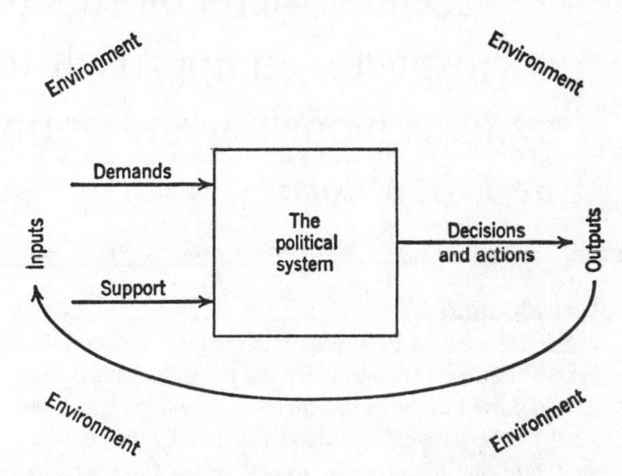

Figure 10.2 A simplified model of a political system.

at least must be in a position to obtain information about what is happening so that they may react insofar as they desire or are compelled to do so.

In *A Framework for Political Analysis* each of these concepts and interrelationships was attended to in varying degrees of detail . . . It will be my task to begin to apply them in an effort to construct a much more elaborate structure for the analysis of political systems.

In doing so, we shall find ourselves confronted with a series of major questions. What precisely are the nature of the influences acting upon a political system? How are they communicated to a system? In what ways, if any, have systems typically sought to cope with such stress? What kinds of processes will have to exist in any system if it is to acquire and exploit the potential for acting so as to ameliorate these conditions of stress? . . .

Reference

1 Easton, D. (1961) *A Framework for Political Analysis*, Prentice-Hall, Englewood Cliffs, N.J.

11 Technosphere, biosphere, and sociosphere: an approach to their systems modeling and optimization*

by J. H. Milsum

Introduction

In the last century, technology and the science underlying it have been primary agents forcing the vast changes that have occurred in the geosphere and sociosphere which we inhabit. The pace of this change has, in addition, accelerated fantastically in more recent years. Consider, for example, that in the lifetime of almost all here present, the following list represents only the most dramatic examples which have been invented, developed, and have reached mass-production with great social effects upon our lives: gas turbine (jet) engines (providing nearly instantaneous transport of people anywhere around the world at a cost that many can afford); rockets and nuclear energy (providing for space travel and potential "over-kill" capability against the whole human population, as well as other less dramatic, but probably more useful activities); radar, television, and solid-state electronics (providing instantaneous and reasonably economic world-wide communication); electronic computers (providing essentially instantaneous computational ability of such great power that we are only beginning to understand its vastly-pervading influence upon our lives).

While there has not been complete agreement in our society that these changes all represent improvement, nevertheless no groups of significant size ever seriously try to operate their societies without utilizing available technological advances. Thus it may be said that the technologist has had society's tacit support, even if he has not formally represented the ideal prototype for its citizens to aspire to. Unfortunately, it has been largely true that during this early technological period when the advantages of technological advances have generally been so obvious to society, in contrast to the disadvantages which tend to become obvious somewhat later and to cost money to avoid, then the technologist does not seem to have felt it his prime duty to anticipate and point out to society the concomitant disadvantages in comparison to the expected advantages. Nor indeed would it seem justified to consider him particularly culpable here, since society has not really encouraged consideration of the aesthetic and ethical aspects which might arise from technological advance if it were to involve larger cost. Consequently, a naïve technologist would seem superficially justified in being puzzled when he finds himself apt to be as much blamed as praised for new technological developments. Clearly this tendency for blame to be attached comes when societies are beginning to realize that things are not going quite according to plan, and yet are not really clear why. Certainly few would now agree with the philosopher, Dr. Pangloss of *Candide*, that "All is necessarily for the best in the best of all possible worlds". In fact, in spite of the invention, development, and widespread production of vast numbers of useful consumer goods and services, which have turned most Western societies into very affluent ones, it is clear that we perceive ourselves to have some very urgent and perplexing problems.

As regards tackling these problems, it now seems clear that the technologist must increasingly take a systems view which is broad enough to include the problems of social values, and that he must be prepared to take more of a lead in pointing out to his less technically-literate confreres in society how some of the different aspects, such as beautification and prevention of pollution, will trade off in the overall optimization of the system under consideration. Then, while it will remain true that simple answers cannot be expected to the complex problems which we are now producing for ourselves as a result of the vast interlocking of our systems, nevertheless society will be encouraged into the dialogue in helping establish how human values should be entered into the complicated systems equations.

Despite such optimization attempts, it will almost undoubtedly remain true that the affluence available through maximum use of technology will bring at the same time many new problems and will also place constraints upon us. Inasmuch as certain of such constraints may be con-

* Presidential address presented at the Annual Meeting of the Society for General Systems Research, New York, December 1967.

sidered unacceptable by societies at any given time, then it is probable that not every technological advance will be automatically accepted. On the other hand, societies are always adapting, if only slowly, and therefore it will seldom be possible to predict with certainty whether any given innovation will remain unacceptable. It is both refreshing and sobering to realize that in the late twenties Freud was very concerned about these same problems (*Civilization and Its Discontents*, Norton, 1929).

In this talk, I hope to make a few useful suggestions towards how we may bring a systems approach to bear on analyzing our problems in the broad areas of geosphere, biosphere, technosphere, and sociosphere, and hopefully towards optimizing them. It should be mentioned that these of course represent only my own opinions, arising from my own background of control engineering, systems research, and biomedical engineering.

Geosphere

Under this heading, we mention some of those vast natural processes in and upon the earth which do not either inherently or necessarily involve living processes. These physical systems involve such complicated interlocking dynamics that they are only now beginning to be understood, in the sense of adequate mathematical models being available.

Perhaps the most obvious dynamic phenomena in the geosphere are those cycles due to the varying exposure of the earth to the sun; namely the annual and daily cycles, and of course, although less noticeably, the lunar cycle. Thus genuine and fairly strict cycles exist in such atmospheric conditions as temperature and precipitation, which have in turn widespread influence upon our biosphere, technosphere, and sociosphere. Other major phenomena depend to a greater or lesser extent upon these cyclic-forcing inputs, in particular the circulation of water in the oceans, the circulation of air in the atmosphere, and the rainwater cycle itself. In turn, the latter is coupled to some extent with an electrical cycle which produces the familiar electrical discharges of lightning, among other things. Actually the word *circulation* may be much better than *cycle* to define these phenomena. The point here is that while a cycle may be traced, namely rain – surface and subsurface water flow – pooling in the seas (both ocean and landlocked) – evaporation – cloud formation – rainfall, the cycling is of given molecules between different phases; and that, from an information flow viewpoint, it is merely a relatively constant circulating flux rather than a true cycle. On the other hand, the circulation shows seasonal cyclicity at any one point on earth, of course; and furthermore, the large spatial coverage of many rain-water systems means that spatio-temporal effects couple-up inextricably.

In these cyclic or circulatory phenomena, some of the effects are unidirectional only, in the sense that what goes in one "direction" is not necessarily returned in the other. A specific important example arising in the rainfall circulation of water is that of erosion of the earth, and specifically of minerals into the sea. As one consequence, the sea cannot itself constitute a constant long-term environment for living species, which has evolutionary implications in the biosphere. It should be noted that the fact that some land also gets thrown back from the sea does not effectively cancel the erosion effect in this biological sense.

At another level of conceptualization, it could be commented that this washing of the land into the oceans is of course in accord with the second law of thermodynamics, namely that closed systems tend towards a state of uniform low order. In fact, the earth is not a closed system; but the geosphere cannot open itself to trap energy from the sun, not at least permanently nor to the same extent as the biosphere can. A further comment is that, in searching for the appropriate physical laws which govern the physical phenomena such as those described, the use of "minimal" principles has often proved important. Minimal principles usually concern conservation of certain properties, that of energy being most generally applicable and valuable, as in the Hamiltonian function.

Biosphere

Since all living organisms essentially exist close to the earth's surface, it is not surprising that many of the relevant major phenomena of the biosphere are closely coupled with those in the geosphere, such as already mentioned. Furthermore, some of the couplings in these combined systems is completely inextricable; for example, in circulatory cycles of such gases as oxygen and carbon dioxide, and indeed of many other materials, both solid and liquid. Again, there are also slow drifts in the sense that, while during the early stages of life on earth the earth may have been considered to provide a largely constant pool of basic materials, this is not necessarily true today when the biomass has become significantly large. Thus, as is obvious to us, the net result of these interactions is that the geosphere itself has been considerably modified from the relevant conditions which would otherwise exist were there no life on earth. In a parallel example to that of the increase of mineral concentrations in the sea of the geosphere, the carbon dioxide content of the atmosphere has increased as a result of life in general, and man's social system in particular. In turn, this is producing significant effects, both as to climate and net electromagnetic radiation conditions at the earth's surface.

The specific major new ability that living things have added over those exhibited in the non-living world is the ability of living organisms to trap high-energy electromagnetic radiations from the sun, on a relatively permanent basis in that the energy is transformed through photosynthesis into other forms useful in life, such as the so-called organic materials, cells, and high order organisms, and thus ultimately, our own societies.

Since apparently it has not proved practical to evolve organisms that live for an indefinitely extended period of time, then an essential feature of maintaining life has been the ability of these organisms to procreate. This ability provides the advantage that evolution can then operate through the competitive process of selection to allow refinement and improvement of the life forms. This expresses what is usually called the adaptive ability in systems theory, which is implemented in the form of some hill-climbing scheme. From a slightly different viewpoint, this ability of evolution to reject the worst results and retain the "best" may be considered a process which is of the positive-feedback, unstable type. Clearly this mechanism can potentially build up regimes of ever-increasing high order, rather than suffering the usual fate, as prescribed by the second law of thermodynamics, of degradation to uniform low order. The manifestly high-order regimes, which living forms have generated and continue to extend, therefore arises solely from the fact that living systems are able to "open" themselves to the high-energy source of the sun.

Inasmuch as the feeding patterns of living organisms represent a fantastically complicated predator–prey situation, a "balance of nature" is usually stated to express the phenomenon that there seems to be some sort of stability principle at work with regard to the populations to be found, both within species and their ecologies. Furthermore, and perhaps rather to the surprise of many, evidence is accumulating that a self-regulation of populations within species and groups exists, in order apparently to match the available food supplies with the size of the group concerned. Apparently this applies not only in the immediate sense of food supply available in a given short time period, but rather in the sense of the minimum amount available when statistical fluctuations are included over a much longer period. Social measures appear at least in part to be responsible for implementing this self-regulation in many different species, notably birds for example (see Wynn-Edwards (1962) *Animal Dispersion in Relation to Social Behaviour*, Hafner.

Technosphere

Historically, it is hardly justifiable to consider that technology started only in the last one, or at most two, centuries. However, such is the nature of exponential growth processes that, for all practical purposes, it suffices to consider that the industrial age started with a small initial condition in the 18th century. Specifically, the so-called first industrial revolution of the 18th–19th century essentially relieved man of the requirement to obtain all his nicely controlled mechanical power through the use of either his own or animal muscles by providing, instead and in succession, steam power, electric power, hydrocarbon fuel power, and finally, nuclear power. This revolution has of course had tremendous social implications, which have been widely discussed, but nevertheless they probably provide only a small component cause in producing the difficult complex problems we face today. The major component arises from the more recent second industrial revolution, which in turn has relieved man of the need to perform all his own data processing by providing, instead, the tremendous capacity of modern electronic (and other) computing machines. This particular revolution has been extremely rapid, for within the course of about twenty-five years from the computer's inception, we now find ourselves affected in a wide spectrum of important ways by this new slave, both helpfully, disturbingly, and frustratingly. Part of the trouble lies in the fact that this slave does accurately and fast so many of the tasks society has found necessary, but which humans themselves have found tedious and therefore do not in fact perform well. Nevertheless, while their work may have generally been decried by these workers, at the same time it has constituted much of the basis for their self-respect. Hence it becomes necessary to define much more carefully what might be considered to constitute the essential characteristics of man, as opposed to a computing machine; in particular, what is necessary to man for him to feel enjoyment in life and to experience emotional and purposeful activity. It is generally presumed that these latter areas are not ones in which a computer excels, and yet, when one tries to define exactly where man does differ in a "superior" way from the computer, the matter becomes very perplexing.

Consequent upon these industrial revolutions, our technosphere now seems capable of providing, at least in principle, large integrated industrial processes which are both automated and computer controlled; as one example, we could cite the possibility of integrating the steel industry through from the mining of metal ores to the inventory control of finished products such as automobiles, almost without human intervention. In other words, given certain sensible and yet fairly unrestrictive constraints on any of man's material needs, he can now aspire to a society in which it is not necessary for the national production of material and information products to be tied essentially to the amount of man power available and its productivity. Theoretically, this should leave him free to develop the higher human qualities with which he believes himself specially endowed. The unfortunate fact that this does not seem to be working out, at present, probably and hopefully relates to the problems of the sociosphere, which at least we may be able to tackle if we can first define them.

Sociosphere

It has been noted that many of today's problems arise partly because the technological advances which have so greatly affected our social environment have occurred much faster than the normal rate at which adaptive changes in social behavior can accommodate to them. Stated slightly differently, the present time constants of technological change are probably at least one order of magnitude larger than that of social change. Some of these problems arise specifically because the result of technological change has been to call into question for needed re-evaluation certain relevant concepts and beliefs that were long thought either to have been handed down to the world by divine inspiration or else to have been adequately proven by long human experience. Indeed, for much of human history, the changes were of a sufficiently slow nature that they could be adapted to without doing great violence to traditional beliefs; that is, their evolution was possible rather than revolution. Today, however, we have seen reflected into society as a result of technological advancement such dramatic changes that it is clear to almost all that certain absurd situations are arising. Some outstanding examples in my view would seem to be:

1. The subject of *human population*, and the need for its rational control by the societies involved. Certainly there existed adequate justification throughout much of human history for attributing sacrosanct status to new human life, inasmuch as all population possible was desirable to offset the high wastage rate due to warfare and disease. Having now tampered very effectively with the death rate side of human population, sufficiently that a runaway condition is in progress, it would seem to represent elementary good husbandry for us to consider some overall objectives for the size of human populations, so as to be able to work towards achieving such conditions before panic measures prove necessary. Relevant criteria in such considerations obviously must relate to matters such as family health, wealth, and security; adequacy of national food supply; natural resources, etc., and finally a satisfactory population density to avoid unsatisfactory, if as yet unknown, psychological pressures.

2. The matter of *work–leisure relations* in an affluent and automated society. A specific basic problem here concerns how we may accomplish change in the deeply-ingrained protestant ethic which makes it difficult for people to feel themselves fulfilled, and thus to maintain their necessary self-pride, if they cannot be working members of the community, no matter how unrewarding the work might seem to be at first glance. It should be noted that this difficulty seems largely independent of whether they are provided adequate sustenance by their society even when they do not work, in fact possibly even *because* they do not work. Again, therefore, the inheritance of a non-technological age, that unremitting toil is good and necessary, has now rapidly become inappropriate, but too rapidly for society to adopt happily to this condition.

3. The problem of *aggression and interpersonal fear* which lead especially to the problems of warfare and ethnic strife within a particular state.

4. The problem of unacceptably *large and probably increasing differences* between the affluence of the developed societies on the one hand, and the developing societies on the other. It should be noted that this is one of the inevitable effects when subsystems having positive feedback pathways are placed in competition with each other. The process is in fact an example of the general competitive exclusion aspect of evolution, and can lead to the same ruthless solutions unless we impose, as civilized societies, some higher performance criteria upon the overall system.

5. The problem of *participation* by the members of society in their society, in such a meaningful way that they are able to feel part of it and therefore to fulfill themselves. This problem is related to the work–leisure problem mentioned in 2 above. The word alienation has been used to describe the general absence of participation that people feel in large remotely-run and apparently inflexible societies.

Let us ask what may in some reasonable sense be considered the purpose of the people within their societies, and then even of their societies themselves. It seems clear that there are many needs which can already be described for individuals: to be closely involved in a "family", and in contributing to the generation and training of the new generation; to be valued by his neighbors for contributions he can make to his immediate community; to be able to call upon his community in times of need; to be able to develop himself in various intellectual, cultural, and other social ways; and finally, in a less well-defined way, to be able to contribute towards the development of his society at large towards some sort of inspiring long-term goal. It is important to notice in this latter regard that people do not need to be in a state of great affluence as a prerequisite to feeling that they are fulfilling themselves satisfactorily in a society, but rather that they are moving towards a better state by their effort. This fact seems again an illustration of the general biological phenomenon that the rate of change seems to be more important than the absolute value itself. A further point is that it is also necessary to have a basis for comparing performance with another outside standard in order to produce an "error signal" for deciding which action should be taken.

A pertinent question follows from these comments in that it is being implicitly suggested that societies can meaningfully be planned with some overall optimization in view. Unfortunately, of course, it is very hard to find successful evidence for this in history, especially since we are necessarily restricted to a history on which many stages of editing have been performed. Furthermore, since societies contain so much positive feedback, there is always the danger that prophecies may become self-fulfilling. Such conditions could considerably obscure the data obtained from historical studies. This emphasizes from another viewpoint that it is not for trivial reasons that the sociosphere is an area of the "soft sciences". However, while the problems are admittedly extremely complex, this fact alone would seem to emphasize how urgent it is that we as societies bring many of our best scientists to work upon the problems.

On the problems

Many of society's problems arise because change is necessary from some established conditions. The codification and implementation of these conditions has generally supported the development of a cadre of professionals, which then inherently tends to resist any changes suggested by its "lay electorate", even though the professionals themselves may generate amendments and even major changes in pursuing their maintenance task. In either case, it follows that changes, such as in amendments of laws, necessarily occur in discrete jumps and at discrete intervals of time, since each change must be brought about by "the pressure to reform" having reached a threshold level after which the resistance may crumble. It may be useful to think of this process as a form of relaxation oscillation, analogous, for example, to the stick-slip friction by which a bow produces a note on the violin string. Furthermore, it should be noted that the decaying of resistance, as the actual change process gathers momentum, is equivalent to positive feedback in the local circuit.

The tendency of the establishment to resist change is not in itself necessarily unhealthy since this procedure represents a reasonable way in which to filter out the high-frequency "noise" for which no change in the rules is really required. On the other hand, society's non-trivial problem in this respect is that of judging how to match the desirable time constant of the filter to the time constant of "DC" or permanent change in the process, so that substantial changes in the "signal" can be adjusted for as rapidly as desirable, while the high-frequency noise is filtered out. Note that use of the word desirable in this sense rather begs the question since, in order to predict this effectively, an observer of godly powers is required.

When changes are made they may fall into either of two general types. In the first type, change is of the amendment or evolutionary type in the sense that the general principles of operation remain the same, but the particular parameter values are changed. The second is of a more revolutionary type in that the whole structure is changed, and obviously therefore this will only occur at a greater threshold of change in the system variables. One classic example of the second type arises when a small one-man business prospers so well that in a relatively short time period it becomes restructured to include professional managers, accountants, salesmen, etc. A frequent by-product of such a change is that the original owner, while still being proud of his success superficially, finds himself increasingly frustrated in that the business is now no longer a simple direct extension of himself, with everything proceeding under his own control. This points to the fact that, during different phases of the life cycle of sociological organisms, staffs of different strengths and weaknesses are required for the different regimes of operation.

Most of man's sociological and industrial enterprises involving many staff have grown more or less successfully to their present stage in a groping manner. Indeed, when the enterprise has been almost entirely focused to achieve the objective of some particular individual, then this type of growth has usually been the most efficient. On the other hand, once its growth reaches the stage where it is essentially a "corporate" business, then it is common experience that the overt objectives of the enterprise can become distinctly modified by these strong but publicly unexpressed motivations of the staff, especially the managers. Thus it has been the well-documented experience of the systems analyst that, when he attempts analysis of such enterprises, this very attempt immediately requires that many questions as to corporate purpose and information-flow details be asked, and that this process itself seems surprisingly revolutionary. The typical consequence is that significant improvements can easily be made in the operation, largely as a result of certain critical questions having been asked for the first time because of the point of view taken by the systems analyst. Stated slightly differently, the first cream of the optimization process can often be gained without resorting to any major refinements. The second stage as a result of the systems study is usually to introduce complicated data processing and hence control schemes for the organization. As the operation becomes closer to optimal, the cost of the analyst and of the computing time necessary for the optimization studies becomes very significant, until finally the diminishing-returns stage is reached in which the cost of the analytical effort exceeds the economy gained as a result.

On the other hand, it should also be noted that the advantages of analytical optimization studies may be realized only after a considerable time delay so that in such cases an immediate economic loss is quite acceptable in view of the profit to be gained later by the experience acquired.

In performing any optimization study, one focal point of the analysis concerns establishing an appropriate performance criterion, also often called cost function or profit function. When the analyst is searching for background knowledge to establish this criterion, he often finds that he must ask questions for which answers are not readily available, while many of the correct answers which are available are unfortunately not answers to useful questions! In defining the performance criterion, one of the most difficult tasks relates to establishing quantitative values for the social aspects, an extreme example being the value of a person's life. In fact, some react with horror to the idea that such an evaluation can even be considered. Such reactions are clearly not rational, since in fact we place great reliance upon insurance schemes, which necessarily quantify such matters; but they do point out the deep-seated nature of reactions that must be considered in any realistic analysis.

The fact that we have accepted a position of such dependence upon our technological processes, in contrast to the condition of being subject only to natural accidents, means that we have to accept the responsibility for setting safety standards, etc., which in turn determine, at least statistically, the risk rate of accidental mortality. To emphasize the case by taking the absurd extreme, almost all new technological innovations would be too expensive for us to utilize if we insisted as a society that under no circumstances could these machines conceivably take or damage human life. As a concrete example, we all recognize that accidents can happen to airplanes, with heavy loss of life, but we go on flying; and furthermore it is doubtful whether we as a society would legislate for significantly lower risk rate, in view of the rapidly increasing cost which would probably be involved. In summary, there is some incompatibility between what might be called the public and private attitudes of society, or the conscious and unconscious attitudes, and all this is borne in mind by the managers of the various important undertakings of our society, for otherwise the society certainly could not function as relatively smoothly as it does.

This consideration suggests another problem concerning inquiries into social beliefs and behaviour, namely that there are strong feedback effects especially on the positive type. To amplify this point, recall that in the physical sciences it is usually possible to design instrumentation with sufficiently high input impedence that no significant power drain is placed upon the system being observed and therefore no significant changes caused upon it. Unfortunately, the position is much worse in the social field, in that it is almost impossible to ask any question of a human being without in some way either "loading the question" or the responder feeling that the question is loaded because of his own subjective reaction. In any case, he may not answer in an entirely frank way, and may even be quite dishonest. Of course, in view of the simultaneously-held, incompatible beliefs that any responder holds, it is probably begging the question to attach any value judgment to the word "dishonest" in this context.

On the phenomena

In this section are considered a number of the most important phenomena concerned in complex systems such as we have been discussing, as follows: growth and positive feedback; oscillations; unidirectionality; psychological and man–machine interactions; minimal principles; and extreme value statistics.

Growth and positive feedback

The expression "positive feedback" conveys some common meaning to most people in a qualitative way, but care must be taken before it is used in a quantative way, especially if any value judgments are to be involved in the matter. For example, the term "vicious circles" is commonly used to connote that a chain of vicious effects circulates in an increasing way. On the other hand, positive feedback is fundamental to all growth processes, whether in living or non-living systems. In this growth, positive feedback is responsible for the rapid growth in the initial stages, although the subsequent decay of growth rate to zero so that some steady state value is reached is due to other feedbacks. Actually, it is very hard to see how social life, as we know it, could be continued without this initiating effect. A few examples are now mentioned from the several domains considered earlier:

Fire, an essential tool in our life, must necessarily start as a small initial condition and then grow rapidly until the equilibrium conditions are approached when a leveling out of growth occurs. Although perhaps trite, it seems worth pointing out that when fire gets out of hand, as in forest and house fire, etc., then the exponential growth stage has merely been able to continue somewhat further than is considered desirable from the viewpoint of our own particular social system. The basic process, however, is the same as that which we consider desirable when we contain the fire within our household grates, oil furnaces, or internal combustion engines to pleasant and/or useful levels.

Growth of organisms, journals, and capital

The very prototype of growth is that in which cells successively divide after relatively constant periods given adequate nutrition, and thus produce the classic exponential growth. To continue the classic picture, however, the usual later growth pattern is modeled reasonably well by the so-called logistic curve or S-curve, in that the decay of growth rate essentially follows a mirror image pattern to the initial increase of growth rate until a steady-state population is reached, typically under a constant nutrition condition (see for example *Positive Feedback*, J. H. Milsum, ed., Pergamon Press, 1968). At present, knowledge, journals, and information are in the stage of exponential explosion. De Solla Price points out that while specific subsets of such knowledge systems, for example, of universities, show more or less the classic S-curve of growth over defined periods, they often set off on a renewed growth curve after some dramatically new condition has appeared (*Science Since Babylon*, Yale University Press, 1961). As regards money, this obviously continues in principle to produce exponential growth if invested with compound interest at constant rate.

Bandwagons, fashions, and vicious cycles

All these terms are used to describe phenomena exhibiting positive feedback in the social domain, the following constituting a few random examples: growth of discussion about a new and important subject, once somebody has first raised the matter; the arousal of passion in mobs; the spread of fashion in such matters as clothing, usage of words, and beliefs.

Another important aspect of growth in complicated systems is that while positive feedback may stimulate the growth of a particular subsystem, there must often be present inhibition for other similar but less successful subsystems in the surrounding domain. For example, there develops a competition between the individual tiny trees in the initial stages of growth of a number of seedlings in the forest so that once some particular examples prevail strongly enough, for example by an appropriate advantage in height, etc., then they can continue to grow preferentially with regard to the others, which latter are effectively inhibited and die off. In turn, of course, further stages of competition arise as these saplings grow larger, until the end result is the few "mighty oaks" for each acre of mature forest. Similar growth examples exist for animal species; for example, in a colony of flour beetles in a jar of flour, the female beetles lay eggs almost continually in the tunnels, but there is an extremely high probability that these eggs will be subsequently eaten by other perambulating mature beetles. Thus, while chance may set the initial conditions by which certain eggs are overlooked, once these newly-hatched beetles have survived to a certain size their probability of survival becomes successively larger. At higher levels of social organization, such as in bee colonies and in political parties, the same initial stage of selecting the queen bee or leader occurs, usually followed by a rallying around or "climbing onto the bandwagon", then elimination of competitors, then a period of comparative stability and unchallenged leadership, until eventually sufficient decay in the leader's performance occurs that the competition starts up again. It is also interesting to note that in certain neuronal data processing, such as in the retina, there is a strong tendency to emphasize contrast and/or movement of perceived objects, which is analogous to the aspect already described of enhancing some central feature and inhibiting the surround.

A common feature, worth noting in this matter of accepting a leader and then inhibiting competition, is that the first important feature in the emergence is to provide a sufficiently superior initial condition for one object to excel over its competition. Once this superiority has been accepted by the surrounding competition, then the survivor essentially loses most of its competition for a period, and usually therefore survives even for some significant period after its performance has fallen below that threshold of superiority which was originally necessary. In our common jargon we say that habits, laws, beliefs, and systems tend to be carried along by their own momentum even after people start to become dissatisfied with them. When we view the matter, however, in terms of the "capital" invested in the system, whether emotional or financial, then such tolerance is not unreasonable. In any case, the result is typically that of the stick-slip friction model already discussed.

An interesting and relevant socio-technological example concerns the circumstances by which the gauge separating the two tracks of railroads became standardized. It provides a classic example of those growth situations in which the apparently unimportant detail decisions at the early stages in fact become inevitably frozen into the design of huge systems. Thus today's railroads apparently developed from their embryonic form of horse-drawn coal carts. In the 19th century these were first carried over wooden planks, and then on iron plateways, and then on crude rails after the invention of the flanged wheel. The essentials of the modern system then started in England at the beginning of the 19th century, and, as the typical explosion of railroad building occurred, there did not seem any compelling reason at any stage to change from the first gauge which had "just happened", namely 4 ft. 8 1/2 in. No compelling reason, that is, for the original inventor Stephenson and subsequently his son, who were responsible for building most of England's railroad trackage. However, there was competition, symbolized by the engineer, Brunel, who realized that this gauge was not technically optimum, and therefore decided to build the Great Western Railway on a gauge of 9 ft. Considerable rivalry grew up over this matter, and technical passions were raised; but ultimately it was resolved on the matter of total capital investment, rather than on technical merit alone. Thus, when it was resolved to standardize in England after many years, there was about 80% of the narrower gauge compared to the remainder of the wider gauge, so the result may have been considered inevitable.

A further pertinent aspect may be pointed out concerning this case history. England, as the originator of railroads, had been more or less strongly involved in developing railroads all over the world. Thus the decision concerning the gauge standardization affected all, but was made only by a few.

Oscillations

Oscillations occur in many systems, and in social systems especially in those where mutual interaction exists. Now, many ecologists consider that an ecology cannot be considered to have reached optimal conditions while considerable oscillation of more or less random patterns continues in the subpopulations of the system. Evidently as ecologies become more mature, then the magnitude of oscillatory behaviour decreases. This comment is probably pertinent for many of our own human social systems also. For example, as Boulding comments, we are no longer interested in merely understanding why our economic systems oscillate, because such oscillations produce profound human misery, but rather we wish to control these oscillations to a satisfactorily low level.

In contrast, it should be noted that, within animals, the perceptual mechanisms seem particularly designed for detection of changes, as characterized rather dramatically by the fact that in the absence of change in the stimuli most of the relevant receptors generally cease to be stimulated. Furthermore, it has been shown that if a large sensory loss is maintained for a considerable period of time, then the human psyche finds it very difficult to operate at all in a stable normal way.

The evidence is increasing that the normal condition of state variables within organisms such as man, animals, and even plants, is often one of continual cycling. This may be anywhere in the range from subcellular systems up to the level of the organism itself. In any one case, the oscillation is in relatively narrow spectral bands, typical examples being the approximately one-second heart cycle; a ten-second "engine" cycle at the capillary level of blood, as noted by Iberall; circadian cycles of temperature, metabolic activity, etc., of order one day; the three- to five-day water cycle; the monthly ovarian cycle; and finally, the various generational cycles of different species.

Most of the technological systems which have been designed to incorporate adaptive and optimizing aspects inherently require a more-or-less continuous cycling action. This is necessary both in theory and practice, because the performance criterion may be time varying, and the fact that the system had once adapted so as to reach maximum performance "on top of the hill" is not satisfactory in general, since the hill may subsequently have moved. A hill-climbing scheme is therefore necessary in which the relevant derivative must be measured so that direction and rate information can be provided for climbing again towards the optimal performance condition. It is therefore at least conceptually possible that much of the cycling found in living systems relates to the possibility of them continually optimizing themselves. However, there are other equally plausible explanations, for example, that the oscillation is difficult to filter out and in itself is not harmful, or that the oscillation may help implement the necessary sharing of materials which are competed for in such economies as those of cells.

Unidirectionality

Asymmetry is a particular form of non-linearity which arises widely in the operation of systems, but perhaps especially those of the biological and social type. It is noteworthy that the relative infrequency of non-linearity in technological systems merely reflects the purpose of the designers, is that there is at least a fairly coherent linear system theory, which unfortunately is not true for the non-linear domain. However, with the advent of more powerful theories, and of the computer, the engineer is increasingly beginning to design his systems to be non-linear, which often has advantages as far as optimization is concerned.

A major reason for asymmetry in many living systems is that negative quantities cannot exist for many variables. Examples are the concentrations of active chemicals such as hormones in the endocrine system; that the force which a muscle can exert is restricted to pulling, rather than pushing; that the firing frequency of neurons can only be at a positive rate, since negative frequencies are meaningless; that such behavioral variables as hostility can have only one polarity, unless an anti-hostility effect is definable; and that the strength of an army is only significant for positive quantities. In the detailed design of biological systems, this fundamental unidirectional nature of certain of the components is overcome by providing another parallel channel which utilizes the same operating principle, but is connected in the reverse sense so as to provide effectively a variable for the negative polarity requirement of the variable. Particularly obvious examples are agonist–antagonist pairs of muscles, and the stimulatory–inhibitory connections of neurons. Note that in obtaining efficient operation, reciprocal-type information networks must be provided so that the two energetically-expensive variables cannot fight each other for no net effect in the system. However, as often is the case in biological systems, more subtle uses can also be made of the system in which in fact such apparently useless action is implemented; for example, a pair of muscles will be provided with "tone" in order to maintain a certain desired stiffness in the system, and furthermore, when the muscles are required to shiver in order to produce heat for the thermoregulation system, then it is clearly necessary for both muscles to shiver at the same time.

The above comments have essentially concerned static unidirectionality, but some very interesting and advantageous system characteristics arise from dynamic unidirectionality of the type specially pursued by Clynes ("Rein Control, or Unidirectional Rate Sensitivity, A Fundamental Dynamic and Organizing Function in Biology", to be published in *Annals of the N.Y. Acad. of Sci.*, 1968). The general effect is that dynamic asymmetry tends to emphasize changes of condition, a process noted earlier as being ubiquitous in living systems. Further, it emerges that such dynamic unidirectionality provides especially good response characteristics at the high-frequency end of the spectrum, but unfortunately this cannot be pursued further here.

There is another steady-state effect of static unidirectionality that is worth mentioning, which is that, in systems exhibiting oscillation or statistical variation, the rectifying effect of unidirectionality can result in an irrevocable loss of system capacity. The prototype for this is perhaps the service capacity of a facility given a waiting line. In particular, when the vehicles arrive with some statistical variation or if the service time is similarly distributed, there is an irrevocable loss of service capacity as compared to the situation when both arrivals and servicings are regular. The classic result is that as the mean arrival time reduces until it approaches the mean service time, then the length of the waiting line extends theoretically to infinity. The physical basis for this result is that there is a basic asymmetry operating, in that whenever chance decrees that no vehicles are waiting for service, then this otherwise available service time is lost forever. This condition applies equally for a neuron which must cut off information flow at zero frequency. Of course, effective bidirectionality can be introduced by incorporating a high spontaneous level of neural firing for which the information transfer is zero, but such a technique is metabolically expensive.

Other similar examples arise wherever material is pooled in some capacitor but tends to escape when the pressure rises above a certain amount; for example, if a bathtub has an overflow system, then the result of a statistical variation in the height of water due to waves will be that the mean height will be less than this threshold value at which overflow can first occur.

Psychological and man–machine interactions

In man–machine systems, and indeed in any systems where the psychological characteristics of man contribute significantly, any satisfactory analysis depends upon obtaining good experimental data regarding the psychological aspects. This point is well illustrated by the study of how the flow of vehicles through a single-lane facility can be maximized. In the case of one of the New York tunnels, a well-defined curve was obtained relating the number of cars passing through the tunnel per hour to the steady-speed at which the cars were constrained to travel. To the untrained intuition, it may seem rather surprising that the maximum throughput rate occurred when the controlled speed was close to 20 mph, and that the hill characterizing the throughput was fairly steep sided. The explanation, why the throughput falls above the optimal speed, relates to the psychological considerations that, with very good reason, the driver is not prepared to remain as closely behind the car in front as he would at a lower speed. One base measure for his spacing would be if it increased as the square of the speed, therefore being proportional to kinetic energy, so that the driver would be assured of being able to brake to a full stop without an accident even if the car in front came to an instantaneous halt, such as could occur if he sustained a head-on collision. In practice, drivers can generally see conditions somewhat ahead of the car in front and, given certain other social pressures, the experienced driver then generally chooses some spacing relation intermediate between the quadratic type and a linear spacing with velocity, which in turn gives rise to the optimal condition already stated.

It should perhaps be pointed out that this particular optimal condition was not recognized as such until the appropriate experimental work had been done; indeed, most of the tunnels mentioned are frequently marked "maintain 40 mph". Furthermore, while the systems analyst can obtain such answers concerning optimality, it is most unlikely that any set of subsystems such as drivers themselves would ever arrive at this solution if left to their own resources. Indeed, as members of the social system, we have all experienced frustrating situations especially in traffic jams, of seeing the overall system degenerate to an obviously non-optimal condition in consequence of people operating individually and apparently in their own best interests; the resultant breakdown of the complete system, however, shows that this is not so; but on the other hand that the subsystems are equally powerless to do anything about it.

While man must necessarily set his own performance criteria, he may not necessarily be able to or even wish to reveal their nature to an experimenter. For example, if a person is asked to move a heavy weight with his arm as rapidly as possible across a smooth surface between two fixed points, then observation of the dynamic performance will reveal quite a lot about the subject's strategy. In fact, this does turn out to be generally similar to the "bang–bang" control which does produce the action in minimum time. However, the subject still has under his control the arbitrary choice of what strength of muscle force he shall utilize, and this never seems to be the very maximum that could be obtained. Furthermore, carefully detailed observation may reveal that the man does not in fact accept the verbally stated criterion as appropriate to him, in the light of his actual performance.

Minimal principles and optimality

A number of minimal principles, that is, equivalently optimality principles, are recognized in physics. A generalized verbal formulation of such principles may be given roughly as follows: The particular solution "selected" by a process in a physical system will be that one out of the possible family of solutions, all consistent with given constraints, which minimizes a certain "cost". Thus a ray of light moves through media of arbitrary refractive indices in least time (Fermat's Principle), while a mechanical system which conserves total energy moves along that path which minimizes the quantity defined as "action" (Maupertuis' Principle of Least Action). When slightly generalized to minimize "virtual displacement" rather than action, Maupertuis' principle becomes consistent with that most general minimal principle of physics called Hamilton's Principle. The major drawback to universal application is that only conservative systems can be so treated. Other minimal principles concern lines of minimum length joining points on a given surface, and generalizations of this. There are implications for the latter principle in biology, for example, concerning the deduction that thin films enclosing material should assume configurations which minimize surface area.

Minimal principles have been of considerable philosophical importance in physics, but have not become necessary working tools for normal problems, such as those of the atmospheric and oceanic circulation mentioned earlier. In part this is because the mechanical, hydrodynamic, electrical and other laws which are derivable from the above principle are already well known (and indeed were discovered first usually), but also importantly because the ubiquity of dissipative phenomena render them ineligible.

With the recent development of a comprehensive control theory (see, for example, Bellman, *Adaptive Control Pro-*

cesses: a Guided Tour, Princeton University Press, 1961) the Hamiltonian and its derivative Euler–Lagrange equations have become of renewed importance. In this case the Hamiltonian represents a function of the performance criterion, being in detail comprised not only of the cost function itself but also of the constraints on the system behaviour. This has significant implications for equally direct optimization in the general types of systems we have been considering, if and when it becomes possible to define quantitatively both the performance criteria and the constraints.

The situation in biology is somewhat different and more promising than in physics, as Rosen points out in his recent book (R. Rosen, *Optimality Principles in Biology*, Butterworths, 1967) "... it is possible to give at the outset a coherent, if qualitative, argument which will to some extent justify a search for such principles in biology, and help in their formulation. This argument is based on the phenomenon of natural selection, and the pressure which selection exerts on nearly every aspect of the structure, function, and performance of biological individuals. . . . It is now possible to make the fundamental hypothesis that biological structures, which are optimal in the context of natural selection, are also optimal in the sense that they minimize some cost functional derived from the engineering characteristics of the situation. This most natural assumption has been called the Principle of Optimal Design (N. Rashevsky, *Mathematical Biophysics*, 3rd edition, Dover, New York, 1960; see also D'Arcy Thompson, *On Growth and Form*, Macmillan, New York, 1945). . . . Usually, however, the appropriate cost functional is not immediately obvious; therein lies the art and difficulty of the entire subject." We might add furthermore that it is not indeed likely that most cost functionals will prove simple enough to allow this direct analytical attack even when they can be plausibly defined. It is of interest that Rosen then considers the vascular system and, by assuming a reasonable cost functional related to the metabolic cost of maintaining the system plus the hydrodynamic power loss in the system, is able to show that the results of an optimality analysis are consistent with those obtained for dogs as regards blood vessel size and the geometry of branching.

The last example illustrates a "design optimization" in the sense that this is achieved at the genetic and growth level, presumably just once for any organism. On the other hand, there is evidence for on-going "operating optimization" in the sense that many organ system functions such as respiration, locomotion, posture, and swimming in fishes seem to be optimized on an energy related criterion (J. H. Milsum, *Biological Control Systems Analysis*, McGraw-Hill, 1966; also J. H. Milsum in *Advances in Bio-Medical Engineering and Medical Physics*, edited by S. N. Levine, Wiley-Interscience, 1967).

There is some first rather qualitative evidence that optimality principles apply also in social systems, where the interacting variables are probably more related to information flow than to energy flow (for example, G. K. Zipf, *Human Behaviour and Principle of Least Effort*, Harvard Univ. Press, 1949).

Extreme value statistics

Since nature's various populations and evolutionary experiments involve very large numbers, some of the small samples constituting our experimental research will inevitably differ from the mean by extremely large numbers of standard deviations, that is, falling into the area of extreme value statistics. For example, if the probability of an event such as a particular genetic combination is less than, say, ten to the minus nine, nevertheless such events may occur with reasonable regularity in some of nature's populations. In particular cases sufficient numbers of these extremely improbable events have happened, so that some form of evolution has even operated upon them. The simple animal hydra provides an example here in its evolution of unlikely methods of escaping either in space or time from unfavorable conditions such as by ballooning, etc. (see Slobodkin in *Positive Feedback*, J. H. Milsum, editor, Pergamon Press, 1968).

A trade-off situation is certainly involved in systems involving extreme value statistics, and some rather elementary examples will be considered; but in general it is certainly extremely difficult to attach quantitative values to the relevant variables. In these situations at least one new component is added to the performance criterion, representing the expected value or cost of the unlikely events. An *expected value* or *cost* equals the probability of the event multiplied by the value or cost of the event, should it happen. Numerically this value can be significant, and even comparable to other cost components of the performance criterion, even if the probability of the event is very small, provided that the cost attached to the event is large. Consider a simplified example of deciding what strength should be designed into buildings to combat the possibility of collapse due to earthquakes. If it is assumed that earthquakes are distributed in some reasonable and known statistical manner, then a probability value can be assigned concerning the occurrence of earthquakes of a given strength or larger during an appropriate period of time.

Clearly the construction cost will increase as the design strength against earthquakes is increased, and for an optimal design this must be traded off against the expected cost due to the finite possibility of building collapse. Thus, as design strength is increased, construction cost increases whereas the expected cost decreases because of the decreasing probability that a sufficiently strong earthquake will occur to cause building collapse. On the other hand, if construction strength is decreased, the latter cost is decreased whereas the expected cost of earthquake collapse is increased. It may be expected that an economic optimum point exists when mathematically the derivatives of the two costs, plotted as functions of construction strength, are equal in magnitude although opposite in sign.

Such an analysis forces us to confront the difficult issue of whether or not we are prepared to accept a possibility of building collapse "in cold blood". Clearly though, there is nothing wrong with the thinking of carrying out the analysis as long as we in society are prepared to say whether we are satisfied with the results, following upon incorporating into the performance criterion all values that concern us. Specifically, the difficult problem revolves about the cost of people's lives potentially lost in a collapse. This is certainly not an easy matter to resolve, but on the other hand, our society's nominal claim that we must not allow any risk of accidental death at all is clearly not carried out either individually or en masse in our normal affairs. As we increasingly generate new systems where such considerations will be forced upon us, as in universal systems of medical care, rational, but ethical guidelines will have to be developed. For example, in medical systems there is not only the problem of optimizing the system with regard both to the cost of medical care and the expected value of extended survival of life or mitigation of pain, but also whether or not the positive feedback dynamics of social pressures may allow any rational thinking about the value of prolonging life of individuals, and so on. Here the considerations move rapidly and increasingly confusingly into the technological, legal, ethical, moral, and religious domains.

On tampering and optimizing; some conclusions

It is a basic concept that systems comprise collections of subsystems, and that the problem in optimizing a social system is especially difficult in that it is as yet almost impossible to define quantitatively a relevant performance criterion. The difficulty resides especially in the fact that hard facts regarding costs of concrete, etc., have to be compared with "soft facts" concerning such social aspects as aesthetics and fulfillment. Furthermore, a satisfactory overall criterion must consist of a weighted collection of criteria which comprise the preferences of all the subsystems, namely ourselves who constitute the electorate for the overall systems planners and controllers, and therefore hopefully the ultimate controllers of the latter. It should be noted that the criteria we then choose will certainly be different from the ones we would otherwise be forced to choose if we were in a "dog-eat-dog" condition where each would have to optimize his own subsystem without regard to his neighbors. As Freud points out, in the work quoted, this is why societies have successively adopted the otherwise undesirable constraints of a civilized system. However, resulting from the energy and information revolutions, such constraints can become increasingly irksome while at the same time we enjoy the advantages, and therefore we must arrange that our legal and social structures, and our systems planning, are adequately flexible and fairly rapidly adaptable. In particular, they must try to anticipate the impact of technological change upon the social scene. This requires a vastly greater acceptance by the technologist and scientist of responsibility for reporting to society their intelligent predictions regarding social effects of technological change, and a vastly greater acceptance by the social scientists and humanists of their need to understand something not only of the general systems approach, but also of technological principles. Morison, in his book *Men, Machines, and Modern Times* (MIT Press, 1966) suggests that it is important for society to get into a mood of continuous, small-scale but ubiquitous experimentation, to be carried on by all intelligent members of society. This would encourage a sense of participation by individuals, would enable multiple-parallel experiments to be carried out on a small scale to predict effects of change rather in the same way that evolution operates, and would be a wonderfully fruitful way of realizing the dream of "intelligence amplification".

If such an experimental mood could be widely cultivated in society, then an atmosphere could be created for multidisciplinary systems analysis on a large number of large-scale technological projects with important social implications. For example, recent suggestions of this type include damming the Bering Straits, and widespread modifications of the North American fresh-water systems. A number of other vast projects are interestingly presented in *Engineer's Dreams* by Willy Ley (Viking, 1954).

Along with these new projects, society will also naturally tackle the pressing socio-technological problems such as that of air and water pollution due to pollutants of all types such as chemicals, noise, visual objects, and so on. A further aspect merging more towards the pure biological domain will be that of improving our attempts to control or eradicate what we conceive of as hostile species of animals, insects, etc., by use of the evolutionary process (by attracting and then sterilizing males, for example) rather than making it work continually against us, as in current pesticides. Furthermore, because of the huge and delicate balances of nature of the ecologies with which we tamper, we shall be forced to study the whole problem in a systems light. Finally, what can we say of man himself? We largely believe that we have provided the technological capability for fulfilling ourselves, but we have yet to define what this consists of, and how we must plan our society so as to be able to achieve it.

Reprinted from Milsum, J. H. (1968) "Technosphere, Biosphere and Sociosphere", *General Systems*, **13**.

12 Understanding the counterintuitive behaviour of social systems*

by Jay W. Forrester

System dynamics has demonstrated how companies and how urban systems behave in ways that run against most of what man would do to correct their ills. Now the same obtuse behaviour can be assigned to the largest social issues which confront the world. Although this article is written with the United States in mind, the analytical techniques described and the conclusions derived from them can be applied to all industrialized countries.

This paper addresses several issues of broad concern in the United States: population trends; the quality of urban life; national policy for urban growth; and the unexpected, ineffective, or detrimental results often generated by government programmes in these areas.

The nation exhibits a growing sense of futility as it repeatedly attacks deficiencies in our social system while the symptoms continue to worsen. Legislation is debated and passed with great promise and hope. But many programmes prove to be ineffective. Results often seem unrelated to those expected when the programmes were planned. At times programmes cause exactly the reverse of desired results.

It is now possible to explain how such contrary results can happen. There are fundamental reasons why people misjudge the behaviour of social systems. There are orderly processes at work in the creation of human judgment and intuition that frequently lead people to wrong decisions when faced with complex and highly interacting systems. Until we come to a much better understanding of social systems, we should expect that attempts to develop corrective programmes will continue to disappoint us.

The purpose of this paper is to leave with its readers a sense of caution about continuing to depend on the same past approaches that have led to our present feeling of frustration and to suggest an approach which can eventually lead to a better understanding of our social systems and thereby to more effective policies for guiding the future.

A new approach to social systems

It is my basic theme that the human mind is not adapted to interpreting how social systems behave. Our social systems belong to the class called multiloop nonlinear feedback systems. In the long history of evolution it has not been necessary for man to understand these systems until very recent historical times. Evolutionary processes have not given us the mental skill needed to properly interpret the dynamic behaviour of the systems of which we have now become a part.

In addition, the social sciences have fallen into some mistaken "scientific" practices which compound man's natural shortcomings. Computers are often being used for what the computer does poorly and the human mind does well. At the same time the human mind is being used for what the human mind does poorly and the computer does well. Even worse, impossible tasks are attempted while achievable and important goals are ignored.

Until recently there has been no way to estimate the behaviour of social systems except by contemplation, discussion, argument, and guesswork. To point a way out of our present dilemma about social systems, I will sketch an approach that combines the strength of the human mind and the strength of today's computers. The approach is an outgrowth of developments over the last 40 years, in which much of the research has been at the Massachusetts Institute of Technology. The concepts of feedback system behaviour apply sweepingly from physical systems through social systems. The ideas were first developed and applied to engineering systems. They have now reached practical usefulness in major aspects of our social systems.

I am speaking of what has come to be called industrial dynamics. The name is a misnomer because the methods apply to complex systems regardless of the field in which they are located. A more appropriate name would be *system dynamics*. In our own work, applications have been made to corporate policy, to the dynamics of diabetes as a medical system, to the growth and stagnation of an urban area, and most recently to world dynamics representing the interactions of population, pollution, industrialization,

* This paper is based on testimony for the Subcommittee on Urban Growth of the Committee on Banking and Currency, US House of Representatives, on October 7, 1970.

natural resources, and food. System dynamics as an extension of the earlier design of physical systems, has been under development at MIT since 1956. The approach is easy to understand but difficult to practice. Few people have a high level of skill; but preliminary work is developing all over the world. Some European countries and especially Japan have begun centres of education and research.

Computer models of social systems

People would never attempt to send a space ship to the moon without first testing the equipment by constructing prototype models and by computer simulation of the anticipated space trajectories. No company would put a new kind of household appliance or electronic computer into production without first making laboratory tests. Such models and laboratory tests do not guarantee against failure, but they do identify many weaknesses which can then be corrected before they cause full-scale disasters.

Our social systems are far more complex and harder to understand than our technological systems. Why, then, do we not use the same approach of making models of social systems and conducting laboratory experiments on those models before we try new laws and government programmes in real life? The answer is often stated that our knowledge of social systems is insufficient for constructing useful models. But what justification can there be for the apparent assumption that we do not know enough to construct models but believe we do know enough to directly design new social systems by passing laws and starting new social programmes? I am suggesting that we now do know enough to make useful models of social systems. Conversely, we do not know enough to design the most effective social systems directly without first going through a model-building experimental phase. But I am confident, and substantial supporting evidence is beginning to accumulate, that the proper use of models of social systems can lead to far better systems, laws, and programmes.

It is now possible to construct in the laboratory realistic models of social systems. Such models are simplifications of the actual social system but can be far more comprehensive than the mental models that we otherwise use as the basis for debating governmental action.

Before going further, I should emphasize that there is nothing new in the use of models to represent social systems. Each of us uses models constantly. Every person in his private life and in his business life instinctively uses models for decision making. The mental image of the world around you which you carry in your head is a model. One does not have a city or a government or a country in his head. He has only selected concepts and relationships which he uses to represent the real system. A mental image is a model. All of our decisions are taken on the basis of models. All of our laws are based on the basis of models. All executive actions are taken on the basis of models. The question is not to use or ignore models. The question is only a choice among alternative models.

The mental model is fuzzy. It is incomplete. It is imprecisely stated. Furthermore, within one individual, a mental model changes with time and even during the flow of a single conversation. The human mind assembles a few relationships to fit the context of a discussion. As the subject shifts so does the model. When only a single topic is being discussed, each participant in a conversation employs a different mental model to interpret the subject. Fundamental assumptions differ but are never brought into the open. Goals are different and are left unstated. It is little wonder that compromise takes so long. And it is not surprising that consensus leads to laws and programmes that fail in their objectives or produce new difficulties greater than those that have been relieved.

For these reasons we stress the importance of being explicit about assumptions and interrelating them in a computer model. Any concept or assumption that can be clearly described in words can be incorporated in a computer model. When done, the ideas become clear. Assumptions are exposed so they may be discussed and debated.

But the most important difference between the properly conceived computer model and the mental model is in the ability to determine the dynamic consequences when the assumptions within the model interact with one another. The human mind is not adapted to sensing correctly the consequences of a mental model. The mental model may be correct in structure and assumptions but, even so, the human mind – either individually or as a group consensus – is most apt to draw the wrong conclusions. There is no doubt about the digital computer routinely and accurately tracing through the sequences of actions that result from following the statements of behaviour for individual points in the model system. This inability of the human mind to use its own mental models is clearly shown when a computer model is constructed to reproduce the assumptions held by a single person. In other words, the model is refined until it is fully agreeable in all its assumptions to the perceptions and ideas of a particular person. Then, it usually happens that the system that has been described does not act the way the person anticipated. Usually there is an internal contradiction in mental models between the assumed structure and the assumed future consequences. Ordinarily the assumptions about structure and internal motivations are more nearly correct than are the assumptions about the implied behaviour.

The kind of computer models that I am discussing are strikingly similar to mental models. They are derived from the same sources. They may be discussed in the same terms. But computer models differ from mental models in important ways. The computer models are stated explicitly. The "mathematical" notation that is used for describing the model is unambiguous. It is a language that is clearer, simpler, and more precise than such spoken languages as English or French. Its advantage is in the clarity of meaning and the simplicity of the language syntax. The language of a computer model can be understood by almost anyone, regardless of educational background. Furthermore, any concept and relationship that can be clearly stated in ordinary language can be translated into computer model language.

There are many approaches to computer models. Some are naïve. Some are conceptually and structurally inconsistent with the nature of actual systems. Some are based on methodologies for obtaining input data that commit the models to omitting major concepts and relationships in the psychological and human reaction areas that we all know to be crucial. With so much activity in computer models and with the same terminology having different meanings in the different approaches, the situation must be confusing to the casual observer. The key to success is not in having a computer; the important thing is how the computer is used. With respect to models, the key is not to computerize a model, but to have a model structure and relationships which properly represent the system that is being considered.

I am speaking here of a kind of computer model that is very different from the models that are now most common in the social sciences. Such a computer model is not derived statistically from time-series data. Instead, the kind of computer model I am discussing is a statement of system structure. It contains the assumptions being made about the system. The model is only as good as the expertise which lies behind its formulation. Great and correct theories in physics or in economics are few and far between. A great computer model is distinguished from a poor one by the degree to which it captures more of the essence of the social system that it presumes to represent. Many mathematical models are limited because they are formulated by techniques and according to a conceptual structure that will not accept the multiple-feedback-loop and nonlinear nature of real systems. Other models are defective because of lack of knowledge or deficiencies of perception on the part of the persons who have formulated them.

But a recently developed kind of computer modelling is now beginning to show the characteristics of behaviour of actual systems. These models explain why we are having the present difficulties with our actual social systems and furthermore explain why so many efforts to improve social systems have failed. In spite of their shortcomings, models can now be constructed that are far superior to the intuitive models in our heads on which we are now basing national social programmes.

This approach to the dynamics of social systems differs in two important ways from common practice in social sciences and government. There seems to be a common attitude that the major difficulty is shortage of information and data. Once data is collected, people then feel confident in interpreting the implications. I differ on both of these attitudes. The problem is not shortage of data but rather our inability to perceive the consequences of the information we already possess. The system dynamics approach starts with the concepts and information on which people are already acting. Generally these are sufficient. The available perceptions are then assembled in a computer model which can show the consequences of the well known and properly perceived parts of the system. Generally, the consequences are unexpected.

Counterintuitive nature of social systems

Our first insights into complex social systems came from our corporate work. Time after time we have gone into a corporation which is having severe and well-known difficulties. The difficulties can be major and obvious such as a falling market share, low profitability, or instability of employment. Such difficulties are known throughout the company and by anyone outside who reads the management press. One can enter such a company and discuss with people in key decision points what they are doing to solve the problem. Generally speaking we find that people perceive correctly their immediate environment. They know what they are trying to accomplish. They know the crises which will force certain actions. They are sensitive to the power structure of the organization, to traditions, and to their own personal goals and welfare. In general, when circumstances are conducive to frank disclosure, people can state what they are doing and can give rational reasons for their actions. In a troubled company, people are usually trying in good conscience and to the best of their abilities to solve the major difficulties. Policies are being followed at the various points in the organization on the presumption that they will alleviate the difficulties. One can combine these policies into a computer model to show the consequences of how the policies interact with one another. In many instances it then emerges that the known policies describe a system which actually causes the troubles. In other words, the known and intended practices of the organization are fully sufficient to create the difficulty, regardless of what happens outside the company or in the marketplace. In fact, a downward spiral develops in which the presumed solution makes the difficulty worse and thereby causes redoubling of the presumed solution.

The same downward spiral frequently develops in government. Judgment and debate lead to a programme that appears to be sound. Commitment increases to the apparent solution. If the presumed solution actually makes matters worse, the process by which this happens is not evident. So, when the troubles increase, the efforts are intensified that are actually worsening the problem.

Dynamics of urban systems

Our first major excursion outside of corporate policy began in February 1968, when John F. Collins, former mayor of Boston, became Professor of Urban Affairs at MIT. He and I discussed my work in industrial dynamics and his experience with urban difficulties. A close collaboration led to applying to the dynamics of the city the same methods that had been created for understanding the social and policy structure of the corporation. A model structure was developed to represent the fundamental urban processes. The proposed structure shows how industry, housing, and people interact with each other as a city grows and decays. The results are described in my book *Urban Dynamics*, and some were summarized in *Technology Review* (*April, 1969, pp. 21–31*).

I had not previously been involved with urban behaviour or urban policies. But the emerging story was strikingly similar to what we had seen in the corporation. Actions taken to alleviate the difficulties of a city can actually make matters worse. We examined four common programmes for improving the depressed nature of the central city. One is the creation of jobs as by bussing the unemployed to the suburbs or through governmental jobs as employer of last resort. Second was a training programme to increase the skills of the lowest-income group. Third was financial aid to the depressed city as by federal subsidy. Fourth was the construction of low-cost housing. All of these are shown to lie between neutral and detrimental almost irrespective of the criteria used for judgment. They range from ineffective to harmful judged either by their effect on the economic health of the city or by their long-range effect on the low-income population of the city.

The results both confirm and explain much of what has been happening over the last several decades in our cities.

In fact, it emerges that the fundamental cause of depressed areas in the cities comes from *excess* housing in the low-income category rather than the commonly presumed housing shortage. The legal and tax structures have combined to give incentives for keeping old buildings in place. As industrial buildings age, the employment opportunities decline. As residential buildings age, they are used by lower-income groups who are forced to use them at a higher population density. Therefore, jobs decline and population rises while buildings age. Housing, at the higher population densities, accommodates more low-income urban population than can find jobs. A social trap is created where excess low-cost housing beckons low-income people inward because of the available housing. They continue coming to the city until their numbers so far exceed the available income opportunities that the standard of living declines far enough to stop further inflow. Income to the area is then too low to maintain all of the housing. Excess housing falls into disrepair and is abandoned. One can simultaneously have extreme crowding in those buildings that are occupied, while other buildings become excess and are abandoned because the economy of the area cannot support all of the residential structures. But the excess residential buildings threaten the area in two ways – they occupy the land so that it cannot be used for job-creating buildings, and they stand ready to accept a rise in population if the area should start to improve economically.

Any change which would otherwise raise the standard of living only takes off the economic pressure momentarily and causes the population to rise enough that the standard of living again falls to the barely tolerable level. A self-regulating system is thereby at work which drives the condition of the depressed area down far enough to stop the increase in people.

At any time, a near-equilibrium exists affecting population mobility between the different areas of the country. To the extent that there is disequilibrium, it means that some area is slightly more attractive than others and population begins to move in the direction of the more attractive area. This movement continues until the rising population drives the more attractive area down in attractiveness until the area is again in equilibrium with its surroundings. Other things being equal, an increase in population of a city crowds housing, overloads job opportunities, causes congestion, increases pollution, encourages crime, and reduces almost every component of the quality of life.

This powerful dynamic force to re-establish an equilibrum in total attractiveness means that any social programme must take into account the eventual shifts that will occur in the many components of *attractiveness*. As used here, attractiveness is the composite effect of all factors that cause population movement towards or away from an area. Most areas in a country have nearly equal attractiveness most of the time, with only sufficient disequilibrium in attractiveness to account for the shifts in population. But areas can have the same composite attractiveness with different mixes in the components of attractiveness. In one area component A could be high and B low, while the reverse could be true in another area that nevertheless had the same total composite attractiveness. If a programme makes some aspect of an area more attractive than its neighbour's, and thereby makes total attractiveness higher momentarily, population of that area rises until other components of attractiveness are driven down far enough to again establish an equilibrium. This means that efforts to improve the condition of our cities will result primarily in increasing the population of the cities and causing the population of the country to concentrate in the cities. The overall condition of urban life, for any particular economic class of population, cannot be appreciably better or worse than that of the remainder of the country to and from which people may come. Programmes aimed at improving the city can succeed only if they result in eventually raising the average quality of life for the country as a whole.

On raising the quality of life

But there is substantial doubt that our urban programmes have been contributing to the national quality of life. By concentrating total population, and especially low-income population, in urban locations, undermining the strength and cohesiveness of the community, and making government and bureaucracy so big that the individual feels powerless to influence the system within which he is increasingly constrained, the quality of life is being reduced. In fact, if they have any effect, our efforts to improve our urban areas will in the long run tend to delay the concern about rising total population and thereby contribute directly to the eventual overcrowding of the country and the world.

Any proposed programme must deal with both the quality of life and the factors affecting population. "Raising the quality of life" means releasing stress and pressures, reducing crowding, reducing pollution, alleviating hunger, and treating ill health. But these pressures are exactly the sources of concern and action aimed at controlling total population to keep it within the bounds of the fixed world within which we live. If the pressures are relaxed, so is the concern about how we impinge on the environment. Population will then rise further until the pressures reappear with an intensity that can no longer be relieved. To try to raise quality of life without intentionally creating compensating pressures to prevent a rise in population density will be self-defeating.

Consider the meaning of these interacting attractiveness components as they affect a depressed ghetto area of a city. First we must be clear on the way population density is, in fact, now being controlled. There is some set of forces determining that the density is not far higher or lower than it is. But there are many possible combinations of forces that an urban area can exert. The particular combination will determine the population mix of the area and the economic health of the city. I suggest that the depressed areas of most American cities are created by a combination of forces in which there is a job shortage and a housing excess. The availability of housing draws the lowest-income group until they so far exceed the opportunities of the area that the low standard of living, the frustration, and the crime rate counter-balance the housing availability. Until the pool of excess housing is reduced, little can be done to improve the economic condition of the city. A low-cost housing programme alone moves exactly in the wrong direction. It draws more low-income people. It makes the area differentially more attractive to the poor who need jobs and less attractive to those who create jobs. In the new population equilibrium that develops, some characteristic of the social system must compensate for the additional attractiveness created by the low-cost housing. The counterbalance is a further decline of the economic condition for the area. But as the area becomes more destitute, pressures rise for more low-cost housing. The consequence is a downward spiral that draws in the low-income population, depresses their condition, prevents escape, and reduces hope. All of this is done with the best of intentions.

My paper, "Systems Analysis as a Tool for Urban Planning" from a symposium in October 1969, at the National Academy of Engineering, suggests a reversal of present practice in order to simultaneously reduce the aging housing in our cities and allocate land to income-earning opportunities. The land shifted to industry permits the "balance of trade" of the area to be corrected by allowing labour to create and export a product to generate an income stream with which to buy the necessities of modern life from the outside. But the concurrent reduction of excess housing is absolutely essential. It supplies the land for new jobs. Equally important, the resulting housing shortage creates the population-stabilizing pressure that allows economic revival to proceed without being inundated by rising population. This can all be done without driving the present low-income residents out of the area. It can create *upward economic mobility* to convert the low-income population to a self-supporting basis.

The first reaction of many people to these ideas is to believe that they will never be accepted by elected officials or by residents of depressed urban areas. But some of our strongest support and encouragement is coming from those very groups who are closest to the problems, who see the symptoms first-hand, who have lived through the failures of the past, and who must live with the present conditions until enduring solutions are found.

Over the last several decades the country has slipped into a set of attitudes about our cities that are leading to actions that have become an integral part of the system that is generating greater troubles. If we were malicious and wanted to create urban slums, trap low-income people in ghetto areas, and increase the number of people on welfare, we could do little better than follow the present policies. The trend towards stressing income and sales taxes and away from the real estate tax encourages old buildings to remain in place and block self-renewal. The concessions in the income tax laws to encourage low-income housing will in the long run actually increase the total low-income population of the country. The highway expenditures and the government loans for suburban housing have made it easier for higher-income groups to abandon urban areas than to revive them. The pressures to expand the areas incorporated by urban government, in an effort to expand the revenue base, have been more than offset by lowered administrative efficiency, more citizen frustration, and the accelerated decline that is triggered in the annexed areas. The belief that more money will solve urban problems has taken attention away from correcting the underlying causes and has instead allowed the problems to grow to the limit of the available money, whatever that amount might be.

Characteristics of social systems

I turn now to some characteristics of social systems that mislead people. These have been identified in our work with corporate and urban systems and in more recent work that I will describe concerning the worldwide pressures that are now enveloping our planet.

First, social systems are inherently insensitive to most policy changes that people select in an effort to alter the behaviour of the system. In fact, a social system tends to draw our attention to the very points at which an attempt to intervene will fail. Our experience, which has been developed from contact with simple systems, leads us to look close to the symptoms of trouble for a cause. When we look, we discover that the social system presents us with an apparent cause that is plausible according to what we have learned from simple systems. But this apparent cause is usually a coincident occurrence that, like the trouble symptom itself, is being produced by the feedback-loop dynamics of a larger system. For example, as already discussed, we see human suffering in the cities; we observe that it is accompanied (some think caused) by inadequate housing. We increase the housing and the population rises to compensate for the effort. More people are drawn into and trapped in the depressed social system. As another example, the symptoms of excess population are beginning to overshadow the country. These symptoms appear as urban crowding and social pressure. Rather than face the population problem squarely we try to relieve the immediate pressure by planning industry in rural areas and by discussing new towns. If additional urban area is provided it will temporarily reduce the pressures and defer the need to face the underlying population question. The consequence, as it will be seen 25 years hence, will have been to contribute to increasing the population so much that even today's quality of life will be impossible.

A second characteristic of social systems is that all of them seem to have a few sensitive influence points through which the behaviour of the system can be changed. These influence points are not in the locations where most people expect. Furthermore, if one identifies in a model of a social system a sensitive point where influence can be exerted, the chances are still that a person guided by intuition and judgment will alter the system in the wrong direction. For example in the urban system, housing is a sensitive control point but, if one wishes to revive the economy of a city and make it a better place for low-income as well as other people, it appears that the amount of low-income housing must be reduced rather than increased. Another example is the world-wide problem of rising population and the disparity between the standards of living in the developed and the underdeveloped countries, an issue arising in the world system to be discussed in the following paragraphs. But it is beginning to appear that a sensitive control point is the rate of generation of capital investment.

And how should one change the rate of capital accumulation? The common answer has been to increase industrialization, but recent examination suggests that hope lies only in reducing the rate of industrialization. This may actually help raise quality of life and contribute to stabilizing population.

As a third characteristic of social systems, there is usually a fundamental conflict between the short-term and long-term consequences of a policy change. A policy which produces improvement in the short run, within five to ten years, is usually one which degrades the system in the long run, beyond 10 years. Likewise, those policies and programmes which produce long-run improvement may initially depress the behaviour of the system. This is especially treacherous. The short run is more visible and more compelling. It speaks loudly for immediate attention. But a series of actions all aimed at short-run improvement can eventually burden a system with long-run depressants so severe that even heroic short-run measures no longer suffice. Many of the problems which we face today are the eventual result of short-run measures taken as long as two or three decades ago.

A global perspective

I have mentioned social organizations at the corporate level and then touched on work which has been done on the dynamics of the city. Now we are beginning to examine issues of even broader scope.

In July 1970 we held a two-week international conference on world dynamics. It was a meeting organized for the Club of Rome, a private group of about 50 individuals drawn from many countries who have joined together to attempt a better understanding of social systems at the world level. Their interest lies in the same problems of population, resources, industrialization, pollution, and world-wide disparities of standard of living on which many groups now focus. But the Club of Rome is devoted to taking actions that will lead to a better understanding of world trends and to influencing world leaders and governments. The July meeting at MIT included the general theory and behaviour of complex systems and talks on the behaviour of specific social systems ranging from corporations through commodity markets to biological systems, drug addiction in the community, and growth and decline of a city. Especially prepared for this conference was a dynamic model of the interactions between world population, industrialization, depletion of natural resources, agriculture, and pollution. A detailed discussion on this world system will soon appear in my book *World Dynamics*, and its further development is the purpose of the "Project on the Predicament of Mankind" being sponsored by the Club of Rome at MIT for a year under the guidance of Professor Dennis Meadows. The plan is to develop a research group of men from many countries who will eventually base their continuing efforts in a neutral country such as Switzerland. The immediate project will re-examine, verify, alter, and extend the preliminary dynamic study of the world system and will relate it to the present world-wide concern about trends in civilization.

The simple model of world interactions as thus far developed shows several different alternative futures depending on whether population growth is eventually suppressed by shortage of natural resources, by pollution, by crowding and consequent social strife, or by insufficient food. Malthus dealt only with the latter, but it is possible for civilization to encounter other controlling pressures before a food shortage occurs.

It is certain that resource shortage, pollution, crowding, food failure, or some other equally powerful force will limit population and industrialization if persuasion and psychological factors do not. Exponential growth cannot continue forever. Our greatest immediate challenge is how we guide the transition from growth to equilibrium. There are many possible mechanisms of growth suppression. That some one or combination will occur is inevitable. Unless we come to understand and to choose, the social system by its internal processes will choose for us. The natural mechanisms for terminating exponential growth appear to be the least desirable. Unless we understand and begin to act soon, we may be overwhelmed by a social and economic system we have created but can't control.

The diagram * shows the structure that has been assumed. It interrelates the mutual effects of population, capital investment, natural resources, pollution, and the fraction of capital devoted to agriculture. These five system "levels" are shown in the rectangles. Each level is caused to change by the rates of flow in and out, such as the birth rate and death rate that increase and decrease population. As shown by the dotted lines, the five system levels, through intermediate concepts shown at the circles, control the rates of flow. As an example, the death rate at Symbol 10 depends on population P and the "normal" lifetime as stated by death rate normal DRN. But death

* All figures are taken from *World Dynamics* by Jay W. Forrester, Wright–Allen Press, 238 Main Street, Cambridge, Mass. 02142.

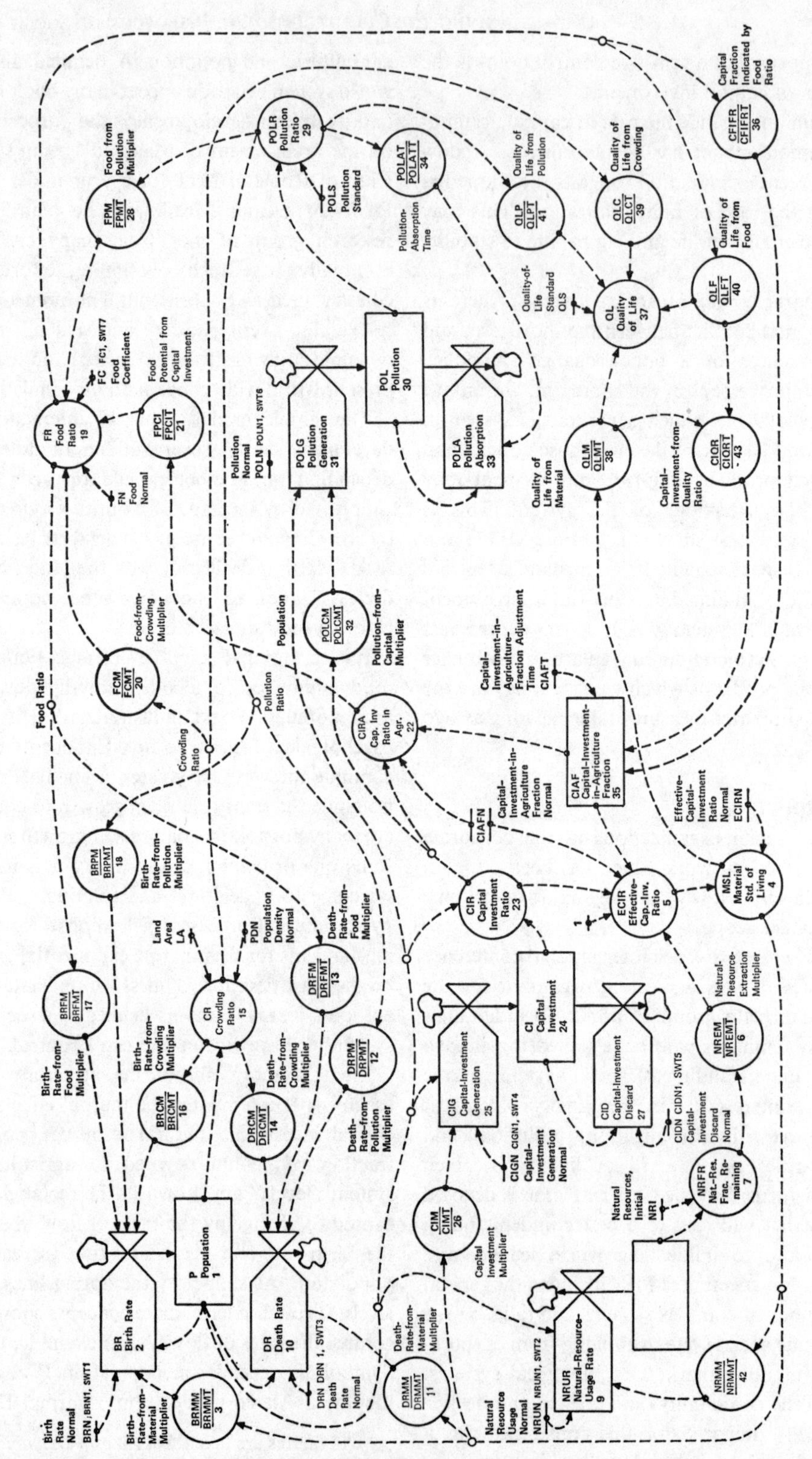

Figure 12.1 Flow diagram of world. Levels are represented by rectangles; rates by valves; auxiliaries (significant components of rates) by circles; dotted lines indicate directions, shown by arrows.

rate depends also on conditions in other parts of the system. From Circle 12 comes the influence of pollution that here assumes death rate to double if pollution becomes 20 times as severe as in 1970; and, progressively, that death rate would increase by a factor of 10 if pollution became 60 times as much as now. Likewise from Circle 13 the effect of food per capita is to increase death rate as food becomes less available. The detailed definition of the model states how each rate of flow is assumed to depend on the levels of population, natural resources, capital investment, capital devoted to food, and pollution.

Individually the assumptions in the model are plausible, create little disagreement, and reflect common discussions and assertions about the individual responses within the world system. But each is explicit and can be subjected to scrutiny. From one viewpoint, the system of Figure 12.1 is very simplified. It focuses on a few major factors and omits most of the substructure of world social and economic activity. But from another viewpoint, Figure 12.1 is comprehensive and complex. The system is far more complete and the theory described by the accompanying computer model is much more explicit than the mental models that are now being used as a basis for world and governmental planning. It incorporates dozens of nonlinear relationships. The world system shown here exhibits provocative and even frightening possibilities.

Transition from growth to equilibrium

With the model specified, a computer can be used to show how the system, as described for each of its parts, would behave. Given a set of beginning conditions, the computer can calculate and plot the results that unfold through time.

The world today seems to be entering a condition in which pressures are rising simultaneously from every one of the influences that can suppress growth – depleted resources, pollution, crowding, and insufficient food. It is still unclear which will dominate if mankind continues along the present path. Figure 12.2 shows the mode of behaviour of this world system given the assumption that population reaches a peak and then declines because industrialization is suppressed by falling natural resources. The model system starts with estimates of conditions in 1900. Adjustments have been made so that the generated paths pass through the conditions of 1970.

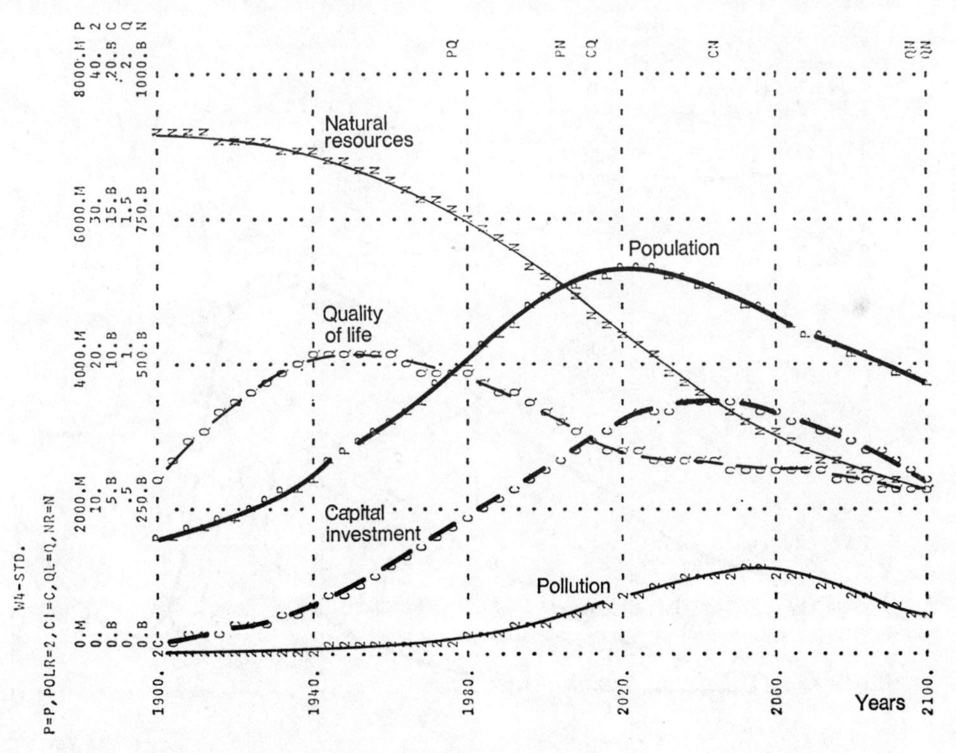

Figure 12.2 Population reaches a peak and then declines because industrialization is suppressed by falling natural resources.

In Figure 12.2 the quality of life peaks in the 1950's and by 2020 has fallen far enough to halt further rise in population. Declining resources and the consequent fall in capital investment then exert further pressure to gradually reduce world population.

But we may not be fortunate enough to run gradually out of natural resources. Science and technology may very well find ways to use the more plentiful metals and atomic energy so that resource depletion does not intervene. If so, the way then remains open for some other pressure to arise within the system. Figure 12.3 shows what happens within this system if the resource shortage is foreseen and avoided. Here the only change from Figure 12.2 is the usage rate of natural resources after the year 1970. In Figure 12.3 resources are used after 1970 at a rate 75 per cent less than assumed in Figure 12.2. In other words, the standard of living is sustained with a lower drain on the expendable and irreplaceable resources. But the picture is even less attractive! By not running out of resources, population and capital investment are allowed to rise until a pollution crisis is created. Pollution then acts directly to reduce birth rate, increase death rate, and to depress food production. Population which, according to this simple model, peaks at the year 2030 has fallen to one-sixth of the peak population within an interval of 20 years – a world-wide catastrophe of a magnitude never before experienced. Should it occur, one can speculate on which sectors of the world population will suffer most. It is quite possible that the more industrialized countries (which are the ones which have caused such a disaster) would be the least able to survive such a disruption to environment and food supply. They might be the ones to take the brunt of the collapse.

Figure 12.3 shows how a technological success (reducing our dependence in natural resources) can merely save us from one fate only to fall victim to something worse (a pollution catastrophe). There is now developing throughout the world a strong undercurrent of doubt about technology as the saviour of mankind. There is a basis for such doubt. Of course, the source of trouble is not technology as such but is instead the management of the entire technological–human–political–economic–natural complex.

Figure 12.3 is a dramatic example of the general process

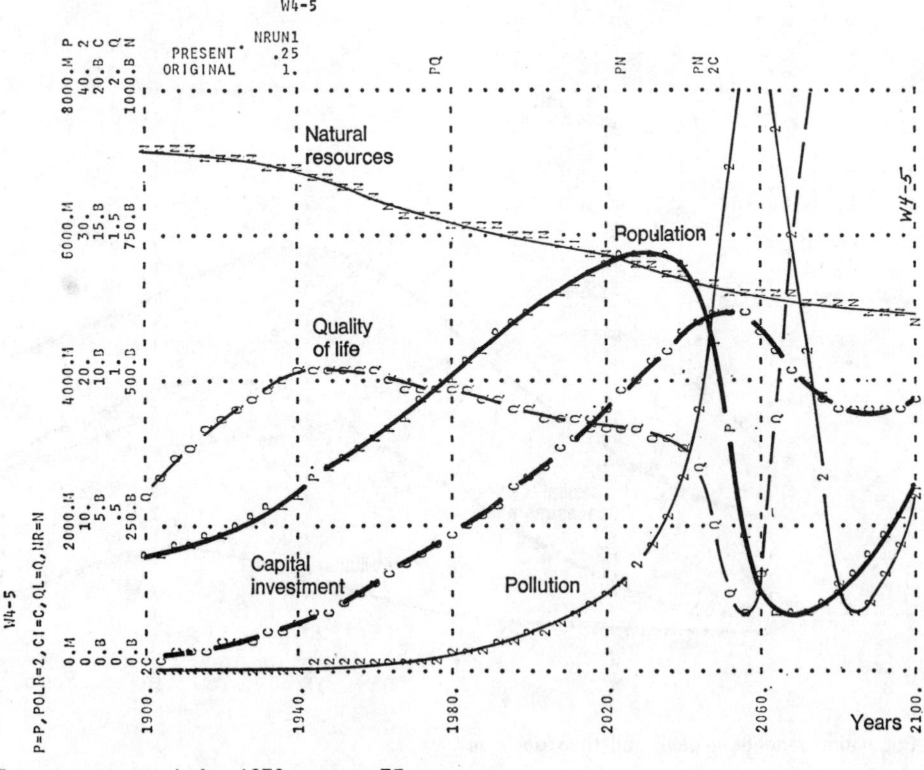

Figure 12.3 Resources are used after 1970 at a rate 75 per cent less than assumed in Fig. 12.2.

discussed earlier wherein a programme aimed at one trouble symptom results in creating a new set of troubles in some other part of the system. Here the success in alleviating a natural resource shortage throws the system over into the mode of stopping population caused by industrialization which has been freed from natural resource restraint. This process of a solution creating a new problem has defeated many of our past governmental programmes and will continue to do so unless we devote more effort to understanding the dynamic behaviour of our social systems.

Alternatives to decline or catastrophe

Suppose in the basic world system of Figures 12.1 and 12.2 we ask how to sustain the quality of life which is beginning to decline after 1950. One way to attempt this, and it is the way the world is now choosing, might be to increase the rate of industrialization by raising the rate of capital investment. Models of the kind we are here using make such hypothetical questions answerable in a few minutes and at negligible cost. Figure 12.4 shows what happens if the "normal" rate of capital accumulation is increased by 20 per cent in 1970. The pollution crisis reappears. This time the cause is not the more efficient use of natural resources but the upsurge of industrialization which overtaxes the environment before resource depletion has a chance to depress industrialization. Again, an "obvious" desirable change in policy has caused troubles worse than the ones that were originally being corrected.

This is important, not only for its own message but because it demonstrates how an apparently desirable change in a social system can have unexpected and even disastrous results.

Figure 12.4 should make us cautious about rushing into programmes on the basis of short-term humanitarian impulses. The eventual result can be anti-humanitarian. Emotionally inspired efforts often fall into one of three traps set for us by the nature of social systems: The programmes are apt to address symptoms rather than

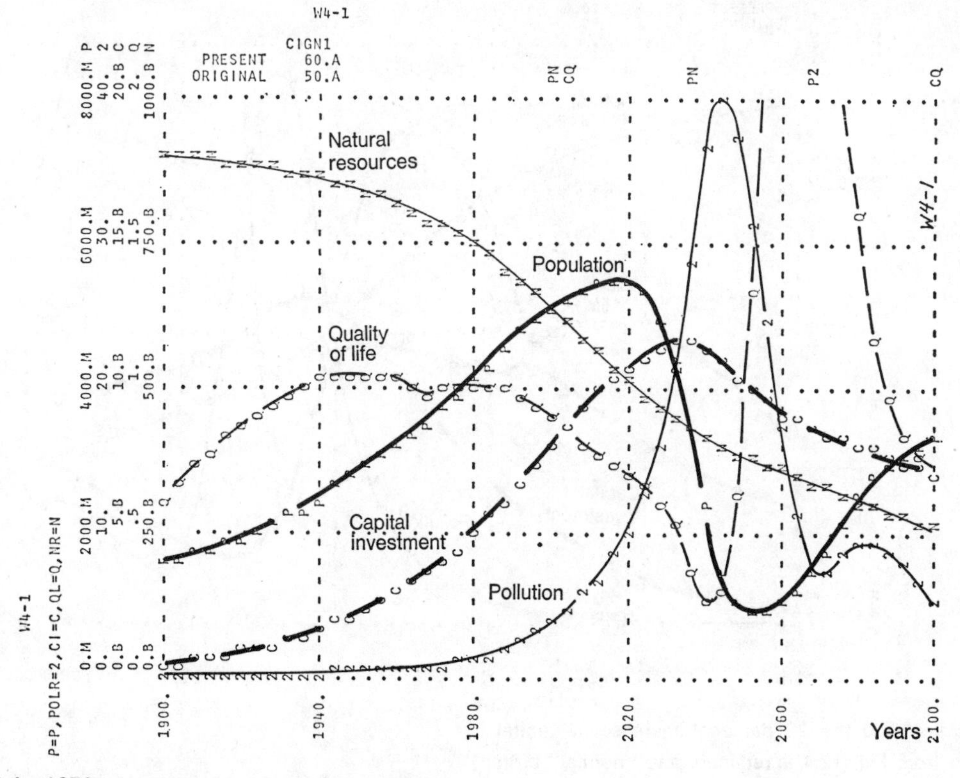

Figure 12.4 In 1970, the rate of capital accumulation is increased by 20 per cent in an effort to reverse the beginning decline in quality of life. The pollution crisis occurs before natural resources are depleted.

causes and attempt to operate through points in the system that have little leverage for change; the characteristic of systems whereby a policy change has the opposite effect in the short run from the effect in the long run can eventually cause deepening difficulties after a sequence of short-term actions; and the effect of a programme can be along an entirely different direction than was originally expected, so that suppressing one symptom only causes trouble to burst forth at another point.

Figure 12.5 retains the 20 per cent additional capital investment rate after 1970 from Figure 12.4 but in addition explores birth reduction as a way of avoiding crisis. Here the "normal" birth rate has been cut in half in 1970. (Changes in normal rates refer to coefficients which have the specified effect if all other things remain the same. But other things in the system change and also exert their effect on the actual system rates.) The result shows interesting behaviour. Quality of life surges upward for 30 years for the reasons that are customarily asserted. Food-per-capita grows, material standard of living rises, and crowding does not become as great. But the more affluent world population continues to use natural resources and to accumulate capital plant at about the same rate as in Figure 12.4. Load on the environment is more closely related to industrialization than to population and the pollution crisis occurs at about the same point in time as in Figure 12.4.

Figure 12.5 shows that the 50 per cent reduction in "normal" birth rate in 1970 was sufficient to start a decline in total population. But the rising quality of life and the reduction of pressures act to start the population curve upward again. This is especially evident in other computer runs where the reduction in "normal" birth rate is not so drastic. Serious questions are raised by this investigation about the effectiveness of birth control as a means of controlling population. The secondary consequence of starting

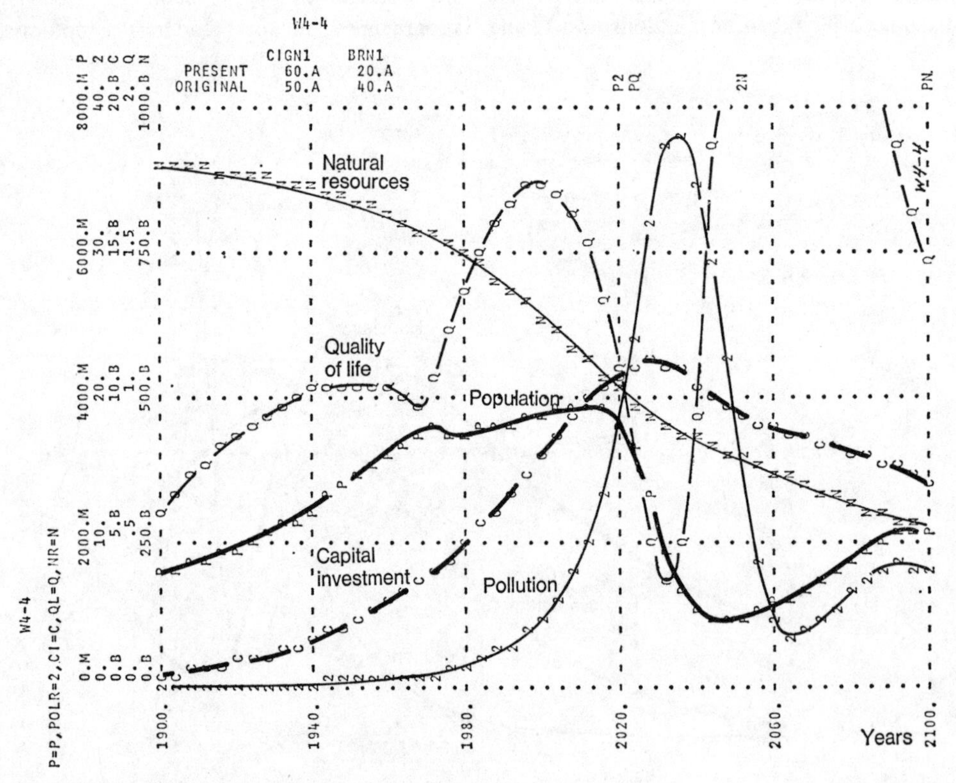

Figure 12.5 In 1970 the 20 per cent increase in capital accumulation from Fig. 12.4 is retained, and "normal" birth rate is reduced 50 per cent. Capital investment continues to grow until the pollution crisis develops. After an initial decline, population is again pushed up by a rapid rise in quality of life that precedes the collapse.

a birth control programme will be to increase the influences that raise birth rate and reduce the apparent pressures that require population control. A birth control programme which would be effective, all other things being equal, may largely fail because other things will not remain equal. Its very incipient success can set in motion forces to defeat the programme.

Figure 12.6 combines the reduced resource usage rate and the increased capital investment rate of Figures 12.3 and 12.4. The result is to make the population collapse occur slightly sooner and more severely. Based on the modified system of Figure 12.6, Figure 12.7 then examines the result if technology finds ways to reduce the pollution generated by a given degree of industrialization. Here in Figure 12.7, the pollution rate, other things being the same, is reduced by 50 per cent from that in Figure 12.6. The result is to postpone the day of reckoning by 20 years and to allow the world population to grow 25 per cent greater before the population collapse occurs. The "solution" of reduced pollution has, in effect, caused more people to suffer the eventual consequences. Again we see the dangers of partial solutions. Actions at one point in a system that attempt to relieve one kind of distress produce an unexpected result in some other part of the system. If the interactions are not sufficiently understood, the consequences can be as bad or worse than those that led to the initial action.

There are no utopias in our social systems. There appear to be no sustainable modes of behaviour that are free of pressures and stresses. But there are many possible modes and some are more desirable than others. Usually, the more attractive kinds of behaviour in our social systems seem to be possible only if we have a good understanding of the system dynamics and are willing to endure the self-discipline and pressures that must accompany the desirable mode. The world system of Figure 12.1 can exhibit modes that are more hopeful than the crises of Figures 12.2 through 12.7. But to develop the more promising modes will require restraint and dedication to a long-range future that man may not be capable of sustaining.

Figure 12.8 shows the world system if several policy changes are adopted together in the year 1970. Population

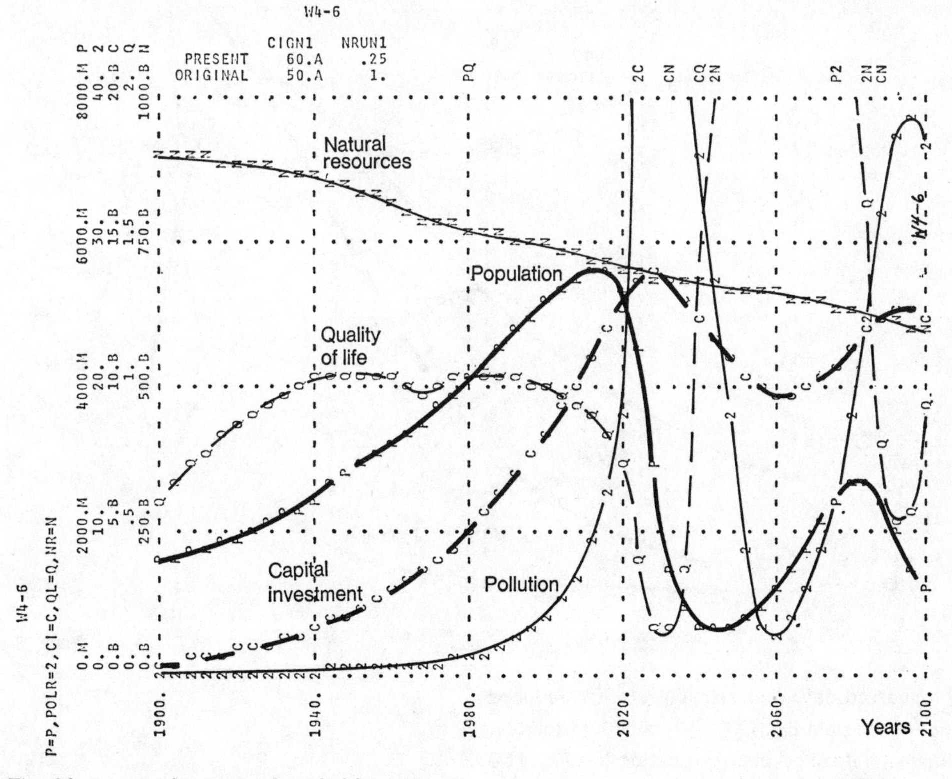

Figure 12.6 The 20 per cent increase of capital investment from Fig. 12.4 and the 75 per cent reduction of natural resource usage from Fig. 12.3 are combined.

is stabilized. Quality of life rises about 50 per cent. Pollution remains at about the 1970 level. Would such a world be accepted? It implies an end to population and economic growth.

In Figure 12.8 the normal rate of capital accumulation is *reduced* 40 per cent from its previous value. The "normal" birth rate is reduced 50 per cent from its earlier value. The "normal" pollution generation is reduced 50 per cent from the value before 1970. The "normal" rate of food production is *reduced* 20 per cent from its previous value. (These changes in "normal" values are the changes for a specific set of system conditions. Actual system rates continue to be affected by the varying conditions of the system.) But reduction in investment rate and reduction in agricultural emphasis are counterintuitive and not likely to be discovered or accepted without extensive system studies and years of argument – perhaps more years than are available. The changes in pollution generation and natural resource usage may be easier to understand and to achieve. The severe reduction in worldwide birth rate is the most doubtful. Even if technical and biological methods existed, the improved condition of the world might remove the incentive for sustaining the birth reduction emphasis and discipline.

Future policy issues

The dynamics of world behaviour bear directly on the future of the United States. American urbanization and industrialization are a major part of the world scene. The United States is setting a pattern that other parts of the world are trying to follow. That pattern is not sustainable. Our foreign policy and our overseas commerical activity seem to be running contrary to overwhelming forces that are developing in the world system. The following issues are raised by the preliminary investigations to date. They must, of course, be examined more deeply and confirmed

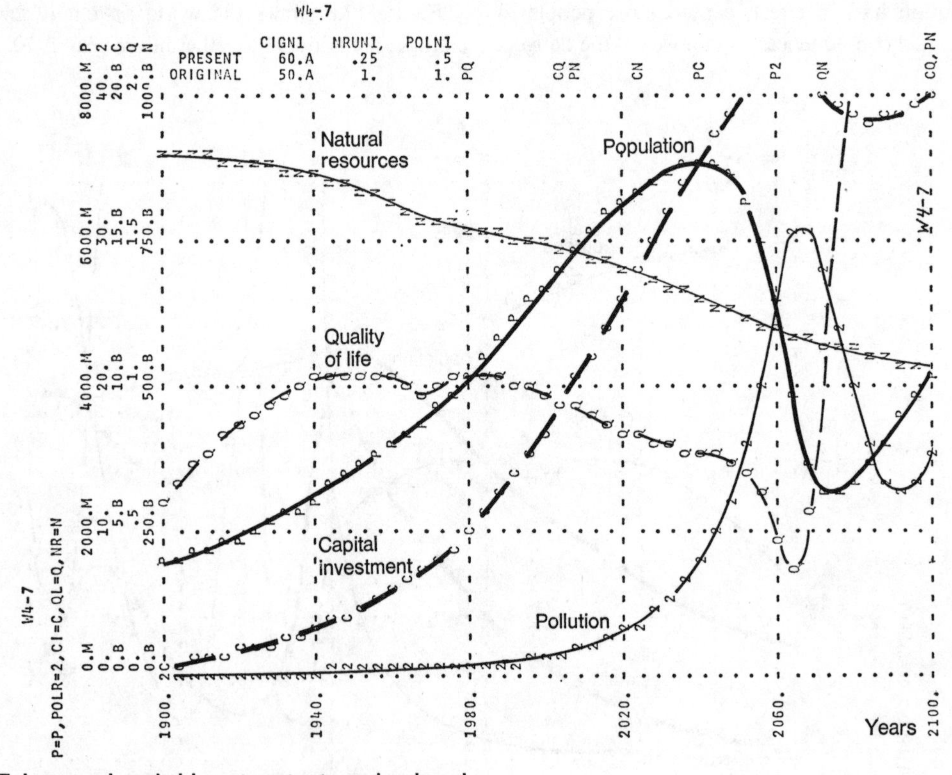

Figure 12.7 Increased capital investment rate and reduced natural resource usage from Fig. 12.6 are retained. In addition in 1970 the "normal" rate of pollution generation is reduced 50 per cent. The effect of pollution control is to allow population to grow 25 per cent further and to delay the pollution crisis by 20 years.

by more thorough research into the assumptions about structure and detail of the world system.

Industrialization may be a more fundamentally disturbing force in world ecology than is population. In fact, the population explosion is perhaps best viewed as a result of technology and industrialization. I include medicine and public health as a part of industrialization.

Within the next century, man may be facing choices from a four-pronged dilemma – suppression of modern industrial society by a natural resource shortage, collapse of world population from changes wrought by pollution, population control by war, disease, and social stresses caused by physical and psychological crowding.

We may now be living in a "golden age" where, in spite of the world-wide feeling of malaise, the quality of life is, on the average, higher than ever before in history and higher than the future offers.

Efforts for direct population control may be inherently self-defeating. If population control begins to result as hoped in higher per capita food supply and material standard of living, these very improvements can generate forces to trigger a resurgence of population growth.

The high standard of living of modern industrial societies seems to result from a production of food and material goods that has been able to outrun the rising population. But, as agriculture reaches a space limit, as industrialization reaches a natural-resource limit, and as both reach a pollution limit, population tends to catch up. Population then grows until the "quality of life" falls far enough to generate sufficiently large pressures to stabilize population.

There may be no realistic hope for the present underdeveloped countries reaching the standard of living demonstrated by the present industrialized nations. The pollution and natural resource load placed on the world environmental system by each person in an advanced country is probably 10 to 20 times greater than the load now generated by a person in an underdeveloped country. With four

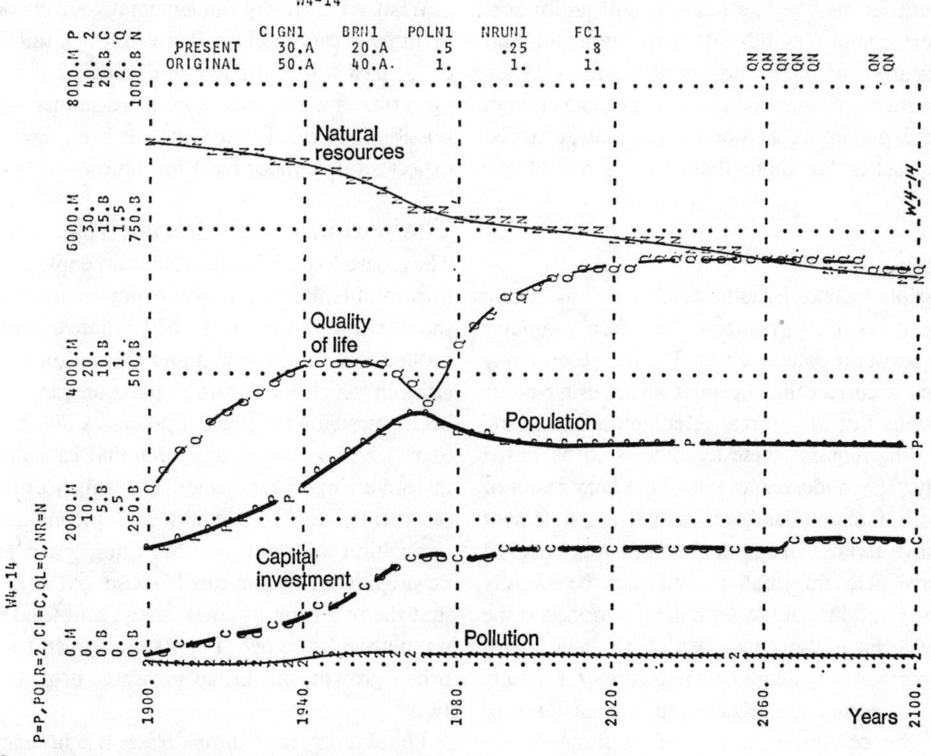

Figure 12.8 One set of conditions that establishes a world equilibrium. In 1970 capital investment rate is reduced 40 per cent, birth rate is reduced 50 per cent, pollution generation is reduced 50 per cent, natural resource usage rate is reduced 75 per cent, and food production is reduced 20 per cent.

times as much population in underdeveloped countries as in the present developed countries, their rising to the economic level of the United States could mean an increase of 10 times in the natural resource and pollution load on the world environment. Noting the destruction already occurring on land, in the air, and especially in the oceans, no capability appears to exist for handling such a rise in standard of living for the present total population of the world.

A society with a high level of industrialization may be nonsustainable. It may be self-extinguishing if it exhausts the natural resources on which it depends. Or, if unending substitution for declining natural resources is possible, the international strife over "pollution and environmental rights" may pull the average world-wide standard of living back to the level of a century ago.

From the long view of a hundred years hence, the present efforts of underdeveloped countries to industrialize along Western patterns may be unwise. They may now be closer to the ultimate equilibrium with the environment than are the industrialized nations. The present underdeveloped countries may be in a better condition for surviving the forthcoming world-wide environmental and economic pressures than are the advanced countries. When one of the several forces materializes that is strong enough to cause a collapse in world population, the advanced countries may suffer far more than their share of the decline.

A new frontier

It is now possible to take hypotheses about the separate parts of a social system, to combine them in a computer model, and to learn the consequences. The hypotheses may at first be no more correct than the ones we are using in our intuitive thinking. But the process of computer modelling and model testing requires these hypotheses to be stated more explicitly. The model comes out of the hazy realm of the mental model into an unambiguous model or statement to which all have access. Assumptions can then be checked against all available information and can be rapidly improved. The great uncertainty with mental models is the inability to anticipate the consequences of interactions between the parts of a system. This uncertainty is totally eliminated in computer models. Given a stated set of assumptions, the computer traces the resulting consequences without doubt or error. This is a powerful procedure for clarifying issues. It is not easy. Results will not be immediate.

We are on the threshold of a great new era in human pioneering. In the past there have been periods characterized by geographical exploration. Other periods have dealt with the formation of national governments. At other times the focus was on the creation of great literature. Most recently we have been through the pioneering frontier of science and technology. But science and technology are now a routine part of our life. Science is no longer a frontier. The process of scientific discovery is orderly and organized.

I suggest that the next frontier for human endeavour is to pioneer a better understanding of the nature of our social systems. The means are visible. The task will be no easier than the development of science and technology. For the next 30 years we can expect rapid advance in understanding the complex dynamics of our social systems. To do so will require research, the development of teaching methods and materials, and the creation of appropriate educational programmes. The research results of today will in one or two decades find their way into the secondary schools just as concepts of basic physics moved from research to general education over the past three decades.

What we do today fundamentally affects our future two or three decades hence. If we follow intuition, the trends of the past will continue into deepening difficulty. If we set up research and educational programmes, which are now possible but which have not yet been developed, we can expect a far sounder basis for action.

The nation's real alternatives

The record to date implies that our people accept the future growth of United States population as preordained, beyond the purview and influence of legislative control, and as a ground rule which determines the nation's task as finding cities in which the future population can live. But I have been describing the circular processes of our social systems in which there is no unidirectional cause and effect but instead a ring of actions and consequences that close back on themselves. One could say, incompletely, that the population will grow and that cities, space, and food must be provided. But one can likewise say, also incompletely, that the provision of cities, space, and food will cause the population to grow. Population generates pressure for urban growth, but urban pressures help to limit population.

Population grows until stresses rise far enough, which is to say that the quality of life falls far enough, to stop further increase. Everything we do to reduce those pressures causes the population to rise farther and faster and hastens the day when expediencies will no longer suffice. The United States is in the position of a wild animal running from its pursuers. We still have some space, natural resources, and agricultural land left. We can avoid the question of rising population as long as we can flee this bountiful reservoir that nature provided. But it is obvious that the reservoirs are limited. The wild animal usually flees until he is cornered, until he has no more space. Then he turns to fight, but he no longer has room to manoeuvre. He is less able to forestall disaster than if he had fought in the open while there was still room to yield and to dodge. The United States is running away from its long-term threats by trying to relieve social pressures as they arise. But if we persist in treating only the symptoms and not the causes, the result will be to increase the magnitude of the ultimate threat and reduce our capability to respond when we no longer have space to flee.

What does this mean? Instead of automatically accepting the need for new towns and the desirability of locating industry in rural areas, we should consider confining our cities. If it were possible to prohibit the encroachment by housing and industry onto even a single additional acre of farm and forest, the resulting social pressures would hasten the day when we stabilize population. Some European countries are closer to realizing the necessity of curtailing urban growth than are we. As I understand it, farm land surrounding Copenhagen cannot be used for either residence or industry until the severest of pressures forces the government to rezone small additional parcels. When land is rezoned, the corresponding rise in land price is heavily taxed to remove the incentive for land speculation. The waiting time for an empty apartment in Copenhagen may be years. Such pressures certainly cause the Danes to face the population problem more squarely than do we.

Our greatest challenge now is how to handle the transition from growth into equilibrium. Our society has behind it a thousand years of tradition that has encouraged and rewarded growth. The folklore and the success stories praise growth and expansion. But that is not the path of the future. Many of the present stresses in our society are from the pressures that always accompany the conversion from growth into equilibrium.

In our studies of social systems, we have made a number of investigations of life cycles that start with growth and merge into equilibrium. There are always severe stresses in the transition. Pressures must rise far enough to suppress the forces that produced growth. Not only do we face the pressure that will stop the population growth; we also encounter pressures that will stop the rise of industrialization and standard of living. The social stresses will rise. The economic forces will be ones for which we have no precedent. The psychological forces will be beyond those for which we are prepared. Our studies of urban systems demonstrated how the pressures from shortage of land and rising unemployment accompany the usual transition from urban growth to equilibrium. But the pressures we have seen in our cities are minor compared to those which the nation is approaching. The population pressures and the economic forces in a city that was reaching equilibrium have in the past been able to escape to new land areas.

But that escape is becoming less possible. Until now we have had, in effect, an inexhaustible supply of farm land and food-growing potential. But now we are reaching the critical point where, all at the same time, population is over-running productive land, agricultural land is almost fully employed for the first time, the rise in population is putting more demand on the food supplies, and urbanization is pushing agriculture out of the fertile areas into the marginal lands. For the first time demand is rising into a condition where supply will begin to fall while need increases. The crossover from plenty to shortage can occur abruptly.

The fiscal and monetary system of the country is a complex social–economic–financial system of the kind we have been discussing. It is clear the country is not agreed on behaviour of the interactions between government policy, growth, unemployment, and inflation. An article by a writer for *Finance* magazine in July 1970, suggests that the approach I have been discussing be applied in fiscal and monetary policy and their relationships to the economy. I estimate that such a task would be only a few times more difficult than was the investigation of urban growth and stagnation. The need to accomplish it becomes more urgent as the economy begins to move for the first time from a history of growth into the turbulent pressures that will accompany the transition from growth to one of the many possible kinds of equilibrium. We need to choose the kind of equilibrium before we arrive.

In a hierarchy of systems, there is usually a conflict between the goals of a subsystem and the welfare of the broader system. We see this in the urban system. The goal of the city is to expand and to raise its quality of life. But this increases population, industrialization, pollution, and demands on food supply. The broader social system of the country and the world requires that the goals of the urban areas be curtailed and that the pressures of such curtailment become high enough to keep the urban areas and population within the bounds that are satisfactory to the larger system of which the city is a part. If this nation chooses to continue to work for some of the traditional urban goals, and if it succeeds, as it may well do, the result will be to deepen the distress of the country as a whole and eventually to deepen the crisis in the cities themselves. We may be at the point where higher pressures in the present are necessary if insurmountable pressures are to be avoided in the future.

I have tried to give you a glimpse of the nature of multi-loop feedback systems, a class to which our social systems belong. I have attempted to indicate how these systems mislead us because our intuition and judgment have been formed to expect behaviour different from that actually possessed by such systems. I believe that we are still pursuing national programmes that will be at least as frustrating and futile as many of the past. But there is hope. We can now begin to understand the dynamic behaviour of our social systems. Progress will be slow. There are many cross-currents in the social sciences which will cause confusion and delay. The approach that I have been describing is very different from the emphasis on data gathering and statistical analysis that occupies much of the time of social research. But there have been breakthroughs in several areas. If we proceed expeditiously but thoughtfully, there is a basis for optimism.

Suggested readings

Forrester, Jay W. (1961) *Industrial Dynamics*, The MIT Press, Cambridge, Mass.

Forrester, Jay W. (1968) *Principles of Systems*, Wright-Allen Press, Cambridge, Mass.

Forrester, Jay W. (1969) *World Dynamics*, The MIT Press, Cambridge, Mass.

Forrester, Jay W. (1971) *World Dynamics*, Wright-Allen Press, Cambridge, Mass.

Meadows, Dennis L. (1970) *Dynamics of Commodity Production Cycles*, Wright-Allen Press, Cambridge, Mass.

 The material for this article was drawn from the author's book *World Dynamics* published by Wright-Allen Press, Cambridge, Massachusetts, U.S.A., 1971.

Section IV

Introduction

Although it is true that we are creating complex systems with technology and in our organizations, we are nowhere near approaching the complexity and behavioural richness of living systems. The highly evolved higher animals, and in particular their nervous systems, are systems of a complexity well beyond anything we can at present construct. Two different systems can be identified: the internal physiological and biochemical control systems, and the system of inter-relations of animals and plants with their ecological environment.

By now, a reader will be familiar with the ideas stemming from von Bertalanffy about the special nature of open biological systems. One of the most important concepts here is that of "stability" in organisms, and we have reproduced a small portion of a classic work by Cannon on homeostasis. The two following articles each deal with one aspect of the two different systems mentioned above. The first, by Priban, discusses modelling the respiratory system, and we have chosen this because it deals with a typical physiological function which is the subject of one module in the Open University course. The second article by Dale suggests how the ideas of systems analysis can be applied to ecology.

Associated reading

Biological control systems have become of great interest to engineers in recent years, partly because engineering analysis can be hoped to reveal more of the functioning of important physiological systems and partly because engineers learn much from these extremely "well-engineered" systems. There are a number of fairly detailed books on control theory and systems ideas applied to biological systems. Milsum's book *Biological control systems analysis* (McGraw-Hill) is a comprehensive treatment; *The application of control theory to physiological systems* by H. T. Milhorn is another. Work on the respiratory system is covered by J. H. Comroe in the *Physiology of respiration* in his article "The Lung" in *Scientific American* (February 1966). The respiratory system is also dealt with in Units 13 and 14 *The human respiratory system* of *Systems Behaviour* T241 (Open University). A second module of this course analyses the sheep-grassland eco-system in Units 11 and 12 of T241 *Systems Behaviour*.

Descriptions and analyses of many failed or failing systems can be found in the Open University's course: TD342 *Systems Performance: Human Factors and Systems Failures* which is described on page 5.

13 Self-regulation of the body

by Walter B. Cannon

I

Our bodies are made of extraordinarily unstable material. Pulses of energy, so minute that very delicate methods are required to measure them, course along our nerves. On reaching muscles they find there a substance so delicately sensitive to slight disturbance that, like an explosive touched off by a fuse, it may discharge in a powerful movement. Our sense organs are responsive to almost incredibly minute stimulations. Only recently have men been able to make apparatus which could even approach the sensitiveness of our organs of hearing. The sensory surface in the nose is affected by vanillin, 1 part by weight in 10,000,000 parts of air, and by mercaptan 1/23,000,000 of a milligram in a liter (approximately a quart) of air. And as for sight, there is evidence that the eye is sensitive to 5/1,000,000,000,000 erg, an amount of energy, according to Bayliss, which is 1/3,000 that required to affect the most rapid photograph plate.

The instability of bodily structure is shown also by its quick change when conditions are altered. For example, we are all aware of the sudden stoppage of action in parts of the brain, accompanied by fainting and loss of consciousness, that occurs when there is a momentary check in the blood flow through its vessels. We know that if the blood supply to the brain wholly ceases for so short a time as seven or eight minutes certain cells which are necessary for intelligent action are so seriously damaged that they do not recover. Indeed, the high degree of instability of the matter of which we are composed explains why drowning, gas poisoning, or electric shock promptly brings on death. Examination of the body after such an accident may reveal no perceptible injury that would adequately explain the total disappearance of all the usual activities. Pathetic hope may rise that this apparently normal and natural form could be stirred to life again. But there are subtle changes in the readily mutable stuff of the human organism which prevent, in these conditions, any return of vital processes.

When we consider the extreme instability of our bodily structure, its readiness of disturbance by the slightest application of external forces and the rapid onset of its decomposition as soon as favoring circumstances are withdrawn, its persistence through many decades seems almost miraculous. The wonder increases when we realize that the system is open, engaging in free exchange with the outer world, and that the structure itself is not permanent but is being continuously broken down by the wear and tear of action, and as continuously built up again by processes of repair.

II

The ability of living beings to maintain their own constancy has long impressed biologists. The idea that disease is cured by natural powers, by a *vis medicatrix naturae*, an idea which was held by Hippocrates (460–377 B.C.), implies the existence of agencies which are ready to operate correctively when the normal state of the organism is upset. More precise references to self-regulatory arrangements are found in the writings of modern physiologists. Thus the German physiologist, Pflüger, recognized the natural adjustments which lead toward the maintenance of a steady state of organisms when (1877) he laid down the dictum, "The cause of every need of a living being is also the cause of the satisfaction of the need." Similarly, the Belgian physiologist, Léon Fredericq, in 1885, declared, "The living being is an agency of such sort that each disturbing influence induces by itself the calling forth of compensatory activity to neutralize or repair the disturbance. The higher in the scale of living beings, the more numerous, the more perfect and the more complicated do these regulatory agencies become. They tend to free the organism completely from the unfavorable influences and changes occurring in the environment." Again, in 1900, the French physiologist, Charles Richet, emphasized the remarkable fact. "The living being is stable," he wrote. "It must be so in order not to be destroyed, dissolved or disintegrated by the colossal forces, often adverse, which surround it. By an apparent contradiction it maintains its stability only if it is excitable and capable of modifying itself according to external stimuli and adjusting its response to the stimulation. In a sense it is stable because it is modifiable – the slight instability is the necessary condition for the true stability of the organism."

Here, then, is a striking phenomenon. Organisms, composed of material which is characterized by the utmost inconstancy and unsteadiness, have somehow learned the methods of maintaining constancy and keeping steady in the presence of conditions which might reasonably be expected to prove profoundly disturbing. For a short time men may be exposed to dry heat at 115 to 128 degrees Centigrade (239 to 261 degrees Fahrenheit) without an increase of their body temperature above normal. On the other hand arctic mammals, when exposed to cold as low as 35 degrees Centigrade below freezing (31 degrees below zero Fahrenheit) do not manifest any noteworthy fall of body temperature. Furthermore, in regions where the air is extremely dry the inhabitants have little difficulty in retaining their body fluids. And in these days of high ventures in mountain climbing and in airplanes human beings may be surrounded by a greatly reduced pressure of oxygen in the air without showing serious effects of oxygen want.

Resistance to changes which might be induced by external circumstances is not the only evidence of adaptive stabilizing arrangements. There is also resistance to disturbances from within. For example, the heat produced in maximal muscular effort, continued for twenty minutes, would be so great that, if it were not promptly dissipated, it would cause some of the albuminous substances of the body to become stiff, like a hard-boiled egg. Again, continuous and extreme muscular exertion is accompanied by the production of so much lactic acid (the acid of sour milk) in the working muscles that within a short period it would neutralize all the alkali contained in the blood, if other agencies did not appear and prevent that disaster. In short, well-equipped organisms – for instance, mammalian forms – may be confronted by dangerous conditions in the outer world and by equally dangerous possibilities within the body, and yet they continue to live and carry on their functions with relatively little disturbance.

III

The statement was made above that somehow the unstable stuff of which we are composed had learned the trick of maintaining stability. As we shall see, the use of the word "learned" is not unwarranted. The perfection of the process of holding a stable state in spite of extensive shifts of outer circumstance is not a special gift bestowed upon the highest organisms but is the consequence of a gradual evolution. In the eons of time during which animals have developed on the earth probably many ways of protecting against the forces of the environment have been tried. Organisms have had large and varied experience in testing different devices for preserving stability in the face of agencies which are potent to upset and destroy it. As the construction of these organisms has become more and more complex and more and more sensitively poised, the need for more efficient stabilizing arrangements has become more imperative. Lower animals, which have not yet achieved the degree of control of stabilization seen in the more highly evolved forms, are limited in their activities and handicapped in the struggle for existence. Thus the frog, as a representative amphibian, has not acquired the means of preventing free evaporation of water from his body, nor has he an effective regulation of his temperature. In consequence he soon dries up if he leaves his home pool, and when cold weather comes he must sink to its muddy bottom and spend the winter in sluggish numbness. The reptiles, slightly more highly evolved, have developed protection against rapid loss of water and are therefore not confined in their movements to the neighborhood of pools and streams; indeed, they may be found as inhabitants of arid deserts. But they, like the amphibians, are "cold-blooded" animals, that is, they have approximately the temperature of their surroundings, and therefore during the winter months they must surrender their active existence. Only among the higher vertebrates, the birds and mammals, has there been acquired that freedom from the limitations imposed by cold that permits activity even though the rigors of winter may be severe.

The constant conditions which are maintained in the body might be termed *equilibria*. That word, however, has come to have fairly exact meaning as applied to relatively simple physico-chemical states, in closed systems, where known forces are balanced. The coordinated physiological processes which maintain most of the steady states in the organism are so complex and so peculiar to living beings – involving, as they may, the brain and nerves, the heart, lungs, kidneys and spleen, all working cooperatively – that I have suggested a special designation for these states, *homeostasis*. The word does not imply something set and immobile, a stagnation. It means a condition – a condition which may vary, but which is relatively constant.

It seems not impossible that the means employed by the more highly evolved animals for preserving uniform and stable their internal economy (i.e., for preserving homeostasis) may present some general principles for the establishment, regulation and control of steady states, that would be suggestive for other kinds of organization – even social and industrial – which suffer from distressing perturbations. Perhaps a comparative study would show that every complex organization must have more or less effective self-righting adjustments in order to prevent a check on its functions or a rapid disintegration of its parts when it is subjected to stress. And it may be that an examination of the self-righting methods employed in the more complex living beings may offer hints for improving and perfecting the methods which still operate inefficiently and unsatisfactorily. At present these suggestions are necessarily vague and indefinite. They are offered here in order that the reader, as he continues into the concrete and detailed account of the modes of assuring steady states in our bodies, may be aware of the possibly useful nature of the examples which they offer. . . .

Reprinted from Cannon, W. B. (1939) *The Wisdom of the Body*, W. W. Norton & Co. Inc.

14 Models in medicine

by Ian Priban

Although medical diagnosis has been analysed frequently in the hope of improving the treatment of disease, doctors still often find themselves obliged to treat symptoms rather than causes of illnesses. Consider, for example, the task facing a physician who is trying to identify the cause of clinical symptoms which could result from a malfunction in one of any number of different physiological mechanisms. He may be able to make key diagnostic tests but the process of detection is generally slow, routine and possibly tedious. What the physician needs is some fault finding strategies akin to those used by the computer engineer which will enable him to test the "biological circuitry" quickly and conveniently. And this is where mathematical models representing physiological functions can help. Equipped with the relevant model, the physician could systematically examine the suspect mechanisms and select the most effective treatment for the patient.

A mathematical model of any system can be constructed only after the relevant variables and their inter-relationships have been determined. This fundamental information can then be incorporated into a model which simulates the real system – anatomical structures and all. Unfortunately building such models inevitably involves some kind of compromise; any living system has so many variables that, even with modern powerful computers, only a small part of it can be dealt with at any one time. Nevertheless, I feel that some models are becoming sufficiently sophisticated and analogous to the natural activities which they represent to help in tracking down the cause of observed symptoms and even, in some cases, to forecast failures in health.

On the other hand, many details of the physiology of even such intensely studied tissues as muscle still have to be worked out in the context of the whole body. To be successful medical models cannot be restricted to the isolated systems so familiar in traditional physiological studies; the data used to construct the model must reflect the influence of other relevant systems in the body.

In order to indicate the kind of problem inherent in providing medically valuable models, I will confine my discussion to just one topic: the control of respiration. In this way I hope to impress upon the reader the innumerable controlled reactions and restraints which together maintain a stable internal environment within the body in spite of sudden fluctuations in both internal and external conditions. I also hope to emphasize that it is not beyond scientific ingenuity to provide a model which will help to relieve the physician of routine tasks and enable his effort to be diverted to areas more rewarding both for him and for his patient.

Most mathematical descriptions of living systems are expressed in the language of the control systems engineer, since this is more amenable to quantitative manipulation than that of the biologist. The late Norbert Wiener pointed out some 20 years ago the similarities between automatically controlled systems and living ones. In fact, control engineers have investigated and modelled certain aspects of human activity for some time – particularly during and since World War 2. Now biologists also are beginning to appreciate what Wiener was saying and as a result they are starting to draw more freely on the experience, methods and theories of the control engineer in their analysis of living systems.

One of the most important concepts in control theory is that of feedback, which is familiar to most people through the control of heating systems by a thermostat. The thermostat compares the ambient temperature with a preset value and an error signal based on the difference is fed back to the heater. The objective of this feedback is to keep the temperature steady at a predetermined level even in the face of large changes in the surroundings. When applied to biological mechanisms, such as the control of body temperature and the level of blood glucose, this concept greatly enhanced the physiologist's understanding; but, paradoxically, by enabling him to look at the control of living systems in more detail, it soon exposed a major problem. How could he explain the close simultaneous control of many variables even when some of the parameters of the system were being subjected to large changes?

Use of the theory of self-adaptive, multilevel control

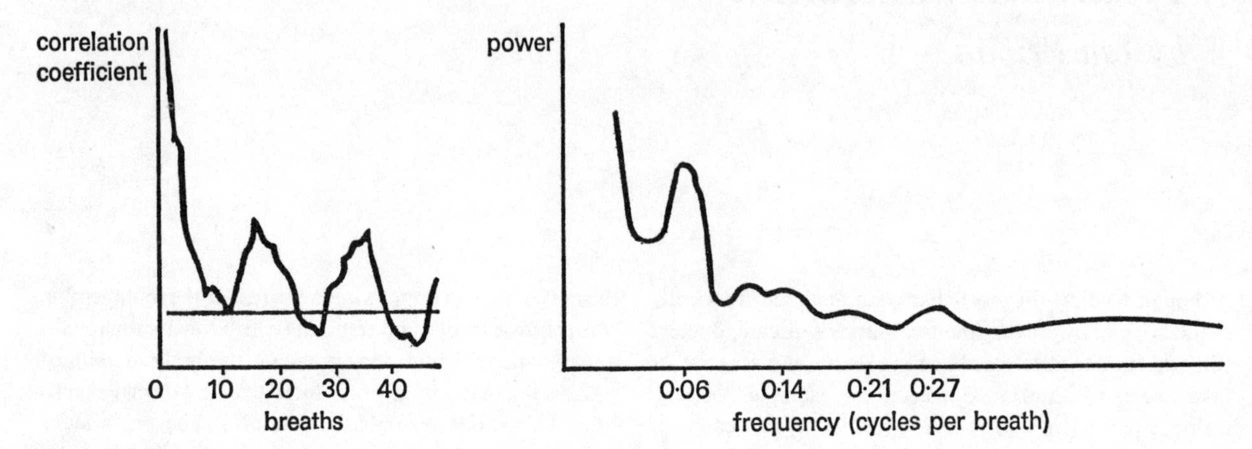

Figure 14.1 Statistical techniques reveal the relationships between different breaths in the respiratory pattern. The graph (left) reveals a direct relationship between the amplitude of any breath and those removed in time by 16 breaths or multiples of this. The prominences seen in the power density spectrum (right) based on two closely related variables, the amplitude and duration of breaths, indicate that four control loops each with a different time delay are concerned with the mutual adjustment of these two variables

systems and of predictive control can go a long way to finding a solution. A classical example of adaptive control – also called self-optimizing control – is the minimizing of the fuel consumption of aircraft piston engines by controlling ignition timing and the composition of the fuel/air mixture. In this adaptive system the "performance index" in kilometres per litre is measured continually, compared with the ideal index, and the error signal used to adjust the mixture and timing to achieve the optimum value. The important feature with adaptive control is that the operating state is determined by the existing physical or physical–chemical characteristics of the process, which may be changing, rather than by the properties of the controller. The best possible performance is achieved by continually evaluating the operation of the system with a test signal which is used to measure the deviation from the optimum. The result of the evaluation is a control signal which adjusts the inputs to the process to attain the optimum performance.

To represent living systems in terms of adaptive control requires that these systems have some unique optimum value and a physiological equivalent of the test signal. Considerable evidence exists for the former in many biological processes. In fact, many so-called "normal" values are the optimum ones in terms of the energy expenditure for the particular processes concerned. The evidence for the existence of test signals, however, is accumulating only slowly; this is due mainly to their being extremely small and difficult to measure.

To build up a model it is necessary also to know how to dissect living systems in a mathematical sense. The respiratory system is suited to such treatment because it is relatively accessible for investigation and making measurements causes little interference to subjects. One of the aims of the work in our laboratory is to show how parts of the respiratory system interact and are co-ordinated. Many of the mechanisms are being investigated individually in detail in laboratories through the world and provide much of the information needed for an operational model.

Breathing is rhythmical but each breath is different from the previous one and the next. The pattern varies continuously even when a person is in a steady state – as, for example, during sleep. Applying "random process" analysis to the breathing pattern, breaking this down into definite amplitude and time or frequency components, yields information about the system generating the pattern which can then be related to anatomical structures.

To do this the true signals must be separated from the random variations which have no explicit mathematical relationship and therefore, on average, cancel one another out. The results of the analysis indicate whether mathematical relationships exist between an event at one time and an event or events at other times. For instance, a recording of a sequence of several hundred to a thousand breaths can be fed into a computer to obtain a series of values of amplitude – the tidal volume. If each value is simply multiplied by itself the average of the products gives a single value. If the original sequence is copied and the copy is moved along by one breath, by multiplying the two, now out of phase, series and again averaging the products another single value is given. This process is then repeated many times to give a series of average values. The illustration Fig. 14.1, shows the final result. There is a direct relationship between any breath and those removed in time by 16 breaths and multiples of this. More simply, this means that the tidal volume of a particular breath is determined partly by what happened 16 breaths earlier.

The "power density spectrum" is a related method for extracting signals and highlighting relationships – in this case in terms of frequency components. The relationship illustrated is between the amplitude of a breath and its duration – two closely related variables. These results were selected from a large number of records. Usually a sequence of breaths shows fewer prominences at any one time, but these features are found when results from different sequences are combined. They indicate that breathing is controlled by signals fed back through at least four control loops with different time delays.

Before relating these results to anatomical structures it is necessary to decide exactly what process is being controlled. In normal man the mean values of the respiratory variables in the arterial blood – the *p*H and the partial pressures of carbon dioxide and oxygen – are nearly constant. It seems that the rate of exchange of both oxygen and carbon dioxide between the environment and the body is controlled, and that the controls ensure that respiration keeps pace exactly with metabolism. The important constituent in this exchange process is blood. Its chemical state depends upon the relative exchange flow rates of carbon dioxide and oxygen from the blood; carbon dioxide moves into the blood and oxygen out as it passes the body tissues and *vice versa* as the blood passes through the lungs.

The respiratory function of blood depends largely on the red pigment, haemoglobin. Its most important property, pointed to by Professor F. J. W. Roughton in Cambridge, is that it can carry the maximum number of molecules of carbon dioxide in exchange for molecules of oxygen and *vice versa* at the partial pressure and *p*H values normally found in arterial blood. These are the values at which the blood accomplishes its respiratory function with the minimum expenditure of energy with respect to the circulation, the ventilation of the lungs and with respect to its own metabolism. Since the purpose of feedback systems is to maintain the stability of a process in face of major disturbances, the gas exchange is obviously being controlled by keeping the blood as nearly as possible in its optimal state. Why have more than one feedback loop to control this

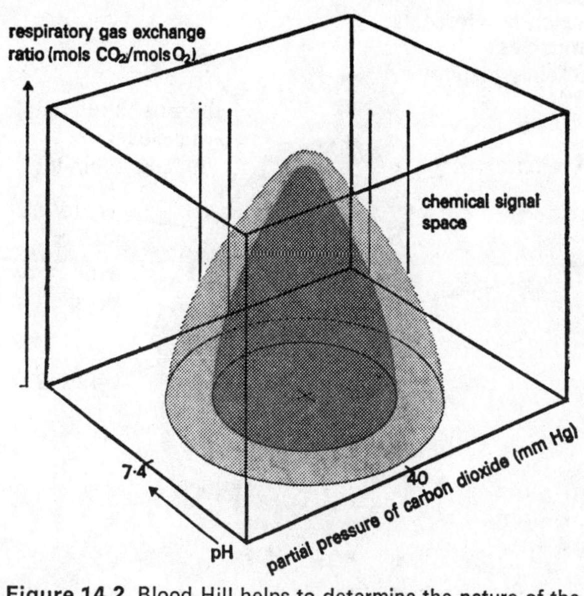

Figure 14.2 Blood Hill helps to determine the nature of the chemical signal which is passed on to the sensors responsible for continually monitoring the blood. The chemical signal is determined by the slope and shape of the hill. Obviously at its peak all the factors combine to give the optimum conditions but as the slope is descended changes in one variable necessitate substantial changes in the others to maintain the respiratory function of the blood. The inner hill shows the effect of reducing the partial pressure of oxygen from about 90 millimetres of mercury to about 50. Clearly such a reduction leads to a considerable increase in the chemical signal and hence a greater change in the pattern of respiration.

process? A simple answer is that the faster a control cycle is, the more rapid the response to small disturbances and the more efficient is the use of available energy. Such a rapid response is important in any condition involving more than resting muscular effort. In rest and sleep the slower loops alone can maintain an adequate respiratory gas exchange. Many of the structures which are part of the fastest control loop are also part of the slower loops – such as the circulatory system. The main differences between the loops are their transport times, which depend on the structures that modify and delay the signals before they arrive at the parts of the brain controlling breathing.

The fastest control loop is a most interesting one. To keep the functional state of blood at the optimum values three processes have to be controlled simultaneously, so that three sub-loops are involved. One of these almost exactly matches the level of pulmonary ventilation to the body's oxygen and carbon dioxide exchange requirements;

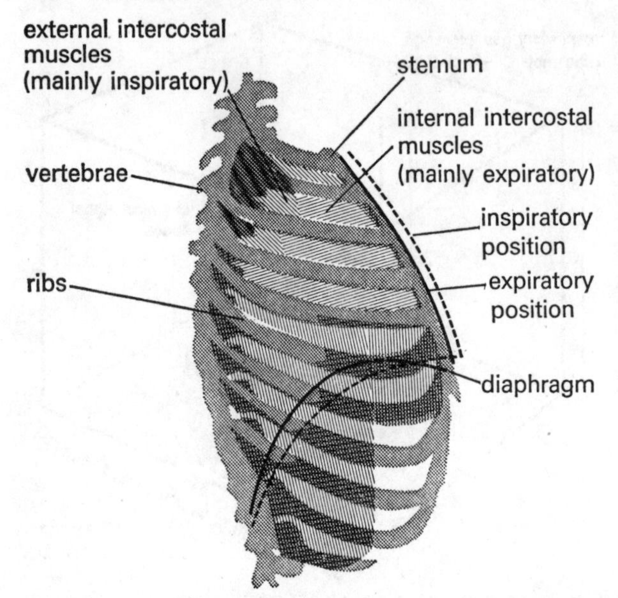

Figure 14.3 Control of the chest muscles which determine the change in volume of the chest and hence the potential gaseous exchange is fundamental to any respiratory system. In fact these respiratory muscles are subject to the finest control which is continually updated by feedback from receptors within the muscles. This feedback also provides additional information for the control of the overall pattern of respiration.

the second ensures that the respiratory muscles accomplish this ventilation with a minimum of energy expenditure, and the third makes sure that the energy used in ventilating the dead space – the conducting part of the lung airways – is kept at a minimum.

A signal which plays a part in evaluating the effectiveness of the gas exchange is caused by the rhythmical nature of breathing. The two way gas exchange associated with each breath causes the state of the blood to fluctuate about a mean value. Recently, W. S. Yamamoto at the University of Pennsylvania suggested that this fluctuation might be a signal controlling breathing. This is highly likely because, when blood is in the optimal state, the gas exchange can be accomplished with the smallest fluctuations or variations of state. By adjusting the level of ventilation in the direction that will minimize the magnitude of these variations, which would appear to be the error signals, the controller ensures that the level of ventilatory gas exchange will tend to keep pace exactly with that required by metabolism. The objective of this control system is very similar to that of a control engineer regulating an electrical power supply network so that generators are brought into action to meet the demands of the consumer.

In order that the ventilatory gas exchange can occur efficiently the activity of numerous respiratory muscle units has to be finely controlled and co-ordinated. In exercise these may have to pump air in and out of the lungs at more than ten times the resting rate. In addition, many of these muscles also have other important functions associated, for example, with posture, movement and speech.

One of the oldest observations about the efficiency of the energy cost of breathing was made by F. Rohrer in 1925. He showed that the natural frequency of breathing is that at which the average power requirement for ventilation is at a minimum. For breaths of shorter duration and small tidal volumes, on the one hand, high input power is required to produce the relatively large changes in velocity of muscle shortening; on the other hand, for larger tidal volumes more sustained muscle action is required and the power needed to distend the chest wall rises steeply as the tidal volume increases.

An examination of the structure and physical properties of the chest respiratory apparatus, which is made up of many different muscles and bones as shown above, provides the important clue to how the control objective – the minimization of the muscular energy cost – can be accomplished. Professor A. V. Hill and his colleagues at University College, London, have established that the efficiency of energy conversion by a muscle depends on its average length and its load or rate of shortening. At any time, therefore, depending on an individual's posture and movement, some muscle units must be able to bring about ventilation of the lungs more efficiently than others. The objective of the control must be to select these units. And to visualize how this is accomplished I have represented the structural arrangement of the muscles by a two dimensional grid, each sector of which contains muscle units with particular performance characteristics (see illustrations). The obvious goal of the controller is to match the "contours" of the muscle activity with contours of the highest value of the performance index (defined as the maximum number of units of ventilation obtained per unit energy expenditure in the muscles).

The following explanation of the mechanisms of control is new and partly hypothetical. It is also part of the operational model which itself provides the basis for new experiments aimed at validating or rejecting the model. The control is seen as being accompanied by a sequence of control actions involving a hierarchial system with three levels. The organization of the respiratory system may be compared with that of a large industrial organization – a

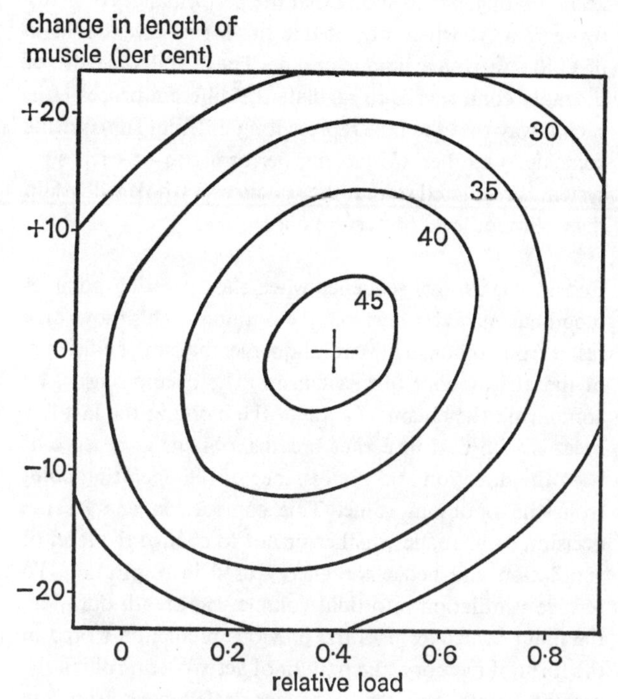

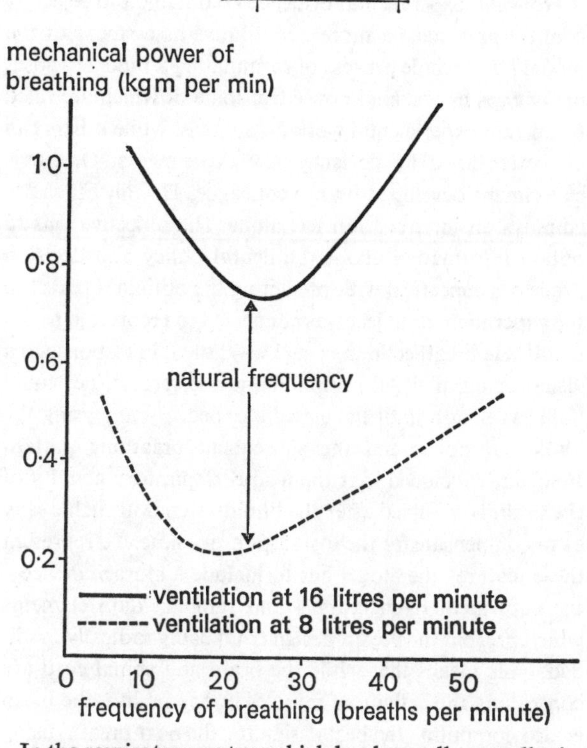

Figure 14.4 Performance contours (above) express the working efficiency of the respiratory muscles – the contour values give the percentage efficiency. The objective of the control system is to call into action those muscles which will bring about the required ventilation of the lungs with the least expenditure of muscular effort. Since the chest muscles are also concerned with posture and other functions, different levels of ventilation and different postures generally involve different contours. The fundamental goal of the body's control system is to keep the balance of energy used by it to the lowest possible level.

The natural frequency (right) of breathing is found to require the least mechanical power. For the same ventilation of the lungs any increase or decrease in the rate of breathing increases the power required for respiration and therefore the amount of work. For example, short rapid breaths demand extra power for the rapid rate of change in the muscle length whereas in slower, deeper breaths the power needed to distend the chest wall rises steeply.

kind of microeconomy. Here information is evaluated, decisions are made in the planning and predicting process and the management takes steps to ensure that the factory, the suppliers and the sales department all fulfil their jobs as predicted.

In the respiratory system a high level overall controller in the brain predicts the total amount of energy to be expended in a breath and the time over which it is to be dissipated in order to keep ventilation in step with metabolism. The prediction is conditioned by the responses of the system during previous breaths and, because living systems are evolutionary systems, also by its own past history as well as that of its ancestors. An intermediate level controller in the brain, also containing memory of past activity, takes the orders from the higher level and produces a more detailed temporal distribution of the pattern of activity. Finally, a lower level controller in the spinal cord takes the orders from the intermediate level and acts by selecting in detail the appropriate respiratory muscle units. Such selection can be accomplished only by a continual evaluation of muscle performance relative to the input signal as determined by the receptors or "spindles" in the muscles. The output from these muscle spindles also feeds back to the highest level of control. If the muscles shorten and effect ventilation as predicted, no correction is required of the higher control activity. But if the load on the muscles has changed the predicted activity will differ from the actual activity. This difference, the error signal, is fed back to the brain via the relay stations and enables a new prediction to be made.

How can experimental evidence be obtained to reject or confirm and make a more quantitative assessment of the model? The whole process of formulating a model exposes many gaps in existing knowledge, some of which are filled by current experimental work on animals, while others can be better filled by devising new experiments. One new experiment developed by my colleague, Dr Bill Fincham, employs an interrogation technique. His objective was to obtain information about the neural policy and decision processes concerned with predicting the optimal breath. In the experiment a subject was connected to recording instruments. He breathed in the usual way; then, in response to a visual or aural signal lasting about one second, he would hold his breath until the signal stopped. By analysing the "hold" response and the subsequent breathing pattern Fincham concluded that the neural respiratory activity of the brain is modified after the "hold" signal in such a way as to compensate for the lost respiratory time. To represent these features the model has to include a storage device – the equivalent of memory – and sampled data elements which prevent the breath design from being radically modified. This means that while the brain and spinal cord are controlling the activity of muscles in one breath, the brain is also computing the best design for the next breath, using both the information stored from previous breaths and that being fed back from the periphery associated with the present one. Breathing is therefore a discontinuous process – a sequence of phasic events.

The efficiency of ventilation also depends on the control of the airway dimensions. It is well established that the airway diameter is actively controlled by means of so-called "smooth" muscles in its walls. If the nerves which normally convey the information to and from the airways are blocked using a local anaesthetic, or are cut, a pattern of breathing is produced which shows that the energy consumption for flushing the airway is increased for the same average level of ventilation in the minute sac-like air spaces – the alveoli – in the lungs.

The overall efficiency of the respiratory system as a whole is undoubtedly due to the well co-ordinated control of the different individual mechanisms. For example, the airway dimensions determine the resistance to air flow and the rate of filling of the alveolar space in which the gas exchanges occur, and both of these are governed also by the rate of shortening of the respiratory muscles. The overall mechanism can be visualized by drawing a grid of any respiratory sub-system and representing on this the operational characteristics in the form of a performance index of the parts of the system, as was done previously for the respiratory muscles. Evidence from different laboratories is now accumulating, which shows that the performances of many living systems, when presented in this form, have the shape of a hill – they have unique optima. The overall objective of the main controller is to regulate the different processes in such a way that the hills representing different sub-systems coincide. In other words, the performance of each sub-system is regulated so that the system as a whole will attain the optimum level of performance.

Consider a control sequence when the operating point of blood has moved away from the optimum. This move may result from a change in ventilation, metabolism or efficiency of the airway control mechanism. The chemo-sensor, by comparing the chemical state of the blood in the last few breaths with that in earlier breaths, obtains a measure of both the direction and the distance of the operating point from the optimum value. This comparison enables the decision to be made whether or not to change the level of ventilation and hence the energy used in respiration. To resolve ventilation into tidal volume and breath duration, the main controller uses the previous prediction stored in the form of memory, the pattern of activity controlling the present breath and the error signals fed back from the muscular and airway sub-systems. These error signals contain the information about the mismatch in performance with respect to the best predicted input to the process in the sub-system. The information content of these signals is reduced at the low and intermediate level controllers, so that only the information concerning the accuracy of the prediction or change in state in the periphery is fed back to the main controller. Part of the information is stored in the lower levels and is then used in the next breath to increase the detail of the neural signal controlling the muscle activity.

I have not used the term "reflex" because this is usually associated with non-anticipatory systems. Instead predictive control has been introduced which is more suited for explaining simply and accurately the behaviour of the neural mechanisms and of naturally operating systems such as those which are part of the respiratory control system.

The fastest control cycle occupies, on average, 3·8 breaths. Most of this time is taken up by the signal, the state of blood being transported in the circulation from the lungs to the chemo-sensitive region in the brain. The transmission and evaluation of the signals in the nervous system is fast and takes relatively little time. The fastest control loop involves three simultaneous sub-processes, resulting in the high efficiency needed by an active working organism.

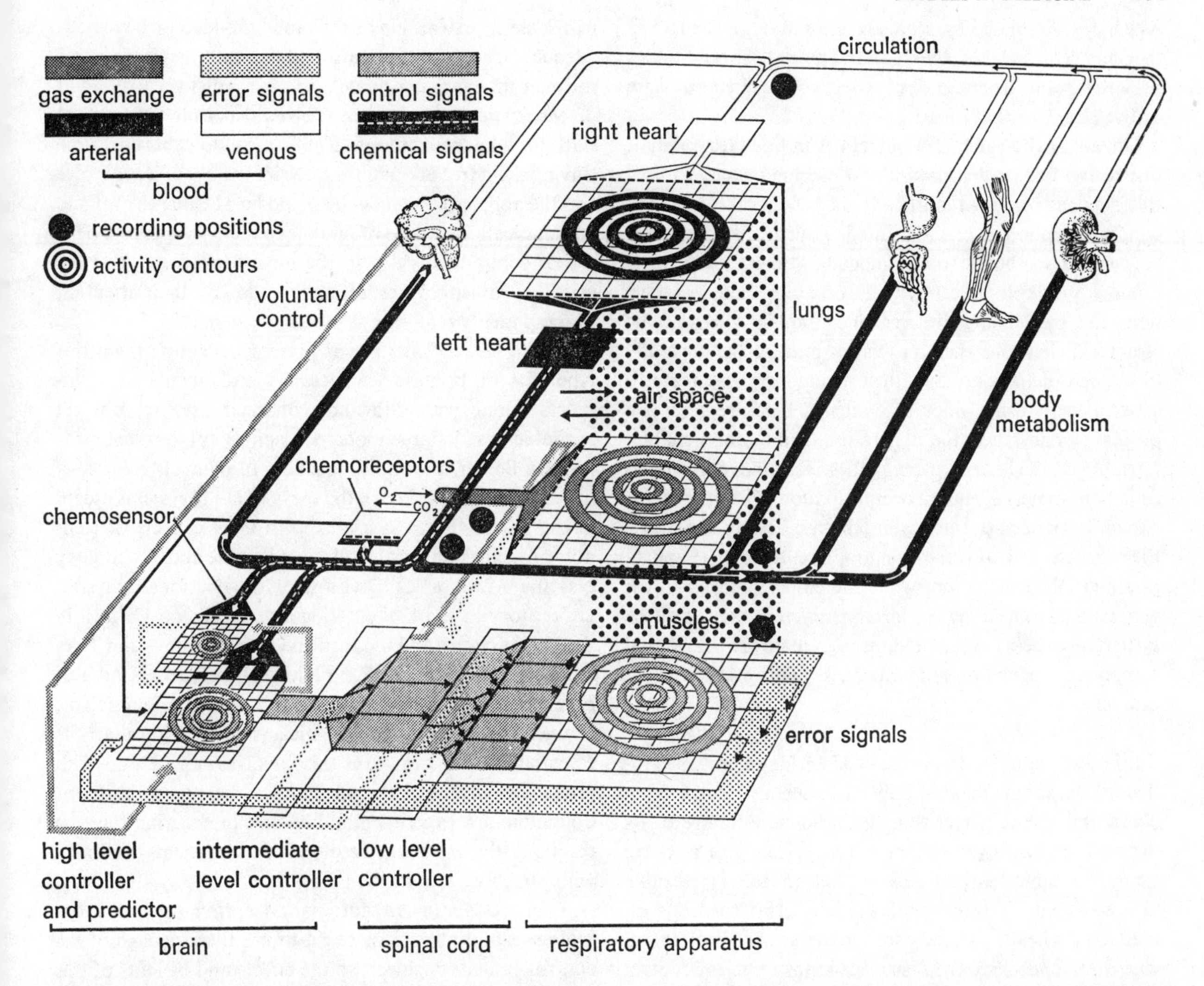

Figure 14.5 Control of respiration, mapped out schematically here, is carried out at three levels with an over-riding "executive" level in the brain itself. The entire system is geared to maintaining the most efficient exchange of respiratory gases and the different levels of control are required both for sensitivity of control and to avoid any violent fluctuations which might occur locally in the system. The different levels also show a gradation in control with the fine control residing in the lowest level which sends signals to the respiratory muscles. Since respiration is a dynamic process, conventional graphic methods cannnot really explain the relationships between the various subsystems such as the pulmonary circulation and the respiratory muscles but data can be recorded and processed to give maps showing "activity contours". Briefly, the diagram shows how the performance of the lungs is assessed by comparing the chemical state of the blood in the arteries with the predicted performance of the respiratory system. The difference between predicted and actual performance gives rise to an error signal which is used by the predictor in controlling the next breath. This coarse control receives additional information from an error signal arising in the respiratory muscles which ultimately determines the potential gaseous exchange within the lungs. As this error signal feeds back it adjusts the control at successive levels continuously modifying the respiratory pattern.

During inactivity or sleep high efficiency is no longer necessary and then the slower loops can adequately control the respiratory gas exchange – but it is still necessary to protect the organism from large disturbances. Important structures concerned with this are the so-called chemo-receptors which are stimulated by more abrupt and large changes in the chemical state of the blood. The still slower cycles probably result from the direct effect of the chemical state of the blood on that of the brain.

Obviously the value of a model lies in how accurately it can be used to predict naturally occurring events. In turn this depends on how realistically the model represents the conditions found in real life. Life is governed by innumerable variables, but to make modelling manageable their number and level of accuracy must be restricted. We have done this by defining the area of interest in terms of the structural elements, their properties and by the generally observable behaviour, and then arranging these features systematically using block diagrams. A block represents a process (for instance, the illustration of the blood characteristic) and the lines connecting these variables correspond to flow of material, energy or information. Thus the whole system is mapped out using, in our case, nearly 50 blocks, the characteristic of some being represented by the appropriate graph or equation, while the empty blocks call for experiments to be done. So far only different aspects of the system are being simulated, using analogue or digital computers depending on which is appropriate to the problem.

The body is complex and because of the innumerable adaptive feedback systems it is difficult to separate cause from effect. But the very fact that the systematic nature of its dynamic organization has been recognized and is being examined quantitatively means that in the foreseeable future we will be using mathematical terms to define an individual's health. Already some patients have been found who apparently lack the fastest and most efficient control loop. Moreover, prominences associated with the control loops can usually be found more easily in the spectra of certain types of patient than in those of normal subjects. The implications are that in these patients the impaired structures tend to exert a more marked effect on breathing. In normal subjects the functioning of the component structures is well matched and controlled so that no one structure or mechanism significantly influences breathing on its own. These tests are therefore of use in assessing the health of individuals.

A particular application of models is in the field of intensive care. With a model it becomes possible to assess automatically certain aspects of a patient's health or even to predict that a patient is moving towards a critical state. For example, some patients suffer from a severe asthmatic condition in which the attacks can be fatal and may occur with little or no warning at any time. Provided we have both adequate means for measuring the patient's respiration and the relevant computer model, we can predict the likelihood of occurrence of an attack. This will permit the medical staff to take precautions at the most appropriate time leaving them free at other times. With the use of a computer and the appropriate model it should be possible to evaluate automatically the state of any one patient in a group. This use of computers will help the medical staff to solve the priority problem by permitting them to give their attention to the patients requiring it most.

Only a few patients are at present benefiting from this type of work because the necessary instruments and computers, along with know-how on their use, are not yet available even in large modern hospitals and medical institutions. So far, the rapid advances in control technology have not been exploited in the study of living systems except in the few institutes specializing in these aspects of technology such as Imperial College of Science and Technology and the Autonomics Division of the National Physical Laboratory in the United Kingdom, MIT and Caltech in the United States and the Institute of Automatics and Telemechanics in Moscow. For biological systems research to be successful the close association between physicians, engineers, scientists, mathematicians and programmers is essential. In my own case this involves association with physicians in a number of London hospitals, both for obtaining new experimental data and in ensuring that the results of the work can be of benefit to patients, and also with my colleagues at the laboratory.

As more centres of automation spring up and more biologists and physicians begin to use the language of the control systems engineer, so the effects and benefits of this type of work must spread more widely. The extent and speed with which this occurs, especially in the clinical application, depends particularly on how readily the medical profession becomes more technologically orientated.

Bibliography

Meethan, A. R. and Hudson, R. A. (eds.) (1968) *Encyclopaedia of Linguistics, Information and Control*, Pergamon Press, Oxford.

Comroe, J. H. (1965) *Physiology of Respiration*, Year Book Med. Pub. Inc.

Milhorn, H. T. (1966) *The Application of Control Theory to Physiological Systems*, W. B. Sanders, London.

Hammond, P. H. (1967) "Living Control Systems", in *Electronics and Power I.E.E.*, **13**, pp. 338–42.

Priban, I. P. and Fincham, W. F. (1965) "Self-adaptive Control and the Respiratory System", in *Nature*, **208**, pp. 339–43.

Yamamoto, W. S. and Raub, W. F. (1967) "Models of the Regulation of External Respiration in Mammals", in *Computers and Biomedical Research*, **1**, pp. 65–104.

Reprinted from Priban, I. (1968) "Models in medicine", *Science Journal*, June, 1968.

15 Systems analysis and ecology

by M. B. Dale

Systems analysis is defined as the use of scientific method with conscious regard for the complexity of the object of study. It has strong relationships with problem solving, in that the same four phases – lexical, parsing, modelling, and analysis – are identifiable in both. Examination of each of these phases reveals some of the problems involved in the use of systems methods in ecology. A model of the precipitation–evaporation system is presented as an example Problems in experimenting with models of systems and with control, optimization, and comparison of such models are considered.

Introduction

Systems analysis has been presented as a desirable framework on which the investigation and comparison of ecosystems can be hung. This approach has been especially emphasized by the productivity subgroups of the International Biological Program (IBP). Claims of the importance of systems analysis are not restricted to ecology, for in other fields the results of employing these methods have been claimed to give additional insight and clarity (see e.g. Halmos and Vaughan 1950, Bush and Mosteller 1955, Glanzer and Glaser 1959, Orcutt 1960, Harary and Lipstein 1962, Keeney, Koenig and Zemach 1967). Examples of explicit use of systems methods in ecology are few (Olson 1963, Patten 1965, Holling 1966, Watt 1968), and it is by no means clear from these examples what systems analysis is, what it does, what restrictions it imposes, nor how the variety of ecology (or more precisely ecological methodology) can be attached to this framework. This paper attempts to clarify some of the questions an ecologist must answer and the problems he must resolve before using systems methods, and to introduce some of these methods in the context of a general systems approach. It does not provide the mathematical, statistical, and other details of the use of the methods, although it is hoped that sufficient references are included to enable the interested ecologist to obtain this information. Necessary definitions are provided and the general nature of systems considered. The relationship between systems analysis and problem solving is established, and the ecosystem is examined in the framework of the problem-solving processes. An example of a systems model is presented and the problems of investigating and manipulating systems and of organizing ecosystem descriptions are considered.

Systems analysis, systems, and ecosystems

Systems analysis

Systems analysis has rarely been defined when introduced into ecological studies. Watt (1968) suggests that it is the determination of those variables which are important in a system, and further adds that systems simulation, systems optimization, and systems measurement are other facets of the systems approach. Others, such as Priban (1968),* view model building as the essence of the systems approach. Morton (1964) has suggested that systems analysis is no more nor less than scientific method itself, and that the distinguishing feature of the systems approach is the conscious application of scientific method to complex organizations in order that no important factor be overlooked, a view expressed by Pascal as "error comes from exclusion". These viewpoints are not necessarily mutually exclusive. Systems analysis is the application of scientific method to complex problems, and this application is further distinguished by the use of advanced mathematical and statistical techniques and by the use of computers. The computers are used as "number-crunching" calculating machines and as convenient tools for modelling systems too complex for analytic solutions to be presently possible. This modelling function is of great importance in studies of complex natural systems, for, provided the model can be treated as representing the real system for the purposes of the investigation, experiments can be performed on the model with a consequent gain in control and rapidity of response. A good model will obviously contain the important variables, so Watt's comments are pertinent. Equally, if natural systems are complex than the modelling phase of the systems analysis will be emphasized and Priban's emphasis accepted. Morton's more general view, since it includes both the others, seems the most acceptable since it does not presuppose some a priori emphasis on certain parts of the analysis.

Systems

A system is a collection of interacting entities, or alternatively it is a collection of parts, together with statements

* [Ed.] Article 14 in this collection of readings.

on the relationships, of some kind, between these parts. The interpretation to be given to the entities is the choice of the investigator, but the entities need not, and in general are not, in one-to-one correspondence with "real" things. They can represent classes of things, or classes of processes if this seems necessary. The state of the system at some point in space and time is described by the values of properties of the entities, and all properties used to so describe the system are termed endogenous. Variables which affect the interrelationships between entities, but which are not included in the state description, are called exogenous and form the environment in which the system acts. If endogenous properties are interchanged with other systems outside the defined one, then the system is said to be "open" for these properties. If there is no import or export, the system is closed. Representations of systems can take a variety of forms. Perhaps the commonest is as a network (Ford and Fulkerson 1962, Harary, Norman, and Cartwright 1965) or as a matrix derived from such a network. An alternative mathematical representation is given by Rosen (1958), and an ecological example is the structural description diagrams of Dansereau, Buell, and Dagon (1966). Such general descriptions permit discussions about systems, but a computer program modelling a system is equally a representation of that particular system.

Any system is composed of subsystems defined for subsets of the entities. Each of these subsystems can be treated as a system in its own right, so that the definition of a system is recursive.* An open system, that is, one open for at least one property, can be considered as a subsystem of some "higher" order system (Cooper 1969), and since each subsystem can be decomposed into sub-subsystems, a hierarchy of systems is produced. A familiar example of such a hierarchy is . . . -organism–organ–tissue–cell–organelle– . . . Obviously some means is required to prevent infinite regress, and in practical work this termination depends on the fidelity of the model of the system to the "real" system. This fidelity requirement will be discussed later.

Finally, it is necessary to define an ecosystem. An ecosystem is a system open for at least one property, in which at least one of the entities is classed as living. This definition is very broad, but restrictions imposed by ecologists to limit this definition for particular studies have not received much consideration. It must be remembered that an ecosystem is a special case of the general system and will possess all properties of the general system. Thus there is no restriction on the number of properties which may be used to describe the system, although many ecologists have so restricted themselves, with the consequent introduction of difficulties with an excessively large number of exogenous variables. There is certainly no restriction to studies of productivity or energy transfer, although many applications of systems analysis in ecology have been on these problems. Population models are systems models and so is the physiognomic description of vegetation. However, since the preponderance of systems studies in ecology have been studies of productivity, it will be convenient to phrase examples in these terms.

* A simple example of recursive definition is the factorial of an integer number written $n!$. This can be calculated as follows:

$$n! = n \times (n - 1) \times (n - 2) \times (n - 3) \times \ldots \times 3 \times 2 \times 1$$

Equally the value can be calculated from the following rule.

$$n! = \textit{if } n = 1 \textit{ then } 1 \textit{ else } n \times (n - 1)!$$

Systems and problem solving

The recursive hierarchical nature of systems is closely paralleled in some theories of human problem solving (e.g., Simon and Newall 1962, Feigenbaum and Feldman 1963). Here an attempt is made to decompose insoluble problems into sub-problems. Any subproblems remaining insoluble are further decomposed, until hopefully all subproblems and their derivatives are soluble, when an attempt is made to reintegrate the solutions into a single solution of the original problem. The parallel between problem solving and systems can be drawn more closely, however. Ross (1967) distinguishes four phases in problem solving: (i) lexical, (ii) parsing, (iii) modelling, and (iv) analysis (see also Morton 1964). In systems analysis these same four phases can also be identified: (i) delimination of the entities or parts; (ii) the choice of relationships between entities which are of interest; (iii) the specification of the mechanism by which these interrelationships take place; and (iv) validation of the model of the system so produced and investigation of its properties.

Ross points out that the rules under which these phases are carried out must be agreed upon a priori, which is not a simple task. An obvious example of changing rules is given by the diversity of human language, which has an additional complication due to the possible existence of several scripts for one spoken language. Some of the phytosociological arguments on vegetation description appear to be arguments regarding rules of procedure, although the situation is complicated here because it is not clear that each system is intended to contain the same information. Ecologists must therefore agree on the rules to be used, otherwise comparison of systems will not be possible. Much of the difficulty lies in the choice of entities, and this will be discussed later.

The four phases of problem solving and systems analysis are used in the following sections as a framework in which to discuss problems in systems analysis.

The lexical phase

One of the most neglected problems in systems studies is the choice of the entities or parts which compose the system. It is commonly assumed that these are self-evident; yet the arguments which have taken place in areas such as the classification of organisms or vegetation concerning sampling, description, and measures of similarity suggest that this is not true. In taxonomy a hierarchy similar to the systems hierarchy is apparent – family, genus, species, etc., and taxonomists have agreed that while studies at any level are possible, the species level is in some way more important. It is by no means clear that the species level is a consistent level: the occurrence of "difficult" genera such as *Hieracium*, *Rubus*, or *Quercus*, and the varied degree of subtlety in characters used to describe and distinguish species in different families such as the Umbelliferae and the Magnoliaceae, attests some inconsistency. Yet the taxonomist has a distinct advantage over the ecologist in that there exists a generating system (Williams 1967), the genetic system, which constrains the possible variation, so that the lexical phase in taxonomy rests on the interpretation of genetic event patterns; that much of the genetic information available is not at the species level but within it is a practical problem though an unfortunate one. The ecologist has no such system presently available, and in the opinion of some ecologists there is no such system.

The choice of entities for the ecosystem is in part determined by the parsing phase, that is, by the nature of the relationships with which the system is concerned. The commonest choices have been between taxonomic, structural, and functional entities. Taxonomic is a convenient adjective to describe entities based on individual organisms, populations, and the commoner taxonomic categories of species, genera, and so on. Structural entities are based on life-form criteria, trees, shrubs, herbs, and bryoids providing a simple botanical example. Life-form criteria are in general more responsive to local environmental fluctuations than taxonomic criteria, since these latter employ characters selected to be invariant within taxa, wherever possible. Functional entities have perhaps received more attention in animal ecology, e.g., herbivore, carnivore, omnivore, although a variety of similar units exists in plant ecology, though less precisely defined, e.g., xerophyte, halophyte, and saprophyte. The definition of entities is not of course concerned with the ease of identification of these parts, although it may well be essential to provide common means of identifying the entities if different systems are to be compared. Of more consequence is the possibility of conversion from one set of entities to another. If one description employs structural categories and another taxonomic categories, how can the two be compared? *Liriodendron tulipifera* is a taxon which could certainly fall into the categories of shrub or tree, and on some definitions the seedlings would be classed as herbs. Even restriction to species as entities fails to resolve the problem, for this ignores all ecotypic and ontogenetic variation and the inconsistencies in the species level noted above.

It may be true that some ecosystems can only be compared at gross levels such as autotroph and heterotroph, for example marine and terrestrial systems. Yet because the United Kingdom has some 1,700 species of vascular plants, 900 bryoids, and various numbers of lichens and fungi, whereas Oak Ridge, Tennessee, has some 2,000 species of vascular plants alone, does not imply that comparisons of the two areas are only possible at some very gross level, even though the species complements are widely different. To demand that comparisons be possible with both very similar and very different ecosystems places severe restrictions on the possible choices of entities which can be employed to describe the systems. It may also be possible to describe systems in terms of a few simple ratios, such as the efficiencies which have been proposed, but this could equally reflect the well-known half-truth that biologists, when given two numbers, divide one by the other.

Functional entities do not resolve the problems of choice any more than structural or taxonomic entities. Omnivores, for example, are both herbivorous and carnivorous, while insectivorous plants are both autotrophic and heterotrophic. Nonliving materials within the ecosystem are less well served with possibilities, while still having problems of ontogeny and chemical equilibrium, such as that between the various forms of nitrogen in the soil. Perhaps a distinction between solid, liquid, and gaseous phases is possible, but is this all?

Since the first phase of a systems analysis is the choice of entities, it is very necessary that an ecologist give considered thought to these problems. In a study of a single system, the problems may well be less acute than when comparison of systems is necessary. But the choice of entities is the ecologists' task and must not be given to the systems analysts by default.

The parsing phase

The second phase is concerned with the definition of the relationships between the selected entities. These relationships can be of any kind and need not be restricted to materials. If has been common practice in ecology to assume that the relationships are those relating to material which the system can reorder or reallocate among its parts. The relationships can, however, be spatial or temporal and need not concern materials at all. Such relationships are important in physiognomic description of vegetation. In view of the present great interest in productivity, however, attention will be concentrated on ecosystems models produced by such studies, and the properties relevant to them. These properties include energy, biomass, carbon, mineral nutrients, populations, individuals, water, and possibly information (in the form of genetic material). It does not include diversity in the sense in which this has been commonly used in ecology (Margalef 1947), which is a measure of the distribution of some property or properties over the entities, or some subset of them, in a single state description. Changes in diversity can provide useful indices of changes in this distribution caused by exogenous variables.

There is no restriction in the definition of the ecosystem given earlier on the use of several properties to describe the state of the system. The description can be multivariate. The importance of interaction between properties in such a multivariate system can be seen in work on mineral nutrient interactions and their effects on yield (Figure 15.1). Such interactions are not of course limited to these particular variables. The difficulties of modelling and experimenting with multivariate systems do impose practical constraints on the investigator (Jacoby and Harrison 1962). The problem can be reduced to a univariate one by treating other properties as exogenous, although this increases the experimentation required. In comparing two systems which have been made univariate, it is essential to ensure that differences between systems do not become confounded with differences in treatment of the relocated exogenous variables.

In the context of the International Biological Program the property most favored has been energy fixed as carbon, usually as total biomass. This implies a preference against edible, palatable, or otherwise desirable biomass such as protein, which preference may not always be desirable. Thus wool growth in sheep may well be related to amounts of sulfur-containing protein rather than total biomass

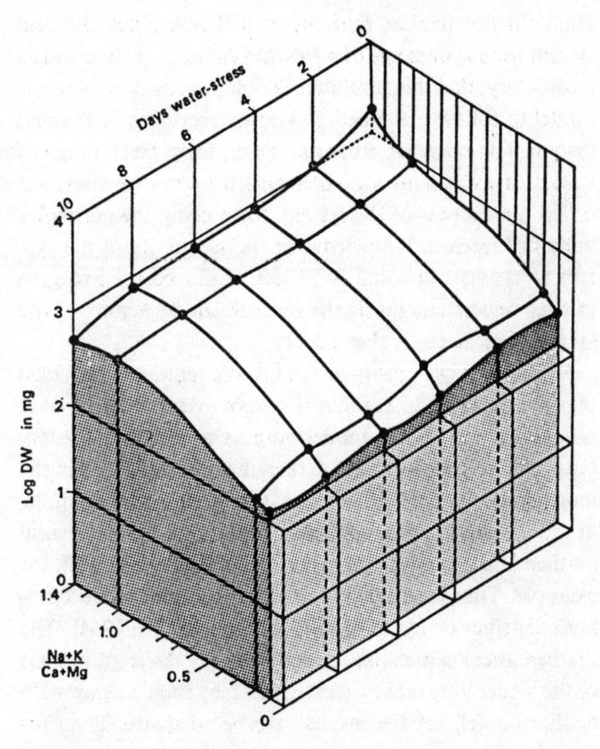

Figure 15.1 Dry weight response of *Atriplex inflata* after 8 weeks in relation to days of withholding water and the ratio (Na + K)/(Ca + Mg).

consumed. This is not to deny the importance of energy transfer as measured by total biomass, but merely to indicate that it alone will be but a partial representation of the "real" system and may not always be the most desirable. Other choices have been made including mineral nutrients, water, radioactive contaminants, and, with growing emphasis in the United States, pollutants in general. In all these cases, however, the possibility that a multivariate system might be more useful than the univariate one must be accepted and consideration given to the requirements of such a system model.

Modelling

Fidelity. Having fixed the entities and the properties, the next phase is the specification of the mechanisms by which changes in the system, that is in the distribution of the properties across the entities, take place. In choosing these processes, an attempt is made to make the model of the system "mimic" the real system, either to increase understanding of the system or to attain control of the system over some range of states. This difference in possible objective is characterized by differing degrees of fidelity.

High fidelity implies that the model resembles the real system for a wide range of states and changes in state and as a corollary, that this similitude is obtained by designing the model to follow presumed or known processes of the real system. The range of property values, for a given range of values of exogenous variables and for some entities, are called the outputs of the system, these being the particular values of interest. Knowledge of the sensitivity of the system to changes in some processes would be used to gain greater understanding of the mechanism of action of the system, and hence higher fidelity.

However, high fidelity is not always required. Provided that the model mimics the real system over some restricted range, that is, that the model outputs and the real system outputs are highly correlated, then the processes used in the model need not reflect the real system at all. As an example, in the description of spatial distributions of plants, several mathematical expressions may fit the data equally well, for example, Thomas' double poisson and the negative binomial distribution (Archibald 1948, Greig-Smith 1964). The mathematical expressions may well imply different models of the underlying real system, which may in fact agree with neither model, yet the results may be adequate. The simplest and most common ecosystem model is

Input–Ecosystem–Output

where the system itself is treated as a closed "black" box. This has been widely used in ecology (Van Dyne, Wright, and Dollar 1968), since it is the basis of univariate multiple regression. The attainment of high fidelity is expensive in time and in the effort required to obtain the precise and accurate data on which to build and validate the model. An analogy with sound reproduction is reasonably drawn. Telephone voice communication neither requires nor uses equipment necessary for the high-quality reproduction of music.* The complexity of model required to attain high fidelity must be matched by the quality of the data. To continue the sound-reproduction analogy, a scratched recording is still scratched on the best equipment. The collection of adequate and relevant data in studies of ecosystems will often be difficult if not impossible. For example, in studies of the interaction of radiant energy and a plant canopy, account must be taken of the spatial distribution of the stems and leaves. The collection of precise and detailed information on this feature is extremely difficult.

The processes to be defined must obviously depend on the choice of properties and entities. Changes in state of plant entities, for example, will require processes defining fixation of energy, carbon, nitrogen, and water, and other processes defining the reallocation of these properties among the entities. Some of the processes operate in sequence, the results of one forming an input to the next. They may also of course operate in parallel, that is, over the same time interval. This parallelism can be troublesome in some methods of investigating systems, such as simulation on digital computers which are essentially serial in operation. Special processing techniques (including special languages such as the SIMULA extension to ALGOL (Dahl and Nygaard 1966)) may be required.

The processes only change endogenous variables although they may employ both previous values of endogenous variables and exogenous variables in the calculation. It must also be realized that high fidelity in the definition of the processes does not guarantee high correlation between model and real outputs. The choice of exogenous variables also constrains the fidelity of the model. For example, consider two models of photosynthesis, one using mean day length to predict amount of carbon dioxide fixed, the other being more sophisticated and employing temperature, carbon dioxide, humidity, and radiation fluxes, together with data on spatial distribution of leaves to estimate the same value. The second might be expected to be of higher fidelity, yet by introducing appropriate stochastic variation into the first model it might be possible to make it of higher fidelity. This prediction requires less data, but the selection of the appropriate stochastic inputs would be troublesome. This emphasizes the importance of considering data-collection techniques when choosing the form of the processes (Watanabe and Abraham 1960). More than this, however, it reinforces the comment made earlier that the definition of the system in the ecologists' problem and is the result of interaction between available, or potentially available, information and the purposes for which the model is required. The modelling process may suggest areas where data-collection techniques might be improved so that a more faithful model becomes possible. One indication can be gleaned from economic models. The choice of interesting and practical models appears to be those models with 30–300 variables, with the experienced worker reducing the number (Forrester, *personal communication*).

Practical considerations. – Watt (1968) has presented a variety of approaches to the problem of defining processes; these by no means exhaust the possibilities. A very simple model of an ecosystem can be constructed consisting of four

* In statistics the problem of fidelity appears in the use of one distribution to provide an approximation to another. One example is the use of the normal distribution to approximate others such as the binomial, poisson, or Mann-Whitney U (see Siegel 1956).

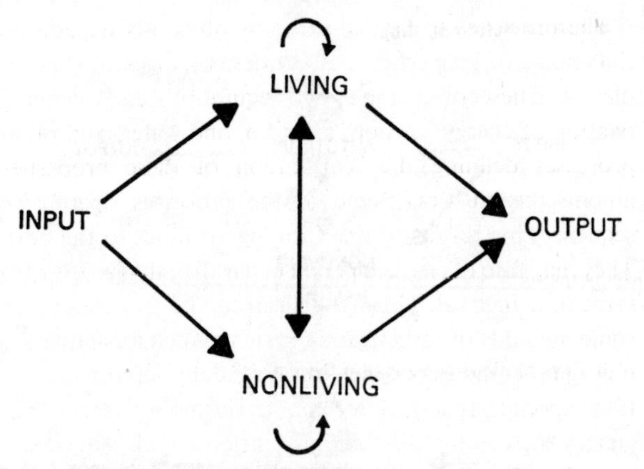

Figure 15.2 The primitive model.

	INPUT	LIVING	NONLIVING	OUTPUT
INPUT	0	a_{IL}	a_{IP}	0
LIVING	0	a_{LL}	a_{LP}	a_{LO}
NONLIVING	0	a_{PL}	a_{PP}	a_{PO}
OUTPUT	0	0	0	0

Figure 15.3 Transition matrix for the primitive model.

entities each of which represents a class and the properties and processes remain unspecified (Figure 15.2). The same system can be presented in the form of a transition matrix, where each a_{jk} represents the probability of transfer of a property between the jth and k^{th} entity (Figure 15.3). Similar matrices could be constructed for all properties, so that the entries can be interpreted as arrays of coefficients. To each of these there must be attached a corresponding process giving the next value of the a_{jk} in terms of the present and previous values of the whole matrix and any additional exogenous variables. The nature of these functions is, of course, of great interest to ecologists. The model does not include some features of human information transfer where questions of the value, reliability, and credibility of information are involved. It would include demographic or population models which form the bulk of ecological work on systems (see Kerner 1957, 1959, Whittle 1962, Bellman, Kagiwada, and Kalaba 1966, Garfinkel 1967*a*, *b*, Watt 1968, and others). Demographic models conveniently illustrate the duality between continuous and discrete models of systems. Many population models employ systems of differential equations which provide a continuous model of the system, including fractional values for the population total. But the population in most cases is discrete, being an integer number of individuals. Of the references given above only Whittle employs a discrete model (a discrete branching Markov process), probably because of the extra effort involved in the mathematics if restriction to integer solutions is imposed.

While demographic models can certainly be included in the ecosystem concept, it is also common to restrict the definition of ecosystem to models of the movement of materials or energy (e.g., Golley 1960, Olson 1963, Witherspoon, Auerbach, and Olson 1964, Patten 1965). If this restriction is accepted, then a slightly more complex model, as shown in Figure 15.4, hopefully would improve in fidelity over the primitive model of Figure 15.2. Autotroph in this model includes both energy and chemical fixation. Minor variants of this model in Figure 15.5 show its generality.

While the network representations of Figures 15.2, 15.4 and 15.5 are convenient visual models, mathematically the transition matrices corresponding to them are more easily handled. The meaning of such matrices can be considered geometrically. Consider a system with two entities and a single property. This can be represented as a point on a graph for any state. If the system changes due to change of an exogenous variable, then the point representing the system is displaced. A series of changes would trace a line, and the transition matrix contains the information describing this line. Of course in most models the graph is not in two dimensions, but the properties of the transition matrix still hold (see Keeney *et al.* 1967 for an extended description of this "state space" model).

The state of the system is a static description and the dynamics of changes in state are incorporated into the model by the processes. While some systems may only show changes in response to changes in exogenous variables, in many systems and certainly in ecosystems the changes in state are partly determined by the previous states of the system, that is, by its history, by means of "feedback" or "memory". This is of course also included in the processes by making these employ previous values of the endogenous variables in the calculations. If these processes themselves employ parameters which change with time, the system is evolutionary, whereas if the parameters do not change with time, the process is stationary. The difficulties introduced by considering evolutionary processes are such that the majority of models employ station-

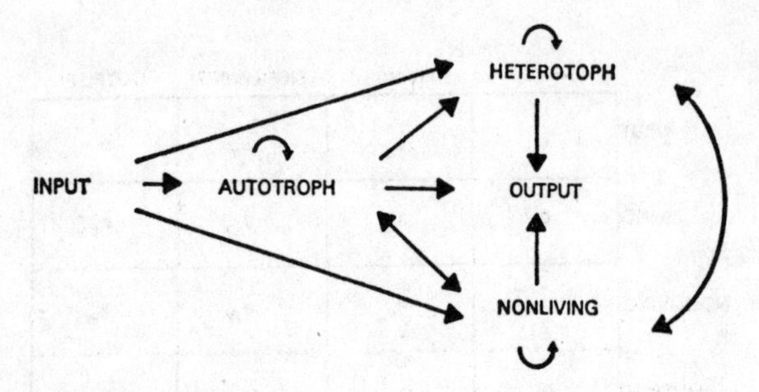

Figure 15.4 Developments of the model for non-demographic systems.

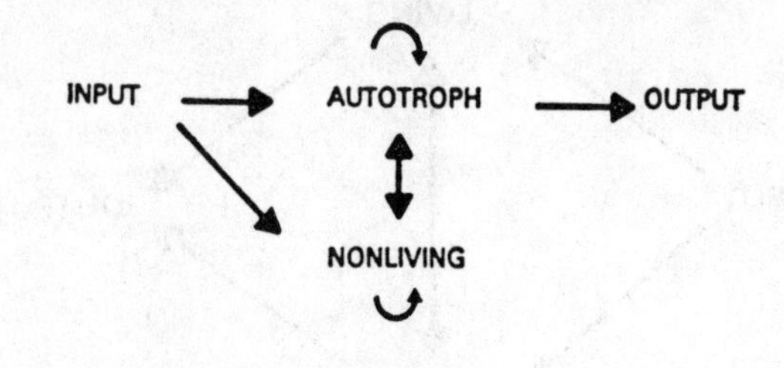

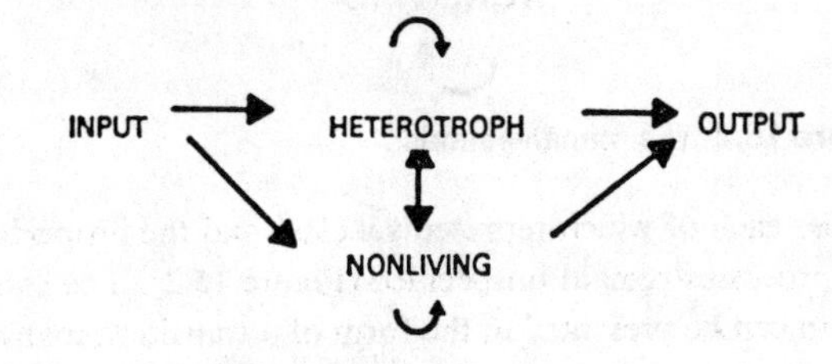

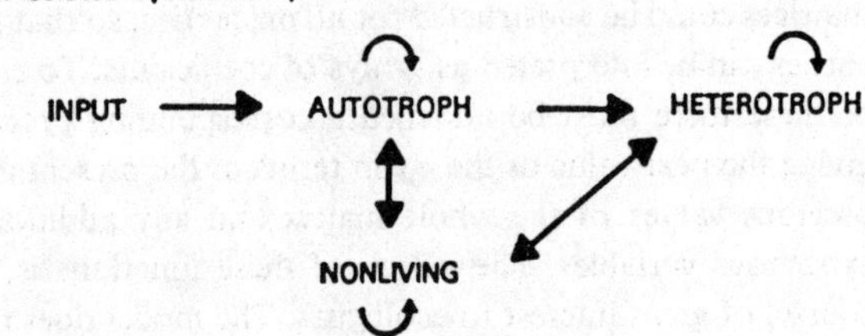

Figure 15.5 Examples of restricted systems.

ary processes, although it is fairly clear that real ecosystems are strictly evolutionary. There is little work on the mathematics of evolutionary processes, and most of this is recent and at a somewhat advanced mathematical level. It would certainly be possible to permit evolutionary processes in simulation models, but this would involve a large increase in time and effort in an already time-consuming method, since the initial state of the system must be specified very carefully for evolutionary processes. For many practical purposes over moderate time intervals the assumption of stationarity may be justified, although, as with multivariate models, the possibility of increasing fidelity by employing evolutionary models must be considered.

Analysis

The final stage of the systems approach is the analysis proper. This involves the solution of the model, in some sense, and the validation of the model outputs by comparing them to the real system outputs. In a few simple cases the model may be solved analytically using standard mathematical techniques. Models employing linear differential equations, for example, may be soluble, and in this case the sensitivity of the model to small changes in parameter values can also be calculated (Wilkins 1966). In general, however, no analytic solution will be available, and recourse must be made to the somewhat time-consuming simulation approach.

The likelihood that high fidelity will be desired suggests that stochastic models, incorporating random processes, will be preferable to deterministic models. This is due to the more realistic incorporation of variability in stochastic models and to the availability of estimates of the expected variability of the outputs. As an example of the greater realism of stochastic models, consider the spread of an infection through a population. Deterministic models suggest the existence of a critical population size at which there is a change from "no epidemic" to "epidemic" and the epidemic is of a fixed size. Stochastic models not only permit the epidemic to be variable in size, but also provide that, whatever the population size, there is an estimable chance of an epidemic occurring, and conversely of its not occurring (Bartlett 1960).

Simulation methods have been widely used both with analogue and with digital computers (Clymer and Graber 1964, International Business Machines Corporation 1966). Most of the ecological applications have used constant time increments, calculating the state of the model periodically. Less commonly event-orientated models have been attempted (Holling 1966). These essentially calculate the time interval between changes in state, so that periods when no change in state occurs require a constant computational effort independent of the length of the interval. Event orientation emphasizes the importance of recurrence intervals, which are ecologically important in determining survival times, where the event of "successful reproduction" and the event of "death" mark the intervals. Recurrence intervals are also important in migration and have been found useful in sampling vegetation (Williams, *personal communication*). Even if there are strongly periodic phenomena such as diurnal or annual cycles, an attempt to define events may force the modeller to consider his system in greater detail.

After a model of the system has been established and some means of investigating its responses has been provided, the crucial problem of the validation of the model remains. Validation may involve the functional form of the processes and the parameters supplied as constants to these processes, but primarily the interest lies in how well the model outputs mimic those of the real system, that is, in the fidelity of the model over the range of interest. If high fidelity is required there will usually be a process of successive approximation, with the model being progressively altered until the desired fidelity is obtained. This requires, of course, some measure of fidelity to assess the disparity between model and reality. Since the processes usually involve subsystems of the model, validating the processes is essentially also a process of measuring the fidelity of a system, in that the outputs of the subsystems to the complete system should presumably also be of high fidelity.

The difficulty of validating outputs depends on the features which it is desired to mimic. If only mean values must be estimated, the disparity can be measured by a test akin to Students' t, and various techniques are available to increase the precision of the comparison, mostly developed in Monte-Carlo studies (Hammersley and Morton 1956). These include such methods as Russian roulette, antithetic variables, and regression. If, however, the variance of the outputs or features of the transient response of outputs to particular changes in exogenous variables is required, the problem is more complex. The outputs form a correlated series of observations, and the comparison and investigation of such series present considerable statistical problems (see Quenouille 1957, Robinson 1967, Jenkins and Watt 1968). Rarely do ecological models specify which features of the output are to be reproduced by the model. Watt (1961) has provided examples of functions which produce outputs of given forms, and the use of least-squares surface fitting can also aid in the selection of possible functions to provide specific output forms. Perhaps the most general techniques are those of Wiener (1949), though these require a large amount of data.

Validation of the parameters of the processes involves searching the response surface of the model to obtain "best" estimates. A variety of techniques might be of use here, including those due to Hooke and Jeeves (1961), Spang (1962), and Marquardt (1963). The general problem of estimation in simulation studies has received most attention in engineering and management studies (e.g., Burdick and Naylor 1966, Fishman 1967, Fishman and Kiviat 1967), but it must be remembered that even if the model is validated, extrapolation beyond the limits of such validation is the responsibility of the ecologist. It would be foolish to say that such extrapolation is never justified, but the justification is not mathematical or statistical.

An autotroph system

A diagrammatic representation of a precipitation–evapotranspiration (PET) system is used as an example of the models employed in systems analysis (Figure 15.6). This is not the only model of this system since both Crawford and Linsley (1966) and Hufschmidt and Fiering (1966) incorporate simple expressions in their larger models to represent the whole of the PET system. Equally, more detailed models might be built up from the equations describing the transfer of heat and water vapor between leaf and atmosphere, and corresponding detailed study of the distribution of stem and leaves. The fidelity of the model will depend on the specifications of the processes by which the transfers of the property, here water, between entities is to be made. Such specifications have been provided and the resulting model converted to a computer program. The diagram shows only the connections between entities which were considered in the modelling.

Two subsystems are easily identified, one modelling the entry of water into the soil, the other modelling its return to the atmosphere. These interact at two points, since changes in the amount of leaves (and their distribution) will affect interception, and changes in the amount of roots will affect water extraction from the soil by the plant. Conversely, both leaf and root growth will be dependent on water availability.

Three other systems are explicitly included. The photosynthetic system requires inputs from the leaves and will feedback to both leaves and roots. This feedback is accomplished by the growth system which is responsible for the partition of photosynthate between the parts of the plant. The atmospheric system provides the source of water and the radiation which finally controls the loss of water. A fourth system could easily be added to introduce the effects of topography on input and output of water as runoff to or from other areas, and possibly erosion effects of such runoff. All the systems here operate in parallel in that they all operate simultaneously.

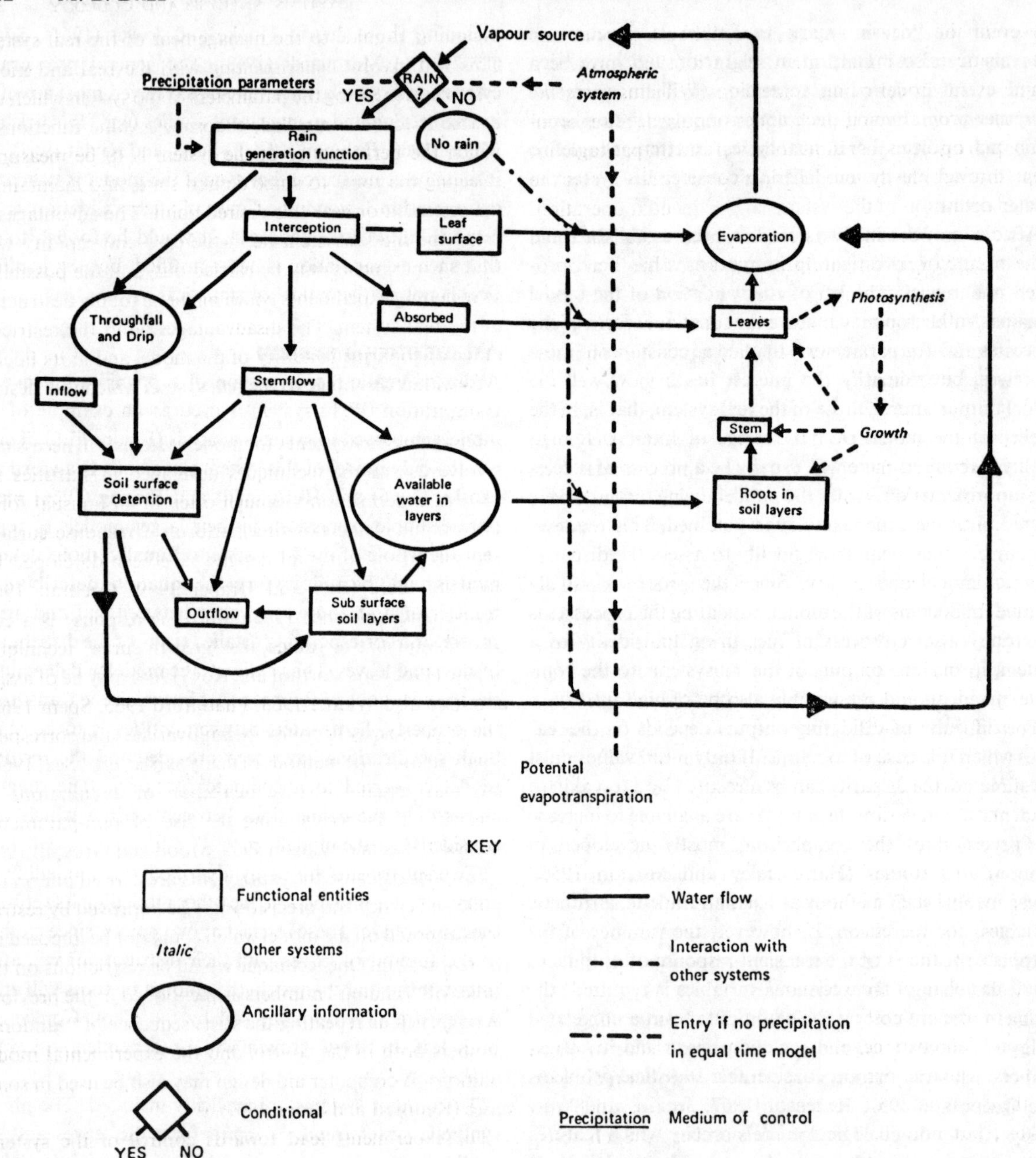

Figure 15.6 The PET system.

An equal time model of this system would review the description of the entities, which in this case would be the water contents and size of plant parts, at periodic intervals recalculating when necessary. During these periods the state is assumed to be constant, but the interval can be arbitrarily small at the expense of more computation. Given the periodicity of instrumental recording this might be acceptable, but the existence of continuous recorders permits the event-orientated model to be investigated.

In the event-orientated model time is variable, but during the interval between events the processes are assumed to proceed in a determinate manner. The events here would mostly be effects on the rates of water movement and on the growth rates. The systems and subsystems need not operate synchronously, each having its own event timing. Thus the photosynthetic system would show no events during darkness, and the growth system might show seasonal and ontogenetic effects. Dahl and Nygaard (1966) present a simple event model of an epidemic which illustrates the computer-programming techniques requir^d. The event technique operates as if it followed small packages of water through the system and in this way transforms the parallel operation of the systems to a sequential operation. The choice between the two models, event or periodic time, will finally depend on the information available and the user's preferences. The PET system has in fact been programmed in both periodic and event-time forms (Cooper, *personal communication*), with the latter proving computationally more efficient.

With either approach to modelling the models may be deterministic or stochastic. The stochastic model effectively replaces certain constants in the deterministic model with random variables drawn from appropriate statistical distributions. This drawing need not in fact be strictly random since by careful manipulation of the technique of drawing, the precisions of comparisons between the performance of the model under varying conditions may be increased. The technique is related to stratified sampling. While the appropriate distributions to use are relatively specific to individual problems and, indeed, form one of the most difficult parts of model building, for very rare events it may be possible to make use of the fact that extreme values have only three possible distributions (Gumbel 1958). This might appear a marginal advantage as there remains some choice to be made. However, this particular problem is usually solved by an automatic choice of the exponential distribution probably because of its ease of computation, and without regard for the alternatives. The systems analyst may indeed accept the simple exponential distribution, but he should be aware of the alternatives.

One of the interesting possibilities of controlling the "random" numbers depends on the nonexistence of such numbers. Random numbers are, in fact, pseudorandom in that they pass some of the tests of randomness which are possible, but not all. The infinite number of such tests makes it impossible to know if any set of numbers is random. Provided that the numbers are random for the tests employed, nonrandomness can be incorporated to reduce the effects of unimportant sources of variation. Tocher (1963) considers the possibilities in some detail.

Experimentation control and optimization

The techniques and considerations of the previous sections will hopefully lead to a valid model of the ecosystem. We will now consider means of using such a model of a single system as a guide to the management of the real system. This will involve experimenting with the real and model systems, identifying the parameters of the system which will enable it to be controlled, choosing a value function by which the performance of the system is to be measured, selecting the route to some desired state, and maintaining the system at or near this desired point. The advantages of using the model system lie in the ease and rapidity with which experiments may be carried out, and the possibility of including experiments which might be totally destructive in the real system. The disadvantages lie in the restricted range of confirmed validity of the model and in its fidelity even within this range to the real system which it is desired to control.

Designing experiments for model systems will necessarily involve use of the techniques developed in statistics for efficient experimenting, though often in an unusual form. The output of the experiments forms a response surface, and special experimental designs have been developed for studying there (Box and Draper 1959, Cochran 1963, Draper and Lawrence 1965). Since the response is a correlated sequence of values, the "growth curve" techniques of Rao (1965) and Potthof and Roy (1964) may be of assistance (see also Whittle 1963, Phattaford 1965, Spent 1967). The response may of course be multivariate, and correspondingly so must the methods of analysis (see, e.g., Seal 1964), and may equally involve relaxation of assumptions of normality thus necessitating the use of non-parametric methods (Box and Watson 1962, Mood and Graybill 1963, Tiku 1964). Because the experiments are carried out on the model, efficiency and precision may be improved by restrictions imposed on the model which could not be imposed on the real system. One technique would be restrictions on the choice of "random" numbers as mentioned in the previous section, such as repeating the same sequence of "random" numbers both in the control and the experimental model solutions. A computer aid design may also be used in some cases (Kennard and Stone 1969).

The experiments lead towards control of the system, enabling a manager to manipulate the system towards some point and to maintain the system in the neighborhood of this point. Almost always there will be constraints on the actions available to the manager, such as avoiding certain states, restrictions on materials, and so on. General mathematical control principles are known (e.g., Pontryagin's continuous maximization principle, for which see Fan 1966), but these have proved difficult to apply in practice. Control implies the existence of a desired state and some means of assessing the importance and hence the size of any deviation from this state. Related to this is the need to measure the effects of any control operation. These subjects have been studied in detail in operations research (Bellman 1961, Muhzam 1963, Box and Tiao 1965). Maintenance of the system near the desired point means the control of variations in the outputs of the system. For example, management of a watershed to have flow proportional to demand would be ideal for hydroelectric power generation, where overproduction is worth little and underproduction is extremely costly. The quest for high fidelity models seems to result from the detailed control necessary, coupled with an assumption that the control of endogenous variables will be more selective, more efficient, and less expensive than control of exogenous variables. The truth of this assumption is debatable since the cost of obtaining the required detail in the model must also be considered. Both endogenous and exogenous variables can be manipulated in many systems.

Efficient control will usually imply the selection of important variables for which several techniques have been developed. Sensitivity analysis (Radanovic 1966, Wilkins 1966) is widely used to study the effects of small perturbations where the effects can be assumed to be nearly linear. Other approaches for isolating important variables exist, such as stepwise multiple regression, canonical correlation analysis (Kendall 1957), multiple predictive analysis, and two-parameter numerical taxonomy (Macnaughton-Smith 1965) or factor analysis (Lawley 1940, Harman 1966). Box and Jenkins (1962) have considered some statistical aspects of control.

The development of optimal control policies is one portion of the overall control process. The goal is provided by a "supersystem", and in this case the individual optimality of subsystems does not ensure the optimality of the system as a whole. An ecosystem might be evaluated in turn (in an appropriately organized society) by an administrative system, a political system, and a social system which employs a judicial system to enforce its control measures (Price 1965, Bulkley and McLaughlan 1966). The goals of all these evaluating systems must be defined and may often be conflicting. For the ecologist it is an evaluating function which is required rather than the goal itself, and this function may constrain the operation of the model and the operations of the managers. The existence of such an evaluating function is crucial, but two further problems are also apparent.

The first of these concerns the existence of "local" optima, which makes the search for the overall optimum more complex. Methods such as linear programming and its extensions to integer, quadratic, stochastic, and dynamic programming (Bellman 1957, Churchman, Ackoff, and Arnoff 1961, Wolfe, 1962, Dantzig 1963, Watt 1963) have proved useful initial guides even if the system models do not always precisely fit the mathematical specifications (see Serck-Hanssen 1963, Watt 1963, Petrini 1964, Heady and Egbert 1964). Other workers have used statistical decision theory in efforts to determine optimal policies for action (e.g., Dillon and Heady 1960, Findler 1966).

The second problem is that of moving from the present state to the optimal one. Here the techniques of network analysis as planning aids are useful, including critical path analysis, resource allocation scheduling, program evaluation, and review techniques and transportation methods (Hein 1967, Hasse 1960, Davis 1965, Martino 1965, Davis 1968). Since the majority of these methods require computer assistance with the calculation and the systems model itself will often be in the form of a computer program, it is interesting to speculate on what additional information is required to enable the computer to design its own experiments, and after analysis to report both the optimal point and the method of reaching it. Certainly cost functions, value functions, and constraints are required, but whether this is sufficient information is not known. As yet the evaluation of ecosystems is at a fairly gross level, and the ecologist is educating himself and others in the extent and degree of complexity inherent in ecosystem management, while avoiding the grosser catastrophes.

Comparison and organization of ecosystems

Some of the problems of comparing ecosystems will be considered briefly. Such comparisons are desirable partly because of the spatial and temporal variation between systems, and partly because as a "pure" science ecology will include the study of patterns in ecosystems.

While it would be possible to extend the description of a system to include those with which it interacts, this will often be impracticable. The pattern of ecosystems with respect to environmental factor, and the processes of successional change are both areas where the comparison of ecosystems is desirable. Such comparisons have for the most part been made by comparing the diversity of the systems as measured by a single property. In vegetation studies this has commonly meant comparison of species lists. The emphasis placed on functional entities by Lindeman (1942) and the increasing use of indicator species has not replaced the taxonomic comparisons, and the success of floristic methods such as those of Heikurainen (1964) suggests that there is strong relationship between functional and taxonomic classes. For some purposes it may be necessary to reconsider presently unfashionable entities, such as the synusiae of Lipmaa (1939).

The process of comparison and the organization of the resulting information to exhibit the patterns of ecosystem structuring is itself a systems process. Clowes (1967), in discussing similar problems in the computer processing of pictures, again distinguishes the four phases: the definition of parts, the provision of a grammar of parts (parsing phase), the representation of this part in relationship to structure in the machine (modelling), and the final analysis of the picture representation (analysis phase). A formal process of comparing ecosystems will itself involve these four phases, although the last analytic phase will be some numerical organization method such as classification, ordination, or spectral analysis (Robinson 1967, Jenkins and Watt 1968). The ecological difficulties all lie in the selection of the entities or parts and the selection of the relationships between the parts which are of interest in the particular study. Simple examples of relationships important in some areas of ecology are the concept of "epiphyte", which involves the relationship "growing on", and the concept of stratification of vegetation involving the spatial relationships "above" and "below". Selection of the appropriate relationships from the many available is a major ecological problem.

While the processing of the ecosystem description is possible, this is not the place to discuss the means available to represent ecosystems and the techniques necessary to compare the complex structures. The majority of the problems so far encountered in this area have been solved, in the sense that something can be done, although the ecological implications of the available solutions is not always clear.

Conclusion

The questions to be asked of this brief account of systems analysis fall into three categories. First, what additional knowledge must the ecologist acquire before he can use systems methods? Second, what ecological questions must be answered before he can apply the methods? Third, what can he hope to gain by using such methods? These will be considered in order.

It is apparent that systems analysis includes a wide variety of mathematical and statistical techniques and borders many areas, including computation, picture processing, language processing, and problem solving. The ecologist need not be fluent in all these areas, but some means of communicating between them seems desirable. The methods used in systems analysis are rarely phrased in ecological language, and the ecologist will certainly have to phrase his questions in non-ecological terms if the developers of the methods are to assist him.

The ecological questions rest on the need for this translation, for the ecologist defines the problems in which he is interested and must interpret them to the assisting workers. For systems he must specify the parts and the relationships and be prepared to modify these definitions in the light of data-collection problems and the fidelity requirement. If he is attempting to control or modify an ecosystem in the light of his models, he must have the desired objective stated, some means of evaluating departures from this state, and some idea of the external constraints imposed on the system and its managers. As an example where the objective function has been variously interpreted, consider the problem of controlling fire in forests. Australian foresters are at present recommending frequent controlled burning as a means of reducing fire hazards. Such a solution has one disadvantage, i.e., the frequency of burning increases due to the selection of rapidly recovering and fire-tolerant species. It also ignores the problems raised by loss of nutrients due to burning and the effects of such losses on the productivity of the trees, since the environment is already nutrient poor. As a solution to the problem of reducing fire risks immediately, controlled burning is probably acceptable, but this is in fact only part of the system.

The gains to be expected from a systems approach come from the precise statement of the problems and the discipline imposed by an ordered approach to the complexities of the real system. It is unlikely that an optimal solution to any problem will be attained directly, a process of successive approximation being likely. That the discipline is helpful can be seen from experience with one technique of management, the program evaluation and review technique (PERT, see Davis 1968). This technique has been credited with saving large sums of money, yet on closer inspection the method consists of little more than an explicit statement of what goes on and in what order! It should also be clear, however, that systems analysis is not a panacea, and its use will involve the ecologist in extending his knowledge, biological and other, before gaining much reward. Hopefully the use of systems methods will prevent ecologists from joining those "who saw the effect but not the cause".*

Acknowledgments

It is a great pleasure to acknowledge the advice and aid of Professor C. F. Cooper under whose auspices I worked at

* St Augustine Contra Pelagium IV 60.

the University of Michigan as a participant in a research project sponsored by Cooperative Research, U.S. Department of Agriculture. It is also a pleasure to record my thanks to Dr. J. Olson of Oak Ridge National Laboratory and other members of the discussion group on systems analysis. My thanks to Professor Cooper again for permission to use the PET model and to Dr. D. J. Anderson of the Australian National University for the response surface diagram.

Bibliography

Archibald, E. E. A. (1948) "Plant populations I. A new application Neyman's contagious distribution", *Ann. Bot.* (London) N.S. **12**, 221–35.

Bartlett, M. S. (1960) *Stochastic population models in ecology and epidemiology*, Methuen, London.

Bellman, R. (1957) *Dynamic programming*, Princeton Univ. Press, Princeton, N.J.

— (1961) *Adaptive control processes: a guided tour*, Princeton Univ. Press, Princeton, N.J.

Bellman, R., Kagiwada, H. and Kalaba, R. (1966) "Inverse problems in biology", *J. Theor. Biol.* **11**, 164–67.

Box, G. E. P., and Draper, N. R. (1959) "A basis for the selection of a response surface design", *J. Amer. Statist. Ass.* **54**, 622–54.

Box, G. E. P. and Jenkins, G. M. (1962) "Some statistical aspects of adaptive optimisation and control", *J. Roy. Statist. Soc., Ser. B*, **24**, 297–343.

Box, G. E. P., and Tiao, G. C. (1965) "A change in level of a nonstationary time series", *Biometrika* **52**, pp. 181–92.

Box, G. E. P., and Watson, G. S. (1968) "Robustness to non-normality of regression tests", *Biometrika* **49**, 93–106.

Bulkley, J. W., and McLaughlan, R. T. (1966) *Simulation of political interaction in multiple purpose river basin development*, Mass. Inst. Technol., Dep. Civil Eng. Hydrodyn. Lab. Rep. 100.

Burdick, D. S., and Naylor, T. H. (1966) "Design of computer simulation experiments for industrial systems", *Commun. Ass. Comput. Mach.* **9**, 329–39.

Bush, R. R., and Mosteller, F. (1955) *Stochastic models for learning*, John Wiley and Sons, Inc., New York.

Churchman, C. W., Ackoff, R. L. and Arnoff, E. L. (1961) *Introduction to operations research*, John Wiley and Sons, Inc., New York.

Clowes, M. B. (1967) "Perception, picture processing and computers". Collins, N. L. and Mitchie, D. (ed.) *Machine intelligence I*, Oliver and Boyd, Edinburgh & London.

Clymer, A. B. and Graber, G. F. (1964) "Trends in the development and applications of analog simulations in biomedical systems", *Simulation* **4**, 41–58.

Cochran, W. G. (1963) *Sampling techniques*, John Wiley and Sons, Inc., New York.

Cooper, C. F. (1969) "Ecosystem models in watershed management". Van Dyne, G. M. (ed.) *The ecosystem concept in natural resource management*. Academic Press, New York.

Crawford, N., and Linsley, R. (1966) *Digital simulation in hydrology. Stanford Watershed Model IV.* Stanford Univ., Dept. Civil Eng. Tech. Rep. 39.

Dahl, O. J., and Nygaard, K. (1966) "SIMULA: an ALGOL-based simulation language", *Commun. Ass. Comput. Mach.* **9**, 671–78.

Dansereau, P., Buell, P. F., and Dagon, R. (1966) "A universal system for recording vegetation", *Sarracenia* **10**, 1–64.

Dantzig, G. S. (1963) *Linear programming and extensions*, Princeton Univ. Press, Princeton, N.J.

Davis, E. W. (1965) "Resource allocation in project network models: a survey", *J. Ind. Eng.* **14**, 177–88.

David, J. B. (1968) "Why not PERT for your next resource management problem", *J. Forest* **66**, 405–8.

Dillon, J. L., and Heady, E. O. (1960) "Theories of choice in relation to farmer decision", *Iowa State Univ. Agr. Exp. Sta. Res. Bull.* **485**.

Draper, N. R., and Lawrence, W. E. (1965) "Designs which minimise model inaccuracies: cuboidal regions of interest", *Biometrika* **52**, 111–18.

Fan, Liang-Tseng (1966) *The continuous maximum principle*. John Wiley and Sons, Inc., New York.

Feigenbaum, E. A., and Feldman, J. (1963) *Computers and thought*, McGraw-Hill Book Co., Inc., New York.

Findler, N. V. (1966) "Human decision-making under uncertainty and risk: computer based experiments and a heuristic simulation program", *Proc. A.F.I.P.S. 1965 Fall Joint Computer Conf.*, **1**, 737–52.

Fishman, G. S. (1967) "Problems in the statistical analysis of simulation experiments: the comparison of means and the length of sample records", *Commun. Ass. Comput. Mach.* **10**, 94–9.

Fishman, G. S., and Kiviat, P. J. (1967) "The analysis of simulation generated time series", *Manag. Sci.* **13**, 525–57.

Ford, L. R., Jr., and Fulkerson, D. R. (1962) *Flows in networks*, Princeton Univ. Press, Princeton, N.J.

Garfinkel, D. A. (1967*a*) "A simulation study of the effects on simple ecological systems of making rate of increase of population density dependent'', *J. Theor. Biol.* **14**, 46–58.

—— (1967*b*) "Effect on stability of Lotka-Volterra ecological systems of imposing strict territorial limits on populations", *J. Theor. Biol.* **14**, 325–27.

Glanzer, M., and Glaser, R. (1959) "Techniques for the study of group structure and behaviour. 1. Analysis of structure", *Psychol. Bull.* **56**, 317–32.

Golley, F. B. (1960) "Energy dynamics of a food chain of an old field community", *Ecol. Monogr.* **30**, 187–206.

Greig-Smith, P. (1964) *Quantitative plant ecology*, Butterworth, London.

Gumbel, E. L. (1958) *Statistics of extremes*, Columbia Univ. Press, New York.

Halmos, P. R., and Vaughan, H. E. (1950) "The marriage problem", *Amer. J. Math.* **72**, 214–15.

Hammersely, J. M., and Morton, K. W. (1956) "A new Monte Carlo technique: antithetic variables", *Proc. Camb. Phil. Soc.* **52**, 449–75.

Harary, F., and Lipstein, P. (1962) "The dynamics of brand loyalty: a Markovian approach", *Oper. Res.* **10**, 19–40.

Harary, F., Norman, R. Z., and Cartwright, D. (1965) *Structural models: An introduction to the theory of directed graphs*, John Wiley and Sons, Inc., New York.

Harman, H. H. (1966) *Modern factor analysis*, 2nd ed. Univ. Chicago Press, Chicago.

Hasse, M. (1960) "Über die Behandlung Graphen theoretischer Probleme unter Verwendung der Matrizenrechnung", *Wiss. Z. Tech. Univ. Dresden* **10**, 1313–16.

Heady, E. O., and Egbert, A. C. (1964) "Regional programming of efficient agricultural patterns", *Econometrika* **32**, 374–86.

Heikurainen, L. (1964) *Suptyypien Ojituskelpoisus: metsänkasvatusta silmälläpitäen*, Kirjayhytyma, Helsinki.

Hein, L. W. (1967) *The quantitative approach to managerial decision*, Prentice-Hall, New York.

Holling, C. S. (1966) "The functional response of invertebrate predators to prey density", *Mem. Ent. Soc. Can.* **48**, 1–85.

Hooke, R., and Jeeves, T. A. (1961) "Direct search solutions of numerical statistical problems", *J. Ass. Comput. Mach.* **8**, 212–29.

Hufschmidt, M. M., and Fiering, M. B. (1966) *Simulation techniques for design of water resource systems*, Harvard Univ. Press, Cambridge, Mass.

International Business Machines Corporation (1966) *Bibliography on simulation*, Report 320-0926-0, White Plains, N.Y.

Jacoby, J. E., and Harrison, S. (1962) "Multivariable experimentation and simulation models", *Naval Res. Log. Quart.* **9**, 121–36.

Jenkins, G. M., and Watt, D. G. (1968) *Spectral analysis and its applications*, Holden Day, San Francisco, Calif.

Keeney, M. G., Koenig, H. E., and Zemach, R. (1967) *State space models of educational institutions*, Michigan State Univ. Div. of Engineering Research, East Lansing, Mich.

Kendall, M. G. (1957) *A course in multivariate analysis*, Griffin, London.

Kennard, R. W., and Stone, L. A. (1969) "Computer aided design of experiments", *Technometrics* **11**, 137–48.

Kerner, E. H. (1957) "A statistical mechanics of interacting biological species", *Bull. Math. Biophys.* **19**, 121–46.

—— (1959) "Further considerations on the statistical mechanics of biological association", *Bull. Math. Biophys.* **21**, 217–55.

Lawley, D. N. (1940) 'The estimation of factor loadings by the method of maximum likelihood", *Proc. Roy. Soc. Edinb.*, a, **60**, 64–82.

Lindeman, R. L. (1942) "The trophic dynamic aspect of ecology", *Ecology* **23**, 399–418.

Lippmaa, T. (1939) "The unistratal concept of plant communities', *Amer. Midland Natur.* **21**, 111–45.

Macnaughton-Smith, P. (1965) *Some statistical and other techniques for classifying individuals*. Home Office Res. Unit Rep. 6, H.M.S.O., London.

Margalef, D. R. (1947) "Information theory in ecology". *Mems. R. Acad. Barcelona* **23**, 373–440. (*Trans. in Gen. Systems* **3**, 36–71. 1958.)

Marquardt, D. W. (1963) "An algorithm for least squares estimation of nonlinear parameters", *J. Soc. Ind. Appl. Math.* **11**, 431–41.

Martino, R. L. (1965) "Advances in network techniques: an introduction to MAP", *Data Process.* **8**, 231–57.

Mood, A. M., and Graybill, F. A. (1963) *Introduction to the theory of statistics*, McGraw-Hill Book Co., Inc., New York.

Morton, J. A. (1964) "From research to industry", *Int. Sci. Technol.*, May 1964, 82–92, 105.

Muhzam, H. (1963) *On multivariate trends*, Paper presented to the 5th Int. Biometric Conf., Cambridge, England.

Olson, J. S. (1963) "Energy storage and the balance of producers and decomposers", *Ecology* **44**, 322–31.

Orcutt, G. H. (1960) "Simulation of economic systems", *Amer. Econ. Rev.* **50**, 893–907.

Patten, B. C. (1965) *Community organization and energy*

relationships in plankton. Oak Ridge Nat. Lab. Rep. ORNL-3634.

Petrini, P. (1964) "Competition between agriculture and forestry under Swedish conditions", *Lantbrukshägskolansannalar* **30.**

Phattaford, R. M. (1965) "Sequential analysis of dependent observations", *Biometrika* **52**, 157–65.

Potthof, R. F., and Roy, S. N. (1964) "A generalized multivariate analysis of variance model useful especially for growth curve problems", *Biometrika* **51**, 313–26.

Priban, I. P. (1968) "Forecasting failure of health", *Sci. Cult.* **34**, 232–5.

Price, D. K. (1965) *The scientific estate*, Harvard Univ. Press, Cambridge, Mass.

Quenouille, M. H. (1957) *Analysis of multiple time series*, Griffin, London.

Radanovic, L. (ed.) (1966) "Sensitivity methods in control theory", *Proc. Int. Symp. Dubrovnik, Yugoslavia*, Pergamon Press, New York.

Rao, C. R. (1965) "Theory of least squares when the parameters are stochastic and its application to the analysis of growth curves", *Biometrika* **52**, 447–58.

Robinson, E. A. (1967) *Multichannel time series analysis*, Holden Day, San Francisco, Calif.

Rosen, R. (1958) "The representation of biological systems from the standpoint of the theory of categories", *Bull. Math. Biophys.* **20**, 317–41.

Ross, D. T. (1967) "The AED approach to generalized computer-aided design", *Proc. Ass. Comput. Mach. National Meeting* **1967**, 367–85.

Seal, H. L. (1964) *Multivariate statistical analysis for biologists*, Methuen, London.

Serck-Hanssen, J. (1963) "A programming model for a fishing region in northern Norway", *Regional Science Association Papers* **12**, 107–18, Lund Congress.

Siegel, S. (1956) *Nonparametric statistics for the behavioural scientist*, John Wiley and Sons, Inc., New York.

Simon, H. A., and Newall, A. (1962) "Simulation of human thinking", 95–131, in M. Greenberger (ed.), *Computers and the world of the future*, Mass. Inst. Technol. Press, Cambridge, Mass.

Spang, H. A. (1962) "Review of minimization techniques for non-linear functions", *Soc. Ind. Appl. Math. Rev.* **4**, 363–5.

Spent, P. (1967) "Estimation of mean growth curves", *J. Theor. Biol.* **17**, 159–73.

Tiku, M. L. (1964) "Approximating the general non-normal variance ratio sampling distribution", *Biometrika* **51**, 83–95.

Tocher, K. D. (1963) *The art of simulation*, English Universities Press, London.

Van Dyne, G. M., Wright, R. G., and Dollar, J. F. (1968) *Influence of site factors on vegetation productivity.* ORNL-TM 1974 Contract No. W-7405-eng-26. Oak Ridge National Laboratory.

Watanabe, S., and Abraham, C. T. (1960) "Loss and recovery of information by coarse observation of stochastic chain", *Information and Control* **3**, 248–78.

Watt, K. E. F. (1961) "Mathematical models for use in insect pest control", *Can. Entomol. Suppl.* **19**, 1–62.

—— (1963) "Dynamic programming, 'Look Ahead' programming and the strategy of insect pest control", *Can. Entomol.* **95**, 525–36.

—— (1968) *Ecology and resource management: a quantitative approach*, McGraw-Hill Book Co., Inc., New York.

Whittle, P. (1962) "Topographic correlation, power-law covariance functions and diffusion", *Biometrika* **49**, 305–12.

—— (1963) *Prediction and regulation*, English Universities Press, London.

Wiener, N. (1962) *The extrapolation, interpolation and smoothing of stationary time series*, Mass. Inst. Technol. Press, Cambridge, Mass.

Wilkins, R. D. (1966) "General time varying systems error sensitivity analysis", *Commun. Ass. Comput. Mach.* **9**, 855–9.

Williams, W. T. (1967) "Numbers, taxonomy and judgement", *Bot. Rev.* **33**, 379–86.

Witherspoon, J. P., Auerbach, S. I., and Olson, J. S. (1964) "Cycling of caesium-134 in white oak trees", *Ecol. Monogr.* **34**, 403–20.

Wolfe, P. (1962) "Recent developments in non-linear programming", 156–87, in Alt, F. L., and Ruhinoff, M. (eds.), *Advances in computers* **3**.

Section V

Introduction

This last section contains four papers which show how systems ideas can be extended to different areas. Martin Davies applies systems ideas to social work and although his article repeats many of the points made in other articles in this collection, it is instructive to compare the examples he uses with those used elsewhere. The application of systems thinking to a different, but related, social area to that of Buckley (Article 9) helps to show how the sociologist can interpret systems concepts in his field and the reader is recommended to compare Davies' exposition with Buckley's.

Van Court Hare Jr. presents an illuminating series of examples which reveal many of the problems involved in implementing systems thinking in practice; these speak for themselves. The final two papers deal with applications in geography. In Haggett's contribution, which is a chapter from a book, the extract only briefly discusses systems ideas and spends more time on the role and use of models. Modelling is an essential system technique and Haggett summarizes their use and value succinctly. Mabogunje's article is a detailed study of a specific area, migration. We have included it because it shows how systems ideas can alter the perspective which an academic discipline normally takes of a problem or area and how system thinking enables a subject to extend its range and to cross the boundaries between academic subject areas.

Associated Reading

Case studies of systems applications can be found in the various issues of the Journal of Systems Engineering published at Lancaster University. Application of systems ideas in geography and planning are developed by McLoughlin in *Urban and Regional Planning*, Faber, 1972.

16 Systems theory and social work

*by Martin Davies**

1 The background to systems theory

Introduction

Despite the frequent use of the phrase "casework theory" the social work profession has never been wholly persuaded of the appropriateness of any one theoretical approach. Even during the years when Freudian thinking was at its most pervasive, there is evidence that *practising* social workers employed a more pragmatic approach to their work than was usually admitted; in the areas of child care, probation and psychiatric and medical social work (i.e. in the pre-Seebohm mainstream) the psychoanalytic foundations of casework teaching were often undermined or at least amended by the realities of the here-and-now situation in practice. Moreover, in all other areas of social work – with the handicapped and the elderly, in the schools and in residential care – the influence of Freudian theory was minimal. And it is one of the ironies of British social work in the 1970's that, although one of the expectations of Seebohm was that the mainstream of social work theory should be given the opportunity to exercise its influence over the less enlightened areas of practice, the effect of re-organization has been at least as powerful in the reverse direction: many traditionally trained social workers have had to come to terms with the fact that much of their work now involves social relations with clients and their environments of such a kind that casework theory derived principally from ego-psychology must take its place alongside other influential frameworks drawn from behaviourism, sociology, organizational theory and penal theory. When the role of the social worker is increasingly caught up in decision-making, the allocation of scarce resources, and the exercise of social control, it is apparent that ego-psychology, though relevant, is only one aspect of the theoretical armour required. Even in probation, which had looked as though it might continue to enjoy its reputation as the last bastion of mainstream traditional casework, it is now clear that the pressures of parliament and public opinion are requiring officers to consider the equally important contributions to be made from behaviourism and learning theory.

In the light of this, and in order to overcome the dangers of adopting a purely cafeteria approach to social work theory – "take what you like and use it as you wish" – a number of tentative attempts have been made in recent years to devise a more all-embracing theoretical framework. As yet, these attempts have to be recognized as being at a relatively premature stage of development; despite their interest and usefulness, none of them has yet convincingly provided a comprehensive theoretical alternative. Indeed, such is the variety of social work settings now, and such is the range of objectives and of functions, that it seems increasingly less appropriate to maintain the belief that there is any such thing as a social work theory per se. There are theoretical contributions to social work knowledge, but their appropriateness may not be of equal weight in different situations or with different clients; thus the work of a private agency family counsellor in San Francisco, a British probation officer responsible for supervising a recidivist-parolee, a local authority social worker allocating aids to the handicapped, and a school counsellor recommending residential care for a persistent truant, covers a wide range of roles, in all of which a variety of theoretical perspectives are likely to be relevant and influential.

Recently, however, *systems theory* (sometimes *system theory*) has been put forward as representing a framework of more than limited usefulness in each of these situations and others; indeed, the case for systems theory is not just that it contains lessons applicable to all aspects of social work, but even of validity throughout society. It will be argued here that, although the term *systems theory* is perhaps over-ambitious (a preferable one is *systems thinking*), the ideas which have emerged from the systems literature are valuable both to the social work practitioner as he plans his programme of intervention with the client, and to the social work administrator whose focus of attention is increasingly on strategic planning and on the need to maintain dynamic control over the relationship

* Director of the social work programme, University of East Anglia at Norwich.

between his department's objectives and resources on the one hand and the demands of its environment on the other.

In this paper, I shall refer en passant to *The Support Project*; in this experimental exercise, the Manchester Council for Voluntary Service recruited volunteers for allocation to the families of children at Manchester's Special Schools (Davies, 1974). The original aim was to help the children over the hurdle of leaving school and starting work, but the Project quickly broadened its perspective and the volunteers were for the most part engaged in providing general family support. Some 100 volunteers were recruited overall, but only one-third operated for twelve months or longer. This small group of volunteers may seem a far cry from the thousands of professional social workers and ancillaries and the heavily capitalized residential sector that together make up the welfare state's provision of personal social services; but the simplicity of the model provides a useful and relevant context within which some elements of systems thinking can be explored; where it becomes necessary, the wider spectrum of the social services will be drawn in, but I have found that the Support Project is quite suitable to provide illustrative material for an understanding of both the strengths and weaknesses of systems theory. Indeed, much of the previous literature relevant to systems theory is excessively abstruse, and there is much to be said for trying to relate complex ideas about systems to simple aspects of social reality; after all, that is the main objective of systems theory – to provide a conceptual framework within which the multi-dimensional qualities of social relations can be studied and understood. There isn't much point in learning about systems theory unless it helps us to make sense out of interactions which had been previously ignored or which had defied analysis.

It has to be acknowledged that social work writers to date have failed to take us very far. Hearn (1958) and Janchill (1969) are the best-known early references to the attractions of systems theory for social work, and more recently Goldstein, Kahn, Mullen and Dumpson, and Pincus and Minahan have made repetitive use of the term *system*, and have lent emphasis to the need to clarify its meaning and its potential contribution. The student embarking upon such a task of clarification might look to the professional literature in vain; like Meyer he may well conclude that it is "a pretentious term at best", or, with Sainsbury, say that "the systems approach suffers at present from being newly fashionable, from the risks of gimmickry and from its attendant jargon". Even if that were the initial reaction – and there would be some justification for it – it does not follow that the ideas which have led these writers to espouse aspects of systems theory are worthless, or that more carefully tested, they might not become of practical value to the profession.

The best brief introduction to systems theory's potential usefulness comes in the contributions made by Carol Meyer to Kahn and to Mullen and Dumpson, and the value of these pieces derives, as always in good academic writing, from the questioning approach she adopts. In one of these essays she pin-points three of the topics that will emerge as central to much of our present discussion: (1) *the identification of goals*, (2) *the recognition of environmental factors*, and (3) *the need to learn to live with uncertainty*. We'll look at each of these in turn:

1 While reviewing a number of research studies, Meyer highlights the lack of precision in social work's objectives. "What are appropriate goals for direct service practice in social work? Who determines them – the agency, the worker, the client? Is client need necessarily correctly assessed by the community? Only as we grapple with the question of goals will we achieve clarity in our specification of casework practice". Meyer reports the familiar research conclusion: "client-worker interactions were demonstrated and in large measure found wanting as a major solution to deviance in school behaviour, dependence on public assistance, the condition of being a multiproblem family in a socio-economically deprived area, and the state of unfulfilled promise", and she says that we must ask if casework has promised more than it can offer. Meyer is quite clear that the business of social work is not with achieving change, nor cure, nor control, but with giving help. "If we choose the helping rather than the socializing goal, then we will be freer to attend to the improvement of services – to socialize *them*, if you will".

2 The goals of casework are closely related to one's view of the relationship between the client and his environment, and Meyer argues that social work cannot be viewed in isolation from the social conditions which foster the problems which the worker is aiming to treat. "Is it possible", she asks, "that we still do not comprehend the role of social casework in the overwhelmingly complex psychosocial systems in which it is involved?" And, unusually in the literature, Meyer recognizes that there are two parts to the environment in social work – that surrounding the client, of course, but also that within which the worker operates. Even those writers who have come to recognize the power of the environment over the

client and the applicability of systems theory to diagnosis and assessment tend to be naively unaware that the social worker is similarly influenced by external factors; Vickery, for example in one of the few British contributions to the literature, writes as if the worker operated from a wholly independent base, and many writers on community work seem to have an idealized view of the essential neutrality of the worker committed to galvanizing underprivileged urban societies into action. Nevertheless even Meyer's main emphasis is on the client's environment and its significance for planned programmes of social work practice. The worker is not restricted to individual or group methods; the systems view takes into account "the interlocking salient systems that radiate from the person", compelling the worker to operate in whatever fashion is appropriate to the client's needs. "The anchor concept is still person-in-situation; only the systems framework has given us the conceptual tools to really make use of both sides of the hyphens instead of bowing to the situation and addressing the person".

3 Management craves for certainty: one of the expressed objectives of applied research in social administration is to reduce the area of uncertainty, and both planners and politicians can fall into the trap of expecting research to provide absolute answers. Social workers, in contrast, are too close to their clients' lives to share these illusions, but the theories of the past, suggests Meyer, have not been equally sophisticated. Now "the world has opened up, loosened up". We can no longer trust psychiatric diagnoses or moral imperatives; the days of linear cause-and-effect are over; "we have to learn to live with generalities as best we can". Meyer herself fails to acknowledge the origins of these arguments in systems theory, but Stein is more precise when she recognizes the inevitability of inconsistencies in social work. If we try to be *totally* consistent, we shall be closing our eyes to new evidence; compromises and syntheses are the way to survive in a complex social system. "Inconsistency is simply a hidden awareness of the contradictions of this world . . . In so far as inconsistency is an individual attitude, it is nothing but a collection of uncertainties which conscience keeps in reserve, a continuous awareness that one may be mistaken or that the enemy may be right". (Stein, quoting Kolakowski).

Values may be absolute, but truths cannot ignore present or potential external factors. The social worker's diagnosis, his planned treatment, his committal to care, his allocation of financial aid: all are uncertain acts in a complex environment, and experience and knowledge can only marginally reduce the area of uncertainty. The social policy that aims at caring, punishing, controlling, educating is similarly vulnerable to factors beyond, far beyond, the policy-maker's control, and necessarily contains within it elements of inconsistency and imprecision. Perhaps if we were to recognize this, we might become more sympathetic to our own failures, especially perhaps in the field of residential care where good intentions come unstuck with monotonous frequency. More important, perhaps, than anything else is the way in which systems theory has taught us not to put our faith in simple legislation as a cure for society's ills.

Systems theory has, then, proved to be attractive to a number of distinguished social work writers, and a prima facie case is established for seeking to extend it beyond the academic frontier in order to test out its potential contribution to practice, even if, as Meyer again suggests, "general system theory turns out to be not a theory but a framework for viewing interrelated phenomena". Research has demonstrated that there are serious reasons for doubting the validity of the linear model of cause-and-effect so far as it relates to most aspects of social problem situations, but it therefore follows that social work must somehow come to terms with the problems of analyzing and conceptualizing the dynamic and multivariable aspects (properties) of the social reality within which its workers practise. In research, both qualitative methods and developments in electronic data processing have brought about some movement in our ability to handle more than two variables, but theory has been more conservative. At a time when it is clear that the present and future functioning of social work practitioners – be they professionals, ancillaries or volunteers – is both questioned and yet encouraged, the need for some clarification of the helping process in our society is self-evident. In order to attempt this, it is necessary to introduce a range of concepts and ideas from systems theory, many of which may be unfamiliar and even alien to some readers. However, with a theoretical framework drawn variously from sociology, biology and physics, it is quite impossible to plunge straight into a discussion of its relevance for social work. Indeed, the literature to date has erred in its incomprehensibility to social workers or in its obvious failure to come to terms with some of the systems concepts. In this paper I shall try to bridge the gap; two risks are run – systems concepts may be over-simplified in order to translate them into the social work situation, and their employment may take us too far from the social work data. Notwithstanding the risks, the effort seems worth making, especially if, in the process, we are able to identify some significant issues for social work in the 1970s.

*The Contribution from Sociology**

Sociologists have continued to debate in the twentieth century one of the central issues that has troubled social philosophers throughout history: is man a product of his society, or does he create, control and determine its development? "Due to the historical circumstances in which it arose, the sociological perspective has always presented a paradoxical vision of man in society: man creates society, but man is also subject to his own creation. As Gouldner puts it:

> The modern concepts of society and of culture arose in a social world that, following the French Revolution, men could believe they themselves had made. They could see that it was through their struggles that kings had been overthrown and an ancient religion disestablished. Yet, at the same time men could also see that this was a world out of control, not amenable to men's designs. It was therefore a grotesque, contradictory world: a world made by men but, despite this, not *their* world". (Thompson, 1972, p. 63.)

These alternative views of man's relationship with society have been expressed in the "social system" approach which emphasizes the ways in which *society makes and controls man*, and in the "social action" approach which "focuses on the ways in which *man constructs and controls aspects of society*". As Dawe indicates, the social system approach reflects a concern with order: man is not to be trusted, but must be made subject to socially determined constraints; man is wild, selfish, greedy, and his instincts have to be curbed by processes of socialization, law, punishment, supervision, control. Conversely, the social action approach expresses a fundamentally optimistic view of human nature, and argues instead that man needs to exercise control over the dangerous tendencies of society to become repressive and dysfunctional.

The significance of these alternative prescriptions needs little emphasis in a social work context. The social system approach, the concern with order, is mirrored in the greater part of social legislation which encompasses the social work profession: the Mental Health Acts, the Children Acts and the Criminal Justice Acts are all explicit about the need for society to control its deviants, and implicitly involve social workers as agents of the law maintaining order in the community.

The social action approach, on the other hand, is also espoused – and usually more willingly – by social workers: clients are encouraged to develop their potential abilities, to claim their rights, to confront elements in society held to be constraining them, and even to rebel against the repressive sanctions of society itself where the social worker feels these to be unjust. Thompson discusses whether the two views are reconcilable or not; conceptually it would seem to be difficult to regard them as being in any way complementary, but in practice there is little doubt that the relationship between the two approaches provides a clear indication of the political complexion of any regime. Dahrendorf has discussed (in his Reith Lectures 1974) the continuing tension that characterizes an open society, and Thompson suggests that "the problems of maintaining social order and of exercising human control over social forces are but two aspects of the general problem of securing the conditions necessary for the development and expression of man's capacities". Because of this, he argues that "the sociological perspective needs to be sufficiently comprehensive to encompass both types of sociological theorizing": action theory which incorporates an essentially optimistic concern for human creativity and a belief in the importance of allowing him the power to control the social world; and systems theory, which reflects the apparent need to restrict the individual's freedom, and to ensure the maintenance of an orderly pattern in society.

It would be wholly inappropriate to argue that systems theory, of the type just described, has much to offer the social work profession; and it would almost certainly be inaccurate to suggest that such a conceptual presentation could successfully be employed to analyze the social worker's role and function. However, Thompson not only argues for the compatibility of the two theoretical approaches (and hence for their possible combination in the social worker's setting), but suggests that the social system is continually affected by, changed by the interactions that occur within it and the information that enters it. "The concrete 'organization' is a resultant both of actors following out rules *plus* the interactions of these actors with each other and with an environment whose constraints or exigencies are usually much too rich to be covered by the rules". (Buckley, 1967, p. 94, quoted in Thompson, 1972, p. 72.)

It thus becomes important in any discussion about

* The subject matter of this section is discussed in the important paper by Dawe, and also in Thompson (1972) and the latter paper has provided the base on which the present author has built.

systems theory in social work to clarify at the outset what is meant by the term; the conceptually pure form of systems theory in sociology is of little help in practice, because its acceptance as valid implies an immediate admission of the wholly determinist nature of society: man has no influence over the social order, so there is little point in either social worker or client exercising any initiative or making any effort to change the status quo. All is as it is meant to be, and the individual is utterly subject to external social forces.

But the systems theory that Thompson acknowledges as apparently valid, and that I shall be describing in these pages is in fact an amalgam of the action perspective and the system approach, and their juxtaposition in reality and the tension that characterizes their relationship seems to be an accurate reflection of human society; in particular for our purposes it provides a helpful context within which we can understand some of the stresses and strains that affect the social worker as he acts at the pivotal point between the individual and society. There is inconsistency in the relationship, there is an unpredictable element, and there are contradictions; there is room for unexpected growth, and there is often breakdown and conflict: these are the dynamic qualities inherent in an open system.

General system theory

The social worker's most pressing problems, like those of the manager in local government, "concern the need to maintain some sort of tolerable relationships between a whole lot of potentially conflicting people, conflicting duties, unforeseen and unavoidable demands on his time, changes of policy and changes in the outside world. He knows, or at least suspects, that the total situation he lives in is complex and uncertain . . ., and he might welcome a more flexible and sophisticated framework of thought within which to cope with all this complexity and uncertainty". (Baker, Systems Theory and Local Government, Local Government Studies, January 1975, p. 30). One of the main attractions of systems theory, as it has been developed outside of theoretical sociology, is its apparent recognition of the confused realities of social life and its willingness to incorporate essentially unpredicatable elements. It does not have the conceptual purity of the best sociological theories, but, as a *framework of ideas*, there is no doubt that it encourages a process of thinking in practitioners and administrators that has greater prima facie validity than many more rigorously defined theories allow.

The literature of systems theory is extensive, and many leading authors have made important contributions; but here my main emphasis will be on the approach taken by one of the leading figures in the systems theory movement, Ludwig von Bertalanffy; only peripheral reference will be made to the work of other theorists, but it is important to keep in mind the necessarily controversial nature of some of the material.

General System Theory – the form of words used by von Bertalanffy to identify his ideas – is interdisciplinary and wide-ranging. It is derived from, and in turn has contributed to, developments in physics and biology, and, more recently, psychology and sociology. It is conceived as a general science of "wholeness", valid for all systems whatever the nature of their component parts – electrons, cells, ants, men as individuals or in groups – and whatever the nature of the relations between them.

Bertalanffy dates his interest in it back to 1937 when he was working as a biologist trying to explain the behaviour of living matter in mechanistic terms; gradually he was led to an *organismic viewpoint*. Living beings had to be recognized as organized, dynamic objects, capable of growth; they do not conform to physical laws which apply to *closed* systems, systems independent of and isolated from their environment, systems which in effect "run down" with the process coming to an end in some form of static equilibrium, the shape of which is determined only by chance.

Human systems, social systems are enormously complicated, but they are not in any logical sense unpredictable – though, in common sense terms, they may appear to be so. General System Theory emerged out of the need to conceptualize, if not to explain, the complexity of the dynamically interacting variables that affect living organisms, and it can be applied in such diverse areas as the relationship of wall lichens to industrial pollution, of disease bacteria to antibiotics, of poor families to the offer of financial aid, and of hostel wardens to unruly residents; in all these examples, a simple cause-and-effect unidimensional approach is known to provide an unsatisfactory account of the true position. (My own interest in systems theory was given practical encouragement when, in 1969, having been engaged in social work research for the Home Office for five years, I had become dissatisfied with what seemed to be oversimplified theories of criminal behaviour and naive policies of penal response. Reference to the influence of poverty, broken homes, cultural deprivation, and the like as determinants of deviance was leading to policies, both conventional and revolutionary, which postulated that programmes of action, involving either social work or political upheaval, could attack the causes of crime and so produce beneficial effects for society. Social workers, perhaps more than most people, are only too well aware that such simple diagnostic assessments are rarely adequate to explain their client's behaviour. Social work and social administration seemed to me to be a fruitful area to test out the lesson of systems theory; it might or might not lead to improvements in practice, but it could hardly achieve less than conventional paradigms had effected in areas like crime, personal relations and mental health in the public sector).

Systems theory also emerged out of the felt need to allow for growth and spontaneous development in groups of living objects, and this emphasizes its closeness to many of the more interesting recent ideas in social work – those of the therapeutic community, of client responsibility, of family dynamics and of creative movement in groups. "Modern systems research can provide the basis of a framework more capable of doing justice to the complexities and dynamic properties of the socio-cultural system. (Buckley, 1967, quoted in von Bertalanffy, pp. 5–6.) It provides an ideal framework for the analysis of what Bertalanffy calls "*organized complexity*", as distinct from the unorganized complexity that is rooted in the laws of chance and probability.

Katz and Kahn's definition of systems theory as being "basically concerned with problems of relationships, of structure, and of interdependence, rather than with the constant attributes of objects" (in Emery, p. 90) could be applied without amendment to social work practice; and Bertalanffy's definition of systems as "sets of elements standing in interrelation" is of self-evident relevance to the psychosocial context of the client's life; it also, of course, incorporates the social worker.

We shall look in rather more detail in Part 2 of this paper at some of the key concepts in systems theory, but it is worth reiterating the main points without delay.

1 *Systems theory is organismic, not mechanistic:* it is concerned with growth, dynamism, spontaneous development; the constituent parts of a system are not passive elements incapable of reacting to external forces, and subject only to mechanical laws of movement; they interact with each other and with elements in other systems; no pattern of events, no process of interaction between component parts will ever be quite the same as any other pattern or process. (Hence the conventional social work riposte to the unduly mechanistic approach of much research: we deal with individuals, and you can't reduce them to statistics. You can; but, as the social worker well knows, you lose something in the process, and, if you take it too far, you may end up with a travesty of the truth).

2 Linear causality is the notion used to explain processes by postulating conceptually simple relationships: A causes B, and B causes C. While there is clear justification for breaking down complex relationships in this way, it has to be recognized that the approach enables the scientist to represent only *a part of* the truth. In particular, it overlooks the importance of interacting relationships, and underestimates the value of viewing a complex pattern of causal relationships as a whole; the separate investigation of isolated variables is often essential and the results can be valid, but taken out of the context of the total situation, they *need* not be an accurate reflection of the truth, and they can sometimes be quite false in their implications. One area where this is now recognized is in regard to maternal deprivation and its effect on childhood behaviour; much of the critique of Bowlby's early work hinges on the argument that the relationship between the two variables (cause and effect) depends on the circumstances surrounding them, and it is therefore unwise to make a simple assertion about the inevitable results of separating a child from its mother at a given point in time; moreover, as with the broken home hypothesis, it is less likely that the causal factor is the simple act of separation or deprivation than that it is rooted in the associated circumstances that preceded it or followed on afterwards. In other words, what is sometimes postulated as a simple matter of a statistical association between two isolated variables is in fact a rounded situation in which a veritable galaxy of factors may have to be taken into account. The importance of this assertion for practice is apparent, for it suggests that a social policy based on an acceptance of the validity of the simple relationship between maternal separation and childhood behaviour is likely to be, at best ineffective, and at worst, damaging, if it fails to take into account other relevant factors. (An even more striking example of the naivety of linear causality, again in criminology, is to be seen in the erstwhile postulated link between poverty and theft; this too failed to take into account relevant intervening variables, such as the concept of relative deprivation – a concept which demands a systems framework if it is to be capable of analysis). *Systems theory is not tied down to linear causality*, but has been able to absorb concepts of wholeness, gestalt, holism. It accepts that problems and circumstances, men and groups, are mutually interdependent,

and that we must accordingly think of systems of elements in interactionist terms. "Many problems particularly in biology and in the social sciences are essentially multivariable problems for which new conceptual tools are needed". (Bertalanffy, p. 99.) (The notion of *organized complexity* helps us to explain why it has proved to be so difficult to develop successful prediction techniques in social work – with regard to foster-care, for example, or absconsion, or mental illness. Such techniques pre-suppose the power of linear causality as a way of explaining, and thence predicting, human behaviour. Generally speaking, only when the causal variables are unambiguous and multiple is it possible to use prediction techniques with any confidence. In other cases, too many additional variables can alter the balance of probabilities over time; and in any case, the very existence of the prediction technique can itself play an important part in either bringing about or preventing the event that was prophesied – a perfect example of the interactionist element at work).

3 The rejection of mechanistic principles enables *systems theory to allow for goal-seeking and purposiveness within the organization or social situation.* Thus one cannot presume that the social worker, for example, will have an undeniable or wholly predictable influence on the client; and although social work theory has paid due regard to notions of client self-determination, practice has not always been so sophisticated. In any case, it is not simply a matter of encouraging client autonomy, for, as Miller and Gwynne show with reference to residential care, the goal-seeking and purposiveness that emerges may prove at times to be disruptive to the *status quo*. Hence the application of systems theory to social work does not necessarily lead to an egalitarian stance; rather does it require the participants, especially perhaps those with administrative or financial control, to clarify their expectations of the system within which they operate; systems theory, by recognizing the respective power of the various members of the group, facilitates an accurate analysis of the group process; it does not make any value judgement about the appropriateness or otherwise of any particular distribution of power (although it might suggest whether or not specific patterns are viable under different external conditions).

4 *The environment surrounding the system plays a critical part in its existence*, and indeed is the key to its organismic, dynamic nature. ". . . for open systems, there exists a relationship to the environment which is based on interconnection with the environment, and hence there exists an interdependence between each system and its environment". (Kremyanskiy in Emery, p. 136). Every organization is conceived as an *open system*, maintaining itself in a continuous inflow and outflow, its components being built up and broken down; the open system is never, so long as it is alive, in a state of equilibrium (which is a static concept), but always maintained in a so-called steady state – that is, it may appear to be stable when in fact it is in a continuous process of growth, and may even be developing towards a state of increased order and organization. The characteristics of an open system involve notions like wholeness, differentiation, growth, hierarchical order, dominance, control, competition, and so on. The human body is, of course, an open system maintaining a steady state, but so too are many social groupings: a family, an office, a student class, an Intermediate Treatment group, a hostel, a shop . . . What about the social worker and client in tandem? Much depends on the power of the relationship, on its longevity, and on the symbolic significance of each party to the other; in many such casework situations, the relationship may be simply too superficial, too fleeting to enable one to speak of it as an open system in its own right, although any analysis of the relationship must take into account the relevance of the *other* systems to which each party belongs.

5 Although mathematical terms are commonly to be found in the systems theory literature, the value of the approach is not dependent on a mathematical model. "A verbal model is better than no model at all . . . Models in ordinary language have their place. The system idea retains its value even where it . . . remains a "guiding idea" rather than being a mathematical construct". (Bertalanffy, p. 22.) In the area of social science, we are content to aim at "explanation in principle" (p. 35).

That, then, is a brief account of some of the main arguments from General System Theory – postulated as a way of reconciling the sociologist's theoretical choice between action and system, process and structure, dynamism and reification, control over society and the imposition of social order. It "incorporates equally maintenance and change, preservation of system and internal conflict" (Bertalanffy, 207–8), although it will doubtless leave dissatisfied thoese who see an impenetrable barrier between the pure sociology of organizations and applied organization theory. However, social work's concern is, as always, to use such theoretical concepts and such tools as emerge from the social sciences, and there is enough in general system theory to make further investigation worthwhile.

Aids to rational analysis in social work

Traditionally, social work's objectives have either been taken as self-evident or been defined in relatively simple terms; students applying for courses leading to a social work qualification often say that their aim is to be able "to help people", thus implying that that is what social work is about. In recent years however, more attention has been paid to the wide range of roles which every serving social worker finds himself fulfilling: from correction and even punishment to community development and social reform. Many observers have found it difficult to reconcile what appear to be sometimes incompatible objectives, and one of the attractions of system theory is that it allows for a recognition of conflicting elements in any social situation. However, if human interactions involve forms of organized complexity, social planning (of the kind which social administration and social work are necessarily committed to) requires a willingness to embark upon rational analyses of system elements, whether one's immediate interest is in family life, residential institutions or a social work intervention organization. Such analyses can be used in the process of designing new systems, or they can be used for reviewing the function of existing systems. (In this section, as in Part 2, we shall return to the Support Project in order to relate the often rather abstract ideas behind systems thinking to the concrete realities of everyday life as it is seen by the social worker).

Two authors, Jaffe and Chapman, have each spelt out the important elements that must be taken into account by social planners and operators. By linking together their work, we find five basic elements identified: they can be presented logically but not necessarily sequentially. "Rather as one proceeds in thinking about the system, in all likelihood it will be necessary to re-examine the thoughts one has already had in previous steps. Logic is essentially a process of checking and rechecking one's reasoning". (Churchman in Optner, 284).

Jaffe (in Optner, 230) suggests that the key to system design is the identification and mapping of system *tasks*, and argues that "this is accomplished by creating a sequence of means-end relationships which link the more abstract objectives of the system user to successively more concrete and specific goals", hence, in social work, if the aspiring students are right and *helping people* is a central objective of the profession, Jaffe would argue that such a broad, all-encompassing and possibly ambiguous objective is of little use in practice and must be reduced to more meaningful, more practical goals.

The five steps in systems analysis are: I The identification of objectives and sub-objectives

These are usually structured in large part by the values and judgement of men; they reflect political, pragmatic, material or religious aims – for example the objectives of the National Front, a darts team, a firm of estate agents or the Society of Friends may be of quite different kinds (although there may be surprising areas of overlap too); objectives are not immutable – they are always liable to change during the operation of the system; for example, in the Support Project, one of the original aims was to help school-leavers into stable employment, but in practice this became a peripheral objective only. Moreover, in Mertonian terms, there may be a need to distinguish between purpose and function, between latent objectives and manifest objectives. What people *say* is the objective of an organization may in fact have been replaced by another; it is this fact which leads Churchman to argue that "the scientist's test of the objective of a system is the determination of whether the system will knowingly sacrifice other goals in order to attain the objective", and his example is worth quoting. "If a person says that his real objective in life is public service and yet occasionally he seems quite willing to spend time in private service in order to maximize his income, then the scientist would say that his *stated* objective is not his *real* objective. He has been willing to sacrifice his stated objective at some time in order to attain some other goal" (p. 284). In professions like teaching and medicine, the objectives are *relatively* clear-cut, but the social worker, whether he is operating in a Family Service Unit, a probation office or prison, a local authority department or a Salvation Army hostel, is not so fortunate; to speak of *teaching* or *practising medicine* itself implies broad objectives, but in the 1970s and 1980s, it is unlikely that *the practice of social work* is so self-explanatory.

Having identified a broad objective, there is a crucial second step to take: its reduction into sub-objectives, in order eventually to identify tasks (which emerge in the fourth step towards systems analysis). For example, once it was decided to launch a Support Project with the overall aim of helping specially identified children and families, a large number of sub-objectives could be defined: to appoint an organizer, to recruit a sufficient number of competent volunteers, to identify the right kind of families, to maintain the support of the headmasters without which access to the schools would be impossible, to secure volunteer-satisfaction . . . Failure to achieve any one of these sub-objectives would have meant at least partial failure to achieve the main objective. Of particular significance in this case, as in many others, is the importance that must be placed on the achievement of worker-satisfaction in the system; to a greater or lesser extent, in both the voluntary and the professional sector, this is a critical factor in the ultimate achievement of the client-focussed objectives. Thus there is a very narrow border-line between the *bad* system in which worker-satisfaction actually becomes the prime objective to the detriment of the client, and other systems in which worker-satisfaction remains an essential but necessarily secondary sub-objective to be satisfied only as a means of achieving the client-focussed end.

In this way, systems analysis requires a constant re-assessment of the balance between objectives and sub-objectives.

II The recognition of the system's environment

Every system has an identifiable environment. The environment of the system is what lies *outside* of it. This is not primarily a matter of boundaries: "when we say that something lies 'outside' the system, we mean that the system can do relatively little about its characteristics or its behaviour. Environment, in effect, makes up the things and people that are 'fixed' or 'given', from the system's point of view". But "not only is the environment something that is outside the system's control, it is also something that determines in part how the system performs"; in other words, environment represents *constraints* on the system.

It is crucial to emphasize that environment is *not* everything outside the system. In deciding whether an element is environmental to our given system, we must ask two questions: Can the system do anything about it? Does it matter relative to the system's objectives? Only if the answer to the first question is "No" and to the second "Yes", does Churchman recognize that "it" is in the environment. (In the Support Project, the fixed grant awarded by the DHSS constituted an obvious fixed constraint of great importance to the system but over which it had no control).

III The assessment of the system's resources

If the answer to the first question had been "Yes", then Churchman would define that element as constituting not an environment exercising constraint but a resource available for use by the system. Resources are *inside* the system, under its control, the means by which it attempts to achieve its objectives. In the Support Project, resources included the paid staff, the volunteers, the efforts of the headmasters and their teacher-colleagues, the goodwill of the local press and the support of other welfare agencies; all of these were available to be nurtured or abused by the system, and all contributed to its evolution over a short or a long period.

The conceptual distinction between environment and resources is an important one in social work, especially when we turn from the organized social work system to the client-family system. Would one normally describe the social worker as a part of the problem-family's environment or as one of its resources? Clearly the answer will vary from case to case, but in a great many situations, it is likely that the family can exercise little or no control over the social worker, and that therefore he must be defined as environmental – but only if he has significance for the family. If he is peripheral to them, he is neither a part of their environment nor a resource. In examining individual cases, the answer arrived at – whether the social worker is environment, resource, or neither? – is a significant indication of the nature of the client-worker relationship.

In the Support Project, it took a long time, and it remained relatively rare for families to perceive volunteers as resource-agents; the majority of volunteers seemed to keep firm control over their own behaviour vis-à-vis the clients, and hence remained, at best, within the external environment. (The same kind of analysis can be applied to many inter-personal relationships: for example, in teacher-student relations, to what extent are teachers a part of the uncontrollable environment of students, and to what extent are they a resource to be influenced and employed by the students to their own ends?)

And again the environment-resource distinction can be used to throw light on the difference between state provisions in the public sector of the welfare state and services offered for cash in the private sector: those who have money to spend can convert environmental elements into resources subject to their own control. The client of the Health Service, Social Security, Housing Departments, and the rest perceives these organizations as environmental: they are relevant to his well-being, sometimes crucial, but he has little or no control over them and must take or leave what they offer; they are not a part of his social system, not a resource under his jurisdiction; when they benefit him, as they may certainly do, he may feel gratitude, but he will not feel personal satisfaction at having contributed to his own development. The possession of money, however, changes this to some extent: it gives the individual real influence – albeit often quite slight – in regard to the receipt of private medicine, perhaps, or over the choice of a house in which to live. Of course there may be perfectly valid reasons why private medicine or private property are better abandoned, so that all are equally denied access to welfare resources, and all are required to make themselves subject to the environmental allocation of facilities subject not to their own choice but to entirely external determinants. But whether or not that is so on political grounds, this analysis gives us a glimpse of a key element in the organization of welfare provisions, universal or selected: they are environmental to the recipient, but there is reason to believe that benefits will accrue if some means can be found for converting them at least partially into resources over which the recipient feels some control and for which he feels some responsibility. The psychology of living in a socially planned welfare state is a remarkably underexplored area; if man has little or no control over those elements necessary to his own welfare, what implications does this carry, if any, for the way in which he uses them or builds on them?

Of course, the distinction between environment and resources can be a conceptual one, rather than a physical one. In many systems, some elements may be identified with both environment and resource; the best example occurs in many different guises – in the life of the child, perhaps, or in the situation faced by the resident in institutional care. In both such cases, there are others – the parent or the residential staff – who can be both environmental in that they exercise control and are not subject to system influence, and agents of resource, in that they give benefits and are open to pressure. It is known that many authority-relationships only function properly when the agent is able to balance the two component parts in appropriate proportions.

IV The refinement of tasks

Churchman talks of components, missions, jobs, activities – these are all the things the system must perform in the course of achieving its objectives. And for Jaffe, the refinement of tasks is the crucial step after moving from objectives to sub-objectives: at this point, inputs, outputs and processing can be inferred, planning can be completed, and the multi-variable organized complexity is brought under control. For example, in the Support Project, early tasks included:

1 The heads delegated responsibilities to teacher-counsellors;
2 Teachers picked suitable names from the school register;
3 Teachers were introduced to selected volunteers;
4 Teachers conveyed the volunteer to the chosen family;
5 The organizer placed advertisements in the paper;
6 She replied to applications received;
7 She clarified the volunteers' role to applicants;
8 She sent for references;
9 She made selections from the applications;
10 She took volunteers along to the heads or teachers;
11 She went with them to meet the family;

. . . and that is just the very beginning. In any organized system, tasks may be innumerable, but Jaffe argues that systems thinking requires their rational analysis: they represent the final step in the means-end sequence, and ultimately the achievement of objectives is dependent on the satisfactory completion of appropriate tasks. Hence the reasons for each task may have to be identified and considered. However, systems design is still under-developed; the early steps in system design are the hard ones, and we "do not yet have ready-made analytic tools for the job". The main need is to make the tasks more and more specific, continually to relate them back to the system objectives (which hopefully are clearly defined), and to relate them forward to the material and personal resources available for use, and to the environmental constraints likely to be influential.

V The management of the system

All systems are "managed", more or less consciously, more or less successfully. Each individual manages his own personal system as it affects different aspects of his life – at home, at work, in his leisure-time. Each family-system is managed, whether it be under a patriarchal, matriarchal, democratic or any other model. Industrial and commercial systems are managed; armies, political parties and football teams are managed; so too are central and local government departments. Only relatively recently has much explicit attention been paid to the different ways in which social service systems can be managed, but it has been quickly realized how inadequate were many of our past assumptions about social work, and how complex are the variables that have to be taken into account when considering future plans.

As Churchman points out (p. 291), most of the people

who have paid attention to systems analysis like to think that their work is outside of the very system they are describing; they outline the preferred process of planning logic but do not really understand how they themselves are a part of the system being observed. You cannot study a system without being a part of it, however obliquely; you cannot make recommendations for action without becoming an integral part of the action yourself: nowhere is there a better, more vivid demonstration of the dynamic nature of systems theory – the theorist, the thinker is necessarily a part of the action.

Nevertheless, bearing in mind his own unanticipated effects on the system, the manager must *generate* plans for development, *control* operations in order to avoid deviations, *evaluate* the effect of the system's efforts (the output) according to prescribed criteria, and, in the light of such evaluations, continually *provide for change* within the system: "no-one can claim to have set down the correct overall objectives, or a correct definition of the environment, or a fully precise definition of resources, or the ultimate definition of the components (tasks)" (p. 292). The recognition of such imperfections in any system highlights the primacy of information (feedback) being conveyed to the manager; only if deficiencies and shortcomings are communicated to those with the power to effect changes can reforms be instituted. Hence the great concern in much management literature with information flow – especially in organizations like social service systems, where shortcomings may not be highlighted by measures of commercial output like productivity, turnover, profit, and so on.

In social work, however, the pattern of generating plans, controlling operations, evaluating effects and providing for change (G-C-E-P) is not only applicable at the level of top management, as in industry. For each social worker in the field and in residential care is, or ought to be, concerned with managing the system of individual clients; I have argued elsewhere that, in many cases, where contact is limited to 15–30 minutes per week, such an objective is unrealistic: in probation, for example, the most that can be said in many instances is that the probation officer is providing a form of oversight as an end (an objective) in itself. However, there is little doubt that the employers of social workers have more ambitious expectations in many cases, and if the job of the social worker is with therapeutics, rehabilitation, corrections, reform, re-education, community development, environmental stimulation, then greater consideration will have to be given to improving the techniques employed by social workers to achieve their specific objectives. Despite lip service paid by textbooks and teachers to social work effectiveness, the lesson of most recent research is that few social workers operate efficiently or successfully when their results are gauged by measures of client-change.

Hence systems theory is applicable at the level of top management, but it necessarily requires much greater clarification of objectives in the context of the social services; it is also applicable at ground level, both in the field and in residential care, but it too demands a re-assessment of aims in social work and associated areas. And, in any case, it cannot be forgotten that systems theory, although it savours of omnipotence and omiscience, and hence tends to legitimize power relationships, nonetheless places equal weight on the shoulders of the client. To the client, the social worker is either an environmental constraint or a resource to be used for the benefit of his own system; client autonomy and client influence are as yet ill-developed concepts, but in a welfare state system operating under democratic principles, it is crucial that they be encouraged; unfortunately, like many liberal virtues, they run the risk of undermining the smooth running machinery of the larger system.

Systems theory is ultimately, like all social theory, about the relationship between the one and the many, between Man and Society, between the minority and the majority, between the deviant and the conformist. It is a helpful tool both for management, social work practice, and sociological theory; but it does not resolve the perpetual political issues of freedom and human rights.

2 The language of systems theory

Introduction

The exploration of systems theory carried out in Part 1 enables us to draw together seven points which highlight its attraction to those involved in the continuing debate about social work's role in modern society; it does not and cannot resolve all the issues that often seem so perplexing, but it casts doubt on some of the more naive dogmas that characterize the social work literature, and it compels practitioners and administrators to bear in mind the complex variables that impinge on even their most modest aspirations.

1 Systems theory confirms the psychosocial focus of social work, but it accords greater emphasis to the contemporary environment as both an enabling and a constraining force; systems theory argues that, if social work is concerned with human behaviour and personal welfare,

then it is crucial to accept as given the notion of client-in-system; only in this way can either social work or social administration plan effective programmes of intervention or development.

2 Systems theory avoids the pitfalls of the monocausal error; it recognizes, not so much the alternative notion of multiple causation, but the force of interactionist effects in social life.

3 Systems theory precludes the necessity of committing oneself to any single objective in social work (for example, that of *change*), while at the same time compelling practitioners and administrators to undertake rational analyses of their objectives in all situations.

4 Systems theory does not demand the overthrow of many of the most important lessons from casework theory and practice. Within a systems approach, the social worker must necessarily employ a variety of techniques determined only by their anticipated likelihood of achieving specified objectives. It accepts that there are different routes to the same goal, and therefore that the personal qualities of individual social workers are important components in the treatment relationship.

5 Systems theory removes from the social worker the danger of ascribed omnipotence. His professional skills, his experience, his agency identity, his personal charisma are still important to him in his work, but the primacy of his position vis-a-vis the client-in-system is not confirmed; the social worker merely becomes either a part of the client's environment or a resource for his selective use, and his impact on the client-as-individual is wholly dependent on the success with which he can achieve absorption into the client's system. He may improve his chances of influence by removing the client from his home situation and placing him in a new system (a prison, a hostel, a hospital), but even here the evidence suggests that the client retains an impressive degree of autonomy in determining his own future.

6 Systems theory not only emphasizes the legitimacy and the essential strength of the client-in-system, it also suggests that the client might come to be seen as an influential agent within the treatment agent's system – for example, in residential care situations, in detached social work, or even in relatively structured fieldwork settings operating at a fairly intensive level (like intermediate treatment, day training centres for offenders, day care centres for the elderly, and so on).

7 Systems theory is clearly not confined to the one-to-one relationship, but has equal applicability to groupwork and community work. In these fields, as in casework, it helps to clarify the role and function of the professional or voluntary social worker.

Seven systems concepts

Systems thinking, it will now be apparent, is not a unified body of knowledge, though it draws on an accumulation of research findings over a considerable period of time; it represents a way of looking at objects in inter-relation, and appears to be of particular value in social situations where agencies are seeking to exert influence, effect change or provide support according to preconceived plans. In this part of the paper, there will be presented seven concepts which are central to systems thinking, and which, taken together, provide a framework for further study. Following the outline of each concept, an illustration will be provided of the way in which it can be applied to a social work situation; all the examples are drawn from the Support Project, although it should be emphasized that the relevance of the theoretical material is not any way restricted to the work of volunteers; the challenge to the reader that is inherent in this paper is to relate the concepts and the thinking to other social work situations with which he is familiar, and to consider whether they throw light on many of the problems and confusions that so bedevil much of social administration and community care.

1 Open and closed systems

An essential distinction is drawn in General Systems Theory between open and closed systems. *Closed systems* are totally isolated from and independent of their environment; they are static, predictable, and ultimately tend towards a state of equilibrium, stillness, inactivity. Some of the cruder theoretical models of society appear to imply a degree of closedness, of immutable structure which absurdly misrepresents the true nature of human relationships, and it is obvious to all that no living organism can be represented as a closed system: all are open systems.

An *open system* is defined as a system in exchange of matter with its environment, importing and exporting energy, building up and breaking down its own component parts. Although stable, open systems are always changing, always evolving; although identifiable, classifiable, they present differences over time and in changing circumstances. The concept of the open system defies many of the laws of conventional physics, despite its claimed applicability in the natural as in the social sciences; and to the physicist it presents many as yet unsolved problems of analysis. Little wonder then that the sociologist, with his crude instruments and under-developed theories, is in some difficulty handling the rapidly evolving interactions of social life. "Open systems maintain themselves in a fantastically improbable state, preserve their order in spite of continuous irreversible processes and even proceed toward ever higher differentiations." (Von Bertalanffy)

A valuable discussion of open systems in a social work context is to be found in Miller and Gwynne. They use the concept of the open system "to help us to understand and explain the ways in which different aspects of functioning are connected with one another". They talk, as we shall in the following pages, of the open system's input, throughput and output. Organisms, including social groups or institutions, can only survive by exchanging materials with their environment; in industry, the materials are usually inanimate objects, but in a school or in a residential home, they are more likely to be people or ideas which may themselves be a part of semi-independent systems. The materials go towards the maintenance of the open system, and the output ensures further input: "these import, conversion, and export-cum-exchange processes are the work the enterprise has to do if it is to live". No enterprise is more than partially independent of its environment.

There are many reasons why social work groups are particularly complex in their dynamics, but Miller and Gwynne emphasize two in particular. The first is the fact that both resources and throughput are human, and their interactions are therefore capable of greater versatility and are more unpredictable than interactions between humans and objects. And second, Miller and Gwynne draw attention to the fact that the individual is himself an open system in his own right. "He has an inner world of thoughts and feelings that are derived from his biological inheritance and from what he has learnt and mis-learnt from his lifetime of experience; he lives in an environment to which he has to relate himself in order to survive; and it is the function of the ego – the conscious, thinking mind – to regulate transactions across the boundary between inside and outside." In the course of their book describing the sequence of events at a Leonard Cheshire Home for the Chronically Sick and Disabled, Miller and Gwynne emphasize that unconscious energy drawn out of the past experiences of each resident can be a vital factor in determining the nature, not only of each individual, but of the group of residents as a whole or of the institution as a system in its own right. Hence, they argue that, firstly, the residents' repressed memories of inter-personal experiences in the past (often extremely stressful and traumatic), and secondly, the residents' suppressed awareness of their own present position, poised uneasily between social death and physical death, *both* supply distorted forms of energy input which the Cheshire Home as an open system has to deal with in some way. Such energy is often not allowed for in the more simplistic models of social work and residential care, or is regarded as an aberration which sound planning and good management will overcome. But Miller and Gwynne's analysis is an important demonstration that such attitudes will not do; not only does systems theory compel administrators to pay attention to the need for clearer objectives, it also requires them to be honest about the very nature of the open systems which they have created to meet those objectives. In the Cheshire Home, the input of unconscious energy from often disturbed residents was rarely predictable, and it had to be considered along with other more straightforward inputs: money, voluntary help from the community, administration, the caring role of the paid staff, political developments, and so on. The unconscious forces, then, are an undoubtedly important energy input which no social work setting can ignore; of course, it is not necessarily or in all cases the most dominant input, but perhaps it is most likely to predominate if its potential force is underestimated – as in the Cheshire Home it did when it led to the scapegoating of sympathetic members of staff.

The concept of "open system" represents a major challenge to those critics who accuse systems theory of being static. The "open system" model combines action theory and the sociologist's version of systems theory; it is dynamic, and allows for interactions of unlimited influence.

Example In the Support Project, the concept of *open system* can be applied to any of the organizations involved in the Project – the schools, the Council for Voluntary Service, etc. It is not applicable to the volunteers *as a group*; they never met all together, very rarely interacted with each other, had no corporate identity; it could however be applied to the Project as a whole, in which the volunteers were separately recruited resources, and, more surprisingly, it could be applied to each volunteer's *own* family, with a potentially significant question being: What input did *that* family-system receive from the volunteer's activity in the Project? Most obviously, however, and most importantly for the practice of professional social work as well, the concept is relevant to the client-family: contained generally within physical boundaries, the family incorporates multiple interactions between its various members, each of whom is the vehicle for continuing input and output; moreover, the Support Project itself hinges on the idea that an additional intervention in the family system – the support offered by the volunteer – will have predictably beneficial effects on the behaviour or welfare of one identified member – the child at the special school. The family is the single most dominant system in most peoples' lives (although it too interacts with and is influenced by other systems, especially the family-head's work-system and the neighbourhood systems): up to and beyond adolescence, and then from marriage until, in one way or another, the individual is left alone, the family is the one system in which the majority of people participate.

The family interacts with its environment, draws resources into its range, uses them for its own maintenance and development, and in turn contributes to the wider community; it changes, as its members age, but still remains distinctive; at times when there are sudden breaks in continuity – the arrival of a new baby or an ageing in-law, the death or desertion of a spouse, the departure of a child-become-adult – the system goes through a period of crisis, but, depending on the intensity of the changes, the component parts will re-form and perpetuate the system so that it remains still identifiable.

II The importation of energy

Energy is essential for the survival of any organism; the cell and the body need nutrients in continuous supply; so too the personality is dependent on a steady inflow of stimulation from its external environment (from people, from animals, from visual images, literature, sound, and so on), and social organizations cannot live unless they have constantly renewed supplies of energy from other organizations, from individuals, and from the material environment. "No social structure is self-sufficient or self-contained." (Katz and Kahn.) Groups, both corporately and through the medium of their constituent members individually, receive energy inputs from a variety of sources: for example, they derive stimulus from educational and socialization agents (hence the emphasis placed in probation and social services departments on the provision of in-service and post-experience training); they derive strength from conscious and unconscious memories influencing the members (even to the extent of reifying the past in anniversaries, founders' days, etc.); and they are enabled to continue by the encouragement and expectations of others outside. All these factors lead to a raising of morale and a consequent enthusiasm for the activities of the group; they lead to a pride in membership, and a consequent demand by non-members to be admitted; they lead to an unquestioning acceptance of the group's norms; and they lead to individuals within the system making adaptive and creative contributions to the group's development. All these are the visible signs of a system that remains alive, that remains open; they may not be, and often are not, always present together, but only when all are absent will the system die and the component parts drift apart.

Vickery recently presented a family case-study within a systems framework; she describes the circumstances surrounding a mother of four who is being treated for depression at a psychiatric outpatient clinic. She and her family appear to be isolated within the community. "If Mrs. X were living in a bright, cheerful neighbourhood among people who were outgoing and friendly, the social worker might take the view that Mrs. X's depression stemmed from intra- and inter-personal conflicts and inadequacies. However the lack of rewarding experiences to be derived from people outside the family means that the family X is deprived of sufficient positive transactions with their environment for the maintenance of psychosocial health." Vickery rightly does not exclude the possibility that the problems may be quite independent of the environment, though it seems unlikely. But nonetheless she uses the case to demonstrate that social workers need to give far greater weight to ways in which they might intervene in community social systems in such a way as to benefit the family systems whose survival and welfare depend on adequate inputs from outside. It may be, says Vickery, that the community worker would view the impoverished community "as a client-system in its own right". But a family worker might retain a focus on Mrs. X and her family, and intervene in the community as a way of influencing indirectly the welfare of the family. "It is easy to see how social group work with Mrs. X . . . might be the method of choice and that it might be more effective if based on a 'social goals' rather than a 'remedial' model. Becoming a member of a resident's association or a play group scheme might do more to relieve Mrs. X's depression than several hours of social work interviewing or even of home-help time. Both the psychological benefits of social interaction and the benefits accruing from the achievement of goals, such as improved play and leisure facilities, would make a contribution to this family's well-being. This does not imply that domiciliary services and a relationship with a social worker might not be an essential bridge in helping Mrs. X to transfer from her role as 'patient' to that of active group member."

The fact that many questions are necessarily begged in Vickery's analysis – for example, the fact that, even in mainstream therapeutic casework, we are still appallingly ignorant of the ways and means by which the social worker moves into the client's system and successfully exerts influence of the kind described in the case-study; and the fact that the effectiveness of such systems oriented techniques is much less well tested than is the effectiveness of casework intervention in the traditional mould – does not undermine its interest and its attraction. Of course, it has to be said before indignant social workers say it themselves, the kinds of activity described by Vickery are by no means revolutionary; many an enterprising social worker in the past as in the present will have worked with his client in much the way that Vickery describes; but it is certainly true that such efforts have rarely been set down in writing; they are not practised on a wide scale; and the theoretical justification for them has tended to be intuitive. What Vickery is suggesting is that systems theory now compels us all to cease seeing family systems as isolated units, and to try to relate our prescribed treatments to the wealth of inputs which the community has to offer. The poverty of the social worker and the inadequacy of his tools have always been matters of concern to theorists and teachers; systems theory is now suggesting that we should not look to the social worker to provide the rich input or to manipulate his blunt instruments for the therapeutic benefit of the average client, but that we should rather see ourselves as enabling agents putting the client in touch with *other* riches, *other* tools which may well be more culturally appropriate to his needs; the social worker becomes a kind of entrepreneur bringing together systems in need of input and energies suitable for these needs. Of course, as Vickery concludes in her case-study, "if social workers are to be sensitive to the need for improved transactions between individuals and the social network outside the family, at least two requirements have to be met. The first relates to the availability of knowledge about *the community* and the second to the knowledge about *the means of effecting change within the community*". Such requirements imply radical developments, not just in the content of teaching for social work, but in the availability of valid material to be employed in such teaching, for our knowledge of how individuals effect change in the community is, to say the least, rather sketchy.

Example In the Support Project, the input is simply defined, and accurately reflects millions of similar inputs provided by social workers everywhere: it comprises the physical, mental and material energy given by the volunteer to the family or to specified members of the family. Inputs included visits to the home and verbal exchanges, work carried out for the clients (for example, form-filling, telephone calls made), the provision of presents for the children, money and clothes for the family, car-rides, organizing picnics and outings; words and personal contact predominated, but some practical and material aid also occurred. The very title of the exercise, *The Support Project*, is focused uncompromisingly on the intended input, and there was of course no thought in the minds of those planning it that it might be the schools or the volunteers or the Manchester Council for Voluntary Service that would be receiving the support (although each might well have derived benefit from the Project indirectly). The process was envisaged as a one-way process – the provision of energy for those deficient in it.

III Throughput

"Open systems transform the energy available to them." (Katz and Kahn.) The body converts food into heat and action, chemical and electrical stimulation into sensory qualities, and information into thought patterns. So, too, does the family, the group or an organization convert energy input into forms which it considers more suitable to its needs: just so does the problem family of folk-lore convert its new bath into a coal bunker, and just so does the client in the after-care office convert the offer of a chat by the probation officer into a ticket for the Salvation Army (Silverman); but so too do social workers convert new legislation into practice which conforms to their cultural norms, and so too, do all people convert earned or unearned income into items of expenditure which represent the desired throughput for their own personal system. (Indeed it is significant that money is one of the most neutral of all inputs, in that the recipient is almost completely free to convert it according to his choice; but it is worth noting the occasional outcry that arises when a welfare recipient spends *his* money on purposes which are deemed to be inappropriate.)

When the input is other than money, its conversion into throughput may involve system-members in negotiations in order to identify those resource-elements which the system can call on; for if the energy on offer (say, the well-intentioned bonhomie of a Lady Bountiful) is not convertible, it may be of no value, and there cannot then be any effective throughput no matter what the intention of the agent responsible for the input of energy; alternatively, the system may sample the input to see how adaptable it proves to be, for once it is absorbed into the system, it may well become an integral part of that system and then become subject to the processes of interaction that characterize any system on the inside; but that requires that the energy input is willing to be so absorbed, is willing to become a part of the inner system.

Example In the Support Project, the throughput is the use made by the family of the volunteer's presence and his offers of help. Maybe the clothes are the wrong size, so the family receives them, then throws them away; or maybe they are altered by Mum and worn by the children. Perhaps the volunteer is kept on the doorstep, and her words remain wholly irrelevant to the family's condition or maybe she is radically accepted as an integral part of the family system, exerting real influence and having an effect on behaviour, feelings and events both in the here-and-now and in the future. Even in cases where the volunteer may feel that she is not accepted by the family, and where the family describes her in rather superficial terms as "a nice lady, not stuck up, from the school, who comes down now and again to see how we're getting on", it may well be that the input, though not converted into the kind of throughput that might have been hoped for, nevertheless becomes a reality for the family, and has unexpected indirect effects on internal system relationships simply because of the input-agent's presence. This might equally be true perhaps of the formal role played by social workers or probation officers acting as agents of control; in this way, even an apparently cursory casework relationship may be "used" as a systems resource within the context of a family-in-need.

IV Output

Output in an open system is energy of any kind generated from within the system, as a result of the system's reaction to energy input – although, in reality the relationship between the two is neither direct nor linear. The energy output may in turn constitute potential energy inputs for other systems. The simplest examples can be drawn from manufacturing industry where end-products are sold for the benefit of other systems (other industries, families, etc.), thus generating further input for the producing system. Models in the welfare sector are more complex, because many of the input-output relationships are indirect, and some recipients of aid, for example, can offer only the fact of their survival and/or their grateful appreciation as outputs in exchange for further inputs; hence, in the social work profession, as in nursing and teaching, much energy is spent for the benefit of client- (patient-, pupil-) systems, but the effective return comes not principally from the recipient of aid but from the input-agent's employer in the form of cash and other career-rewards.

Example In the Support Project, the complexity of the output concept is immediately apparent, for the *intended* output from the family, following volunteer-input, is to be seen as improved school attendance by the identified child, improved social functioning in the community, or even increased self-confidence among family members leading in turn to their more effective use of welfare agency resources. These are the objectives of the volunteer and the Project. But in this setting, as in many other spheres of social work, both professional and voluntary, the applications of systems thinking should show us that such expectations are naive; it is unfortunate but true that we are quite ignorant about the probable outputs emerging from client systems in receipt of given worker-inputs, although the growing volume of social work research is at least beginning to demonstrate how rarely do we achieve our stated objectives. (Sainsbury's recent work amply illustrates the point.)

It is not that there are *no* outputs from client-systems – such a thing would be unlikely once the imported energy is accepted by the system and transformed into a throughput – but that they are not necessarily the outputs that fit into traditional models of what help, support, casework, treatment, or therapy are intended to achieve. In the Support Project, the most frequent family-system output identified by both families and volunteers was a feeling of gratitude and verbally expressed satisfaction; there were a small number of cases in which improved functioning was reported, but it was generally restricted to personality variables and did not extend to more tangible behavioural factors: "the chats made me feel better". Certainly such outputs, however modest they might seem, were sufficient reward for many of the volunteers and served in turn as inputs to maintain the momentum established. One case with a more visible output was that in which the volunteer's main input was his car; the parents used it to make journeys to their son away at a residential special school; and the output was therefore the more frequent visits that they paid to their son than would otherwise have been the case.

V Systems as cycles of events/feedback

Feedback is one of the terms most commonly associated with systems theory, but it is nonetheless criticized by Von Bertalanffy for its static, mechanistic qualities. The best example of a feedback process is to be seen in a thermostat mechanism within a heating system: fuel-inputs are burnt, the temperature rises to a given point, the thermostat switches off the flow of fuel, the temperature falls and the thermostat switches on the fuel-supply once again.

However even if the *feedback* concept has shortcomings when employed within open systems, the notion of positive and negative responses is of value within the broader perspective of cyclical interactions. Systems, as dynamic organs, are constantly in interaction with other systems, and especially with those systems that make up their own external environment. "The product exported into the environment furnishes the sources of energy for the repetition of the cycle of activities"; or, in the language of cybernetics, it provides positive feedback. Alternatively a failure to convert input into an appropriate end-product will tend to lead to a breakdown in the cycle of events, a cessation of the interactional process; it indicates negative feedback.

Example In some of the Support Project's cases, the families conveyed to the volunteers their rejection of the proffered help, not usually by saying "go away!", but by their disinterest, their coolness, their formality, their failure to respond to requests or invitations; the volunteers were thus given negative feedback, and quickly withdrew, sometimes from the Project altogether; they could not continue because of their failure to establish a satisfactory cycle of interaction. In other cases, the volunteers provided input, the families accepted it, used it as seemed appropriate to their needs, and were privately or explicitly glad of the volunteer's presence; however, at this point, some of the volunteers failed to pick up their families' positive feedback, or discovered that, for them, the families' gratitude or improved morale was insufficient reward for their efforts, and withdrew. In this latter group, it was not that the client-system intended to reject the volunteers – they didn't – but that the volunteers failed to receive the input for *their* own personality or social system that they were looking for; in particular, many volunteers (after a considerable period, and a large number of cyclical interactions) became disillusioned because they had looked for, hoped for and expected much more radical changes in the family-systems than had occurred. Hence, on a reality level, they were receiving positive feedback out of the family-system's response to their efforts, and had they come to terms with the limited objectives inherent in their role, there was no reason why the process of interaction should not continue indefinitely; but, because of their unrealistic expectations, they had developed a fantasized model of their effect on families, with the result that they interpreted positive feedback in negative terms; hence, the failure to achieve exaggerated objectives (reduction of truanting, improved job performance, better marital relations, etc.) became negative feedback in the worker's eyes, despite the fact that it was perceived as positive feedback in the client's.

A critical finding from the Support Project, and one which has obvious application to the use of volunteers generally, as well as relevance to professional social work too, is this: the continuation of the relationship between worker and client seemed to be dependent on *both* parties receiving enough positive feedback to maintain the cycle of interactions. Each party could break off the relationship; each could look for changes in the other's performance so as to improve the level of his satisfaction; but each could refuse to change, such refusal almost certainly constituting negative feedback with a consequent breach in the relationship. The longer the pattern continued, the more the volunteer might be drawn in to the family's system, and the less he would be isolated in the external environment; the closest interactions of all were to be found in those cases where one or more members of the family-system were also drawn tentatively and briefly into the volunteer's own family-system. This step was frequently interpreted by clients as the most vivid form of positive feedback they could ask for from the volunteer.

VI Steady state

"In the open system, continuous decay and synthesis is so regulated that the organism is maintained approximately constant in a so-called steady state. This is one fundamental mystery of living systems." (Von Bertalanffy, p. 165.) And it is argued that this principle, which is most obviously applicable to the human body, is also relevant to social systems. The open system is self-regulating; it maintains itself in a dynamic equilibrium. Any disturbance from the environment tends to be counteracted; but if it persists then such disturbance can lead to adaptations within the system and the establishment of a new steady state. Unlike the more mechanistic concepts of equilibrium or homeostasis, the notion of steady state allows for growth, for transition towards a higher order, more organization, greater heterogeneity.

If the concept is valid, it has obvious implications for social work intervention, and would tend to provide a theoretical explanation for much recent research evidence about the difficulty of making inroads on the client's situation. The idea of steady state also lends credence to the argument that social workers must beware of assuming that even the most deviant of clients is necessarily in some disorganized (anomic) social situation, or that families failing to conform to a hypothesized conventional norm are necessarily in need of therapeutic intervention, for even in such apparently extreme instances a steady state may exist which will be superior to any externally imposed alternative that the social worker may be able to arrange. Thus the concept links up both with deviancy theory in sociology and with the work of the Institute of Marital Studies in social work, and leads to a note of caution for the therapeutic practitioner.

Example The evidence from the Support Project suggested that volunteers had little impact on the total situation of the families to which they were allocated, despite the intensive efforts that some of them made. Those who came closest to being absorbed into the client-system exercised the greatest influence therein; but conversely those whose influence appeared to be potentially the most disruptive (i.e. the most radical in terms of the client's cultural perspective) were met by an insurmountable barrier that precluded them from affecting the family's steady state in any way. It might therefore be suggested that the client's system is open to influence by the social worker only when he comes to be perceived as a part of that system – a rare occurrence under normal circumstances, although there were some cases in the sample where the presence of a social worker – any social worker – appeared to have become an essential part of the family-system anyway, with the result that volunteers were simply absorbed into the system without difficulty. A different kind of influence might occur when the social worker employs techniques derived from his external authority, enabling him to exercise constraints over the client and his system, and in extreme cases, to remove him from it: but this is a totally different role for the social worker to play – at least, in the eyes of the client.

(In a different example, the concept of steady state is crucial to residential settings, in which new arrivals are imposed upon the existing system which is thereby compelled either to adapt to the expectations of the newcomer, or more usually to exert control over him, to socialize him and ultimately to absorb him within the fabric of the existing system. It is apparent that many residential settings now make free use of medication in order to reduce the disruptive behaviour of any newcomers and so induce readier conversion and absorption. On the other hand, it is also apparent that regimes in residential care do change over a period of time as newcomers among staff and residents have influence over the system.)

VII Negentropy

Negentropy is a crucial characteristic of open systems – as distinct from closed systems – and is linked to the concept of steady state. It means simply that open systems do not tend to run down, in the way, for example, that a clockwork motor (which is a closed system) runs down. Because they are *open* systems, and constantly in receipt of energy from the environment, they are continually being stimulated into new growth, new developments, evolutionary movement. Though remaining as a steady state, the open system not only preserves its order, maintains itself, but might also evolve towards even more advanced states.

Example While, at first sight, the concept of steady state may lead to some denigration of the social worker's role – after all, if the family is immune to such environmental intervention, why bother? – the concept of negentropy is more encouraging. Every system is dependent for its survival and growth on receiving energy inputs from the environment; these inputs may not bring about change, but they are essential for maintenance. Hence it is possible to argue that, provided he is offering a form of energy input required by client systems, the social worker is fulfilling a vital role in ensuring the continuation of living systems – whether they are in the community or in residential care. The analogous model would represent the social worker as a provider of basic and essential foods for survival rather than as a therapist offering cures for sickness; and this analogy would apply, not merely to preventive work but in ordinary casework as well.

In the Support Project, the work of the volunteers can be presented as ensuring the negentropic process so long as they remained in contact with the family; of course, as an open system, the client-family is not dependent on the volunteer for survival, but the rationale of the Project is that the availability of a helpful volunteer will make a significant difference in the pattern of survival characterizing each case. In fact, some volunteers made minimal impact and withdrew; but even among these, there were some whose clients expressed greater appreciation of their work than would have been thought likely from what the volunteers themselves told us: often just visiting had been a valuable energy input for families who felt particularly isolated from other systems of community support. Where the volunteer carried on visiting well beyond a year, and in some cases spoke of having made "a friendship for life", the force of the input and its consequent effect on the family's steady state were considerable – but such cases, of course, were exceedingly rare.

Conclusion

The language of systems theory is relatively alien in a traditional social work context, and, because it is employed in order to grapple with social interactions of great complexity, it too reflects something of the multidimensional reality that characterizes interpersonal relations. Within the concept of systems, there is still room for a focus on one-to-one relations, on isolated interview or group occasions, and on the minutiae of client diagnosis; and for the worker, the identification of tasks en route to the achievement of specified objectives is crucial: hence, it is not necessary to think only or all the time in strategic terms. But systems thinking certainly demands a recognition in social work, not only that strategy is essential if tasks are to be both relevant and successfully performed, but also that sometimes the most obvious helping functions are wholly inappropriate given the reality of the client-family-system.

References

Baker, R. J. S., 1975, Systems Theory and Local Government, *Local Government Studies*, January, pp. 21–35.

Bertalanffy, L. von, 1973, *General System Theory*, Penguin.

Bowlby, John, 1953, *Child Care and the Growth of Love*, Penguin Books.

Buckley, Walter, 1967, *Sociology and Modern System Theory*, Prentice-Hall.

Churchman, C. W., 1973, Systems. In Optner, 1973, pp. 283–293, q.v.

Davies, Martin, 1974, *The Support Project*, Department of Social Administration, Manchester University. (Mimeo).

Dawe, Alan, 1971, The Two Sociologies. In Kenneth Thompson and Jeremy Tunstall (Eds.), *Sociological Perspectives*, Penguin.

Emery, F. E., 1969 (Ed.), *Systems Thinking*, Penguin.

Goldstein, Howard, 1973, *Social Work Practice: A Unitary Approach*, Columbia, University of South Carolina Press.

Hearn, G., 1969, *The General Systems Approach*, Council on Social Work Education, N.Y.

Jaffe, J., 1973, The System Design Phase. In Optner, 1973, pp. 228–259, q.v.

Janchill, M. P., 1969, Systems concepts in casework theory and practice, *Social Casework*, 50–52.

Kahn, A. J., 1973, *Shaping the New Social Work*, Columbia U.P.

Katz, D. and Kahn, R. L., 1969, Common Characteristics of Open Systems. In Emery, 1969, pp. 86–104, q.v.

Kolakowski, L., 1964, In Praise of Inconsistency, *Dissent*, pp. 201–209.

Kremyanskiy, 1969, Certain peculiarities of organisms as a 'system'. In Emery, 1969, pp. 125–146, q.v.

Meyer, Carol, 1972. In Mullen and Dumpson, 1972, pp. 158–190, q.v.

Meyer, Carol, 1973. In Kahn, pp. 47–53, q.v.

Miller, E. J. and Gwynne, G. V., 1972, *A Life Apart*, Tavistock.

Mullen, E. J. and Dumpson, J. R., 1972 (Eds.), *Evaluation of Social Intervention*, Jossey-Bass.

Optner, S. L., 1973 (Ed.), *Systems Analysis*, Penguin, 1973.

Pincus, A. and Minahan, A., 1973, *Social Work Practice: Model and Method*, Ithaca, Ill., Peacock Publications.

Sainsbury, Eric, 1975, *Social Work with Families*, Routledge and Kegan Paul.

Silverman, M. and Chapman, B., 1971, After-Care Units in London, Liverpool and Manchester. In Sinclair, Ian, *et al.*, *Explorations in After-Care*, HMSO.

Stein, Irma, 1974, *Systems Theory, Science and Social Work*, Metuchen, N.J., The Scarecrow Press.

Thompson, Kenneth, 1972, *Sociological Perspectives*, Open University Press.

Thompson, Kenneth and Tunstall, Jeremy, 1971, *Sociological Perspectives—a Reader*, Penguin Books.

Vickery, Anne, 1974, A systems approach to social work intervention: its uses for work with individuals and families, *Brit. J. Social Wk*, 4–4, pp. 389–404.

Reprinted by permission of the author.

17 Analysis for implementation

by Van Court Hare Jnr.

Most systems analysts find that they introduce change in the systems they study. They correct the faults of malfunctioning systems, or seek to improve an existing system. Or, they design new systems and seek to implement them, displacing older systems by new ones. Indeed, the systems analyst, probably has "the future in his bones", to use C. P. Snow's phrase. Otherwise, he would not have taken up his occupation.

Often such changes greatly broaden the scope of the systems analyst's problem and call upon his knowledge not only of technical possibilities but also of institution and culture.

When we consider this broader subject, discussions of the nature of man, a subject central to most of the great systems of human thought, eventually confront us. Based on assumptions or ideas, usually simplified to suit the times, the concepts of the rational man, the sinful man, the man of will and power, the economic man, and the Freudian man have all become the basis for arguments and theories of human behavior that are still with us in many forms.

The latest trend in building a picture of human nature is to consider "behavioral man", who is defined by the sum total of his observed actions, rather than by assumptions about his character. For our present discussion, this latter viewpoint provides a more varied picture of the human scene and a number of detailed observational results that can be used by the systems analyst. Because all operating systems at one or more stages of their definition, development, analysis, and use are affected by human individuals, knowledge of actual behavior patterns in different circumstances becomes an essential ingredient in understanding how systems come about and how they will be accepted and work in practice.

In what follows we present a sketch of a major cultural conflict and a corresponding discussion of behavioral science man that has meaning for the systems analyst who hopes to implement system change.[1]

Social science and technological change

Social scientists have long been concerned with the problems of technological change and how they influence the society and the culture in which we live. The reverse problem, of course, is also important because the historical setting, and the tools and ideas presently known, affect the selection of projects considered worthwhile.

In viewing this grand process of technological advancement and social change we generally find a conflict between those who propose change and those who prefer the present state of affairs. In analyzing a given society, for example, it is useful to make a distinction between what Wheelis has called "the instrumental process" and "the institutional process". These are concepts representing two opposing clusters of activities, attitudes, and kinds of criteria for what "good" is. Both processes affect groups within society at a given time, and indeed to a greater or lesser extent they affect the individual at a given time. Thus, our understanding of these processes sheds much light on the ways in which system change may be acceptably introduced, so we shall consider each of them in turn.

The instrumental process

The instrumental process, which is concerned with "the facts", stresses replication, verifiability, and usefulness in social life. "The authority of the instrumental process is rational, deriving from its demonstrable usefulness to the life process. The final appeal is to the evidence."[2]

The scientific method approximates the heart of the matter, but the instrumental process is a larger concept that concerns the development of tools and techniques (both physical and mental) used to solve problems. Electric saws *and* the differential calculus are both instruments for this purpose. It includes art, both fine and applied, because materials and methods are required in their completion. The instrumental process is bound to reality, facts are facts, it seems to say. Ignoring them is of no avail. Reality can be altered, particularly if it is clearly observed. Indeed, the better one understands it and the more tools one has to deal with it, the more radically it can be changed.[3]

We may not like the facts, but there they are – for better or for worse. Our job is to proceed with the project at hand. As we proceed with the instrumental process, we learn more and develop more tools, which, in turn, increase the output of the instrumental process and the number of possible combinations of what is already available.

The instrumental process is respected because it is useful, productive, growing, and bountiful in many, though not all, spheres of human life. Yet, from an individual or social viewpoint, for very personal reasons, the instrumental process "is often disparaged as mere problem-solving; for the security it creates, though real, is limited".[4] It may be respected, but not loved.

The institutional process

The institutional process, on the other hand, builds certainty, not doubt, for the individual. It seeks stability, sure-footedness, a rock of ages. Change, particularly rapid change, is shunned.

Thus, both the individual and society build strong barriers, both conscious and unconscious, for protection against change. Most frequently, these barriers seek an authority, organization, or tradition – in short an institution – larger than self or even everyday reality. This institutional process is diametrically opposed to what we have previously called the instrumental process.

> Everything mundane is subject to change, and hence certainty is not to be found in the affairs of men. The searcher arrives at his goal [of certainty], therefore, in a realm of being superordinate to man. Solomon put it succinctly: "Trust in the Lord with all thine heart; and lean not unto thine own understanding."[5]

Although religion is one example of the institutional process, the concept is broader than that. The institutional process includes customs, taboos, rites, mores, ceremonial compulsions, magic, kinship, status, coercive power systems, and such modern institutions as private property and the sovereign state. As such, "The authority of the institutional process is arbitrary; the final appeal is to force."

> In particular, the institutional process is bound to human desire and fear. Wishing will make it so, it seems to say. It is unbearable that no one should care; so there must exist a heavenly Father who loves us. Activities of the institutional process do not, objectively, gratify any need or guard against any danger; incantation does not cause rain to fall or game to be plentiful. But such activities may engender a subjective sense of security, and this has always been a factor to be reckoned with – and, indeed, to be exploited. Honor and prestige accrue to the institutional process; for the security it creates, though illusory, is unlimited.[6]

Some findings about behavioral science man

The matter is put in a slightly different way by Berelson and Steiner, who, after compiling an inventory of scientific findings in the social sciences, summarize that body of knowledge on "behavioral science man".[7]

> Perhaps the character of behavioral science man can be grasped through his orientation to reality. . . . First, he is extremely good at adaptive behavior – at doing or learning to do things that increase his chances for survival or for satisfaction. . . . But there is another way man comes to terms with reality when it is inconsistent with his needs or preferences. . . . In his quest for satisfaction, man is not just a seeker of truth, but of deceptions, of himself as well as others. . . . When man can come to grips with his needs by actually changing the environment he does so. But when he cannot achieve such "realistic" satisfactions, he tends to take the other path: to modify what he sees to be the case, what he thinks he wants, what he thinks others want.
>
> [In the latter case] he adjusts his social perception to fit not only the objective reality but also what suits his wishes and his needs . . . he tends to remember what fits his needs and expectations . . ., or what he thinks others will want to hear . . .; he not only works for what he wants but wants what he has to work for . . .; his need for psychological protection is so great that he has become expert in "defense mechanisms" . . . he will misinterpret rather than face up to an opposing set of facts or point of view . . .; he avoids the conflicts of issues and ideals whenever he can by changing the people around him rather than his mind . . ., and when he cannot, private fantasies can lighten the load and carry him through . . .; he thinks that his own organization ranks higher than it actually does . . .; and that his own group agrees with him more fully than it does . . .; and if it does not, he finds a way to escape to a less uncongenial world. . . .[8]

The introduction of change, particularly change that seems to the individual beyond his control and which therefore threatens, or reduces perceived security, elicits reactions that are not necessarily logical and that the systems analyst may expect in the course of his work. If the systems analyst proposes change, which is the essence of the instrumental process, he meets the proponents of status quo, for the essence of the institutional process is to stand pat.

Indeed, some institutional processes are so rigid, that the

innovator contests them at his peril. In his *Letters from Earth*, Samuel Clemens wrote,

> We do not know how or when it became custom for women to wear long hair, we only know in this country it *is* the custom, and that settles it. . . . Women may shave their heads elsewhere, but here they must refrain or take the uncomfortable consequences. . . . The penalty may be unfair, unrighteous, illogical, and a cruelty; no matter, it will be inflicted, just the same.[9]

And, some habits are so strongly ingrained that they are impervious to change. Many individuals work at night to avoid change. As one elderly worker, who had worked the night shift for forty years expressed it to the author:

> When I went to work there were no cars on the street, only horses. Wagons and horses. Now there are cars, and too many people. Too many. Why do I work at night? Things never change much at night. In the morning there are cars and people. But, things never change much at night. Who needs those cars and people?

It is interesting that the instrumental and institutional processes described above have been compared to the term *ego* and *superego* used in individual psychoanalysis.

In those terms, the ego represents the executive department of the human personality – the instrumental activities that recognize facts, marshall resources, devise plans of action, and get things done.

The superego is the judicial department, which performs a screening function, directs awareness, vetoes unacceptable proposals, and sets values and effectiveness measures used in goal formation. The superego is institutional in character, and derives its "conscience" from the culture, its customs, habits, and mores.

Most studies show that the conscience so formed is highly relative to the culture or society in which the individual lives.[10]

For many important issues, constraints upon the superego are set by a small group – those near home as it were. Thus, the constraints are greater where families live together for several generations than where they do not, greater in small towns than in large, greater at home than abroad, greater when there are strong religious beliefs or formal institutional ties than when there are none, greater when only one set of values is perceived than when there are many that seem relative to time or place. We are all familiar with acts and common phrases that illustrate these facts.

> Things happen in motels that do not happen in homes, and towels are swiped in distant hotels by persons who would not steal a pin in their hometowns. Some persons, indeed, travel for just this purpose – to lose an unwanted reinforcement of conscience. For them wanderlust is not a lust for wandering but a wandering for lust.[11]

Similarly, methods exist for strengthening the superego or judicial function in the personality (and therefore the institutional function in the culture), and also for reducing its effect. Street lights reduce burglaries, and double-entry bookkeeping reduces embezzlement. Conversely, in surroundings where multiple values are evident and where institutional restrictions are consciously relaxed, the range of acceptable value and goal choices increases, leading to a greater variety of possible actions or considerations and to a greater potential for change. Thus, as institutional restrictions become less (and instrumental efforts are, relatively speaking, more respected), new proposals become more easily accepted by those who must approve and use them and vice versa.*

Note also that the capacity of the investigator or the individual to effect system change, or to alter his perception of the facts to suit his needs, is due to his capacity for the creation and manipulation of *symbols*. Language and abstraction form the concepts, ideas, and instructions that permit learning and the transfer of experience (and the very act of systems definition and analysis).

This capacity, which permits the analyst to generalize, also permits him to change the name of the game to suit his needs.

Although it would appear to be a fact that a rose by any other name would smell as sweet, for individuals who view a scene this invariance of description may not hold. A plain ham-and-cheese sandwich may taste better if it is described as a "wedge of cheddar wedded to a generous portion of prime Virginia ham surrounded by California tomatoes and Florida lettuce and a discrete portion of pure egg

* The converse effect has interesting psychological implications, although we cannot explore it in detail. For example, when the rate of instrumental change is high in a culture, that area of the individual personality controlled by the superego could be expected to diminish. The individual may thus be involved by a search for certainty in his changing world, yet finds few institutional guides acceptable to him. Wheelis, for example, cites the quest for group consensus and the appeal of mass movements for the individual as evidence of this change in the superego in complex societies. He derives many interesting results from this thesis, one of which is that psychiatry as practiced in many cases treats problems no longer relevant to the present scene. See Wheelis, *op. cit.*, pp. 87–9, and Chapter 7.

mayonnaise". Even though a simple yes or no might suffice, a problem solution may seem more impressive and convincing if crouched in mathematical symbols and presented with a slight but correct accent by a man with a Ph.D. from Cambridge.

But, because our symbols, our abstractions, and our ability to conjure up favorable or unfavorable impressions with a word or gesture are products of our culture, and because the participant-listener's symbols and abstractions are formed by his culture, a knowledge of how symbols are formed and held is the key to our understanding of culture.*

Finally, we should note that institutional problems are magnified as the instrumental process advances with time. Although institutions do change under the impact of technology and instrumental advances, they change slowly and reluctantly, ". . . and make peace, finally, with the conditions which altered them".[12]

But institutional change is *slower* than instrumental change. By the time the institutional process, or the culture, has made peace with instrumental change, ". . . technology has moved on, and the laggard is still trailing".[13] Indeed, the discrepancy between instrumental growth and institutional change becomes worse with time. The combinatorial possibilities of instrumental growth are geometric; more tools, more techniques, more facts, and more concepts lead to a cornucopia of new possibilities. The instrumental process is regenerative, but the institutional process does not partake of this bounty and holds steadfast, unless intimidated, coerced, and pummeled into movement.

The resulting effect is a cultural lag – an ever increasing gulf between what is possible and what is acceptable.[14]

Introducing change

Changes that alter no dearly held belief, custom, or mode of habitual operation are often introduced with relative ease in highly technical areas. Tools may be redesigned, new production methods may be introduced, new weapons may be brought out; tactics, competitive goals, and impersonal means may all be changed with relative ease. However, primary group relations, territorial and religious stability, systems of prestige, customs, mores, and habits resist activation.

Change is easier to introduce in matters arranged on a scale with narrow intervals than in those arranged in a sharp dichotomy – i.e., when the only answer is black or white. Change is easier to introduce when the elements of change are congenial to the culture, and the society has roughly equivalent substitutes or existing alternatives. Change is easier to introduce through existing institutions rather than through new, through individuals of high prestige and status rather than low, through a third "disinterested" party rather than directly. Change is easier to introduce if it directly affects only a small segment of society, rather than the mass, if its side effects are imperceptible rather than pervasive, and if secrecy instead of full publicity is the rule. Change is easier to introduce in times of crises and stress than in more tranquil times.

In short, the introduction of change is eased if the symbols of change present no apparent alteration or modification of the culture's widely held symbols. Indeed, change is greatly facilitated if the culture's present symbols and instructions reinforce the proposed alterations in operation.

To exploit the cultural symbols of the time or to create others that are only slightly changed but useful, to present a carefully edited story, to obtain the approval of a high-status group for a project, to associate the new development with values already held dear – all of these activities help bridge the cultural gap and make a new proposal acceptable.

Example: Pharmaceutical firms have the problem of introducing and marketing new drugs as they are developed and tested. Because many new drugs of a specialized type are introduced each year, the physician is deluged with circulars, samples, and "detail men" explaining the virtues of their products, many of which compete for the same type of treatment. Several drug firms have organized this effort to exploit institutional values held by physicians. Extensive mailing lists are maintained with records on each physician (containing, for example, age, school attended, organizational affiliations, and the like). When a new drug is to be introduced, it is often possible to analyze this list and to compile a relatively shorter list of former professors or maestros who represent a higher status group. If the maestros react favorably to the new drug – and their smaller number makes them easier to convince with limited resources – then their approval can have a strong

* A culture is most frequently defined by its community of understanding, and may be measured by the frequency of compatible interactions that occur between its members, as indicated by Deutsch, K. W. (1953) *Nationalism and Social Communication*, Technology Press and Wiley, New York. C. P. Snow expressed the same idea in literary terms: "Without thinking about it, they respond alike. That is what culture means." To speak of a cultural lag is perhaps not so descriptive as to speak of a cultural conflict or gap between the instrumental and institutional community of understanding.

institutional influence in making the technological advance acceptable to the total list.

Example: The elimination of elevator operators (by the substitution of push buttons) to reduce operating costs may well be spoken of as an attempt to improve passenger service, to increase passenger comfort, and to reduce delays in the lobby. The acceptable symbols are stressed, and the less acceptable image of a mercenary landlord throwing old retainers out of work is played down.

Another benefit derives from these seemingly devious devices. Change is easier to introduce in form than in substance: When imposed "from the outside" the forced change may result in overt compliance, but covert resistance. If the proposed change can be made to come "from the inside" (or seem to), the form and the substance of change are more likely to coincide. The change is also more likely to be permanent and not to be a mere verbal acquiescence until the analyst has gone.

Anthropologists also tell us that in the evolution of a culture pattern action comes first and values second. We begin to like what we do. Thus, it is frequently easier to change the values of individuals concerned with a system by alteration of their activities rather than by a direct attack upon their beliefs.

Example: When a system's users are unfamiliar with it they may be afraid of it, and thus hold it in low esteem. However, *after* experiencing success with its use they become more optimistic about their ability to influence their own future, and the values employed by the users in assessment of the system change. Consequently, instead of describing the virtues of a new product or system, a demonstration model is put in the user's hands so that he may convince himself of its value.

Similarly, if behavior can be changed, a change in values usually follows more easily than if the reverse procedure is followed.

Example: These alternate approaches to change are evident in the policies of the Eisenhower and Johnson administrations with regard to racial discrimination. The Eisenhower policy was that social values would change with time, leading to later changes in behavior. The argument for this approach is that when the behavioral change does occur, it will be permanent because it is reinforced by the individual's value structure. The Johnson administration, on the other hand, in urging stronger legislation against discrimination, sought to change behavior from which a change in values would emerge. It is argued that the latter course produces permanent social change more swiftly than the former, although covert resistance may be expected at the outset.

When introducing change in large-scale systems and organizations, policies are much easier to change than procedures. This observation, made by many professional administrators and systems analysts,[15] does not contradict what has just been said, but blends instrumental considerations with it.

First, the detail and variety of specific changes that must be made in procedures to obtain a given result frequently exceed the resources of the controller. And, because procedures usually adapt themselves to the demands of policy, much more can be accomplished in a given time with limited resources by the policy-change approach, leaving resources available for the adjustment of critical procedures when necessary.

But, perhaps more important, a change in policy rather than in detailed procedure, leaves some freedom, although it may be illusory, for the individual to adjust to the proposed change. His values and institutional roadblocks are not directly confronted, and he is more likely to accept "his own" adjustments as good ones.

For similar reasons, making a few large-scale policy changes is often easier and more effective than instituting many small changes. The few major changes can be controlled, and the individual does not develop the frustrations, confusions, and value disturbances that a series of harassing minor alterations will produce.

Problems of implementation

Without going further, it would appear from what has been said that many repetitive difficulties arise when certain kinds of change are proposed and introduced, and that from this experience much has been learned about social and cultural changes. Moreover, numerous techniques and artifices may be used in easing the introduction of instrumental change and in mitigating institutional obstacles. Yet the field is an open one, and much remains to be done.

If this is so, it would appear reasonable to conduct analyses aimed at the implementation of specific systems, or at specific systems improvements – so-called action research – whenever system innovation and improvement are proposed.

For example, the efforts of political scientists and public relations firms to elect a given political candidate, to ease passage of a school bond issue to reduce racial discrimination, or to promote the fluoridation of drinking water represent research of this type in the social sciences. Market and advertising research also has the same "action" flavor. The same may be said for the use of psychological tests and research in personnel selection for given occupations. The research is not "disinterested" in its outcome; it is purposeful and goal directed.

Such activities, which we prefer to call analysis for implementation, are often shunned by the scientist (for reasons we shall mention hereafter), although, for a particular purpose and stated objective, analysis for implementation may greatly increase the success of a technical advance, discovery, or improvement.

For example, even relatively simple system changes may raise questions for which the uninitiated analyst is unprepared. The automobile owner hesitates to spend more funds on his car because his children are sick, so he believes the car will run another year – even though the mechanic knows objectively that the vehicle is potentially dangerous. What is the mechanic to do? The patient hestitates to have the clearly indicated operation and believes he will get better. Should the physician retire after stating the bald facts? Persons who have not experienced the disasters of a tornado, flood, or large explosion tend to deny or to disbelieve warnings that danger is near. They search for more information, and often "... interpret signs of danger as signs of familiar normal events until it is too late to take effective precautions." [16] Should the weather man report his scientific conclusions and make no interpretation or exhortation to his listeners to take care? A system that takes care of today's conditions is installed, but it does not provide for future contingencies or for "updating" the system as conditions change with time. Later the system efficiently performs functions no longer relevant to prevailing conditions, with possibly disastrous consequences. Should the analyst not concern himself with these problems of implementation and use?

Should the analyst concern himself with the problems of cultural change?

There are many scientists, and thus many systems analysts, who would, on serious grounds, take issue with the proposal that the investigator concern himself with the use made of his work. Rational investigation and logical decision cannot take place, they argue, if one has a vested interest in obtaining a given outcome. This position is strongly held by science as an institution because of the historical struggle to free science, and the instrumental process, from the constraints, the dogma, and the myth-making of institutions.

There is yet another reason for the strength of this position. As older institutional constraints have been removed, others have come into play as the individual searches for the universal certainty which the rational process cannot supply.

For example, a vote of the majority or a mass movement may threaten the dispassionate scientific pursuit just as much as constraints imposed by a monarch or the clergy. The mathematician does not arrive at a problem solution by conducting a public opinion poll on the street and averaging the answers obtained, even if those polled are other mathematicians! The test is different. It depends upon the demands of a verifiable procedure, not the beliefs of any individuals, regardless of their reputation, their power, their eminence, or their number.* If in his choice of variables or the development of alternatives (both of which are subtle selection processes) the scientist consciously begins to favor or exclude one group of possibilities for institutional reasons, or if he must shade his thinking, model building, and verification to meet institutional constraints, he has abandoned the basic tenets of science. Surely, he says, many new discoveries that might otherwise be his will be lost, or worse he will be deluded and falsely evaluate what he observes.

The scientist may also argue that his energies are limited. To worry about implementation will not only debase his pursuit of knowledge but will also embroil him in a conflict of personal values that will consume his limited resources in a wasteful fashion. How can he simultaneously be a myth-maker and an iconoclast?

The dilemma presented by arguments for and against research for implementation is well known, and although exaggerated here for effect it is nevertheless real. The scientist–analyst may choose to avoid the problems of implementation altogether, but someone will implement the results of his work, or not, and the job may be done well, or poorly, depending upon how well the analyst and the implementer understand each other.†

* For example, one test for an instrumental versus an institutional process is whether a conclusion is validated by the success of a procedure or the agreement of an individual or group in society. Thus, if a surgeon sets a broken arm, that is first aid. If the nurse performs an appendectomy, that is nevertheless surgery. First aid and surgery are instrumental. If a priest administers the sacrament with wine and wafers, it is a holy ceremony; but if a lay person performs the same act, it is a sacrilege. A marriage performed by an unlicensed individual is invalid. The latter procedures are institutional.

† An interesting sidelight on this problem is the reluctance of professional people to engage in politics, an attitude encouraged by the Federal Tax Policy. "... Federal tax policy forbids

Leadership and organization to bridge the cultural gap

One resolution of this difficulty – which almost always arises in applications of management science and operations research, to name one area of systems analysis – is to conduct a project with mixed teams of investigators, some of whom have instrumental skills, and others who have institutional skills. These skills when shared in the work group bridge the interface between the two worlds and permit a smoother transition between problem definition, analysis, and implementation.

Many organizations use similar devices to bridge the gap between instrumental and institutional requirements. For example, often an organization has two leaders instead of one, although one man may be apparently in charge. The role played by one leader is to institute instrumental change. The role played by the other leader is to represent the institutional requirements of the organization and the component individuals. When the instrumental leader causes friction, the institutional leader smooths it over or rephrases the requirements in more acceptable, warmer, more congenial, or more orthodox terms. It matters little who is the obvious leader, so long as the two cooperate (and can stand each other)! Together they will be able to produce more change in the organization than either could accomplish individually. The family unit, to come closer to home, offers a similar example of dual leader effectiveness in creating change in the habits of the young.

A final example from intelligence operations is interesting because it illustrates the power of the same approach, employed to different ends.

Example: Many of the intelligence services of the world employ a method of interrogation in which two interrogators, each employing a distinctly different role characterization, alternately confront the subject. The first man, for example, may affect a stern air, a military costume with riding boots and crop, a crew cut, an air of efficiency and dispatch, and a stern, cruel disposition. As the perfect martinet, he demands the subject talk at once or be shot at dawn, withdraws all physical comfort, inflicts various apparent physical and mental tortures – just as the subject might anticipate. Finally, in exasperation, ordering the firing squad to be assembled, the first interrogator leaves the room. Very likely he has learned nothing, and expected to learn nothing, from the subject. After some time, the second interrogator enters. He is an entirely different type. Dressed in baggy tweeds, probably smoking a pipe, he seems distressed with the subject's plight, orders some food to be brought at once, produces cigarettes, offers apologies for the abruptness of his colleague, and settles down to commiserate with his unfortunate friend. He may offer some personal experiences of his own, some philosophies and observations on life and his friend's present plight. But then, he is unavoidably called from the room, and the subject is again alone. He did not expect this. As the night wears on, the alternate presences continue. First he is threatened, then he is consoled. In the end, although some time may be required, the subject begins to change: He sees the constraints of his former world as less important than before, his values begin to seem irrelevant to his present state, he may even be convinced of the error of his ways and seek to explain himself in his defense. Then he talks, as was intended, most likely to his tweedy friend who in fact was the instrument of the subject's change.*

Little more can be said here on this topic, but the administrator of systems analysis projects who seeks to blend instrumental and institutional skills for a given study can benefit from further study of the dual role approach to organization.

Organization of data for implementation

To continue, let us suppose the investigator is concerned with the problems of implementation, and that his concern is to anticipate problems that may occur in a specific case, rather than to worry in general.

Many of the examples and points cited in our next few pages follow from underestimates of institutional power to affect routine and emergency operations of instrumental processes.

Why are implementation data scarce?

In most specific cases, the selection of treatments and the anticipation of specific problems and reactions requires high technical competence and experience in a given field. Nevertheless, in a given field, the data of past experience can be organized in an orderly fashion and "what if"

charitable deductions of gifts to organizations engaged in promoting or opposing legislation. Thus, for example, the bulk of foundation money goes to organizations that conduct research, and financial aid for organizations that take public positions, however worthy, is hard to come by. As a result, lobbies representing private interests operate freely and effectively, while those groups that might represent the public interest stay out of politics." Abrams, C. (1965). "The City Planner and the Public Interest", *Columbia University Forum.*

* For an extensive discussion of interrogation and training methods in this form of intelligence operation, see Pinto O. (1952) *Spy-Catcher*, Harper, New York.

questions can be asked for specific analyses of implementation. For example, the collection of symptom–cause relationships described in Chapter 11 might be a typical example for this type of display. (Problem characteristics would then replace symptoms, and probable difficulties would replace causes.)

However, in collecting data for implementation, the purpose of the collection is expanded. The emphasis is not on the apparent difficulties and present symptoms, but on those that *might* appear in the future, or that have been known to occur in the past, given specified side conditions.

In medicine, for example, certain treatments or drugs are known to be "counterindicated" when the patient has a given past history, and these counterindications are reported and publicized in the profession.

In the same way, certain forms of system alteration or modification are known to raise given problems with great regularity. When information systems are installed, or when decision-making processes are analyzed and changed, we encounter the same problems over and over again.

Unfortunately, in the latter case, published warnings and admonitions are uncommon. The investigator who analyzes and installs business systems, for example, is seldom forewarned against probable pitfalls, and he seldom has organized, published data to aid his implementation work. We may have diagnostic aids for hardware maintenance, and for trouble-shooting in many fields. However, there has been little emphasis to date on the diagnosis of conceptual or decision-making failures on the part of the analyst when he implements systems change. Where such data are available, the files are kept secret or transmitted from one worker to another as an art – for institutional reasons which should be obvious from our previous discussion in this chapter.

For example, from the 35,000 or so electronic computer installations made in the United States in the past ten years much has been learned and published about electronic difficulties, hardware reliability, component failures, and design defects. Yet little, if anything, has been reported in an organized form about the many specific failures of the systems that have been installed – if those failures were due to faulty decision-making, inadequate system conception, or lack of individual experience and foresight. Little, if anything, has been published about the specific difficulties those systems have encountered as times have changed. Not only is the nature of such data collection and organization difficult, but institutional barriers also prevent publicity. The physician does not testify against his colleagues except within the instrumental framework and the users and manufacturers of computing equipment are not likely to publicize their failures except when the onus can fall upon the hardware details, and frequently not then. The available data are locked in consultants' files, or in the notebooks of internal investigators. The auditor has his check list of devious practices to look for, but he is not likely to publish a statistical account of his, or his clients', difficulties and shortcomings.

The point of these illustrations is that such compilation of problems, even if maintained privately by the investigator, can be a powerful tool in anticipating and preventing a repetition of implementation difficulties. The construction of such a listing is thus one of the major steps in analysis for implementation. It is a valuable step if the result is only a list. It is an even more valuable tool for the analyst, in a given case, if the broad listing can be classified and cross referenced in a hierarchy of problem types, with specific classifications for the anticipated problems of specific system types. In such a tabulation it would be seen that many of the specified problem areas arise when instrumental and institutional processes meet at an interface, when man-and-machine or man-and-man meet.

A problem anticipation file

To illustrate one such listing, we now present a selected group of implementation problems that are general enough to provide both a manageable list and a set of categories for further development. For each of these categories, we provide an example.

The reader may add his own examples to this outline. Our aim in the following presentation is to suggest a set of major problems to anticipate in a wide variety of systems implementations, and to leave further development to the reader. The categories chosen are taken from a composite of systems analysis problems in several fields, and are presented in the form of questions to stimulate further discussion and thought.

1 *Are the objectives and constraints perceived by the investigator the same as those perceived by the organization?*

Frequently, differences in cultural background and the conflict between professional and administrative interests cause the systems analyst to solve the wrong problem. This most subtle of systems mistakes happen too frequently – even when the problem of analysis is stated in writing and agreed to by the analyst and the user – that great care must be taken to reach a common understanding of what is needed.

The trouble is usually not so obvious as maximizing profit instead of minimizing cost, although that type of error is frequent enough. Usually, an objective is stated formally by the system user, but perhaps incorrectly or incompletely. So, the analyst may not at first perceive constraints that the user imposes upon acceptable solutions, or, conversely, he may anticipate constraints that are not in fact considered important. He may often be purposefully deluded.

Example: A major oil company sought a scientific procedure for locating filling stations as new stations were added to the distribution system. Initially, the objective was to locate stations that would return at least a minimum return on investment, as specified by the firm's management. A procedure that included many factors, among them estimated sales of petroleum products at the proposed stations, was developed. Check of the new procedure indicated that it would locate stations more consistently and reliably than less organized methods. However, the new procedure was rejected by the executives charged with station location. After some time these executives revealed the method was not suitable to them because it included forecasts of station sales, and this limited executive flexibility in acquiring new locations. As it turned out the true objective was to add 200–300 new stations to the system each year. Many proposed station locations could not make the required return on investment, and competition for good locations was severe. Knowing that the computed return on investment was based on internal transfer prices, which also included contributions to profit, the executives often inflated the estimate of station sales to get around what they felt was an unrealistic investment requirement and come up with the required number of new additions (which otherwise would not have been possible). Had the proposed method taken these facts into account, the systems analysis project might have been implemented, rather than rejected. It is not surprising in this instance that proposals for follow-up studies to compare predicted sales and investment return against actual values were also rejected.

The analyst may avoid this pitfall or sidestep it in many cases by investigating in advance alterations in constraints and objectives that may be of possible interest, and by determining how the problem solution is affected by alternate problem statements. He may find that the problem statement contains many noncritical factors, learn which factors are critical, and be prepared for the presentation of alternate proposals should they be required.

Because of his training and knowledge of methodology, the analyst may also tend to frame problem objectives and constraints to make his job easier, unwittingly leading also to the solution of the wrong problem. The work may be professionally competent and workmanlike in every respect, but it may also be irrelevant to the present need. The cliché that the right problem solved approximately is a better result than the wrong problem solved precisely is a fair warning. Most major problems of systems implementation are introduced at the beginning, when the wrong objectives and constraints are assumed in a system definition.*

There is another reason for looking carefully into the goals, objectives, and constraints to be used in a systems analysis project or design. The implications of the chosen goals, objectives, and constraints may not be clearly understood at the outset and a solution which seems desirable at first may in fact be fraught with difficulties that once commenced are difficult to overcome.

For example, when automatic or automated systems are given a goal, it is, as the late Norbert Wiener observed, like invoking a form of magic. The goal is interpreted literally, and the full implications of a given instruction or objective must be understood by the analyst if he is to stave off embarrassment and woe.

Folklore, for example, contains stories – ranging from the *Sorcerer's Apprentice* to the *Monkey's Paw* – in which magic, once begun, was difficult or impossible to turn off. The Apprentice learned the words to bring magic to the alleviation of his workday tasks, but forgot how to stop the magic broom and the magic pail of water from overdoing their appointed tasks. (Fortunately, he was saved by the Master's return.) In the *Monkey's Paw* and many similar tales, three wishes were granted. In each case, the first wish was fulfilled, but with unsuspected side effects. And usually in these stories the last two wishes were used to undo the horrors created by the fulfillment of the first.

Thus, at the outset, the analyst should beware lest he have King Midas' touch.

2 *Are the effectiveness measures used in the analyses appropriate?*

Since goals are formulated from sets of values or effectiveness measures, the measures used may be incorrectly chosen and throw off the analysis.

* "At first it is impossible for the novice to cast aside the minor symptoms, which the patient emphasizes as his major ones, and to perceive clearly that one or two facts that have been belittled in the narration of the story of the illness are in reality the stalk about which everything else in the case must be made to cluster." Herbert Amory Hare (1899) *Practical Diagnosis*, Lea Brothers, Philadelphia.

Example: A classic example, reported by Morse and Kimball[17] from their World War II experience, concerns the installation of antiaircraft guns on merchant vessels. On the one hand, guns installed on these ships were so "ineffective" as to be useless; on the other hand, they made the crews feel safer. Because the guns were expensive and were needed elsewhere, their removal was proposed. Indeed, data on equipped and nonequipped ships showed that only 4% of attacking planes were shot down, a dismal figure that served to indicate the guns were not worth installing or keeping aboard. On second thought, however, it was apparent that the percentage of planes shot down was not the correct effectiveness measure for the guns. Guns were installed on the ship *to protect the ship,* and the proper measure was whether the ship was damaged less with or without a gun. Analyses of the observed data in this light showed the guns definitely increased the ship's chance of survival. Even though the antiaircraft guns did not often shoot down the attacking planes, a gun's use lowered attack accuracy, reduced damage, and often saved the ship. The change of measure changed the decision, and the ships were equipped with guns.

A typical error in selecting effectiveness measures is to scale alternatives by their ability to reach a given objective without regard to the resources consumed, or to seek a resource measure only without regard to effectiveness. The property of "cost-effectiveness", which gives the contribution to effectiveness per unit resource, is often used to remove this problem. Even then, it is essential that the "effectiveness" measure be the one desired.[18]

3 *Are the attention and awareness functions of the system correctly oriented?*

Newly designed systems or organizations and those that have been in operation for some time can suffer from problems of incorrect awareness or goal rigidity, generally described as "fixation" problems. Goal-directed systems have some form of goal setting and holding function. And, because a system's awareness of alternate possible courses of action, forms of organization, and other goal and value possibilities limits the types of goals and values that will be formulated, goal setting and holding functions of a system can be affected by these factors.

Example: If the reader will tightly hold the thumb of his left hand in his closed left fist, and close his eyes, he will after a time "feel" that his thumb is larger than the other four fingers combined. This "perception" of the size or importance of the thumb, were it constant, could easily alter the individual's work habits, selection of desirable jobs, and even his sensory view of the world about him. The pictures drawn by children, which seem distorted to adults but realistic to children, are another example of how awareness and perception affect the organization of behavior. The senses most acutely tuned to a given scene, and the "mix" of a system's sensory input types can greatly affect behavior and goal-setting functions. An analysis of the nerve structure of the pig would cause an analyst to believe that the pig considers his snout a very important information source. The pig thus "views the world through his nose". Several authors have discussed this problem of perception and awareness in different contexts.[19]

Difficulties arise if the data required to solve a given problem are not available, are not sought, or are not perceived "realistically", i.e., with respect to the system's survival and growth.

Similarly, the methods used for combining data, generating new goals and values, and updating presently used objectives may cause trouble. On the one hand, the goal setting process may be too rigid. Like the driver who locks his steering wheel and drives over the cliff, many systems can reach disaster if erroneous goal locking is present. On the other hand, lack of goal stability is also a problem. Then, the system acts like a small child who first picks up one toy, then another, in a succession of unrelated activities.

Example: During World War II, and even today, postmasters throughout the country are ordered to watch the mail of certain individuals. During the war individuals suspected of Nazi leanings were put on "watch lists". Ten years after the war, some of these lists were still in use because the original orders had not been withdrawn. Obsolete reports, procedures, and objectives may also be found in most large-scale systems.

The shortcomings illustrated in this section can be alleviated by improving the quality of system inputs, and by focusing attention on the procedures the system uses for formulating and updating goals to meet changed conditions.

4 *Have operating standards been developed for the system?*

If operating standards are not developed when a system is implemented and installed, particularly in lower level operations, errors or deviations cannot be corrected. As a result, even minor deviations create a crisis. Lack of specifications and documented standards for component operations can cause the system to get out of hand, and to deviate unexpectedly and unpredictably from its over-all purpose.

In short, planning, control, and design are virtually impossible without the use of standards.

Hardware systems analysts take great pains to provide detailed operating specifications and standards for their equipment. Of equal or greater importance are standards in procedure and information handling systems.

Example: The frequent underestimates of computer programing costs, research and development completion times, and the effort required to introduce new products are examples of cases where even approximate standards could have prevented implementation mistakes. Monitoring these processes against standards for segments of the total job can usually reduce the error between projected and actual cost, completion dates, and promised performance specifications. Should deviations be noted early in the procedure, the projected time, cost, or performance estimates can be revised, or corrective action can be taken early enough to prevent serious problems.[20]

5 *Are vital system processes protected against danger or failure?*

To lose a leg is not so serious as decapitation. In one instance you can carry on, in the other you cannot. The problem of guarding vital control and flow processes against loss or failure is equally serious for the human being, the organization, the hardware system, or the information system. Systems that do not protect these vital processes against damage and the vicissitudes of the environment do not survive, or have difficulty retaining stable continuity of operation.

Example: An extreme example is the protection provided for a military communications center. The new combat operations center of the North American Defense Command (NORAD) is buried deep inside Colorado's 9656 foot Cheyenne mountain and protected against any predictable hazards from enemy sabotage to a direct hit by a nuclear bomb. Thirteen computers, each independent, are able to pick up the work of others in case of failure. The installation is shielded against radioactivity and the electromagnetic effects of nuclear explosions. Power, water, food, fuel, and other essential supplies are stored in gigantic reserves. Houses and rooms within the mountain are set on four-foot springs of three-inch steel to protect personnel and equipment from the shock of a blast or an earthquake. The tunnels and chambers that house the central communications and control center of the military defense effort for the nation are 1400 feet under the mountain top.[21]

Because critical, higher-order systems processes usually involve human operators, the continuity and stability of service of these individuals is often guarded or planned for in systems implementation.

Example: In one large order-processing system 90 sales officers were connected by teletype to a central office, where orders, received as punched paper tape, were rerouted and scheduled to plants through what is called a "torn-tape filter center". The operation of this center, which was current technology in the late 1950's, required operators to read, tear, and redispatch messages on short lengths of paper tape. These operators, at the heart of the information system, were clearly critical to the success of the operation. Employee turnover, divided loyalty, and similar problems that might affect the operation of this vital spot were virtually eliminated by recruiting as operators only divorced women with several children who had never worked for the company or with the equipment before. Trained in these special tasks and held together by their common economic need and background, the girls hired remained on the payroll for many years, and turnover, absences, and job dissatisfaction were astonishingly rare.

6 *Has adequate provision been made for updating the system?*

Installed systems often have many components or adjustable elements "set" at the time of installation. As time passes, these settings may no longer be appropriate to the system's condition of operation, so the system fails.

Example: The November 9, 1965, electrical blackout of the eastern seaboard was traced to a relay located at the Ontario Hydroelectric Commission's distribution plant at Queenston, Ontario, near Niagara Falls. According to the Federal Power Commission's report, the Ontario relay was set to operate a circuit breaker if the power load exceed 375 million watts. It was set at this point in 1963 and was not reviewed. Subsequently, the average power load on the line controlled by the relay increased to 356 million watts, and thus an ordinary upward fluctuation in power tripped the relay and started the whole blackout.[22]

Similar difficulties occur in systems and operating organizations when policies have not been reviewed and are no longer appropriate, when decision making is based on obsolete data, when component operation has deteriorated from design values, when system structure has changed with time but without the analyst's knowledge, or when the goals and values of the system are not kept current with present requirements.

7 *Is the system protected against direct falsification or illicit interruption?*

Although information errors can occur in many parts of a system, either from mechanical failure or human mistake, there are usually several ways in which to introduce false information into the system purposefully or to intercept confidential information for competitive purposes.

Example: Inventory control systems depend upon correct reports of sales so that orders may be related to demand. However, when reports of sales are made by distributors or agents who have inventory "on consignment" and who collect from customers before remitting to a central source, some direct or indirect alteration of the data may be expected.

For example, commission agents frequently delay reports of sales and use the collected funds as working capital as long as possible. In other situations commission agents were found to be reporting and paying for bulk sales when in fact package goods of the same item were often sold at a higher price. The difference was eventually detected, but the errors were blamed on the computing and inventory system. Similar shading of input data may be expected if there is a strong conflict of objectives and values at an information interface. For this reason, some manufacturers operate their own distribution facilities so that accurate data inputs to data control systems can be assured.

In a similar way, when the competitive objective of organization and system differ strongly, information protection problems arise.[23]

Example: Communications systems, such as telephone lines, computers, teleprinters, and similar devices, all radiate energy as they work. These signals can often be captured and decoded by an industrial spy, usually with little chance of detection. For example, a teletype machine generates sparks as it prints. Even when located in a shielded room, its signals can be detected from several miles away, and reproduced with suitable equipment if security precautions are not taken. The problem of information protection is increased when information processing facilities are shared, as is the case in real-time computing centers. Planning for the World Trade Center in New York, where computer systems on every ten floors will be shared, includes protection against monitoring systems. Every computer system for defense operations must be protected against monitoring, according to the specification of FED-STD-222 and DCAENS 422-5s, the contents of which are known only to those with clearance. Less exotic forms of data falsification and theft (as well as material spoilage and theft) often present difficulties that are overlooked. For example, material theft, damage, hoarding, and similar activities can cause major data errors in inventory control systems.

8 *Are operators actually performing according to the system definition and plan?*

Many systems have not performed as predicted because an essential operation was not performed by an operator or group of operators according to the plan assumed by the systems analyst.

Example: Electronic computers are often used to calculate minimum-flight-time and minimum-fuel flight plans for commercial aircraft. These plans take into account weather, wind, load, and similar factors, and are prepared for the pilot just before takeoff. When first introduced, the time and fuel reductions predicted by these schemes failed to materialize. The question arose as to whether the optimizing computations were in error or whether pilot adherence to the computer plan was unsatisfactory. An investigation of the latter problem led to increased pilot cooperation, and the predicted savings were at last obtained.

9 *Will the system accept and act upon signs of impending disaster?*

Although the desire for survival and perpetuation is often strong in organizations and individuals, and is often designed into procedural and hardware systems, this survival or continuity of operation is often threatened when the system fails to act or delays action in response to clear signs of danger.

There are cases in which the system is not "aware" of impending danger because such input data are not received or sought. But even when such data are in hand, many technical and institutional blocks can prevent correct evaluation and acceptance in time for adequate action. For example, the required pieces of intelligence information may be on the "desks" of several intelligence analysts who do not individually see the emerging pattern of danger and dismiss the isolated pieces of intelligence as unimportant. The organizational leader may have surrounded himself with weak yes-men who fail to tell him the truth. There may be technical delays in transmission, reception, and decoding and possible mistakes along the way. But even when these faults are overcome, the executive, the operator, or the control system often cannot, or will not, accept and act quickly upon the fact that disaster is at hand.

The following is reported in some detail as a modern and classic example of this often neglected implementation problem.

Example: In the major eastern power failure of November 9, 1965, previously mentioned, the first signs of failure were at 5.16.11 pm when service was interrupted to only portions of upstate New York and Ontario.

There was a period of some 7 minutes to 12 minutes between the initial disturbance at 5.16 pm when the service to the various portions of southern New York and New England finally collapsed. The operators at the various dispatching centers all knew after 5.16 on November 9 that the frequency on their system was going down and that the load had reversed and was placing a large drain on their generating reserves. . . .

The night of the blackout a 62-year-old Con Edison Engineer, Edwin J. Nellis, who has been with the [New York City] utility for 41 years, was on duty at Con Edison's automated Energy Control Center on Manhattan's West Side. At 5.16 pm, the lights dimmed, and Mr. Nellis called for an instrument check, which showed nothing wrong with Con Edison's generating equipment.

Checking his instruments again, Mr. Nellis noticed a surge of power in from the north, then a surge outward. Con Edison officials report that he immediately called the Niagara Mohawk Company in Syracuse to determine what the trouble was. At the same time he was ordering all 12 Con Edison generating plants to peak power output – to handle any extra demand should the trouble be serious.

Meanwhile, three other stations in the power network, or grid – a station in Flushing, Queens, one in Rockland County and one in Orange County – were reporting troubles of their own to Mr. Nellis.

It was then that the lights dimmed a second time, and instruments showed a tremendous surge of power into New York, then out again. Mr. Nellis was now on the phone to Syracuse, which told him "of trouble to the north".

On hearing this, he told Syracuse, "I'm going to cut clear of you," and began pushing the first of eight buttons to cut Con Edison away from the rest of the grid. He also began ordering Con Edison's network stations – 42 in all – to shed their loads. He was too late, and in just 2.5 seconds New York City was blacked out – at 5.28 pm.

. . . In retrospect," the [Federal Power Commission] report said, "it seems likely that a timely shedding of the load in some sections of New York might have avoided a citywide blackout and the breakdown of service elsewhere, as well as facilitating restoration of service. But whether because of lack of clarity in the control room instrumentation or for other reasons, the system operator did not make an immediate clear-cut decision in this emergency."

In reply, Charles E. Eble, president of Con Edison, insisted that the company's operators "followed established procedures and in our opinion made proper and timely decisions in the short space of time available to them".

However, the automatic controls on Con Edison's generators, as T. H. White aptly observed for *Life Magazine*, were quicker "to protect their own". They cut out by themselves when the survival of their generators was threatened by the massive heat of extreme overload.[24]

10 *Have potential difficulties at the boundaries of functions, departments, components or modules have adequately "bridged"?*

The analyst may in general expect difficulty in system implementation when his system crosses functions or boundaries. Although this statement is true of hardware systems – more interfaces or intercommunications between components generally mean more potential trouble – it is particularly true for procedural and information processing systems that cut across organizational boundaries. Additional care in implementation is always needed to bridge these sources of potential system disruption. Severe measures or implementation decisions may be required if the "boundary problem" is aggravated by a man-machine interface.

Example: To avoid such boundary problems at the outset, many systems analysts take the position that a new procedural or information system should be designed either to be completely manual and so simple that any operator can understand it or, at the other extreme, so automated that no human intervention is required in the intermediate steps from input to output. Many failures with combinations of manual and automated steps are the basis for this conclusion.

11 *Has the analyst correctly evaluated his own resources in relation to the task of analysis and implementation?*

Gross underevaluation of the time, cost, and personnel requirements for system development and installation is unfortunately the rule for projects specified by performance standards. Similarly, overoptimism is often the rule when a systems project is proposed with fixed resources. Usually too much is attempted with too little, with the frequent result that the project flounders, and this leads to general disappointment and losses.

Example: The installation of even a modest data processing facility will take about two years, on the average, although many firms, to their regret, frequently make estimates of a year or less. The many steps required in preparing for a changeover to the computer system invariably bring snags that prolong the effort and raise its cost beyond what was expected.

Similar problems occur when the analysts' range of actions is smaller than is required to handle the system he confronts, and an assessment of this relationship – discussed further in Chapter 6 of the original book – is in order, particularly when implementation, with its greater variety of difficulties, is confronted.

Finally, systems implementation is subject to many threshold effects that raise questions of resource availability. Often a little implementation does no good, and a basic minimum of resources must be committed to assure any degree of success. If this minimum is not available, attempts at implementation are usually wasted, an economic fact of life of which the analyst should be aware.

12 *Is the system chosen for analysis and implementation big enough?*

The scope of a system considered may not be large enough to result in any major improvement, or worse, it may lead to unwanted suboptimization. In addition, the costs in-involved in system analyses and implementation are often relatively fixed after a certain threshold of resources is reached, and may not vary thereafter directly with the size of the system considered. Thus, major improvements or large-scale improvements may be no more costly than lesser efforts, although the payoff can be much greater.

Example: The design of an inventory control system for a small company with 1000 items requires almost as much effort as one for a larger company with 10,000 or 100,000 items, although the absolute dollar improvements in the latter case will usually be much greater. The hardware and paper conversion will be somewhat more costly in the latter case, but development problems, personnel problems, and the political problems of installation are essentially the same. This fact usually places the larger firm in a better position to use systems analysis than the smaller firm.

In the same way, the analyst within an organization should look for systems improvements that will produce the largest payoff first, and should shun forms of implementation that offer a small return in relation to the cost of analysis and implementation.

For institutional reasons, the reverse strategy is often unwisely chosen. ("Do a series of small projects to prove the worth of research to the organization.") Although this may appear at first to make good political sense, the result is often not as expected. Both the analysts and the organization become discouraged with the lack of demonstrable and dramatic return from the costs incurred after a time. The bold choice carries the risk of dramatic failure as well as dramatic gain; but the piecemeal approach carries the risk that the results of the smaller efforts are not measurable above the noise inherent in a system's reporting and evaluation procedures.

Thus, there is a minimum size for systems analysis projects set, on the one hand, by the threshold cost of analysis and implementation and, on the other hand, by the need for measurable results, which often must be dramatic to be noticed.

References

[1] This discussion is abstracted basically from two sources: Wheelis, A. (1958) *The Quest for Identity*, Norton, New York, and Victor Gollancz Ltd., London, especially Chapter III, and Berelson, B. and Steiner, G. A. (1964) *Human Behavior: An Inventory of Scientific Findings*, Harcourt, Brace & World, New York. The Wheelis book is a discussion of current psychiatric and cultural problems. Berelson and Steiner report 1045 experimental and observational findings from which general conclusions are drawn in Chapter 17. There, confirming cross references to the literature may be found in abundance.

[2] Wheelis, *op. cit.*, p. 74.

[3] Wheelis, *op. cit.*, p. 75.

[4] Wheelis, *op. cit.*, p. 75.

[5] Wheelis, *op. cit.*, p. 74.

[6] Wheelis, *op. cit.*, p. 75.

[7] Berelson and Steiner, *op. cit.*, Chapter 17.

[8] Berelson and Steiner, *op. cit.*, pp. 663–4.

[9] Clemens, S., *Letters from Earth* as quoted in Berelson and Steiner, *op. cit.*, p. 642.

[10] For example, see the cases summarized by Berelson and Steiner, *op. cit.*, Chapter 17.

[11] Wheelis, *op. cit.*, p. 101. In 10 months of 1965 the Americana

Hotel in New York City is reported to have lost 38,000 demitasse spoons, 20,000 towels, and 475 bibles to its guests. *New York Daily News*, January 24, 1966, p. 23.

[12] Wheelis, *op. cit.*, p. 82.

[13] Wheelis, *op. cit.*, p. 82.

[14] The culture lag was first discussed in length by Ogburn, W. F. (1927) *Social Change*, Dell, New York.

[15] Hitch, C. J. (1965) *Decision-Making for Defense*, University of California Press, Berkeley. (This is a series of four lectures by the former Comptroller of the Defense Establishment.) The reader may also find two other works of interest in the same vein, Kaufman, W. W. (1964) *The McNamara Strategy*, Harper, and Quade, E. S. (1965) *Analysis for Military Decisions*, Rand McNally.

[16] Fritz, C. E. (1961) "Disaster", in Merton, R. K. and Nisbit, R. A. (eds.), *Contemporary Social Problems*, 1st ed., Harcourt, Brace & World, p. 665.

[17] Morse, P. M. and Kimball, G. E. (1951) *Methods of Operations Research*, Technology Press, and Wiley, New York, pp. 52–3.

[18] See also Hitch, C. J. and McKean, R. N. (1960) *The Economics of Defense in a Nuclear Age*, Harvard University Press, Cambridge, Mass., and Hitch, C. J. (1965) *Decision-Making for Defense*, University of California Press, Berkeley. The Operations Research Society of America has a "Cost Effectiveness" Section, which concentrates on problems of measure selection and use.

[19] See Adrian, E. D. (1928) *The Basis of Sensation: The Action of the Sense Organs*, Norton, New York, and the same author's *The Physical Background of Perception*, (1947) Clarendon Press, Oxford. Also, Hebb, D. O. (1949) *The Organization of Behavior*, Wiley, New York, and Penfield, W. and Rasmussen, T. (1950) *The Cerebral Cortex of Man*, Macmillan, New York. In a completely different vein, the works of Marshall McLuhan are based upon changes in the ratio of sensory inputs from different media, such as the printed page versus television, and their effects upon the individual and society. See *The Gutenberg Galaxy* (1962) University of Toronto Press, Toronto, and *Understanding Media* (1964) McGraw-Hill, New York.

[20] As one example, see Brandon, D. (1963) *Management Standards for Data Processing*, Van Nostrand, Princeton, N.J. Critical Path Methods and PERT Networks are planning tools often used to introduce project control.

[21] *Time*, June 28, 1966, pp. 52–3.

[22] *New York Times*, December 7, 1965, p. 41.

[23] Sarafin, E. E. (1965) "Information Protection", *Control Engineering*, pp. 105–7.

[24] From the Federal Power Commission Report, reprinted in the *New York Times*, December 7, 1965, p. 40, and the report by *Times* reporter Thomas O'Toole describing the sequence of events in New York, p. 41. Copyright 1965 by The New York Times Company. Reprinted by permission. One may well ask what good is an automated information or control system if the operators who have the final word on system action fail to take action or veto the action clearly indicated by the system itself?

18 Systems approach to a theory of rural-urban migration

by A. L. Mabogunje

In the growing literature on the study of migration, two theoretical issues have attracted the greatest attention, namely, why people migrate and how far they move. A simple model for explaining the reasons why people move has been formulated in terms of the "pull-push" hypothesis.[1] This has been elaborated variously to take account of internal migration movements of the rural-rural, rural-urban, or urban-urban types and international migrations. The issue of how far people move has, in turn, given rise to the formulation of a surprisingly large number of models of varying degrees of statistical or mathematical sophistication. In most of these models the distance covered is treated as either the sole independent variable or as one of many independent variables explaining the number of migrants moving to particular destinations. Morrill[2] has provided a valuable summary of these models and suggests that they can be classified broadly into deterministic and probabilistic models.

Most of these theoretical formulations have been applied to conditions in the developed countries of the world and especially to urban-to-urban migrations. Their relevance for handling migratory movements from rural to urban areas and particularly in the circumstances of underdeveloped countries has hardly been considered. Yet, it is these areas of the world where rural-urban migrations are presently taking place that afford the best opportunity for testing theoretical notions about this class of movements.

It is suggested that Africa in particular is a unique area from which to draw important empirical evidence about this type of movement. Similarly valuable data, however, can also be derived from examining the history of some of the advanced countries of the world. It is, of course, true that in Africa attention to data has been focused to a disproportionate extent on seasonal and other non-permanent transfers of population from rural to urban areas, that is, on what has been referred to as a "constant circulatory movement" between the two areas.[3] But, it will be shown that this type of movement represents a very special case of rural-urban migration. To make the point clear, it is necessary to offer a definition of the latter. Essentially, rural-urban migration represents a basic transformation of the nodal structure of a society in which people move from generally smaller, mainly agricultural communities to larger, mainly non-agricultural communities. Apart from this spatial (or horizontal) dimension of the movement, there is also a socio-economic (or vertical) dimension involving a permanent transformation of skills, attitudes, motivations, and behaviour patterns such that a migrant is enabled to break completely with his rural background and become entirely committed to urban existence. A permanence of transfer is thus the essence of the movement.

Rural-urban migration also represents an essentially spatial concomitant of the economic development of a region. Indeed, it has been suggested that one of the basic goals of economic development is to reverse the situation wherein 85 per cent of the population is in agriculture and lives in rural areas while only about 15 per cent is in non-agricultural activities and lives in the cities.[4] Rural-urban migration represents the spatial flow component of such a reversal. It is a complex phenomenon which involves not only the migrants but also a number of institutional agencies, and it gives rise to significant and highly varied adjustments everywhere in a region.

It can be argued with a great deal of justification that few of the theoretical models provided so far have considered migration, especially rural-urban migration, as a spatial process whose dynamics and spatial impact must form part of any comprehensive understanding of the phenomenon. It is the main contention of this paper that such an understanding can best be achieved within the framework of General Systems Theory.[5] This approach demands that a particular complex of variables be recognized as a system possessing certain properties which are common to many other systems. It has the fundamental advantage of providing a conceptual framework within which a whole range of questions relevant to an understanding of the structure and operation of other systems can be asked of the particular phenomenon under study. In this way, new insights are provided into old problems and new relationships whose existence may not have been appreciated previously are uncovered. In this paper no attempt is made to define major components and relationships in a formal, mathematical manner. The emphasis here is on a verbal analysis of the ways in which the system operates. This, it is hoped, will enable us to identify areas where present knowledge is fragmentary and where future research may be concentrated with some profit.

Defining the system of rural-urban migration

A system may be defined as a complex of interacting elements, together with their attributes and relationships.[6] One of the major tasks in conceptualizing a phenomenon as a system, therefore, is to identify the basic interacting elements, their attributes, and their relationships. Once this is done, it soon becomes obvious that the system operates not in a void but in a special environment. For any given system, this environment comprises "the set of all objects a change in whose attributes affects the system, and also those objects whose attributes are changed by the behaviour of the system". Thus, a system with its environment constitutes the universe of phenomena which is of interest in a given context.

Figure 18.1 indicates the basic elements in the rural-urban migration system as well as the environment within which the system operates. It shows that a systems approach to rural-urban migration is concerned not only with why people migrate but with all the implications and ramifications of the process. Basically, the approach is designed to answer questions such as: why and how does an essentially rural individual become a permanent city dweller? What changes does he undergo in the process? What effects have these changes both on the rural area from which he comes and on the city to which he moves? Are there situations or institutions which encourage or discourage the rate of movement between the rural area and the city? What is the general pattern of these movements, and how is this determined? These, and other such questions, define the problems for which we require a theory of rural-urban migration.

It can be shown theoretically that areas with isolated and self-sufficient villages such as were found in many parts of Africa until recently, are not likely to experience rural-urban migration, since, in any case there would be hardly

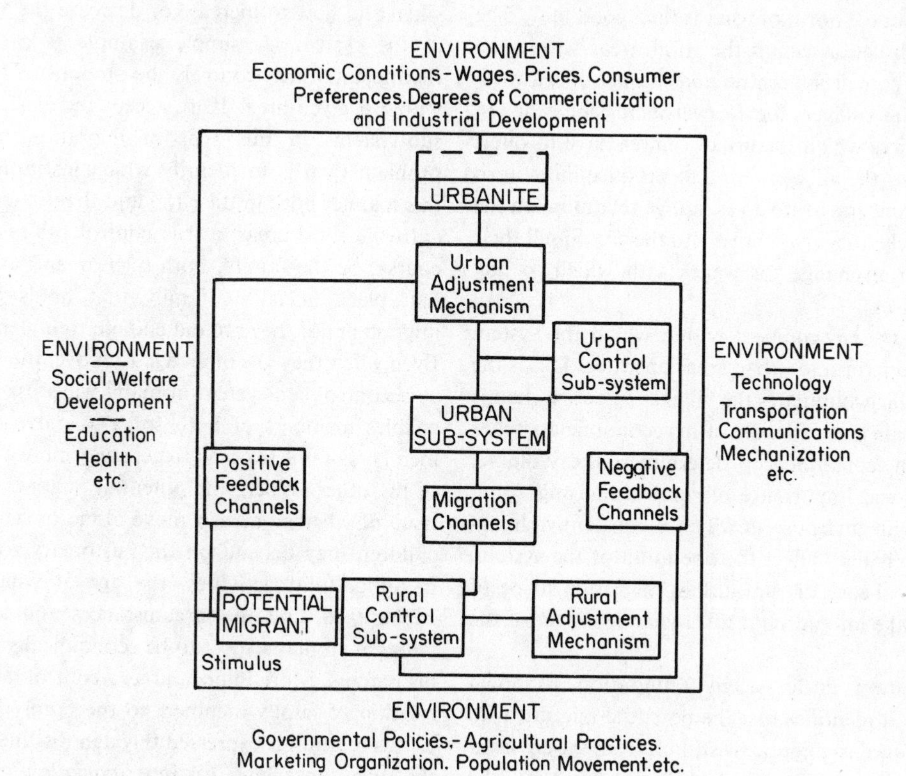

Figure 18.1. A system schema for a theory of rural-urban migration.

any cities in such areas. The fact that today such movements characterize many parts of the continent and are lately assuming spectacular proportions means that rural areas are in general no longer isolated or self-sufficient. It is therefore relevant to ask: what forces have contributed and continue to contribute to the decline in these conditions of isolation and self-sufficiency in the rural areas? They are, in the main, forces set in motion by increasing economic development. In most African countries, this was brought about initially by the colonial administrations and further reinforced in recent years by the activities of the new African governments. Decreasing isolation means not only improvement in transportation and communication links but also greater integration of the rural economy into a national economy. Such integration makes the rural economy more responsive to changes in wages and prices, consumer preferences, and the over-all demand pattern within the country. It also subjects it to a wide range of government legislation of official policies over which, in many cases, the rural community has little or no control. Decreasing isolation also means greater social and cultural integration of rural and urban areas such that levels of expectations in both areas begin to converge towards a recognizable national norm of what is the "good life". The breakdown of isolation brings the rural areas within the orbit of one or more urban centres and sharpens the awareness and desire of villagers for the ever-increasing range of goods and services which the urban centres have to offer. To acquire these, the villagers have to produce more agricultural goods and enter into an exchange relation with the city. Alternatively, they may move into the city to sell their labour direct in exchange for wages with which to buy goods and services.

This then is the environment within which the system of migration from rural to urban areas operates. This is the environment which stimulates the villager to desire change in the basic locale and rationale of his economic activities and which, in consequence, determines the volume, characteristics, and importance of rural-urban migration. Moreover, it is an environment which is constantly changing, and these changes affect the operation of the system. Hence, for any theory of rural-urban migration to be of value it must take into account this dynamic aspect of the problem.

The basic elements in the system of migration are shown in Figure 18.1. It identifies first the potential migrant who is being encouraged to migrate by stimuli from the environment. Few studies have concerned themselves with the universe of potential migrants. More often, the tendency has been to study only those who successfully made the move. Even for these, attention is given mainly to classificatory characteristics such as age, sex, religion, education, and ethnic or racial origin rather than to an analysis or understanding of the background to their move.[8] There is, of course, no doubt that what this variety of information is meant to indicate is the pattern of distribution of the "propensity to migrate" within the rural population. But this is neither explicitly stated nor formulated. Moreover, an equally valuable concept which this variety of information might have been used to explore is that of "migration elasticity".[9] This relates not so much to the propensity to migrate but to how long impulses or stimuli from the environment must be transmitted to a potential migrant before he makes the desired move.

Within the systems framework, attention is focused not only on the migrant but also on the various institutions (sub-systems) and the social, economic, and other relationships (adjustment mechanisms) which are an integral part of the process of the migrant's transformation. The two most important sub-systems are the rural and the urban control sub-systems. A control sub-system is one which oversees the operation of the general system and determines when and how to increase or decrease the amount of flow in the system. A simple example is provided by the thermostat which controls the amount of heat that flows within a given area. If we accept the existence of control sub-systems in this type of migration movement, the problem then is to identify which institutions operate in this manner both in the rural and the urban areas.

In the rural areas, a true control sub-system would, of course, be the family, both nuclear and extended. In the first place, it is the family that holds back potential migrants until they are old enough to undertake the move. Even when they are of an age to move, the family still acts as a control sub-system in many ways. In some places, it enables members of both sexes to move out; in others, members of one sex tend to get away more easily than those of the other. Where the potential migrant is married, the issue of whether he can move alone or with his wife and children may depend on the customary role of the sexes in agricultural activities, the age at which marriage is encouraged, and the circumstances and age at which a young man may expect to be economically independent of his parents. More important as a control mechanism is the relation of family members to the family land, especially as this relation is expressed through the lineage system and the inheritance law. An inheritance law that encourages most of the land to go into the hands of the first child (the primogeniture rule) will tend to stimulate more migration of the other children[10] compared to one based on the equality of access (partible inheritance rule) by all the children. In either case, the size of the farmland, the nature of the major agricultural products, and the prevailing prices for these would also be of decisive significance.

Apart from the family, the village community itself may act as a control sub-system. Its controlling role is not often direct but is obvious in either a positive or a negative way in the various activities which it sponsors or encourages. Thus, a village community which attempts to improve its economic conditions, for instance, through co-operative farming or marketing, may discourage, at least in the short run, permanent migration. On the other hand, a village community which puts emphasis on social betterment, for example, through education, may inadvertently stimulate migration to the city through training the younger generation to be more enlightened and more highly motivated. A pertinent aspect of the study of rural-urban migration is thus to assess how different rural communities react to migration away from the village. Such assessment should involve more than the opinion survey of the older generations. It should include an investigation of village activities and administration, and of the degree of cohesiveness in the community.

The urban control sub-system operates at the opposite end of the migrant's trajectory to encourage or discourage his being absorbed into the urban environment. Absorption at this level is of two kinds: residential and occupational. Basically, the control sub-system here can be identified with the city administration and other employment agencies operating under national laws and statutes. The city administration can ensure availability of relatively cheap and adequate housing in quantities which could make the transition of the rural migrant either difficult or easy. Apart from housing, the activities of the city administration in providing reception centres as well as various amenities and services may be a vital factor in gradually inducing a migrant to commit himself to the urban way of life.

A major factor in this commitment is, of course, the securing of an employment. In the city, there are numerous employment agencies offering, at any one time, very limited opportunities for the migrant. A pressing problem in the control sub-system is how to bring together and make known these disparate, but sometimes impressive, lists of vacancies. In some urban communities, this function of collation is left entirely to the press and their advertising columns. In others, a labour exchange is provided. The effectiveness with which these organizations function can be crucial for the inflow of migrants. However, once the migrant has secured an employment, a number of other factors determine his final commitment. Among these are: the type of job he secures, whether seasonal or permanent, the opportunity the job offers for improvement in his skill and for advancement in his status, the provisions available for security against the normal hazards of industrial life, and his eventual retirement due to old age.

At both the village and the city level, the decision of the migrant to move from or to move into the community sets in motion a series of adjustments. With regard to the village community, the mechanism for these adjustments should operate in such a way as to lead to an increase in the *per capita* income of the community. At least theoretically, the loss of one of the productive units in the village should lead to an increase of the productive capacity available to the remaining units; otherwise such losses from the rural area would eventually lead to a significant drop in agricultural production, to food shortage, and to famine. That these do not occur in many places means that some adjustments do take place to maintain aggregate productivity from these areas. The Ardeners[11] in their study of the Hsu of the Cameroons, for instance, point out that, in spite of the fact that as much as 40 per cent of the adult male population in the village was absent, food production did not show any significant drop. Studies of other communities in Africa have indicated similar observations.[12] However, what is involved in the adjustment to rural-urban migration is more than the minor arrangements by which the farmlands belonging to seasonal or short-term migrants are tended in their absence by their wives, their friends, or other members of their families. What is involved here are the ways and methods by which rural communities permit migrants to renounce partially or wholly their rights to productive resources in the rural areas.

One of the major research frontiers in rural-urban migration studies is the understanding of how this renunciation is accomplished. In Africa, for instance, such renunciation must be seen against the background of a complex land tenure situation and the fact that sale of land is regarded as basically a foreign concept. There is some evidence that one of the implications of rural-urban migration is to encourage a growing individualization of land-holding (with or without enclosuring) and a disposition to treat land as a marketable commodity. In the Eastern Region of Nigeria, for instance, rural-urban migration has been leading to a new pattern of land distribution and ownership. This is especially so in those areas not too far from the major urban centres.[13] It would appear, however, that initially it is the usufruct (or right to beneficial usage) on the land and not ownership that is regarded as negotiable. As a result, leasehold or annual rental of agricultural land has become widespread in many parts of West Africa and serves as a means of reallocating land which would otherwise remain unutilized because its owners have migrated to the cities. In some other cases, it is the right to exploit tree crops such as the oil-palm, cocoa, or rubber that is exchanged for monetary considerations either by outright payments or by share-cropping arrangement. In all cases, the effect of the renunciation of the migrant's claims on land or other resources is to enable some members of the village community to increase their net income by the expenditure of their often under-utilized labour. The more complete the renunciation by the migrant, the greater the acceptance of the idea of outright sale or alienation of land. Renunciation, by providing increased capacity in land or other resources, also encourages attempts at production on individual farms, and a reduction in the subsistence sector of the village economy.

Sometimes, however, this process of adjustment is induced by government and has the effect of widely stimulating migration from the rural areas. This was in fact, what happened in Britian in the eighteenth and nineteenth centuries with the various Enclosure Acts. In Africa, especially in East and Central Africa, the same process can be witnessed today. Thus, in Rhodesia, the Native Land Husbandry Act of 1951 individualized agricultural holdings and occasioned the loss by many farmers of their right to cultivate former family land. This disenfranchisement, as was only to be expected, gave rise to a flood of migrants most of whom had no alternative but to become permanently committed to wage employment and psychologically attuned to surviving in an urban environment.

In the urban areas, the mechanism of adjustment is basically one of incorporating the migrant into a new frame of reference more relevant to his needs in the city. In this respect, a city can be described as an assemblage of interacting interest groups. Part of the process of becoming a member of such a community would thus be to identify closely with one or more of these groups. The mechanism of incorporation in the urban areas, in contrast to the adjustment process in the rural areas, has been the subject of a number of studies. In particular, attention has been called to the role of ethnic unions and various voluntary organizations such as the Church, trade unions, occupational associations, and recreation societies in helping the rural migrant to adjust to his new environment.[14]

Finally, of the various elements of the system, there is the city *qua* city, seen as part of an urban sub-system. What aspects of urban life and activities are relevant for the understanding of rural-urban migration? To answer this question, it is important to visualize the city as comprising a hierarchy of specializations. In other words, a city is a place where everyone is trying to sell specialized skill. The more specialized the skill, the greater the demand for it, and hence the higher the price it commands on the market. Within this conceptual framework, the illiterate, unsophisticated rural migrant is seen as belonging to the lowest level of the hierarchy. A corollary is that the higher a person moves up within the hierarchy, the greater his commitment to the urban way of life and the less the probability of his reversion to rural existence. This is one reason why the type of job which a rural migrant secures in the city can be so crucial to how soon he becomes committed to urban life. This is also one reason why those countries, such as Rhodesia, which are anxious to ensure that the African does not become an urban resident, pursue a discriminatory policy with regard to his acquisition of skill in urban employment. Yet, as Masser pointed out, even in Rhodesia the propensity to return to the village after migrating to the city decreases with the minimal rise in the skill of the migrant.[15]

Another interesting aspect of this concept of the city is that upward mobility within the hierarchy of specializations is often accompanied by changes in residential location within the city. This is no doubt a function of rising income, but it is also closely related to the length of stay in the city and the increasing commitment of a migrant to spend the remaining part of his working life there. There have been many studies of the residential pattern and varying length of stay of migrants in urban areas. Unfortunately, a good number of these studies have been concerned more with indicating the ethnic basis of this pattern than with investigating the dynamic factor of skill differentiation and status advancement which is operating to blur out the importance of ethnicity. As a result, this rather crucial dimension of rural-urban migration has tended to be neglected. Its investigation should yield some rather interesting results.

One final aspect of the examination of cities as a hierarchy of specializations relates to the significance of size. What effect has the size of a city on the type of migrants attracted to it? Clearly, small urban centres have fewer tiers of specializations and more restricted employment opportunities than the larger ones. Yet, competition for positions in them may be less intense. Are certain types of migrants attracted to such centres first and then able to "leap-frog" gradually to bigger and bigger centres? What type of migrants would make direct for the larger cities? Does rural-urban migration into the larger cities take place in the manner of "a series of concentric migratory contractions" suggested by Ashton[16] for Britain in the eighteenth century? According to Ashton, the larger industrial centres attracted a number of workers and their families who were living in the larger market towns on their perimeters. These towns in turn made good their losses from the surrounding villages, the villages from the hamlets, and the hamlets from the farms. In this way, there was no sharp discontinuity in the pattern of life with which the migrant was familiar. Are there conditions, for instance, the nature of the transportation network and development, which would make such a pattern of migration appropriate for Britain of the eighteenth century but not for Africa in the twentieth century? Or, does this pattern reflect stable human reaction to a permanent spatial dislocation of existing networks of social contacts?

The energy concepts in systems analysis

A system comprises not only matter (the migrant, the institutions, and the various organizations mentioned) but also energy. In the physical sense, energy is simply the capacity of a given body to do work. It can be expressed in a number of ways, but two forms of it are relevant here. There is "potential energy" which is the body's power of doing work by virtue of stresses resulting from its relation either with its environment or with other bodies. The second form is "kinetic energy" which is the capacity of a body to do work by virtue of its own motion or activity.

In a theory of rural-urban migration, potential energy can be likened to the stimuli acting on the rural individual to move. What is the nature of these stimuli? As pointed out earlier, a number of studies have tried to identify why people migrate and have come up with a variety of answers generally subsumed under the push-and-pull hypothesis. This suggests that people migrate from rural areas to the cities because of one of two general causes: overpopulation and environmental deterioration in the rural areas (the push factor) or the allurement or attraction of the city (the pull factor or the so-called "bright-light theory"). The push factor, it is claimed, explains migrations directed to earning extra income to pay the annual tax or to take a new bride or to buy a few manufactured articles or to escape oppressive local *mores*. The pull factor, on the other hand, explains migrations undertaken as a modern form of initiation ceremony to adult status or as the basis for later receiving preferential admiration of the village girls or as the product simply of an intense curiosity about the city.

These explanations, to the extent that they have any theoretical validity at all, are relevant only at the aggregate level. These are notwithstanding the results from completed questionnaire surveys requiring individuals to indicate the reason or reasons for their migration into the city. But, as Richards[17] and Gulliver [18] stress, the battery of questions usually asked of migrants hardly ever reveals anything about why they moved. In Africa, the great number of temporary migrants to the cities on whom most studies have been concentrated, are involved in making no major decisions other than on the length of time they can or have to be away from home. The reasons for their migration are very often manifold and usually not easy to articulate in a few, simple sentences. What the questionnaire does, in fact, is to suggest to the migrant a set of equally plausible reasons, besides the obvious one of coming to earn extra income.

Within the systems framework, the explanation of why people migrate must be in terms of differential individual responses to the stimuli both from the environment and from within the system. It differs from the pull-and-push hypothesis in putting the emphasis at the individual level, not on why people migrate from particular areas but why any person from any village would want to migrate to the city. The stimulus to migrate is related to the extent of the integration of rural activities into the national economy, to the degree of awareness of opportunities outside of the rural areas, and to the nature of the social and economic expectations held by the rural population not only for themselves but also for their children. Indeed, the notion of "expectations" or "aspirations" is central to an understanding of the ways in which the stimulus from the environment is transmitted to individuals, and for that reason it is a crucial variable in the theory of rural-urban migration. What determines the variation in the level of individual expectations in rural areas and conditions individual responses to the stimulus to migrate? Clearly, for a given cohort in any rural area, one can, at least theoretically, conceive of individuals who respond promptly to the stimulus and others who take a much longer time to respond. One may in fact ask whether there is a threshold below which the stimulus cannot be expected to act and an upper limit beyond which its impact is no longer felt? How are these limits defined – by age, wealth, natural alertness, or family position? In short, two problems in the theory of rural-urban migration which still require resolution concern the nature and significance of rural expectations and their relation to the differential effectiveness of the stimulus to migrate.

Once an individual has been successfully dislodged from the rural area, we can assume that he is translating his "potential energy" into its "kinetic form". The major issues concern not only the act of moving but also the cost, the distance, and the direction of movement. These three variables clearly determine the crisscross channels of migration as well as their destinations. Again, as already indicated, this aspect of migration studies has received considerable theoretical attention. Starting with Ravenstein's laws of migration[19] which try to establish the relation between distance and the propensity to move, there have been various attempts to seek understanding through using the gravity model,[20] and the intervening opportunities model.[21] There have also been other studies which have tried to understand the pattern of migration channels through probabilistic models.[22]

As soon as a migrant has moved from the rural to the urban area, his role in the system is greatly amplified. Basic to an understanding of this amplified role is the concept of "information", a central notion in the theory of communications. Information can be defined simply as bits of messages in a system which lead to a particular set of actions. Thus, one can easily assume that the first migrant from a village to a city would soon start to transmit back to the village information about his reception and progress in the city. Ignoring for the moment the question of "information content", it can be shown that the level of information can be measured in terms of decisions.[23] A particular set of decisions can be compared with the random choice from a universe of equally probable decisions. Its deviation from the latter becomes a measure of the level of information. It also represents a statement of the level of order or organization existing within the system. Information is thus a crucial feature of the operation of a system since it determines at any point of time the state of organization of the system.

Of equal importance is the notion of "feedback" which has been the focus of the field of Cybernetics. This can be explained quite simply in terms of stimulus-response behaviour. A stimulus affects a receptor which communicates this message to some controlling apparatus and from this to an effector which gives the response. In feedback, the effector's activity is monitored back to the receptor with the result that the system's behaviour is in some way modified by the information. The feedback process can have one of two effects. It can further amplify the deviation (in this case by stimulating further migration), or it may counteract the deviation by encouraging a return to the initial situation. Deviation-amplifying feedbacks are regarded as positive; deviation-counteracting feedbacks as negative.

The notion of a "most probable or random state" is one that needs further clarification. Imagine a situation in which migrants from a village are lost to their communities as soon as they move out and send back no information on their reactions to the cities to which they moved. Later migrants then, not knowing where the first set of migrants went to might choose any city in the system, almost in a random manner. Over time, the distribution of migrants from individual villages may come to approximate a situation in which the number of migrants from any village to a city is proportional to the size of that city. This is the most probable state in which no order or organization is evident in the system. Conceptually, it can be seen as a state of maximum disorder, or a state of maximum "entropy".

Yet, the general experience is that migrants are never lost in this sense to their village of origin but continue to send back information. If the information from a particular city dwells at length on the negative side of urban life, on the difficulties of getting jobs, of finding a place to live, and on the general hostility of people, the effect of this negative feedback will slow down further migration from the village to this city. By contrast, favourable or positive feedback will encourage migration and will produce situations of almost organized migratory flows from particular villages to particular cities. In other words, the existence of information in the system encourages greater deviation from the "most probable or random state". It implies a decrease in the level of entropy (or disorder) or an increase in negative entropy (negentropy). The result is greater differentiation in the pattern of migration which reflects some form of organization. Thus, experience of rural-urban migration in many parts of the world emphasizes this organized nature of the moves. In many North African cities, for instance, it is not uncommon for an entire district or craft occupation in a city to be dominated by permanent migrants from one or two villages.[24] Furthermore, this element of "organization" resulting from the operation of feedback in the system underlies the varying rate of population growth among cities.

A major area of research into rural-urban migration thus concerns the flow of information between the urban and the rural areas. Considerable work on this question has been undertaken in Europe and the United States and some of the results are of great interest. Hägerstrand, for instance, insists that we must distinguish between "active" and "passive" migrants.[25] The former are those who seek out suitable destinations which, in their eyes, guarantee future prosperity; the latter are those who follow impulses (feedbacks) emanating from persons of their acquaintance, primarily those who had made "fortunate" moves. One implication of this distinction is that in a theory of rural-urban migration, the crucial moves which we need to understand and explain are those of the active-migrants. In the aggregate, these moves are likely to be complex and not easily explained in terms of a few choice variables.

A number of other studies have concentrated on the measurement of the information field of a potential migrant as a means of understanding the general pattern of his behaviour in space.[26] Individual fields may be aggregated to produce community mean information fields, and these have been used in studies which attempt to predict the volume and pattern of migratory movements.[27]

Relation between a system and its environment

Systems can be classified into three categories depending on the relationship they maintain with their environment; first, the isolated systems which exchange neither "matter" nor "energy" with their environment; second, the closed systems which exchange "energy" but not "matter"; third, the open systems which exchange both "energy" and "matter". The distinction between the categories, however, is largely one of scale and depends on which elements are regarded as belonging to the system and which to the environment. Thus, if the scale was to be reduced significantly, an open system could become an isolated system.

Given the system in Figure 18.1, it can be seen that rural-urban migration is an open system involving not only an exchange of energy but also of matter (in this case, persons) with the environment. The persons concerned would be defined as all those, who having migrated into cities, have become involved in making local decisions or formulating national policies and legislation on economic and other matters which do affect the volume, character, and pattern of migration. The energy exchange has to do with the increasing economic activities resulting from rural-urban migration and affecting the over-all economic and social conditions of the country.

One major implication of viewing rural-urban migration as an open system is the fact that it enables us to explore the principle of equifinality in so far as it applies to this phenomenon. This principle emphasizes that the state of a system at any given time is not determined so much by initial conditions as by the nature of the process, or the system parameters. In consequence, the same results may spring from different origins or, conversely, different results may be produced by the same "causes". In either case, it is the nature of the process which is determinate, since open systems are basically independent of their initial conditions. This principle is of considerable importance in studying rural-urban migration in different parts of the world since there has been a tendency to regard this movement in countries such as in Africa and Asia as a special kind different from elsewhere in the world. There is, of course, no doubt that initial conditions in Africa today are vastly different from what they were in countries such as Britain and the United States at the times of the massive migrations there of people from the rural areas into the cities. But, according to the principle of equifinality, as long as we keep in mind the particular system's parameters, an understanding of the migration process as it affected and continues to affect those developed countries may throw considerable light on what is currently happening in many parts of the underdeveloped world.

Growth process in the system

From what has been said so far, it must be assumed that one of the concomitants of the continued interaction between the system and its environment will be the phenomenon of growth in the system. This will be indicated by, among other things, a rise in the volume of migration from the rural to the urban areas. Within a system framework, this phenomenon involves more than a simple growth or increase in the number of people moving from one area to another. It is much more complex, involving not only the individual components of the system but also the interaction between them and the system as a whole.

Boulding[28] has identified three types of growth processes that may occur in a system. The first is "simple growth" and involves the addition of one more unit of a given variable such as a migrant, a farm, a vehicle, or a retail establishment. The second type is "population growth", a process which involves both positive and negative additions. In general, this type of growth depends on the surplus of births (positive additions) over deaths (negative additions) and applies to variables which have an age distribution and regular rates of births and deaths. The third type is "structural growth", the growth process of an aggregate with a complex structure of interrelated parts. This process often involves a change in the relation of the components since the growth of each component influences and is influenced by the growth of all other components in the system. Structural growth shades imperceptibly into structural change since, in most cases, it is not only the over-all size of the structure that grows but also its complexity.

In viewing rural-urban migration as a system, growth, in the form of structural growth, is an important dimension for more detailed investigation and study. What effects have an increase in the volume of migration on the character of the cities? What effects have the growth in the size and complexity of the cities on the types of migrants, on villages and their spatial distribution, on farms and their areal extent, on the crops grown and their qualitative importance, on the types of equipment used and on the average income of families in the rural areas? What effects have changes or growth in these variables on the volume and characteristics of migrants and on further growth and complexity in the urban areas?

It may be argued, of course, that to conceive of a theory of rural-urban migration in this broad, systematic framework is to suggest a catchall embracing a wide range of changes taking place in a country at any given time. In a sense, this is deliberate since part of the object of this paper is to call attention to the paramount importance of "flow phenomena" in the spatial processes modifying the character of any country. Thus, just as the flow of water acts as a major sculpturing agent in the physical geography of any area of the world, the flow of persons (migration), of goods and services (trade and transportation), and of ideas (communication) is a crucial agency in shaping the human geography of a country.

More than this, there is the fact that growth in such "flow phenomena" creates form. Growth in the flow of rural-urban migrants affects the pattern of population distribution, the areal size and internal configuration of cities, the types of buildings in rural areas, the size and arrangements of farms, and the number, size, and network density of rural roads. These, in a sense, are simply the results of the way the system tries to adjust to growth processes. However, as Boulding has pointed out, there is a limit to the extent to which the system can go on making these adjustments. "Growth", states Boulding, "creates form; but form limits growth. This mutuality of relationship between growth and form is perhaps the essential key to the understanding of structural growth".[29]

This paper has tried to show how a theory of rural-urban migration can gain in incisiveness and breadth by being construed within a General Systems Theory framework. The conceptualization of the problem in this way emphasizes the structural congruencies or isomorphy with other problems. Further, one of the major attractions of this approach is that it enables a consideration of rural-urban migration no longer as a linear, uni-directional, push-and-pull, cause-effect movement but as a circular, interdependent, progressively complex, and self-modifying system in which the effect of changes in one part can be traced through the whole of the system. Such a circularity gives special prominence to the dynamic nature of rural-urban migration and allows the process to remain as one of considerable interest over an indefinite period of time. In other words, it emphasizes rural-urban migration as a continuous process, occurring in most countries all the time though at different levels of complexity. In this respect, the systems approach also serves as a normative model against which one can seek to explain obvious deviations. If the movement of people from the rural to the urban areas is not generating the set of interconnected effects which the theory leads us to expect, we may ask why. We may then investigate the various elements in the system to ascertain which of them is not functioning in the proper way. Alternatively, we may examine critically the politico-economic environment (such as, for example, the situation in those areas of the world where discriminatory policies exist based on race or caste) in order to appreciate those features that do impair the efficient operation of the system. In either case, the basic systems approach would provide the most important insight to the many dimensions of the problem. More than that, it would emphasize the crucial role of rural-urban migration as one of the most important spatial processes shaping the pattern of human occupation of the earth's surface.

Notes and references

1 Herberle, R., "The Causes of Rural-Urban Migration: a Survey of German Theories", *American Journal of Sociology*, **43** (1938), 923–950; Mitchell, J. C., "Migrant Labour in Africa South of the Sahara: the

Causes of Labour Migration", *Bulletin of the Inter-African Labour Institute*, **6** (1959), 8–16.

2 Morrill, R. L., "Migration and the Spread and Growth of Urban Settlement", *Lund Studies in Geography* Ser. B, **26** (1965).

3 Mitchell, J. C., "Wage Labour and African Population Movements in Central Africa", in *Essays on African Population*, ed. Barbour, K. M. and Prothero, R. M. New York, 1962, p. 232.

4 Lewis, W. A., *Theory of Economic Growth*. London, 1955, p. 333.

5 Bertalanffy, L. von, "An Outline of General System Theory", *British Journal of the Philosophy of Science*, **1** (1950), 134–165; id., "General System Theory", *General Systems Yearbook*, **1** (1956), 1–10; id., "General System Theory – a Critical Review", *General Systems Yearbook*, **7** (1962), 1–20.

6 Hall, A. D. and Fagen, R. E., "Definition of System", *General Systems Yearbook*, **1** (1956), 18.

7 Ibid., p. 20.

8 Diop, A., "Enquête sur la migration toucouleur à Dakar", *Bulletin de l'Institut Français d'Afrique Noire*, Ser. B, **22** (1960), 393–418.

9 Wolpert, J., "Migration as an Adjustment to Environmental Stress", *Journal of Social Issues*, **22** (1966), 92–102.

10 Arensberg, C. M., *The Irish Countryman, an Anthropological Study*. New York, 1937.

11 Ardener, E., Ardener, S. and Warmington, W. A. *Plantation and Village in the Cameroons*. London, 1960, pp. 211–229.

12 Prothero, R. M., "Migratory Labour from North-Western Nigeria", *Africa*, **27** (1957), 250.

13 Ndukwe, N. I., "Migration, Agriculture and Trade in Abriba Town". Unpublished MS., Department of Geography, University of Ibadan, 1964; Onwueke, A. I., "Awka Upland Region, the Land of Migrant Farmers". Unpublished MS., Department of Geography, University of Ibadan, 1966.

14 Banton, M., *West African City*, London, 1957; id., "Social Alignment and Identity in a West African City", in *Urbanization and Migration in West Africa*, ed. Kuper, H., Los Angeles, 1956, pp. 131–147; Little, K., *West African Urbanization*, London, 1965.

15 Masser, F. I., "Changing Pattern of African Employment in Southern Rhodesia", in *Geographers and the Tropics: Liverpool Essays*, ed. Steel, R. W. and Prothero, R. M., London, 1964, p. 229.

16 Ashton, T. S., *An Economic History of England: The Eighteenth Century*. London, 1966, ed., pp. 15–17.

17 Richards, A. I., ed., *Economic Development and Tribal Change*. Cambridge, 1954, p. 66.

18 Gulliver, P., "Nyakyusa Labour Migration", *Rhodes-Livingstone Journal*, **21** (1957), 59.

19 Ravenstein, E. G., "The Laws of Migration", *Journal of the Royal Statistical Society*, **48** (1885). 167–235; **52** (1889), 242–305.

20 Dodd, S. C., "The Interactance Hypothesis: A Gravity Model Fitting Physical Masses and Human Groups", *American Sociological Review*, **15** (1950), 245–256; Stewart, J. W., "Demographic Gravitation: Evidence and Applications", *Sociometry*, **11** (1948), 31–57.

21 Stouffer, S. A., "Intervening Opportunities: A Theory Relating Mobility and Distance", *American Sociological Review*, **5** (1940), 845–867; id., "Intervening Opportunities and Competing Migrants", *Journal of Regional Science*, **2** (1960), 1–26.

22 Kulldorf, G., "Migration Probabilities", *Lund Studies in Geography*, Ser. B, **14** (1955).

23 Bertalanffy, L. von, *op. cit*.

24 Clarke, J. I., "Emigration from Southern Tunisia", *Geography*, **42** (1957), 99–101; Marty, G., "A Tunis: éléments allogènes et activités professionnelles", *Revue de l'Institut des Belles Lettres Arabes*, Tunis (1948), pp. 159–188.

25 Hägerstrand, T., "Migration and Area: Survey of a sample of Swedish Migration Fields and Hypothetical Considerations on their Genesis", in *Migration in Sweden, a Symposium*, ed. Hannerberg, D. *et al.*, *Lund Studies in Geography*, Ser. B, **13** (1957), p. 132.

26 Marble, D. F. and Nystuen, J. D., "An Approach to the Direct Measurement of Community Mean Information Fields", *Papers and Proceedings, Regional Science Association*, **11** (1962), 99–109; Morrill, R. L. and Pitts, F. R., "Marriage, Migration and the Mean Information Field: a Study in Uniqueness and Generality", *Annals of the Association of American Geographers*, **57** (1967), 401–422.

27 Morrill, R. L., *op. cit*.

28 Boulding, K., "Toward a General Theory of Growth", *General Systems Yearbook*, **1** (1956), 66–75.

29 Boulding, K., *op. cit.*, p. 72.

From *Geographical Analysis*, **2**, (1970), 1–18.

19 On systems and models
by P. Haggett

1 Human geography and general systems theory

During the last decade there has been a remarkable growth of interest in the biological and behavioural sciences in *general systems* theory (Bertalanffy, 1951). Some attempts have been made (notably by Chorley, 1962) to introduce its concepts into geomorphology and physical geography, and there seems no good reason why the concept of systems could not be further extended into geography. In this section we explore the possibilities.

a. Nature of systems. What is a system? One loose definition, cited by Chorley, describes it as "... a set of objects together with relationships between the objects and their attributes" (Hall and Fagen, 1956, p. 18). In everyday plumbing parlance we speak of a "hot-water system" in which the set of objects (stove, pipes, cylinders, etc.) are related through circulating water with inputs of energy in the form of heat. In geomorphology we may speak of an "erosional system" in which the set of objects (watersheds, slopes, streams) are related through the circulation of water and sediment with inputs of energy in the form of rainstorms.

In human geography, our nearest equivalent is probably the nodal region in which the set of objects (towns, villages, farms, etc.) are related through circulating movements (money, migrants, freight, etc.) and the energy inputs come through the biological and social needs of the community. This idea is implicit in most central-place theory, though in only a few statements (notably that of Vining, 1953, and Curry, 1964-B) is the description couched in "system" terms.

Clearly then, systems are arbitrarily demarcated sections of the real world which have some common functional connections. Von Bertalanffy (1950) distinguishes two separate frameworks: the *closed system* and the *open system*. Closed systems have definable boundaries across which no exchange of energy occurs, but since they are likely, by this definition, to be rather rare in geographical studies (except in the limiting case of a world-wide study) they are not considered here.

b. Nodal regions as open systems. The view taken in the first half of this book* is that we may regard nodal regions regions as open systems (Philbrick, 1957; Nystuen and Dacey, 1961). Indeed the organization of the chapters (Chaps. 2–6) shows the build-up of such a system; viz., the study of *movements* (Chap. 2) leads on to a considera- tion of the channels along which movement occurs, the *network* (Chap. 3), to the *nodes* on that network (Chap. 4) and their organization as a *hierarchy* (Chap. 5), with a final integration of the interstitial zones viewed as *surfaces* (Chap. 6). This progression, from energy flows to recog- nizable landforms, may be seen more clearly from Figure 19.1, in which more familiar geographical forms may be substituted for their abstract geometrical equivalents, i.e. roads, settlements, the urban hierarchy, and land-use zones. If the sceptic still regards the nodal region as a purely mental construct, then Dickinson (1964, pp. 227–434) has provided a detailed review of city-regions within the United States and western Europe, while Caesar (1955; 1964) has shown strong nodal organization within regions as unlike in scale as the communist block in eastern Europe and northeast England.

If we wish to view nodal regions as open systems we must first look at the typical characteristics of such systems and check their existence in the regional system. Chorley (1962, pp. 3–8) suggests that open systems have some of the following six characteristics: (i) the need for an energy supply for the system's maintenance and pre- servation, together with, the capacity to (ii) attain a "steady-state" in which the import and export of energy and material is met by form adjustments, (iii) regulate itself by homeostatic adjustments, (iv) maintain optimum magnitudes over periods of time; (v) maintain its organiza- tion and form over time rather than tending (as do closed systems) towards maximum entropy, and (vi) behave "equifinally", in the sense that different initial conditions may lead to similar end results.

In our regional systems we certainly find some of these six characteristics. Regional organization needs a constant

* *Locational Analysis in Human Geography* by Peter Haggett.

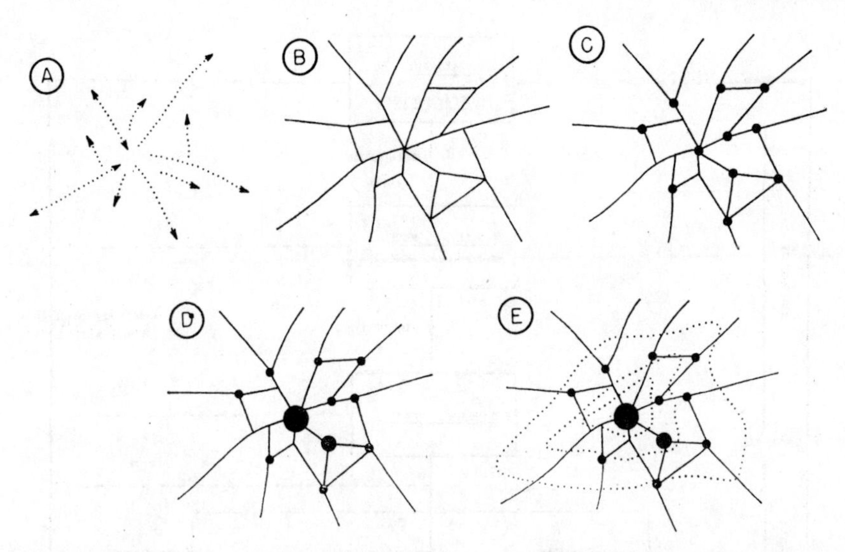

Figure 19.1. Stages in the analysis of regional systems. *A* Movements, *B* Networks. *C* Nodes. *D* Hierarchies. *E* Surfaces.

movement of people, goods, money, information to maintain it; an excess of inward movements may be met by form changes (city expansion and urban sprawl) just as decreased movement may lead to contraction and ghost cities. The first two conditions are clearly met. Similarly, on the third condition the urban region follows Le Châtelier's Principle in that its hinterland may expand or contract to meet increased or decreased flows. Berry and Garrison (1958-c) would also suggest that it meets the fourth and fifth requirements in that the form of the urban rank-size relationships tends to be relatively constant over both space and time. Finally, the growing convergence of the form of the major cities in different continents suggests that the urban open system is capable of behaving equifinally.

The advantages of viewing the region as an open system are that it directs our attention towards the links between process and form, and places human geography alongside other biological and social sciences that are organizing their thinking in this manner. Exchanges between students of "ecosystems" at all scale levels should prove rewarding (e.g. Thomas, 1956, pp. 677–806).

2 Model building in human geography

In everyday language the term "model" has at least three different usages. As a noun, model implies a representation; as an adjective, model implies ideal; as a verb, to model means to demonstrate. We are aware that when we refer to a model railway or a model husband we use the term in different senses. In scientific usage Ackoff (Ackoff, Gupta, and Minas, 1962) has suggested that we incorporate part of all three meanings; in model building we create an idealized representation of reality in order to demonstrate certain of its properties.

Models are made necessary by the complexity of reality. They are a conceptual prop to our understanding and as such provide for the teacher a simplified and apparently rational picture for the classroom, and for the researcher a source of working hypotheses to test against reality. Models convey not the whole truth but a useful and comprehensible part of it (Society for Experimental Biology, 1960).

a. Types of models. A simple three-stage breakdown has been suggested by Ackoff (Ackoff *et al.*, 1962) into *iconic*, *analogue*, and *symbolic* models, in which each stage represents a higher degree of abstraction than the last. Iconic models represent properties at a different scale; analogue models represent one property by another; symbolic models represent properties by symbols. A very simple analogy is with the road system of a region where air photographs might represent the first stage of abstraction (iconic); maps, with roads on the ground represented by lines of different width and colour on the map, represent the second stage of abstraction (analogue); a mathematical expression, road density, represents the third stage of abstraction (symbolic). At each stage information is lost and the model becomes more abstract but more general.

Chorley (1964) carried this classification process further and created a "model of models" (Figure 19.2), illustrating

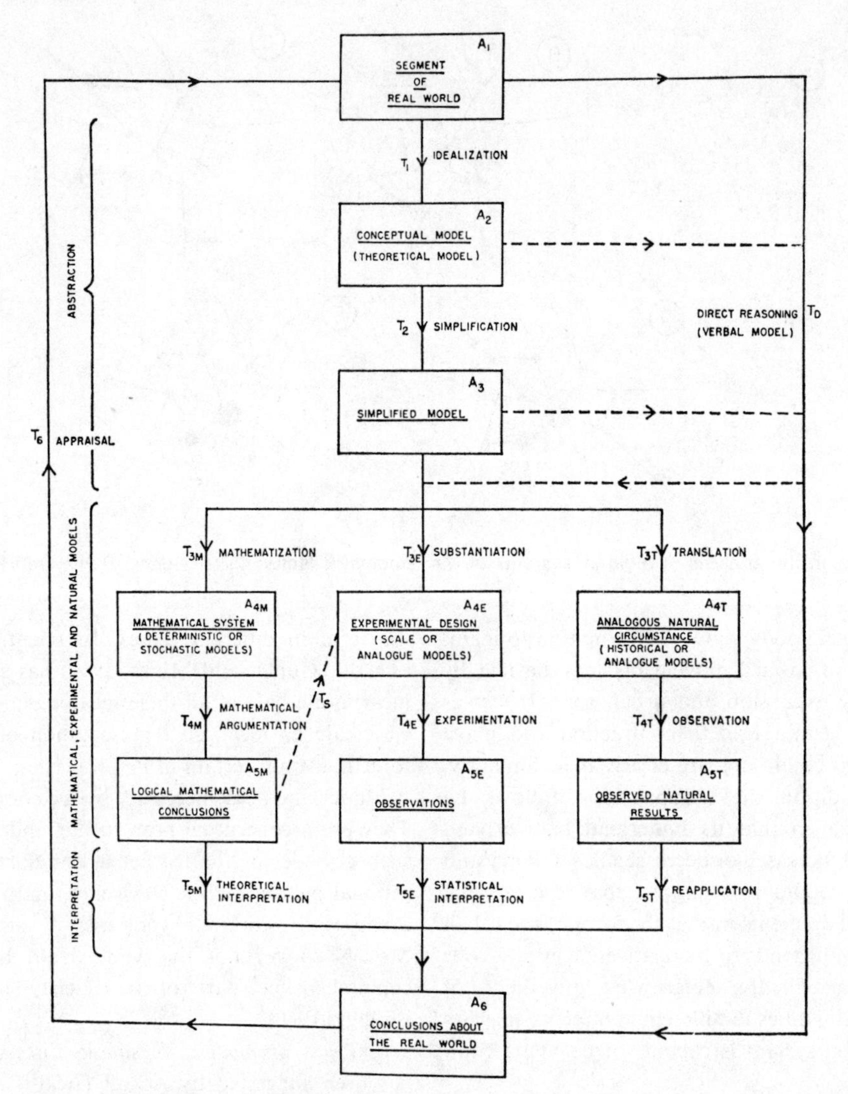

Figure 19.2. A model for models. Source: Chorley, 1964, p. 129.

it with examples from both physical and human geography. His model consists of a flow diagram in which a series of *steps* (A_1 to A_6) are linked by *transformations* (T_1 to T_6). Each step contains some aspect of the real world, model, observation or conclusion; each transformation connects these by some process (idealization, mathematical argument, statistical interpretation, etc.) which advances or checks on the reasoning process.

The first section of Figure 19.2 is concerned with the "abstraction" process in which the complexities of the real world are so simplified that they may become more comprehensible. Chorley argues that this process is difficult largely because as huge amounts of information are lost, extraneous "noise" is introduced; a Cézanne painting represents an abstracted model of a landscape in which the noise level (brush-marks, etc.) is high, while a Van Ruysdael is less simplified but considerably less noisy (Chorley, 1964, p. 132). Successful models are those which manage a considerable amount of simplification without introducing extraneous noise.

The second section of Figure 19.2 breaks into the three main stems of mathematical, experimental, and natural models. *Mathematical models* might be represented in human geography by Isard's distance-inputs equations (Isard, 1956) or Beckmann's "equation of continuity" (Beckmann, 1952) in which features of the system being studied are replaced by abstract symbols and subjected to mathematical argument. *Experimental models* might be represented by Hotelling's use of a heat-flow analogue in migration theory (Hotelling, 1921; cited by Bunge, 1962, p. 115) or Weber's weight and pulley machine in industrial location (Weber, 1909) where tangible structures are used to simulate certain aspects of reality. Finally *natural models* might be represented by the Garrison ice-cap analogy of city growth (cited by Chorley, 1964, p. 136), where some analogous natural circumstance which is believed to be simpler or more readily available is substituted for reality. The problem in each case is to translate the circumstances being studied into some analogous form in which it is either simpler, or more accessible, or more easily controlled and measured; to study it in this analogue or model form; and to reapply the results of this study to the original system. Models then represent idealized parts of systems, just as systems represent an arbitrarily separated segment of the real world.

b. Approaches to model building. In economic geography, model building has proceeded along two distinct and complementary paths. In the first, the builder has "sneaked up" on a problem by beginning with very simple postulates and gradually introducing more complexity, all the time getting recognizably nearer to real life. This was the approach of Thünen (1875) in his model of land use in *Der Isolierte Staat*. In this "isolated state" he begins by assuming a single city, a flat uniform plain, a single transport media, and like simplicities and in this simple situation is able to derive simple rent gradients which yield a satisfying alternation of land-use "rings". But Thünen then disturbs this picture by reintroducing the very things that he originally assumed inert and brings back soil differences, alternative markets and different transport media. With their introduction, the annular symmetry of the original pattern gives way to an irregular mosaic far more like the pattern we observe in our land-use surveys. Nevertheless, Thünen's model has served its point; in Ackoff's terminology it has "demonstrated certain properties" of the economic landscape.

The second method is to "move down" from reality by making a series of simplifying generalizations. This is the approach of Taaffe (Taaffe, Morrill and Gould, 1963) in his model of route development. The study begins with a detailed empirical account of the development of routes in Ghana over the period of colonial exploitation. From the Ghanaian pattern a series of successive stages is recognized. In the first, a scatter of unconnected coastal trading posts; in the last, an interconnected phase with both high-priority and general links established. This Ghanaian sequence is finally formalized as a four-stage sequence, common to other developing countries like Nigeria, East Africa Malaya, and Brazil.

Not all such models have developed inductively from observations within geography. Some of the most successful have come from borrowing ideas from related fields, especially the field of physics. Thus Zipf (1949) attempted to extend Newton's "divine elastic" of gravitation to social phenomena and his P_iP_j/d_{ij} formula for the interaction between two cities of "mass" P_i and P_j at a distance d_{ij} is a direct extension of Newtonian physics. When modified by Isard's refined concept of distance (Isard, 1960) and Stouffer's addition of intervening opportunity (Stouffer, 1962) it has proved a very powerful predictive tool in the study of traffic generation between points. A less widely known borrowing was used by Lösch (1954, p. 184). He has related the "bending" of transport routes across landscapes of varying resistance and profitability to the sine formula for the refraction of light and sound. While such borrowing may have its dangers, it is a most fruitful source of hypotheses that can be soberly tested for their relevance to the problems of economic geography. A book like D'Arcy Thompson's *On growth and form* (1917) illustrates how many subjects find common ground in the study of morphology; there is inspiration still to be found in his treatment of crystal structures or honeycomb formation as Bunge (1964) has illustrated. These models are treated at length in Part One of this book and in Chorley and Haggett (*in press*).

c. Role of models. In his *Novum Organum*, Bacon describes scientific theory as consisting of "anticipations, rash and premature". Certainly we might argue that most of the models put forward in the first half of this book fit this description admirably: all are crude, all full of exceptions, all easier to refute than to defend. Why then, we must ask, do we bother to create models rather than study directly the "facts" of human geography? The answer lies in the inevitability, the economy, and the stimulation of model building:

(i) Model building is inevitable because there is no fixed dividing line between facts and beliefs; in Skilling's terms ". . . belief in a universe of real things is merely a belief . . . a belief with high probability certainly, but a belief none the less" (1964, p. 394A). Models are theories, laws, equations, or hunches which state our beliefs about the universe we think we see.

(ii) Model building is economical because it allows us to pass on generalized information in a highly compressed form. Like rules for the plurals of French adjectives there may be exceptions but the rule is none the less an important ladder in learning the language. This use of models as teaching aids is discussed by Chorley and Haggett (1965–A, pp. 360–364).

(iii) Model building is stimulating in that, through its very over-generalizations, it makes clear those areas where improvement is necessary. The building and testing of models is as important to geography as aeronautics; the test flight of a hypothesis, no less exciting, nor much less dangerous, than the test flight of a prototype "Comet". Each leads on to further research and modifications.

In short the role of models in geography is to codify what has gone before and excite fresh inquiry. To be sure the present stock of models may be unprepossessing enough, but as Lösch asked "... does not the path of science include many precarious emergency bridges over which we have all been willing to pass provided they would help us forward on our road"; certainly his hope that his work on regions would open "... a path into a rich but almost unknown country" (Lösch, 1954, p. 100) has been richly fulfilled.

References

Ackoff, R. L., Gupta, S. K. and Minas, J. S., (1962). *Scientific method: optimizing applied research decisions*. New York.

Beckmann, M. J. (1952). A continuous model of transportation. *Econometrica*, **20**, 643–660.

Berry, B. J. L. and Garrison, W. L. (1958–C). Alternate explanations of urban rank size relationships. *Annals of the Association of American Geographers*, **48**, 83–91.

Bertalanffy, L. von (1951). An outline of general system theory. *British Journal of the Philosophy of Science*, **1**, 134–165.

Bunge, W. (1964). Patterns of location. *Michigan Inter-University Community of Mathematical Geographers, Discussion Papers*, **3**.

Caesar, A. A. L. (1955). On the economic organization of eastern Europe. *Geographical Journal*, **121**, 451–469.

Caesar, A. A. L. (1964). Planning and the geography of Great Britain. *Advancement of Science*, **21**, 230–240.

Chorley, R. J. (1962). Geomorphology and general systems theory. *United States, Geological Survey, Professional Paper*, **500**–B.

Chorley, R. J. and Haggett, P., editors (1965–A). *Frontiers in geographical teaching: the Madingley lectures for 1963*. London.

Curry, L. (1964–B). Landscape as system. *Geographical Review*, **54**, 121–124.

Dickinson, R. E. (1964). *City and region: a geographical interpretation*. London.

Hall, A. D. and Fagen, R. E. (1956). Definition of system. *General Systems Yearbook*, **1**, 18–28.

Hotelling, H. (1921). A mathematical theory of migration. *University of Washington, M.A. Thesis*.

Isard, W. (1956). *Location and space-economy: a general theory relating to industrial location, market areas, land use, trade and urban structure*. New York.

Lösch, A. (1954). *The economics of location*. New Haven.

Nystuen, J. D. and Dacey, M. F. (1961). A graph theory interpretation of nodal regions. *Regional Science Association, Papers and Proceedings*, **7**, 29–42.

Philbrick, A. K. (1957). Principles of areal functional organization in regional human geography. *Economic Geography*, **33**, 299–336.

Skilling, H. (1964). An operational view. *American Scientist*, **52**, 388A–396A.

Stouffer, S. A. (1962). *Social research to test ideas*. New York.

Taaffe, E. J., Morrill, R. L. and Gould, P. R. (1963). Transport expansion in underdeveloped countries: a comparative analysis. *Geographical Review*, **53**, 503–529.

Thomas, W. L., Jr., editor (1956). *Man's role in changing the face of the earth*. Chicago.

Thünen, J. H. von (1875). *Der Isolierte Staat in Beziehung auf Landwirtschaft und Nationalökonomie*. Hamburg.

Vining, R. (1953). Delimitation of economic areas: statistical conceptions in the study of the spatial structure of an economic system. *Journal of the American Statistical Association*, **18**, 44–64.

Weber, A. (1909). *Uber den Standort der Industrien*. Tübingen.

Zipf, G. K. (1949). *Human behaviour and the principle of least effort*. Cambridge.

Reprinted from *Locational Analysis in Human Geography* by Peter Haggett (1965). Used with permission of Edward Arnold (Publishers) Ltd.

Glossary

Amplifier An amplifier is a device or system which increases either the effort, flow, or the power level of a signal or physical quantity.

Amplitude The maximum departure from an equilibrium value. Used to define one of the major characteristics of a wave, its maximum value above or below an equilibrium value.

Anabolism The conversion of simple compounds into living, organized substance. Also called Constructive Metabolism (cf. catabolism).

Artificial Intelligence The use of computers or electro-mechanical or electro-chemical devices to perform operations analogous to the human abilities of learning and decision making.

Autotrophe An organism which is independent of outside sources of organic substances for the provision of its own organic constituents. It can manufacture its own organic substances from inorganic material, e.g. certain bacteria and some plants.

Biosphere The part of the Earth's atmosphere in which life can exist.

Black Box A component of a system that is only considered in terms of its inputs and outputs. Its internal mechanisms may be unknown or ignored.

Catabolism The conversion of living, organized substance into simpler compounds also termed Destructive Metabolism (cf. anabolism).

Closed Loop A system where part of the output is fed back to the input so that the system's output can affect its input or some of its operating characteristics.

Closed System A system which does not take in or give out anything to its environment.

Cluster-analysis A specific type of multi-variate analysis (see multi-variate).

Constraint A requirement that a system has to satisfy, or a limit or set of limits restricting its behaviour.

Core Temperature Internal body temperature.

Cost–Benefit Analysis A method for comparing the outcomes (benefits and disbenefits) of different courses of action in monetary terms.

Cost Function An equation which relates cost to the variables in a given situation.

Cybernetics The study of control and communication in living beings and machines.

Damping Suppression of the degree of movement or change of a variable.

Demography The field of human population analysis, also applied to the study of other animal populations.

Deterministic Having a specific outcome or where outcomes must follow from specific courses of action or inputs (compare stochastic).

D.N.A.–Deoxyribonucleic acid Long thread-like molecules, found in chromosomes and viruses, responsible for storing the genetic code.

Ecology The study of the relation of plants and animals to their environment.

Econometric The branch of economics which expresses theories in mathematical terms in order to verify them by statistical methods.

Ecosystem A natural unit of living and nonliving elements which interact to produce a stable system.

Endogenous Originating within the organism, e.g. endogenous metabolism is concerned with tissue waste and growth.

Entropy The entropy of a system is a measure of its degree of disorder.

Environment The totality of external conditions which affect the behaviour of a system.

Epiphyte Plant attached to another plant, not growing parasitically upon it but merely using it for support, e.g. lichens on trees.

Epistemology The theory of knowledge.

Equilibrium A state of balance.

Exogenous Originating outside the organism, e.g. exogenous metabolism concerned with effector activities and temperature.

Feedback The return of some of the output of a system as input or where the input to a system is affected by the present output (cf. closed loop).

Frequency The rate at which a repeated event or change occurs expressed as the number of occurrences in a standard time interval.

Game Theory A mathematical approach to idealized problems of competitive conflict or games.

Heterotrophe An organism requiring a supply of organic material (food) from its environment from which to make most of its own organic constituents. Depends ultimately on synthetic activities of autotrophic (q.v.) organisms, e.g. all animals and fungi, most bacteria, and a few plants.

Hill-climbing A method of finding the highest or lowest values of a function, by changing individual variables and noting the effect.

Holistic Pertaining to totality, or to the whole.

Homeostasis The maintenance of static or dynamic stability irrespective of external effects.

Isomorphism Similarity or identity in form.

Iterative Repeated application of a procedure or method.

Markov Chains Chains of successive events forming a stochastic process. The particular property of a Markov chain is that the probability of the next event to come depends only on the present state *not* of any previous event.

Metaphysics The branch of speculative philosophy which deals with the ultimate nature of things.

Morphogenesis The origination of morphological characters, i.e. characteristics of form.

Multi-variate Factor Analysis A mathematical method of analysing sets of data to discover which variables are statistically linked, by causative or other factors.

Multi-variate Analysis The branch of statistical analysis concerned with the relationship of sets of dependent variates.

Noise Any disturbance which does not represent any part of a message from a specified source. Usually used to refer to random disturbances.

Nomothetic Relating to the formulation of laws.

Norm A standard, or model, or pattern, or type to which individuals in a group tend to conform.

Onteogeny or Ontogeny The developmental history of the individual organism (compare *phylogeny*).

Open Loop A control system in which corrective action is not automatic, but depends on external intervention. Control actions are made without reference to the present output of the system.

Open System A system that is connected to, and interacts with, its environment.

Optimization Finding the "best" solution or situation in terms of some stated criteria.

Paradigm A pattern or example.

P.E.R.T. (*Program Evaluation and Review Technique*) A method of expressing sets of events and the time between them in a network, and of analysing the network to determine which sequence of events will determine the overall time taken to complete the set.

pH A measure of acidity, hydrion concentration.

Photosynthesis The process by which green plants manufacture their carbohydrates from atmospheric carbon dioxide and water in the presence of sunlight.

Phylogeny The evolutionary development of any organic type or species.

Physiognomic Relating the features of the face or form to a person's character.

Phytosociological Branch of botany comprising ecology (relationship between organism and its surroundings), chorology (biogeography), and genetics (hereditary and variation) of plant associations.

Program(me) A predetermined plan or set of instructions.

Recursive A procedure which includes itself within the procedure.

Retina The light-sensitive layer in the eye.

Role The part or character which a component has to play, undertake, or assume. Chiefly with reference to the part or parts played by a person in society or his life.

Role Set Refers to the set of social roles with which a person interacts by virtue of occupying a particular status. E.g. the status of a schoolteacher has a distinctive role set, relating the teacher to his pupils, to colleagues, to the school principal.

Role-taking Taking the role of the other. Role-taking attempts to explain how social interaction is possible as a sustained activity where each individual is able to take different roles successfully.

R.N.A.–Ribonucleic Acid Long thread-like molecules, found in living cells and viruses.

Saccadic Movement Sudden movement of the eyes from one fixation point to another.

Sensitivity Analysis A method of discovering by how much the estimates used in a model can vary without changing the result.

Shannon's 10th Theorem "If noise appears in a message, the amount of noise that can be removed by a correction channel is limited to the amount of information that can be carried by the channel."

Signal The physical embodiment of a message.

Steady-state A situation where inputs and outputs are constant.

Stochastic Having a probability attached to it. A stochastic process is one where the next event is probabilistically related to previous events.

Syndrome A complex set of various symptoms which are thought to be characteristic of a disease.

Synergy Special correlated action or co-operation resulting in unusual or unexpected results. Sometimes expressed as $2 + 2 = 5$, which is meant to illustrate that co-ordination of objects may give a larger result than their individual characteristics suggest.

System An entity which consists of interdependent parts.

Taxon A category in the taxonomic hierarchy (see *Taxonomy*).

Taxonomy The science of arranging organisms in logical and natural groups, in such a manner as to cast light on their evolution and affinities. Can be applied to the classification of other objects or items.

Teleology The doctrine of final causes or of adaptation to a definite purpose. Loosely, the idea that future states or goals can affect current behaviour.

Thermodynamics: 2nd Law The entropy of a closed system increases with time.

Topology A branch of geometry concerned with the interconnection of objects rather than with their size or shape.

Vitalism The theory that bodily functions are produced by a non-physical inner force called "vital force".

Index